GUIDELINES FOR
RISK BASED
PROCESS SAFETY

This book is one in a series of process safety guideline and concept books published by the Center for Chemical Process Safety (CCPS). Please go to www.wiley.com/go/ccps to see the full list of titles.

GUIDELINES FOR RISK BASED PROCESS SAFETY

Center for Chemical Process Safety

An **AIChE** Industry
Technology Alliance

A JOHN WILEY & SONS, INC., PUBLICATION

A Joint Publication of the Center for Chemical Process Safety of the American Institute of Chemical Engineers and John Wiley & Sons, Inc.

Published by John Wiley & Sons, Inc., Hoboken, New Jersey.
Published simultaneously in Canada.

For general information on our other products and services or for technical support, please contact our Customer Care Department within the United States at (800) 762-2974, outside the United States at (317) 572-3993 or fax (317) 572-4002.

Wiley also publishes its books in a variety of electronic formats. Some content that appears in print may not be available in electronic format. For information about Wiley products, visit our web site at www.wiley.com.

Library of Congress Cataloging-in-Publication Data is available.

ISBN 978-0-470-16569-0

Printed in the United States of America.

10 9

Guidelines for Risk Based Process Safety
is dedicated to the memory of

Lester H. Wittenberg
1929 – 2005

Les was among the first employees of the Center for Chemical Process Safety and was a major contributor to its success. All who knew and worked with Les remember him for his knowledge, his courage, and for his love of life. Although he is gone, his smile is still with us.

It is sincerely hoped that the information presented in this document will lead to an even more impressive safety record for the entire industry; however, neither the American Institute of Chemical Engineers (AIChE), its consultants, the AIChE's Center for Chemical Process Safety (CCPS) Technical Steering Committee and the Risk Based Process Safety Subcommittee members, their employers, their employers' officers and directors, nor ABSG Consulting Inc. and its employees warrant or represent, expressly or by implication, the correctness or accuracy of the content of the information presented in these *Guidelines*. As between (1) the AIChE, its consultants, the CCPS Technical Steering Committee and Subcommittee members, their employers, their employers' officers and directors, and ABSG Consulting Inc., and its employees, and (2) the user of this document, the user accepts any legal liability or responsibility whatsoever for the consequence of its use or misuse.

CONTENTS

I   COMMIT TO PROCESS SAFETY   37

3   PROCESS SAFETY CULTURE   39

III MANAGE RISK   243

10 OPERATING PROCEDURES   245

13 CONTRACTOR MANAGEMENT 365

20 MEASUREMENT AND METRICS

21 AUDITING

LIST OF TABLES

LIST OF FIGURES

ACRONYMS AND ABBREVIATIONS

ACA   apparent cause analysis
ACC   American Chemistry Council
AIChE   American Institute of Chemical Engineers
ALARP   as low as reasonably practicable
ANSI   American National Standards Institute
API   American Petroleum Institute
ASME   American Society of Mechanical Engineers

BLEVE   boiling liquid expanding vapor explosion

CAER   Community Awareness and Emergency Response
CAP   Community Advisory Panel
CBT   computer-based training
CCPS   Center for Chemical Process Safety
CFR   Code of Federal Regulations
CMMS   computerized maintenance management system
COMAH   Control of Major Accident Hazards (UK HSE regulation)
CSB   U.S. Chemical Safety and Hazard Investigation Board

EDMS   electronic document management system
EOC   emergency operations center
EPA   U.S. Environmental Protection Agency
ERT   emergency response team
ESH   environmental, safety, and health
EU   European Union

FDA	Food and Drug Administration
FMEA	failure modes and effects analysis
FMECA	failure modes, effects, and criticality analysis
HAZMAT	hazardous material
HAZOP	hazard and operability analysis
HIRA	hazard identification and risk analysis
HSE	Health and Safety Executive (UK)
IEC	International Electrotechnical Commission
IPL	independent protection layer
ISA	Instrumentation, Systems, and Automation Society
ISO	International Organization for Standardization
ITPM	inspection, testing, and preventive maintenance
JCAIT	Joint Chemical Accident Investigation Team
KSA	knowledge, skills, and ability
LEPC	local emergency planning committee
LOPA	layer of protection analysis
LPG	liquefied petroleum gas
MOC	management of change
MSDS	material safety data sheet
MTTF	mean time to failure
NASA	National Aeronautics and Space Administration
NFPA	National Fire Protection Association
NOHSC	National Occupational Health and Safety Commission
NTSB	National Transportation Safety Board
OEM	original equipment manufacturer
OSHA	U.S. Occupational Safety and Health Administration
P&ID	piping and instrumentation diagram
PDA	personal digital assistant
PHA	process hazard analysis

PPE	personal protective equipment
PSM	process safety management (U.S. OSHA regulation)
QC	quality control
RBPS	risk-based process safety
RC	Responsible Care®
RCA	root cause analysis
RIK	replacement-in-kind
RMP	risk management program (U.S. EPA regulation)
ROI	return on investment
SIF	safety instrumented function
SIL	safety integrity level
SIS	safety instrumented system
SME	subject matter expert
UK	United Kingdom

GLOSSARY

Accident: An incident that results in significant human loss (either injury of death), sometimes accompanied by significant property damage and/or a significant environmental impact.

Accident prevention pillar: A group of mutually supporting RBPS elements. The RBPS management system is composed of four accident prevention pillars: (1) commit to process safety, (2) understand hazards and risk, (3) manage risk, and (4) learn from experience.

Apparent cause analysis (ACA): A less formal investigation method that focuses on the immediate causes of a specific incident.

As low as reasonably practicable (ALARP): The concept that efforts to reduce risk should be continued until the incremental sacrifice (in terms of cost, time, effort, or other expenditure of resources) is grossly disproportionate to the incremental risk reduction achieved. The term *As low as reasonably achievable (ALARA)* is often used synonymously.

Asset integrity: An RBPS element involving work activities that help ensure that equipment is properly designed, installed in accordance with specifications, and remains fit for purpose over its life cycle.

Audit: A systematic, independent review to verify conformance with prescribed standards of care using a well-defined review process to ensure consistency and to allow the auditor to reach defensible conclusions.

Checklist: A list of items requiring verification of completion. Typically, a procedure format in which each critical step is marked off (or otherwise acknowledged/verified) as it is performed. Checklists are often appended to procedures that provide a more detailed description of each step, including information regarding hazards, and a more complete description of the controls associated with the hazards. Checklists are also used in conjunction with formal hazard evaluation techniques to ensure thoroughness.

Chemical reactivity hazard: The potential for an uncontrolled chemical reaction that can result directly or indirectly in serious harm to people, property, or the environment. The uncontrolled chemical reaction might be accompanied by a temperature increase, pressure increase, gas evolution, or other form of energy release.

Competency: An RBPS element associated with efforts to maintain, improve, and broaden knowledge and expertise.

Conduct of operations: The execution of operational and management tasks in a deliberate and structured manner that attempts to institutionalize the pursuit of excellence in the performance of every task and minimize variations in performance.

Continuous improvement: Doing better as a result of regular, consistent efforts rather than episodic or step-wise changes, producing tangible positive improvements either in performance, efficiency, or both. Continuous improvement efforts usually involve a formal evaluation of the status of an activity or management system, along with a comparison to an achievement goal. These evaluation and comparison activities occur much more frequently than formal audits.

Contractor management: A system of controls to ensure that contracted services support (1) safe facility operations and (2) the company's process safety and personal safety performance goals. It includes the selection, acquisition, use, and monitoring of contracted services.

Controls: Engineered mechanisms and administrative policies/procedures implemented to prevent or mitigate incidents.

Core value: A value that has been promoted to an ethical imperative, accompanied with a strong individual and group intolerance for poor performance or violations of standards for activities that impact the core value.

Decommissioning: Completely deinventorying all materials from a process unit and permanently removing the unit from service. Decommissioning normally involves permanently disconnecting the unit from other processes and utilities, and is often followed by removal of the process piping, equipment, and support structures.

Demand for resources: Staff hours, funding, or other inputs needed to support RBPS work activities. The demand per unit time (demand rate) normally fluctuates; the peak demand rate is the maximum demand per unit time, such as the maximum number of changes submitted for approval in a given week, which provides a rough estimate of the maximum resource requirements for reviewing change requests.

Demand rate: The required intensity and/or frequency of RBPS work activities, which determines what types of resources, and the amount of resources, needed to conduct work activities or produce work products in any given timeframe.

Effectiveness: The combination of process safety management performance and process safety management efficiency. An effective process safety management program produces the required work products of sufficient quality while consuming the minimum amount of resources.

Efficiency: The ratio of outputs (work products, such as a risk analysis report) to inputs (e.g., staff hours).

Element: Basic division in a process safety management system that correlates to the type of work that must be done (e.g., management of change [MOC]).

Element owner: The person charged with overall responsibility for overseeing a particular RBPS element. This role is normally assigned to someone who has management or technical oversight of the bulk of the work activities associated with the element, not necessarily someone who performs the work activities on a day-to-day basis.

Emergency management: An RBPS element involving work activities to plan for and respond to emergencies.

Essential feature: A set of activities or actions that help support a key principle of an RBPS element (e.g., *involving competent personnel* is one essential feature that is required to maintain a dependable practice within most management systems).

Facility: The physical location where the management system activity is performed. In early life cycle stages, a facility may be the company's central research laboratory or the engineering offices of a technology vendor. In later stages, the facility may be a typical chemical plant, storage terminal, distribution center, or corporate office.

Hazard: Chemical or physical conditions that have the potential for causing harm to people, property, or the environment. In these *Guidelines*, hazard refers to the first risk attribute: What can go wrong?

Hazard Identification and Risk Analysis (HIRA): A collective term that encompasses all activities involved in identifying hazards and evaluating risk at facilities, throughout their life cycle, to make certain that risks to employees, the public, or the environment are consistently controlled within the organization's risk tolerance.

Implementation: Completion of an action plan associated with the outcome of the process of resolving audit findings, incident investigation team recommendations, risk analysis team recommendations, and so forth. Also, the establishment or execution of RBPS element work activities.

Implementation options (for a work activity): Different ways that a work activity might be executed, depending on the risk reduction desired by the facility or demanded by regulation.

Improvement: See continuous improvement.

Incident: An unplanned sequence of events with the potential for undesirable consequences.

Incident investigation: A systematic approach for determining the causes of an incident and developing recommendations that address the causes to help prevent or mitigate future incidents. See also root cause analysis and apparent cause analysis.

Independent protection layer (IPL): A device, system, or action that is capable of preventing a postulated accident sequence from proceeding to a defined, undesirable endpoint. An IPL is (1) independent of the event that initiated the accident sequence and (2) independent of any other IPLs. IPLs are normally identified during layer of protection analyses.

Inherently safer: A condition in which the hazards associated with the materials and operations used in the process have been reduced or eliminated, and this reduction or elimination is permanent and inseparable from the process.

Inspection: See worksite inspection.

Inspection, testing, and preventive maintenance (ITPM): Scheduled proactive maintenance activities intended to (1) assess the current condition and/or rate of degradation of equipment, (2) test the operation/functionality of equipment, and/or (3) prevent equipment failure by restoring equipment condition.

ITPM program: A program that develops, maintains, monitors, and manages inspection, testing, and preventive maintenance activities.

Integrated corrective action tracking system: A corrective action tracking system that is common to and monitors all relevant RBPS action items, including recommendations from incident reports, risk analyses, emergency drills, audits, and so forth.

Key principle: A part of an RBPS element, which is often generic to all elements because of the nature of how management systems are defined in these *Guidelines*. For example, almost all elements include a key principle called *maintain a dependable practice*, which is further expanded into essential features and work activities that help ensure that appropriate actions are undertaken to provide the required level of dependability for activities related to the particular element.

Knowledge (or process safety knowledge): An RBPS element that includes work activities to gather, organize, maintain, and provide information to other RBPS elements. Process safety knowledge primarily consists of written documents such as hazard information, process technology information, and equipment-specific information.

Knowledge, skills, and abilities (KSAs): Knowledge is related to information, which is often associated with policies, procedures, and other rule-based facts. Skills are related to the ability to perform a well-defined task with little or no guidance or thought. Abilities concern the quality of decision making and execution when faced with an ill-defined task (e.g., applying knowledge to troubleshooting).

Lagging indicator: Outcome-oriented metrics, such as incident rates or other measures of past performance.

Layer of protection analysis (LOPA): A process of evaluating the effectiveness of independent protection layer(s) in reducing the likelihood of an undesired event.

Leading indicator: Process-oriented metrics, such as the degree of implementation or conformance to policies and procedures, that support the RBPS management system.

Life cycle: The stages that a physical process or a management system goes through as it proceeds from birth to death. These stages include conception, design, deployment, acquisition, operation, maintenance, decommissioning, and disposal.

Limiting conditions for operation: Specifications for critical systems that must be operational and critical resources that must be available to start a process or continue normal operation. Critical systems often include fire protection, flares, scrubbers, emergency cooling, and thermal oxidizers; critical resources normally involve staffing levels for operations and other critical functions.

Management review: An RBPS element that provides for the routine evaluation of other RBPS management systems/elements with the objective of determining if the element under review is performing as intended and producing the desired results as efficiently as possible. It is an ongoing "due diligence" review by management that fills the gap between day-to-day work activities and periodic formal audits.

Management system: A formally established set of activities designed to produce specific results in a consistent manner on a sustainable basis.

Metrics: Leading and lagging measures of process safety management efficiency or performance. Metrics include predictive indicators, such as the number of improperly performed line breaking activities during the reporting period, and outcome-oriented indicators, such as the number of incidents during the reporting period.

Near miss incident: An unplanned sequence of events that could have caused harm or loss if conditions were different or if the events were allowed to progress, but actually did not.

Nonroutine activity/operation: Any production or maintenance activity that is not fully described in an operating procedure. Nonroutine does not necessarily refer to the frequency at which the activity occurs; rather, it refers to whether the activity is part of the established (routine) sequence of operating a process.

Normalization of deviance: A gradual erosion of standards of performance as a result of increased tolerance of nonconformance.

Operating mode: A phase of operation during the operation and maintenance stages of the life cycle of a facility. Operating modes include startup, normal operation, shutdown, product transitions, equipment cleaning and decontamination, maintenance, and similar activities.

Operational readiness: An RBPS element associated with efforts to ensure that a process is ready for startup/restart. This element applies to a variety of restart situations, ranging from restart after a brief maintenance outage to restart of a process that has been mothballed for several years.

Operator: An individual responsible for monitoring, controlling, and performing tasks as necessary to accomplish the productive activities of a system. Operator is also used in a generic sense to include people who perform a wide range of tasks (e.g., reading, calibration, maintenance).

OSHA Process Safety Management, 29 CFR 1910.119 (OSHA PSM): A U.S. regulatory standard that requires use of a 14-element management system to help prevent or mitigate the effects of catastrophic releases of chemicals or energy from processes covered by the regulation.

Outreach: See stakeholder outreach.

Performance: A measure of the quality or utility of RBPS work products and work activities.

Performance assurance: A formal management system that requires workers to demonstrate that they have understood a training module and can apply the training in practical situations. Performance assurance is normally an ongoing process to (1) ensure that workers meet performance standards and maintain proficiency throughout their tenure in a position and (2) help identify tasks for which additional training is required.

Performance-based requirement: A requirement that defines necessary results – the "what to do", but not "how to do it." The means for producing the desired results is left up to the discretion of the facility based on an evaluation of its needs and conditions, and on industry practices. For example, the requirement to implement a management of change system that considers the impact of safety and health as part of the review/approval process, and to prevent changes that pose an unacceptable risk to workers, is a performance-based requirement. The implementer must define the process to identify and review risk associated with changes, determine what level of risk is tolerable, and evaluate risk in sufficient detail to demonstrate that they have met the defined standard of care, which in this case may be to provide a safe work environment. (See also prescriptive requirement, which differs from a performance-based requirement in that a prescriptive requirement states how the activity should be performed.)

Performance indicators: See metrics.

Pillar: See accident prevention pillar.

Prescriptive requirement: A requirement that explicitly states both "what to do" and "how to do it." For example, the specifications for a full body harness and the requirement that it be used when working at a certain height or within a specified distance from the edge of a roof are prescriptive requirements. (See also performance-based requirement, which differs from a prescriptive requirement in that a performance-based requirement does not state how the activity should be performed.)

Procedures: Written, step-by-step instructions and associated information (cautions, notes, warnings) that describe how to safely perform a task.

Process safety competency: See competency.

Process safety culture: The combination of group values and behaviors that determines the manner in which process safety is managed. A sound process safety culture refers to attitudes and behaviors that support the goal of safer process operations.

Process safety knowledge: See knowledge.

Process safety management (PSM): A management system that is focused on prevention of, preparedness for, mitigation of, response to, and restoration from catastrophic releases of chemicals or energy from a process associated with a facility.

Quantitative risk analysis (QRA): The systematic development of numerical estimates of the expected frequency and/or consequence of potential accidents associated with a facility or operation based on engineering evaluation and mathematical techniques.

RBPS criteria: Three criteria – risk, demand for resources, and existing process safety culture – that determine the appropriate level of detail and rigor for risk based process safety management systems and practices that support the RBPS management systems. These criteria must be understood well enough to make rational, consistent decisions, but are normally not characterized in a quantitative manner.

RBPS element: See element.

RBPS management system audit: The systematic review of RBPS management systems, used to verify the suitability of these systems and their effective, consistent implementation.

Readiness review: A work activity that occurs prior to initial startup or restarting a process unit to verify that the condition of process equipment and safety systems, the status of limiting conditions for operations, and in some cases, the training and qualification status of personnel conform to predefined conditions.

Replacement-in-kind (RIK): An item (equipment, chemical, procedure, etc.) that meets the design specification of the item it is replacing. This can be an identical replacement or any other alternative specifically provided for in the design specification, as long as the alternative does not in any way adversely affect the use of the item or associated items.

Resolution: Management's determination of what needs to be done in response to an audit finding (and/or associated recommendation), incident investigation team recommendation, risk analysis team recommendation, and so forth. During the resolution step, management accepts, rejects for cause, or modifies each recommendation. If the recommendation is accepted, an action plan for its implementation will typically be identified as part of the resolution. (See implementation.)

Resources: The labor effort, capital and operating costs, and other inputs that must be provided to execute work activities and produce work products.

Risk: The combination of three attributes: What can go wrong?, How bad could it be?, and How often might it happen?

Risk analysis: A study or review of risk associated with a set of activities or list of potential accident scenarios. A risk analysis normally considers all three risk attributes. A risk analysis can provide qualitative or quantitative results.

Risk-based: The adjective "risk-based" is used to portray one or more risk attributes of a process, activity, or facility. In this context, considering any one of the three risk questions can be viewed as a risk-based activity. For example, when considering the hazards of a substance or a process in deciding how much rigor to build into an operating procedure, the term risk-based design is used rather than hazard-based design, even though understanding the hazard attributes was the primary determinant in the design of the procedure. So, for simplicity, rather than use the independent terms hazard-based, consequence-based, or frequency-based, the single term **risk-based** is used to mean any one or a combination of these terms.

Risk-based process safety: The CCPS's process safety management system approach that uses risk-based strategies and implementation tactics that are commensurate with the risk-based need for process safety activities, availability of resources, and existing process safety culture to design, correct, and improve process safety management activities.

Risk control measures: See controls.

Risk significance: The potential impact that an activity has on risk. A control or work activity that greatly reduces risk is more risk significant than one that marginally reduces risk. Activities can also have a higher significance if a deviation from intended practice greatly increases risk.

Risk tolerance criteria: A qualitative or quantitative expression of the level of risk that an individual or organization is willing to assume in return for the benefits obtained from the associated activity.

Root cause analysis (RCA): A formal investigation method that attempts to identify and address the management system failures that led to an incident. These root causes often are the causes, or potential causes, of other seemingly unrelated incidents.

Safe operating limits: Limits established for critical process parameters, such as temperature, pressure, level, flow, or concentration, based on a combination of equipment design limits and the dynamics of the process.

Safe work practices: An integrated set of policies, procedures, permits, and other systems that are designed to manage risks associated with nonroutine activities such as performing hot work, opening process vessels or lines, or entering a confined space.

Safeguards: See controls.

Stakeholder: Individuals or organizations that can (or believe they can) be affected by the facility's operations, or who are involved with assisting or monitoring facility operation.

Stakeholder outreach: An RBPS element associated with efforts to (1) seek out and engage stakeholders in a dialogue about process safety, (2) establish a relationship with community organizations, other companies and professional groups, and local, state, and federal authorities, and (3) provide accurate information about company/facility operations, products, plans, hazards, and risks.

Standards: The RBPS element that helps identify, develop, acquire, evaluate, disseminate, and provide access to applicable standards, codes, regulations, and laws that affect a facility and/or the process safety standards of care that apply to a facility. Standards also refers to requirements promulgated by regulators, professional or industry-sponsored organizations, companies, or other groups that apply to the design and implementation of management systems, design and operation of process equipment, or similar activities.

Standards of care: Established guidelines, standards, or regulations against which judgments regarding conformance to requirements are based (e.g., used by auditors to define acceptable practice). Standards of care can also include the organization's self-imposed requirements.

Technology manual: A document that explains how a process operates and documents the designer's intent. Technical manuals often include or reference engineering calculations, technical reports, and a wide range of other technical information that is relevant to the process covered by the manual.

Technology steward: A person who is formally appointed to be responsible for maintaining the collective knowledge regarding a process, including process safety-related knowledge.

Toolbox meeting: A meeting held with a work crew, typically at the start of the work shift, during which safety topics and other related items are discussed.

Training: Practical instruction in job and task requirements and methods. Training may be provided in a classroom or at the workplace, and its objective is to enable workers to meet some minimum initial performance standards, to maintain their proficiency, or to qualify them for promotion to a more demanding position.

Work activity: A specific action that is typically required to implement or support the ongoing activities of an RBPS management system.

Workforce: A general term used to refer to employees and contractors at a facility. This term is often, but not exclusively, used to refer to operators, maintenance employees, and other employees or contractors who are not in a supervisory or technical role.

Workforce involvement: An RBPS element that consists of a series of work activates that (1) solicit input from the entire workforce (including contractors), (2) foster a consultative relationship between management and workers at all levels of the organization, and (3) help sustain a strong process safety culture.

Worksite inspection: A work activity designed to determine if ongoing work activities associated with operating and maintaining a facility comply with an established standard. Inspections normally provide immediate feedback to the persons in charge of the ongoing activities, but normally do not examine the management systems that help ensure that policies and procedures are followed.

Written program: A description of a management system that defines important aspects such as: purpose and scope, roles and responsibilities, tasks and procedures, necessary input information, anticipated results and work products, personnel qualifications and training, activity triggers, desired schedule and deadlines, necessary resources and tools, continuous improvement, management review, and auditing.

ACKNOWLEDGMENTS

The American Institute of Chemical Engineers (AIChE) and the Chemical Center for Process Safety (CCPS) express their gratitude to all of the members of the Risk Based Process Safety (RBPS) Subcommittee and their CCPS member companies for their generous efforts and technical contributions in the preparation of these *Guidelines*. The AIChE and CCPS also express their gratitude to the team of authors from ABS Consulting.

RBPS Subcommittee Members:

Jack McCavit	*CCPS Emeritus, Committee Chair*
Don Abrahamson	*Celanese Chemical*
Steve Arendt	*ABS Consulting*
Tim Blackford	*Chevron Energy Technology Company*
John Herber	*3M Company*
Dan Isaacson	*The Lubrizol Corporation*
Shakeel Kadri	*Air Products and Chemicals, Inc.*
Greg Keeports	*Rohm and Haas Company*
Jim Klein	*DuPont*
Pete Lodal	*Eastman Chemical Company*
Bill Marshall	*Eli Lilly and Company*
Darren Martin	*Shell Chemical Company*
Neil Maxson	*Bayer Material Science*
Lisa Morrison	*BP*
Karen Tancredi	*DuPont*
Tony Thompson	*Monsanto Company, Retired*
Scott Wallace	*Olin Corporation*
Roy Winkler	*INEOS Olefins and Polymers USA*
Gary York	*Rhodia, Inc., Retired*

CCPS Staff Consultant:

Bob G. Perry	*Center for Chemical Process Safety*

The CCPS wishes especially to acknowledge the many contributions of the principal authors and other staff members of ABS Consulting who contributed to this book.

Principal authors:

Steve Arendt	Don Lorenzo
Bill Bradshaw, *Project Manager*	Lee Vanden Huevel
Walt Frank	

The authors wish to thank the following ABS Consulting personnel for their technical contributions and review: Earl Brown, Myron Casada, Randy Montgomery, and David Whittle. The authors are also greatly indebted to the technical publications personnel at ABS Consulting. Karen Taylor was the editor for the manuscript, Paul Olsen created many of the graphics, and Susan Hagemeyer prepared the final manuscript for publication.

Before publication, all CCPS books are subjected to a thorough peer review process. CCPS also gratefully acknowledges the thoughtful comments and suggestions of the peer reviewers Their work enhanced the accuracy and clarity of these *Guidelines*.

RBPS Peer Reviewers:

Kevin Allars	*Health and Safety Executive, UK*
Jim Belke	*U.S. Environmental Protection Agency*
Michael Broadribb	*BP*
Graham Creedy	*Canadian Chemical Producers Association*
Les Cunningham	*Merck & Company, Inc.*
Ray French	*ExxonMobil, Retired*
Cheryl Grounds	*BP*
Dennis Hendershot	*Center for Chemical Process Safety*
Brian Kelly	*Center for Chemical Process Safety Emeritus*
Murty Kuntamukkula	*Washington Savannah River Company*
Sam Mannan	*Mary Kay O'Conner Process Safety Center*
Bob Ormsby	*Center for Chemical Process Safety*
Tim Overton	*Dow Chemical Company*
Adrian Sepeda	*Center for Chemical Process Safety Emeritus*
Mike Rodgers	*Syncrude Canada Ltd.*
John Shrives	*Environment Canada*

PREFACE

The American Institute of Chemical Engineers (AIChE) has been closely involved with process safety and loss control issues in the chemical and allied industries for more than four decades. Through its strong ties with process designers, constructors, operators, safety professionals, and members of academia, AIChE has enhanced communications and fostered continuous improvement of the industry's high safety standards. AIChE publications and symposia have become information resources for those devoted to process safety and environmental protection.

AIChE created the Center for Chemical Process Safety (CCPS) in 1985 after the chemical disasters in Mexico City, Mexico, and Bhopal, India. The CCPS is chartered to develop and disseminate technical information for use in the prevention of major chemical accidents. The center is supported by more than 80 chemical process industry sponsors who provide the necessary funding and professional guidance to its technical committees. The major product of CCPS activities has been a series of guidelines to assist those implementing various elements of a process safety and risk management system. This book is part of that series.

Process safety practices and formal safety management systems have been in place in some companies for many years. Nevertheless, many organizations continue to be challenged by inadequate management system performance, resource pressures, and stagnant process safety results. To promote process safety management excellence and continuous improvement throughout industry, CCPS created risk-based process safety (RBPS) as the framework for the next generation of process safety management.

This new framework builds upon ideas first published by the AIChE in 1989 in its book titled *Guidelines for Technical Management of Chemical Process Safety* that were further refined in AIChE's 1992 book titled *Plant Guidelines for Technical Management of Chemical Process Safety*. The RBPS approach recognizes that all hazards and risks are not equal; consequently, it advocates that more resources should be focused on more significant hazards and higher risks. The approach is built on four pillars:

- Commit to process safety
- Understand hazards and risk
- Manage risk
- Learn from experience

These pillars are further divided into 20 elements. The 20 RBPS elements build and expand upon the original 12 elements proposed in the 1989 work, reflecting 15 years of process safety management implementation experience and well-established best practices from a variety of industries. The safety record within the chemical and allied process industries is impressive. CCPS member companies, as well as the industry in general, are committed to continually improving on this impressive safety record. It is CCPS's hope that adopting a risk-based approach to managing process safety will become an integral part of this effort.

EXECUTIVE SUMMARY

Process safety practices and formal safety management systems have been in place in some companies for many years. Process safety management (PSM) is widely credited for reductions in major accident risk and in improved chemical industry performance. Nevertheless, many organizations continue to be challenged by inadequate management system performance, resource pressures, and stagnant process safety results. To promote PSM excellence and continuous improvement throughout the process industries, the Center for Chemical Process Safety (CCPS) created *risk-based process safety* (RBPS) as the framework for the next generation of process safety management.

PURPOSE OF THESE *GUIDELINES*

The purpose of the *RBPS Guidelines* is to provide tools that will help process safety professionals build and operate more effective process safety management systems. These *Guidelines* provide guidance on how to (1) design a process safety management system, (2) correct a deficient system, or (3) improve process safety management practices.

This new framework for process safety builds upon the original ideas published by the CCPS in the early 1990s; integrates industry lessons learned over the intervening years; applies the management system principles of plan, do, check, act; and organizes them in a way that will be useful to all organizations – even those with relatively lower hazard activities – throughout the life cycle of a process or operation. For RBPS to work most effectively, companies should integrate its practices with elements of other management systems so that RBPS is totally consistent with manufacturing operations; safety, health, and environmental controls; security; and related technical and business areas.

These RBPS *Guidelines* are neither a compliance obligation nor a prescription; they do not define a sole path for compliance with process safety regulations. However, these *Guidelines* may create a new performance-based expectation for process safety.

The RBPS elements are meant to be evaluated by companies, which may then elect to implement some aspects of these practices, based on thoughtful consideration of their existing process safety management systems. Not all companies, even those with facilities in nearly similar circumstances, will adopt and implement the RBPS activities in the same way. Company-specific and local circumstances may result in very different RBPS activities based on the perceived needs, resource requirements, and existing safety culture of the facility.

RISK-BASED PROCESS SAFETY APPROACH

The RBPS approach recognizes that all hazards and risks are not equal; consequently, it focuses more resources on higher hazards and risks. The main emphasis of the RBPS approach is to put just enough energy into each activity to meet the anticipated needs for that activity. In this way, limited company resources can be optimally apportioned to improve both facility safety performance and overall business performance.

Risk-based process safety criteria. Effective efforts to improve safety must be based upon:

- An understanding of the hazards and risks of the facilities and their operations.
- An understanding of the demand for, and resources used in, process safety activities.
- An understanding of how process safety activities are influenced by the process safety culture within the organization.

The hazards and level of risk associated with the facilities or operations should be the primary consideration in the design and improvement of PSM activities. The demand for the activity, resources needed, and embedded safety culture also influence design and improvement decisions.

Effectiveness measures. Organizations must find ways to measure performance and efficiency so they can apply finite resources in a prioritized manner to a large number of competing process safety needs. Focusing on effectiveness helps ensure that the organization is getting the promised business return by doing the right things in its journey toward process safety excellence.

Accident prevention pillars. The following four accident prevention pillars should be implemented at a risk-appropriate level of rigor:

> **Commit to process safety** – The cornerstone of process safety excellence. A workforce that is convinced the organization fully supports safety as a core value will tend to do the right things, in the right ways, at the right times – even when no one else is looking.
>
> **Understand hazards and risk** – The foundation of a risk-based approach. An organization can use this information to allocate limited resources in the most effective manner.
>
> **Manage risk** – The ongoing execution of RBPS tasks. Organizations must (1) operate and maintain the processes that pose the risk, (2) keep changes to those processes within risk tolerances, and (3) prepare for, respond to, and manage incidents that do occur. A company that uses its risk understanding is better able to deal with the resultant risk and, subsequently, sustain long-term, accident-free, and profitable operations.
>
> **Learn from experience** – The opportunities for improvement. Metrics provide direct feedback on the workings of RBPS systems, and leading indicators provide early warning signals of ineffective process safety results. When an element's performance is unacceptable, organizations must use their mistakes – and those of others – as motivation for action.

If an organization focuses its process safety efforts on these four pillars, then its process safety effectiveness should improve, the frequency and severity of incidents should decrease, and the long term safety, environmental, and business performance should improve. This risk-based approach also helps avoid gaps, inconsistencies, overwork, and underwork that can lead to system failure. For PSM to work most effectively, companies should integrate their RBPS practices with other management systems, such as those for quality, reliability, environmental, health, safety, and security.

RISK BASED PROCESS SAFETY ELEMENTS

The 20 elements listed in Table S.1 expand upon the original CCPS PSM elements to reflect 15 years of process safety management implementation experience, best practices from a variety of industries, and worldwide regulatory requirements.

These elements can be designed and implemented at varying levels of rigor to optimize process safety management performance, efficiency, and effectiveness. The new elements also help eliminate gaps and inconsistencies that have contributed to PSM failures.

TABLE S.1.   Risk Based Process Safety Elements

Commit to Process Safety
- *Process Safety Culture*
- *Compliance with Standards*
- *Process Safety Competency*
- *Workforce Involvement*
- *Stakeholder Outreach*

Understand Hazards and Risk
- *Process Knowledge Management*
- *Hazard Identification and Risk Analysis*

Manage Risk
- *Operating Procedures*
- *Safe Work Practices*
- *Asset Integrity and Reliability*
- *Contractor Management*
- *Training and Performance Assurance*
- *Management of Change*
- *Operational Readiness*
- *Conduct of Operations*
- *Emergency Management*

Learn from Experience
- *Incident Investigation*
- *Measurement and Metrics*
- *Auditing*
- *Management Review and Continuous Improvement*

APPLICATION OF THE *RBPS GUIDELINES*

The RBPS system may encompass all process safety issues for all operations involving the manufacture, use, storage, or handling of hazardous substances or energy. However, each organization must determine which physical areas and phases of the process life cycle should be included in its formal management systems, based on its own risk tolerance considerations, available resources, and process safety culture.

The following technical issues are addressed within, or excluded from, the scope of the RBPS elements:

Total life cycle. The RBPS elements are meant to apply to the entire process life cycle. Some elements may not be active in early life cycle stages; but for some elements, the early life cycle stages provide a unique opportunity to minimize risk, for example, identifying and incorporating inherently safer process characteristics early in project development. In later stages, such as decommissioning, some element work activities may not be as important or may no longer be needed, while others may be simplified.

Fixed facilities, not transportation. The RBPS elements are described herein for fixed facilities. Although the risk-based thought process and many RBPS elements and activities are relevant to transportation or maritime situations, their application in those environments was not considered during the development of these RBPS *Guidelines.*

Processes, not products. Some RBPS elements and activities are relevant to managing product safety and consumer risks; however, these topics were not considered during the development of these RBPS *Guidelines.*

Related technical areas. Any organization can use management system approaches to address complex issues. Some promote the integration of management system activities into a single system to achieve more efficient operation. These RBPS *Guidelines* were written to address process safety as a stand-alone issue and do not explicitly include:

- Occupational health and safety
- Environmental protection
- Product stewardship
- Product distribution
- Security
- Quality

Companies can use the information in this book to help implement new process safety management systems, repair defective systems or elements, or improve mature systems or elements.

ORGANIZATION OF THE *RBPS GUIDELINES*

Chapter 1 provides background information and lays the foundation for this new approach to managing process safety. Chapter 2 defines the risk-based process safety approach for applying the RBPS elements to industrial operations. Chapters 3 through 22 provide the management system framework for each RBPS element. Each element chapter has the same organization:

- Overview
- Key principles and essential features
- Work activities and implementation options
- Performance and efficiency improvement examples
- Possible metrics
- Management review topics

Chapter 23 describes approaches for initial implementation, correction of deficiencies, and ongoing improvement of an RBPS system at a facility. Chapter 24 sets goals for ongoing improvement of process safety management systems.

1

INTRODUCTION

Process safety management is widely credited for reductions in major accident risk and improved process industry performance. Process safety practices and formal safety management systems have been in place in some companies for many years. Over the past 20 years, government mandates for formal process safety management systems in Europe, the U.S., and elsewhere have prompted widespread implementation of a management systems approach to process safety management.

However, after an initial surge of activity, process safety management activities appear to have stagnated within many organizations. Incident investigations continue to identify inadequate management system performance as a key contributor to the incident. And audits reveal a history of repeat findings indicating chronic problems whose symptoms are fixed again and again without effectively addressing the technical and cultural root causes. Table 1.1 lists some of the reasons that process safety management programs may have plateaued or declined.

While all of these issues may not have occurred in your company, they have all happened to some degree in other companies. Left unchecked, such issues can do more than cause stagnation, they can leave organizations susceptible to losing their focus on process safety, resulting in a serious decline in process safety performance or a loss of emphasis on achieving process safety excellence. This is one of the reasons the Center for Chemical Process Safety (CCPS) created the next generation process safety management framework – *Risk Based Process Safety (RBPS).*

TABLE 1.1. Possible Causes of Process Safety Management Performance Stagnation

- In the U.S., process safety management has become synonymous with OSHA's PSM regulation, 29 CFR 1910.119, resulting in a minimum cost, compliance-based approach to managing process safety . . . "If isn't a regulatory requirement, I'm not going to do it!"

- Since worker injuries are much more frequent and are easier to measure, company resources are sometimes disproportionately focused on personal safety instead of process safety.

- Since worker injury rates are steadily declining at most facilities, management assumes this also indicates that the risk of low-frequency, high-consequence process safety incidents must likewise be declining.

- Process safety management was developed by and for big companies. Small companies often do not have the capability to implement similar systems.

- Organizations lack a thorough understanding of recognized and generally accepted good engineering practices and are inconsistent in interpreting and applying them.

- Process safety management was implemented as a separate, stand-alone system that was not integrated into the organization's overall management system.

- Process safety management was implemented as a one-time project instead of an ongoing process.

- Management systems are overemphasized while the technical aspects of process safety, which actually control the hazards and manage risk, are neglected.

- No consistent, widely recognized measurement systems are available for process safety.

- Auditing costs are high and audits have focused on symptoms of problems; they have failed to identify underlying causes.

- Management does not understand or apply risk-based decision processes.

- The legal system inhibits the application of risk-based decision processes.

- Engineering curricula often do not include or emphasize process safety.

- Verbal support for implementation is inconsistent with financial support.

- Diminishing resources are devoted to process safety; facilities face increased pressure to achieve short-term financial objectives.

- Mergers, acquisitions, and divestitures have decreased organizational stability.

- Senior management lacks plant/process operating experience, resulting in a perceived (or real) lack of commitment to process safety management.

- Success has led to complacency – the absence of major accidents lessens a company's sense of vulnerability; statistics continue to demonstrate that worker safety in the process industries is better than almost all other industrial sectors.

- Process safety professionals communicate poorly with senior management, or management does not receive and act on the messages.

1.1   PURPOSE OF THESE GUIDELINES

The purpose of these *RBPS Guidelines* is to help organizations design and implement more effective process safety management systems. These *Guidelines* provide methods and ideas on how to (1) design a process safety management system, (2) correct a deficient process safety management system, or (3) improve process safety management practices. The RBPS approach recognizes that all hazards and risks in an operation or facility are not equal; consequently, apportioning resources in a manner that focuses effort on greater hazards and higher risks is appropriate. Using the same high-intensity practices to manage every hazard is an inefficient use of scarce resources. A risk-based approach reduces the potential for assigning an undue amount of resources to managing lower-risk activities, thereby freeing up resources for tasks that address higher-risk activities.

This approach is a paradigm shift that will benefit all industries that manufacture, consume, or handle hazardous chemicals or energy by encouraging companies to:

- Evolve their approach to accident prevention from a compliance-based to a risk-based strategy.
- Continuously improve management system effectiveness.
- Employ process safety management for non-regulatory processes using risk-based design principles.
- Integrate the process safety business case into an organization's business processes.
- Focus their resources on higher risk activities.

This new framework for process safety builds upon the original process safety management ideas published by the CCPS in the late 1980s, integrates industry lessons learned over the intervening years, applies the management system principles of "plan, do, check, act", and organizes them in a way that will be useful to all organizations – even organizations with relatively lower hazard activities – throughout the life cycle of a process or operation.

An RBPS management system addresses four main accident prevention pillars (Table 1.2).

Authentic *commitment to process safety* is the cornerstone of process safety excellence. Management commitment has no substitute. Organizations generally do not improve without strong leadership and solid commitment. The entire organization must make the same commitment. A workforce that is convinced that the organization fully supports safety as a core value will tend to do the right things, in the right ways, at the right times, even when no one is looking. This behavior should be consistently nurtured, and celebrated, throughout the organization. Once it is embedded in the company culture, this commitment to process safety can help sustain the focus on excellence in the more technical aspects of process safety.

TABLE 1.2. RBPS Management System Accident Prevention Pillars

- Commit to process safety
- Understand hazards and risk
- Manage risk
- Learn from experience

Organizations that *understand hazards and risk* are better able to allocate limited resources in the most effective manner. Industry experience has demonstrated that businesses using hazard and risk information to plan, develop, and deploy stable, lower-risk operations are much more likely to enjoy long term success.

Managing risk focuses on three issues: (1) prudently operating and maintaining processes that pose the risk, (2) managing changes to those processes to ensure that the risk remains tolerable, and (3) preparing for, responding to, and managing incidents that do occur. Managing risk helps a company or a facility deploy management systems that help sustain long-term, incident-free, and profitable operations.

Learning from experience involves monitoring, and acting on, internal and external sources of information. Despite a company's best efforts, operations do not always proceed as planned, so organizations must be ready to turn their mistakes – and those of others – into opportunities to improve process safety efforts. The least expensive ways to learn from experience are to (1) apply best practices to make the most effective use of available resources, (2) correct deficiencies exposed by internal incidents and near misses, and (3) apply lessons learned from other organizations. In addition to recognizing these opportunities to better manage risk, companies must also develop a culture and infrastructure that helps them remember the lessons and apply them in the future. Metrics can be used to provide timely feedback on the workings of RBPS management systems, and management review, a periodic honest self-evaluation, helps sustain existing performance and drive improvement in areas deemed important by management.

Focusing on these four pillars should enable an organization to improve its process safety effectiveness, reduce the frequency and severity of incidents, and improve its long-term safety, environmental, and business performance. This risk-based approach helps avoid gaps, inconsistencies, over work, and under work that can lead to system failure. For process safety management to work most effectively, companies should integrate their RBPS practices with other management systems, such as those for product quality, equipment and human reliability, personnel health and safety, environmental protection, and security.

These *Guidelines* offer two central strategies for how companies can succeed in applying the above principles:

- *Use RBPS criteria to design, correct, or improve process safety management system elements.* Review the work activities associated with each element and update them based on (1) an understanding of the risks associated with the facilities and operations, (2) an understanding of the demand for process safety activities and the resources needed for these activities, and (3) an understanding of how process safety activities are influenced by the process safety culture within the organization.

- *Focus on process safety effectiveness as a function of performance and efficiency.* Use metrics to measure performance and efficiency so that finite resources can be applied in a prioritized manner to the large number of competing process safety needs. Use management reviews to verify that the organization is doing the right things well in its journey toward process safety excellence.

To help companies implement these strategies, these RBPS *Guidelines* offer a set of "new and improved" technical approaches:

- New process safety management elements.
- New activities for traditional process safety management elements.
- New ways to organize and improve process safety management practices.

Companies, whether novices or veterans in process safety management practices, will benefit from examining, adapting, and incorporating the risk-based process safety management approach throughout the entire life cycle of their operations. The RBPS design and implementation process described in this book can be used to develop and implement a practical process safety management system that has a level of detail and effort commensurate with the hazards associated with the facility.

The RBPS management system is not meant to represent the sole path for compliance with process safety regulations, nor is it meant to establish new performance-based requirements for process safety. Nonetheless, in some sense, the RBPS approach does establish new risk-based expectations for process safety management.

The RBPS element guidance is meant to be thoughtfully evaluated by companies, which by using the RBPS criteria, may elect to implement some aspects of these practices while ignoring others. Not all companies, even those with facilities in similar circumstances, will elect to adopt the same elements or implement a given RBPS element or work activity in the same way. Company- and facility-specific circumstances may give rise to very different RBPS activities based on the perceived needs, resource requirements, and the existing process safety culture of the facility.

1.2   BACKGROUND

Causes of chemical process incidents can be grouped in one or more of the following categories:

- Technology failures
- Human failures
- Management system failures
- External circumstances and natural phenomena

For many years, companies focused their accident prevention efforts on improving the technology and human factors. In the mid-1980s, following a series of serious chemical accidents around the world, companies, industries, and governments began to identify management systems (or the lack thereof) as the underlying cause for these accidents. Companies were already adopting management systems approaches in regard to product quality, as evidenced by various Total Quality Management initiatives, with widely reported success (Ref. 1.1). Companies developed policies, industry groups published standards, and governments issued regulations, all aimed at accelerating the adoption of a management systems approach to process safety. Thus, the initial, somewhat fragmented, hazard analysis and equipment integrity efforts were gradually incorporated into integrated management systems. The integrated approach remains a very useful way to focus and adopt accident prevention activities. More recently, inclusion of manufacturing excellence concepts has focused attention on seamless integration of efforts to sustain high levels of performance in manufacturing activities. Done well, manufacturing excellence deeply embeds process safety management practices into a single, well-balanced process for managing manufacturing operations.

The American Institute of Chemical Engineers' Center for Chemical Process Safety was established in 1985 as one of the U.S. chemical industry's reactions to a major chemical accident in Bhopal, India. In 1988, the CCPS published a motivational advertisement for its forthcoming process safety management structure, *Chemical Process Safety Management – A Challenge to Commitment* (Ref. 1.2). This item was intended to educate chief executives in the chemical industry about the importance of implementing process safety management activities into their company operations and to motivate them to adopt a management systems approach.

In 1989, the CCPS began to publish a series of guidelines, beginning with *Guidelines for Technical Management of Chemical Process Safety*, to encourage its members to pursue accident prevention in more integrated, holistic ways (Ref. 1.3). Since then, the CCPS has published more than 100 guidelines, tools, and concepts books covering a wide range of topics related to process safety management. Table 1.3 lists a few of the key guidelines and tools that have paved the way for companies seeking to adopt, implement, and improve process safety management systems for chemical accident prevention.

TABLE 1.3.   CCPS Guidelines and Tools for Chemical Process Safety Management

- *Guidelines for Technical Management of Chemical Process Safety*, 1989 (Ref. 1.3)
- *Plant Guidelines for Technical Management of Chemical Process Safety*, 1992, 1995 (Ref. 1.4)
- *Guidelines for Auditing Process Safety Management Systems*, 1993 (Ref. 1.5)
- *Guidelines for Implementing Process Safety Management Systems*, 1994 (Ref. 1.6)
- *Guidelines for Integrating Process Safety Management, Environment, Safety, Health and Quality*, 1996 (Ref. 1.7)
- *ProSmart: Performance Measurement of Process Safety Management Systems*, 2001 (Ref. 1.8)

Other industry groups and government agencies also developed process safety management frameworks. Tables 1.4 and 1.5 list a sampling of these initiatives. Most of the frameworks are similar in construction, include identical or similar safety management system elements, and promote similar process safety work activities. Differences exist in the frameworks, however, particularly the newer ones. In many cases, the sponsoring country or organization wisely looked around the world and then built its process safety structure on current best practices within the industry.

Prior to publishing these *RBPS Guidelines*, the CCPS published a motivational paper for industry executives similar to the original *Challenge to Commitment*. This paper acknowledges that, while industry has made great progress since the CCPS began publishing its process safety management guidelines series, serious accidents continue to occur. This paper challenges companies to recommit to continuous improvement and process safety excellence.

TABLE 1.4.   North American Industry Process Safety Management Initiatives

- Canadian Chemical Producers Association: Responsible Care program, 1986
- American Chemistry Council (formerly Chemical Manufacturers Association): Responsible Care initiative Process Safety Code of Management Practices, 1987
- AIChE Center for Chemical Process Safety: Technical Management of Chemical Process Safety, 1989
- American Petroleum Institute: Recommended Practice 750 – Management of Process Hazards, 1990
- ISO 14001:1996 and 2001 – Environmental Management System
- Organization for Economic Cooperation and Development: Guiding Principles on Chemical Accident Prevention, Preparedness, and Response, 2003
- American Chemistry Council: Responsible Care Management Systems and RC 14001, 2004

These items are all referenced in Chapter 4.

TABLE 1.5. **Partial List of Worldwide Governmental Accident Prevention and Process Safety Management Initiatives**

• European Commission: Seveso I Directive, 1982 and Seveso II Directive, 1997
• U.S. Occupational Safety and Health Administration: Process Safety Management of Highly Hazardous Chemicals (29 CFR 1910.119), 1992
• U.S. Clean Air Act Amendments: Section 112(r) – Accident Prevention, 1992
• U.S. Environmental Protection Agency: Risk Management Program rule (40 CFR 68), 1996
• Mexico: Integral Security and Environmental Management System (SIASPA), 1998
• United Kingdom: Health and Safety Executive COMAH regulations – The Control of Major Accident Hazards Regulations, 1999
• Australia: Occupational Health and Safety Act 1985 Occupational Health and Safety (Major Hazard Facilities) Regulations 1999 (SR 1999). National Standard for the Control of Major Hazard Facilities [NOHSC:1014(1996)]
• Canada: Canadian Environmental Protection Act – Environmental Emergency Regulation, Section 200 Part 8, 1999
• Republic of Korea: Korean OSHA PSM standard, Industrial Safety and Health Act – Article 20, Preparation of Safety and Health Management Regulations. Korean Ministry of Environment – Framework Plan on Hazardous Chemicals Management, 2001-2005
• Brazil: ANG Oil & Gas industry accident prevention regulations
• Malaysia: Department of Occupational Safety and Health, Ministry of Human Resources, Section 16 of Act 514

These items are all referenced in Chapter 4.

Companies are seeking new ways to improve process safety management activities based on the following strategies:

- Decreasing unnecessary process safety management work, based on risk judgments.
- Performing process safety management activities more efficiently.
- Using the same resources, but using better practices to generate improved results.
- Getting better process safety management results, but with fewer resources.
- Extending existing process safety management practices into new areas.
- Extending existing process safety management practices throughout the life cycle.
- Adding new process safety management activities to existing process safety management elements.
- Creating new process safety management elements.
- Restructuring the process safety management system.

This RBPS *Guidelines* book proposes a management system structure, offers examples of emerging effective practices, and defines a risk-based strategic implementation process that can help companies find effective ways to break through their process safety management barriers to become more effective and to operate safer processes.

TABLE 1.6.   Some Factors that Motivated the CCPS RBPS Project

- Process safety management has become a mature activity for many chemical manufacturing companies with few new drivers for innovation and improvement.
- Innovative practices have emerged from facilities that have been challenged to improve performance despite diminishing process safety resources – achieving better results with fewer resources.
- Much experience, good and bad, has been accumulated on process safety management implementation that should be shared across industry.
- The CCPS process safety management elements are more than 15 years old; many companies and many countries have improved on the CCPS's original structure and contents.
- Some companies have done everything they reasonably could to minimize PSM regulatory coverage, but failed to address their general duty obligations to protect workers, the public, and the environment.
- Many companies are attempting to integrate safety, health, and environmental management systems with security management systems; however, few have succeeded in achieving the efficiency improvements promised by such integration.
- Process safety management costs, and subsequently, value, are often questioned by management.
- Society demands improved process safety performance; serious accidents are not acceptable.

1.3   IMPORTANT TERMINOLOGY

The Glossary defines many terms used within these *Guidelines*. This section emphasizes several terms of particular importance that are used frequently in these *Guidelines*.

Risk. Risk is the combination of: What can go wrong?, How bad could it be?, and How often might it happen? When the term **risk** is used in connection with **evaluating risk,** whether qualitatively or quantitatively, all three questions are typically addressed in some way to generate a risk picture (Ref. 1.9). However, in these *Guidelines*, the term **risk-based** is used more generally to portray one or more **risk attributes** of a process, activity, or facility. In this context, considering any one of the three risk questions can be viewed as a risk-based activity. For example, when considering the hazards of a substance or a process in deciding how much rigor to build into an operating procedure, the term risk-based design is used rather than hazard-based design, even though understanding the hazard attributes was the primary determinant in the design of the procedure. So, for simplicity, rather than use the independent terms hazard-based, consequence-based, or frequency-based, the single term **risk-based** is used to mean any one or a combination of these terms.

Process Safety Management. A management system that is focused on prevention of, preparedness for, mitigation of, response to, or restoration from catastrophic releases of chemicals or energy from a process associated with a facility.

OSHA Process Safety Management, 29 CFR 1910.119 (OSHA PSM). This regulatory standard requires use of a 14-element management system to help prevent or mitigate the effects of catastrophic releases of chemicals or energy from a covered process containing a threshold quantity of specific highly hazardous chemicals.

Risk-based process safety. RBPS is the CCPS's process safety management system approach that uses risk-based strategies and implementation tactics that are commensurate with the demand for process safety activities, availability of resources, and existing organizational culture to design, correct, and improve process safety management activities.

Life cycle. The life cycle consists of the stages that a physical process or a management system goes through as it proceeds from birth to death. These stages include conception, design, deployment, acquisition, operation, maintenance, decommissioning, and disposal.

Facility. Facility, as used in these *Guidelines*, refers to the physical place where the management system activity is performed. In early life cycle stages, a facility may be the company's central research laboratory or the engineering offices of a technology vendor. In later stages, the facility may be a typical chemical plant, storage terminal, distribution center, or corporate office.

Effectiveness. Effectiveness is the combination of process safety management performance and process safety management efficiency. An effective process safety management program produces quality results with minimum consumption of resources.

Measurement and metrics. These measures of process safety management performance include outcome oriented lagging indicators (e.g., incident rates) and predictive leading indicators (e.g., rate of improperly performed line-breaking activities). A combination of leading and lagging indicators is typically needed to provide a complete picture of process safety effectiveness.

Improvement. Improvement means doing better in performance or efficiency, or both, with respect to a starting point or a goal.

1.4   MANAGEMENT SYSTEMS CONCEPTS

In this book, the term **management system** means:

A formally established and documented set of activities designed to produce specific results in a consistent manner on a sustainable basis.

These activities must be defined in sufficient detail for workers to reliably perform the required tasks.

For process safety management, the CCPS initially compiled a set of important characteristics of a management system, which were published in Appendix A of the *Guidelines for Technical Management of Chemical Process Safety*. The CCPS gleaned those important characteristics from interactions with its member companies and traditional business process consulting firms that had significant experience in evaluating management systems. Those guidelines were the first generic set of principles to be compiled for use in designing and evaluating process safety management systems.

Although Appendix A of the *Guidelines for Technical Management of Chemical Process Safety* was groundbreaking, most readers overlooked it as a practical tool because the management systems concept was foreign to them. Since that time, most companies, including their chemical process safety professionals, have accumulated significant practical experience in implementing formal process safety, occupational safety, and environmental management systems.

Table 1.7 lists issues that have proven to be most important when designing, developing, installing, revising, operating, evaluating, and improving process safety management systems. A process safety management framework (such as RBPS) can address one or more or these issues on an element-by-element basis. For example, companies normally define the roles and responsibilities for a particular element within the written program for that element, rather than defining roles and responsibilities for the entire process safety management system within a single discrete system element. On the other hand, a single issue can be the sole focus of an individual element. For example, many companies choose to have a discrete *auditing* element rather than building the activity into each individual system element. In any case, the most important thing is that companies thoughtfully consider all of the issues in Table 1.7 when establishing a new management system, fixing an existing one, or improving a mature system.

TABLE 1.7. Important Issues to Address in a Process Safety Management System

- Purpose and scope
- Personnel roles and responsibilities
- Tasks and procedures
- Necessary input information
- Anticipated results and work products
- Personnel qualifications and training
- Activity triggers, desired schedule, and deadlines
- Necessary resources and tools
- Metrics and continuous improvement
- Management review
- Auditing

1.5   RISK BASED PROCESS SAFETY ELEMENTS

The CCPS RBPS subcommittee reviewed various accident prevention management system structures in place around the world (Tables 1.4 and 1.5), solicited ideas from member companies on new and improved process safety practices, and focused on addressing the process safety management weaknesses and concerns listed in Tables 1.1 and 1.6. The result of that activity was the development of the RBPS elements.

Table 1.8 lists the RBPS elements and compares them to the original CCPS process safety management and OSHA PSM and EPA RMP accident prevention elements (Refs. 1.10 and 1.11). Some of the element names have been changed or expanded to include enhanced activities. Gray shading in the original CCPS PSM or OSHA PSM Element columns indicates that the RBPS element is new.

Chapters 3 through 22 contain a complete description of each element in the management system framework. Because these *RBPS Guidelines* were built upon the original concepts behind the 12-element system described in the CCPS's 1989 process safety management publication, readers need not review the original system.

1.6   RELATIONSHIP BETWEEN RBPS ELEMENTS AND WORK ACTIVITIES

These *RBPS Guidelines* define a structure for the RBPS management system and its elements. Design and implementation of an effective RBPS management system should be based on a company's current risk understanding with regard to the processes to which the RBPS management system applies. Additional factors can influence the design and operation of the RBPS structure. These factors include (1) the rate at which the RBPS management system is used (for example, the number of management of change reviews performed at a facility), placing demand on facility resources and (2) the existing process safety culture at the facility.

Chapter 2 discusses the general application of the risk-based management system design principles to the creation, correction, and improvement of RBPS management systems, to help companies (1) implement RBPS management systems and elements, (2) repair deficient systems and elements, or (3) fine-tune existing systems and elements by continuously improving effectiveness. The information in Chapter 2 can also be used by corporate personnel responsible for establishing company-wide standards or guidelines for process safety management systems. The RBPS design and implementation process described herein allows management to develop and implement process safety management systems that are appropriate and practical at a level of detail and effort that is commensurate with the risk associated with the facility.

TABLE 1.8. Comparison of RBPS Elements to Original CCPS PSM Elements

RBPS Element	Original CCPS PSM Element	OSHA PSM/EPA RMP Elements
Commit to Process Safety		
Process Safety Culture	Accountability: Objectives and Goals	
Compliance with Standards	Standards, Codes, and Laws	Process Safety Information
Process Safety Competency	Enhancement of Process Safety Knowledge	
Workforce Involvement		Employee Participation
Stakeholder Outreach		
Understand Hazards and Risk		
Process Knowledge Management	Process Knowledge and Documentation	Process Safety Information
Hazard Identification and Risk Analysis	Capital Project Review and Design Procedures	Process Hazard Analysis
	Process Risk Management	
Manage Risk		
Operating Procedures	Training and Performance	Operating Procedures
	Human Factors	
Safe Work Practices		
Asset Integrity and Reliability	Process and Equipment Integrity	Operating Procedures
		Hot Work Permits
		Mechanical Integrity
Contractor Management		Contractors
Training and Performance Assurance	Training and Performance	Training
	Human Factors	
Management of Change	Management of Change	Management of Change
Operational Readiness		Pre-startup Safety Review
Conduct of Operations		
Emergency Management		Emergency Planning and Response
Learn from Experience		
Incident Investigation	Incident Investigation	Incident Investigation
Measurement and Metrics		
Auditing	Audits and Corrective Actions	Compliance Audits
Management Review and Continuous Improvement		

Table 1.9 lists the work breakdown structure for each RBPS management system element described in Chapters 3 through 22. This structure is intended to simplify the application of these *Guidelines* when implementing the risk-based approach.

1.7 APPLICATION OF THESE *RBPS GUIDELINES*

In general, the RBPS management system is meant to address process safety issues in all operations involving the manufacture, use, or handling of hazardous substances or energy. Each company must decide which physical areas and phases of the process life cycle should be subject to RBPS, using the risk-based thought process to decide the depth of detail to use in meeting the process safety need. The following paragraphs describe technical issues that are addressed within or excluded from the scope of the RBPS elements.

TABLE 1.9. Generic Work Breakdown Structure for the RBPS System

Item	Description
Element	This basic division in a process safety management system correlates to the type of work that must be done, for example, *management of change* (*MOC*).
Key Principle	Elements are organized according to key principles, which may be generic in nature or specifically defined by the type of element (e.g., *identify potential change situations*).
Essential Feature	Key principles are met by adherence to such essential features as *manage all sources of change.*
Work Activity	Essential features are accomplished by completing activities that are risk-appropriate, for example, developing specific examples of changes and replacements-in-kind for each category of change, and using these in employee awareness training to minimize the chance that the MOC system is inadvertently bypassed.
Implementation Options	Implementation options represent a spectrum of how the work activities can be achieved (e.g., multiple examples of changes and replacements-in-kind are developed for all types of change in different manufacturing areas; they are updated based on MOC performance).

Total life cycle. The RBPS elements are meant to apply to the entire process life cycle. Some elements may not be active in early life cycle stages (e.g., during conceptual design there is little need for developing operating procedures). Other elements may be active, but the information available in early stages may not be very detailed; therefore, the work performed in that element would be more preliminary (e.g., hazard identification and risk analysis).

For some elements, however, the early life cycle stages provide a unique opportunity to minimize risk by identifying and incorporating inherently safer process characteristics. In later stages, such as decommissioning, some work activities may not be as important or may no longer be needed (e.g., maintenance), while others may still be necessary, but might be satisfied using a simpler approach (e.g., hazard reviews of decommissioning activities using checklists).

Fixed facilities, not transportation. The RBPS management system is meant to apply to fixed facilities. Transportation activities are only within the scope of these *Guidelines* when cargo vehicles, such as trucks, rail cars, containers, are connected to a fixed facility during loading and unloading or used as a storage vessel. Although risk-based principles and most RBPS elements and activities are relevant to transportation or maritime situations, the application to those operating environments was not considered when these *RBPS Guidelines* were developed. Thus, readers are cautioned that applying these *Guidelines* to transportation activities may require significant adjustment to, or expansion of, the process safety activities identified in these *Guidelines*.

Processes, not products. The RBPS management system is meant to be applied to process safety-related situations and not product safety issues. Some RBPS elements and activities may be relevant to product safety situations, but such issues were not considered when these *RBPS Guidelines* were developed. Thus, readers are cautioned that applying these *Guidelines* to product safety or consumer risk issues may require significant adjustment to the process safety activities identified in these *Guidelines*.

Related technical areas. Many companies and organizations use management system approaches to address complex issues. In some cases, companies, as well as industry organizations, promote the integration of management system activities into one system to achieve more efficient operation. For example, some companies have established an integrated environmental, safety, and health (ESH) management system. Other companies integrate similar activities across the ESH domain at an element or work activity level; in other words, management of change applies to changes that could impact the environment as well as process safety.

The RBPS management system focuses on process safety issues. Recognizing the potential overlaps, companies may want to consider possibilities for integration. However, these *RBPS Guidelines* were written to address process safety as a stand alone issue and do not explicitly include the following related technical areas:

- Occupational health and safety
- Environmental protection
- Product stewardship

- Product distribution
- Security
- Quality

1.8   ORGANIZATION OF THESE *GUIDELINES*

These *Guidelines* are organized to facilitate their use for any of the following basic needs:

- Implementing the first process safety management policy within a company or process safety management system at a facility.
- Diagnosing and correcting an existing deficient process safety management element or system.
- Determining ways to continuously improve process safety management performance or efficiency.

Chapter 2 defines the risk-based process safety approach advocated in applying the RBPS elements to industrial operations. Chapters 3 through 22 provide the details of the management system framework for each RBPS element. Each element chapter has the same organization:

- Element overview.
- Key principles and essential features.
- Possible work activities and implementation options.
- Examples of ways to improve effectiveness.
- Element metrics.
- Management review.

Section 2.3 applies this roadmap to a spectrum of anticipated user needs and suggests which sections should be reviewed first by readers fitting that user/need category.

Chapter 23 covers approaches for initial implementation, corrective implementation, and ongoing improvement of the RBPS management system at a facility. Chapter 24 describes the current state of process safety practice and areas in which additional development is needed. The appendices provide tools and examples for companies to use in applying the RBPS principles contained in these *Guidelines*.

1.9   REFERENCES

1.1   Feigenbaum, A. V., *Total Quality Control,* McGraw-Hill, Inc., New York, New York, 1983.

1.2   Center for Chemical Process Safety, "Chemical Process Safety Management – A Challenge to Commitment," 1988.

1.3   Center for Chemical Process Safety, *Guidelines for Technical Management of Chemical Process Safety,* American Institute of Chemical Engineers, New York, New York, 1989.

1.4   Center for Chemical Process Safety, *Plant Guidelines for Technical Management of Chemical Process Safety*, American Institute of Chemical Engineers, New York, New York, 1992 (and revised edition, 1995).

1.5   Center for Chemical Process Safety, *Guidelines for Auditing Process Safety Management Systems*, American Institute of Chemical Engineers, New York, New York, 1993.

1.6   Center for Chemical Process Safety, *Guidelines for Implementing Process Safety Management Systems*, American Institute of Chemical Engineers, New York, New York, 1994.

1.7   Center for Chemical Process Safety, *Guidelines for Integrating Process Safety Management, Environment, Safety, Health, and Quality*, American Institute of Chemical Engineers, New York, New York, 1996.

1.8   ProSmart – The Tool You Need to Improve Process Safety, Center for Chemical Process Safety, American Institute of Chemical Engineers, New York, 2001, available at *www.aiche.org/ccps*.

1.9   *Evaluating Process Safety in the Chemical Industry – A Users' Guide to Quantitative Risk Assessment,* Center for Chemical Process Safety, New York, 2004.

1.10   Process Safety Management of Highly Hazardous Chemicals (29 CFR 1910.119), *U.S. Occupational Safety and Health Administration, May 1992*, available at *www.osha.gov.*

1.11   Accidental Release Prevention Requirements: Risk Management Programs Under Clean Air Act Section 112(r)(7), 40 CFR 68, U.S. Environmental Protection Agency, June 20, 1996, Fed. Reg. Vol. 61 [31667-31730], available at *www.epa.gov.*

2

OVERVIEW OF RISK BASED PROCESS SAFETY

Over the years, the process industries have evolved several strategic approaches for chemical accident and loss prevention (Figure 2.1). At any given time, industries, companies, and facilities will not find themselves at the same point along this spectrum. In fact, different departments within a facility, different functions within a department, or the same departmental function at different times, may choose to implement multiple strategies at the same time.

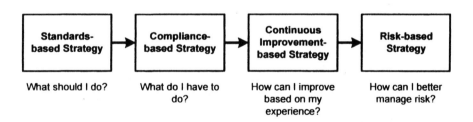

FIGURE 2.1. Evolution of Process Safety and Accident/Loss Prevention Strategies

Standards-based process safety management. For a long time, companies depended solely upon experience-based standards to define their process safety and loss prevention efforts. These standards included both internal company practices and external consensus-based standards, such as standards issued by the ANSI, API, ASME, and NFPA. In a standards-based process safety management strategy, an organization relies on applying proven design, operating, and maintenance practices that have evolved through years of hard lessons (i.e., accidents and other loss events). However, because process safety incidents are typically rare, past experience alone may not adequately alert a company about how to prevent future accidents.

Standards-based process safety management is a prescriptive approach that is fairly easy to implement; however, it can lead to a mixed spectrum of performance. Companies might limit their process safety effort to conformance with consensus codes or standards, even though many process-specific failure issues are not addressed in consensus-based standards or the standards may not address recently discovered failure issues. For many companies not subject to additional federal or state requirements, consensus standards remain the core of their accident prevention and process safety programs. Fortunately, standards organizations have begun to adopt policies that will help ensure that codes and standards are routinely updated to reflect new experience and technology. Some standards organizations have also begun to adopt performance-based and risk-based approaches in their codes and standards to allow companies more flexibility in managing risk.

Compliance-based process safety management. In reaction to public concerns about the actual and potential effects of major accidents involving the chemical process industry, government agencies issued regulations to define minimum levels of accident prevention activity for the protection of workers, the public, and the environment. Many accident prevention regulations, including OSHA's process safety management (PSM) standard and EPA's risk management program (RMP) rule in the U.S., and the Seveso II Directive in the EU, are performance-based regulations that allow companies some degree of latitude to tailor their process safety activities to the appropriate level of risk.

Regulations establish minimum requirements that, in some situations, may not be enough to adequately manage risk while, in other cases, may force companies to overwork process safety issues. The PSM and RMP regulations prompted many companies to implement new activities that have achieved important process safety performance improvements. Although this catalyzed many positive results, these regulatory compliance drivers have also contributed to some companies adopting a compliance-only mindset.

Regulations tend to be experienced-based and take into account national resource constraints. Subsequently, situations deemed hazardous in theory but that have not manifested themselves sufficiently to warrant national attention may not be addressed in regulations. In addition, facilities containing quantities of a hazardous substance that are below the threshold quantity stated in the regulation are not bound by process safety compliance, although risks still exist for these facilities. Thus, a compliance-only approach is not likely to be the optimum strategy for a particular company or facility.

Continuous improvement-based process safety management. With the growth of various total quality management programs, many companies applied the same emphasis on continuous improvement, a hallmark of a quality focused operation, to process safety programs. Companies recognized that merely trying to maintain the status quo in chemical accident prevention was not good enough for three reasons. First, companies have found that "treading water" in process safety can lead to declining performance. Second, global competitiveness demands that companies seek to improve quality and reduce costs. Finally, society's safety expectations are constantly increasing, and most companies intend to meet those expectations to remain welcome members of the community.

This desire led to the incorporation of continuous improvement mantras into the environmental, safety, and health policies of many companies. "Let's learn from our experience" evolved into a more proactive approach, with an expectation to "keep raising the bar." A traditional continuous improvement-based strategy uses lagging indicators to define historical performance and to help guide management system changes based upon the feedback. This model works well when a highly responsive link exists between a business process and one or more lagging indicators. However, this strategy is likely to fail when the lagging indicators are low-frequency, high-consequence events such as catastrophic accidents. The absence of loss events does not reliably indicate that a process safety management system is working well.

Risk-based process safety management. In a risk-based process safety management approach, the organization complies with regulatory requirements, appropriately applies lessons learned from experience across the company or industry, and continues to use lagging indicators to help guide its process safety program. However, risk information and leading indicators are also studied by management to help measure and reliably predict the performance of various aspects of the system, independent of any loss events. Risk information is also used to determine the level of effort and management attention that is appropriate for the risks that have been identified. Management review, a process in which the management team periodically sets aside time to determine how management systems are really performing, supplements risk understanding and metrics. These *RBPS Guidelines* encourage companies to understand and consider risk when making decisions regarding process safety management system resources.

Understanding hazards and risk, monitoring a suite of leading and lagging indicators, and periodically conducting management reviews helps managers highlight strengths, identify weaknesses, and take corrective action in a timely manner. The major challenges in moving to this strategy are (1) developing an adequately detailed and accurate understanding of risk, (2) managing the initial difficulty in selecting the appropriate performance metrics, (3) acquiring the discipline required to maintain these performance metrics, (4) developing the organizational trust and integrity to candidly review performance, and (5) overcoming resistance to making management system changes based on the predictive metrics.

The *Risk Based Process Safety* approach advocated in these *Guidelines* encourages companies to progress toward adoption of accident prevention practices that are risk based. However, companies should continue to use an appropriate blend of the other three strategies as well. The CCPS is certainly not advocating that standards-based codes for structural design, compliance-based approaches for environmental permit compliance, or continuous improvement-based approaches for equipment reliability programs be abandoned in favor of a risk-based approach. Rather, the CCPS proposes a fourth, and sometimes better, way to design process safety management systems. The following section describes how the CCPS believes the risk-based approach applies to the design, correction, and improvement of process safety management systems.

2.1   RISK BASED PROCESS SAFETY SYSTEM DESIGN STRATEGIES

Understanding the risk associated with an activity requires answering the following questions:

- What can go wrong?
- How bad could it be?
- How often might it happen?

Based upon the level of understanding of these answers, a company can decide what actions, if any, are needed to eliminate, reduce, or control existing risk. Early in the life cycle of a process (i.e., conceptual design), limited information typically provides answers to only the first question. Additional information may be available from other facilities, open literature, lessons learned databases, or other sources to improve the company's understanding of risk. Once a process moves into the detailed design stage or is deployed into operation, more detailed answers to all three questions become evident.

Understanding risk helps a company decide how to shape its process safety management activities. Even in a highly regulated environment, process safety professionals can select from a wide range of options when deciding how much technical rigor to incorporate into the process safety management activities at their facilities. Sometimes this flexibility is limited by regulatory constraints, which define a minimum standard of performance for process safety activities. In other cases, an industry consensus standard or internal company requirement may define or limit the range of the process safety professional's options. The options may be further constrained by corporate policies, standards, or guidelines.

These constraints may be written in a prescriptive form or in a performance-based form. Prescriptive requirements state precisely how the process safety activity is to be conducted and what the activity is to produce. Performance-based requirements are more flexible because they specify only what is to be produced and leave the method for generating the desired results up to the facility's management team. Performance-based approaches typically include a series of elements that must be addressed, along with minimum standards of performance. In practice, most performance-based requirements include some blending with prescriptive requirements. All of the process safety management systems listed in Table 1.4 use this blended approach.

The CCPS has identified four accident prevention pillars that form the foundation for risk-based process safety. Each of these pillars can be implemented by companies in a flexible, risk appropriate fashion, based upon company and local circumstances. Table 2.1 lists the accident prevention pillars along with their focal points for process safety assurance. Also listed are the RBPS elements that are aligned with each pillar.

Striving for good process safety results by simply embracing the four accident prevention pillars may be appropriate in a low-hazard situation. For example, a company that simply stores relatively low-hazard materials would likely not require a rigorous approach to process safety. Company policy could state the company's core values and business strategy, describe how the pillars support these elements, and train employees to adopt these ideas and execute their work with appropriate care.

In situations with higher hazards, greater specificity will be required to adequately incorporate the accident prevention principles. In such situations, process safety activities should be explicitly defined in formalized process safety management programs, although some prescriptive requirements can relax into performance-based requirements if a facility has an exceptionally strong process safety culture.

Thus, in the low risk case, the resources needed to properly manage process safety may simply be appropriate management attention supported with consistent messages that foster a sound process safety culture. At the other end of the spectrum, a high risk facility is likely to need a well-integrated system of highly structured policies and procedures, supported by full-time staff positions and, more important, strong management commitment at all levels supporting a mature process safety culture. Section 2.2 describes the RBPS approach for balancing process safety concerns with performance and efficiency goals in a business environment marked by resource constraints.

TABLE 2.1. Process Safety Accident Prevention Principles and Associated RBPS Elements

Process Safety Accident Prevention Pillars and Focal Points	RBPS Elements
Commit to Process Safety	
• Ensure management cares and provides adequate resources and proper environment. • Ensure employees care. • Demonstrate commitment to stakeholders.	• Process Safety Culture • Compliance with Standards • Process Safety Competency • Workforce Involvement • Stakeholder Outreach
Understand Hazards and Risk	
• Know what you operate. • Identify means to reduce or eliminate hazards. • Identify means to reduce risk. • Understand residual risk.	• Process Knowledge Management • Hazard Identification and Risk Analysis
Manage Risk	
• Know how to operate processes. • Know how to maintain processes. • Control changes to processes. • Prepare for, respond to, and manage incidents.	• Operating Procedures • Safe Work Practices • Asset Integrity and Reliability • Contractor Management • Training and Performance Assurance • Management of Change • Operational Readiness • Conduct of Operations • Emergency Management
Learn from Experience	
• Monitor and act on internal sources of information. • Monitor external sources of information.	• Incident Investigation • Measurement and Metrics • Auditing • Management Review and Continuous Improvement

2.2 RISK BASED PROCESS SAFETY DESIGN AND IMPROVEMENT CRITERIA

The main objective of this RBPS approach is to help an organization build and operate a more effective process safety management system. These *Guidelines* describe how to design or improve each process safety activity so that the energy put into the activity is appropriate to meet the anticipated needs for that activity. In this way, limited company resources can be optimally focused to generate improved process safety performance and improved overall business performance.

The RBPS strategic approach is founded on the principle that appropriate levels of detail and rigor in process safety practices are predicated on three factors:

- A sufficient understanding of the risk associated with the processes on which the process safety practices are focused.
- The level of demand for process safety work activity (e.g., the number of change requests that must be reviewed each month) compared to the resources that are available.
- The process safety culture within which the process safety practices will be implemented.

The level of risk associated with the facility or process should be the primary criterion used to guide the design and improvement of process safety management activities. In mathematical terms, risk is the independent variable and resources are a dependent variable. That is, resources should be tailored to risk. Nothing in the RBPS approach advocates the application of insufficient resources, or justifies the replacement of a resource-intensive management system with an ineffective, under-funded system simply because management fails to provide adequate resources. Process safety culture shapes performance at all levels. It shapes human performance, and it can reinforce (or deemphasize) activities that help ensure reliable operation.

Within RBPS management systems, design decisions and implementation actions are guided by the rate of demand for an activity, the level of resources needed, and the process safety culture. Using this risk-based approach, an appropriate suite of practices can be designed and implemented at varying levels of rigor. This approach optimizes process safety management performance and efficiency while avoiding gaps, inconsistencies, overwork, underwork, and associated process safety risks and economic losses.

The following three sections describe why each of the three RBPS design criteria is important and how an organization can develop sufficient insight into its situation to make appropriate decisions regarding process safety activity design and improvement. Each of the RBPS criteria, and the organization's judgment about how the process fits within these criteria, can dramatically affect the appropriate design or improvement approach that should be used. Ignoring these criteria when making adjustments to process safety practices can lead to poor quality or inefficient process safety activities.

2.2.1   Risk of the Process

Why is risk important to the design of process safety management activities?
A company should strive to ensure that process safety activities are completed
in a way that minimizes the risk to employees, the public, and the
environment, whatever the level of risk. Activities with high levels of risk
may warrant greater levels of management attention and higher resource
allocation. On the other hand, activities with a lower risk may require fewer
resources and less management involvement. This approach recognizes that
addressing all risk equally or expecting perfection in every process safety
activity is both unrealistic and a poor use of limited resources. Instead, the
RBPS approach intentionally guides companies to direct their limited
resources to where they are most needed.

*What are some sources of information and ways to develop or refine the
perception of the risk for a specific situation?* The actual hazards or risks of
a process or activity are difficult to know precisely; however, a knowledgeable
process safety professional will become familiar with a variety of chemical,
process, or activity characteristics to gain insight into potential hazards or
risks. To do so involves:

- Learning the intrinsic or extrinsic hazards of a substance, such as flash
  point, toxicity, reactivity, pressure, or temperature. For example,
  propane can burn or explode, chlorine is a toxic exposure hazard, and
  hot oil systems can present a fire or thermal burn hazard.
- Studying the process design or operating conditions. An exothermic
  reaction that normally operates close to the runaway reaction
  temperature usually presents greater risk than a biological reactor that
  operates at ambient conditions.
- Knowing the quantity of hazardous material(s) that could be involved in
  the potential event.
- Knowing the location of people and assets in relation to the hazard.
- Discovering the frequency at which an activity is performed. For
  example, connecting or disconnecting a transfer hose from a small
  cylinder containing toxic material many times a day results in a greater
  likelihood of a leak at a fitting than connecting or disconnecting a
  transfer hose to a larger container once each week (although the
  consequence of a fitting failure involving a small cylinder may be
  lower).
- Learning the operating history of the industry, company, facility, or
  process.

In addition, the precision, accuracy, or completeness of this understanding
is also affected by where in the life cycle the process or activity is when the
risk insight is needed. For example, during the laboratory development stage,

little may be known about the process beyond the basic process chemistry, giving rise to uncertainty about the ultimate hazards and risk the process may pose. Numerous sources of risk insight are available for processes that have operated at the plant scale for years, including operating experience, incidents and near misses, and results of past risk studies. In addition, as a process matures through its life cycle, more information becomes available, and more risk assessment tools can be used, thus increasing opportunities for refining risk perception.

2.2.2   Demand for, and Availability of, Resources for Implementing Process Safety Activities

Why is demand for resources important to the design of process safety management activities? The risk level and the activity level at a facility generally drive the demand for process safety resources. For example, some facilities make only a few changes each year while others make many changes each week. The most effective system for evaluating and controlling changes would not be the same for these two distinctly different types of facilities. A continuous process facility may conduct nearly all inspection and maintenance activities during a very short scheduled shutdown, while a specialty batch facility may space these activities throughout the year, resulting in very different demands on planning and controlling maintenance work. The quantity of activities and the peak rate of activities each present unique challenges that the management systems and the process safety resources must accommodate.

One of the main areas of emphasis for anyone charged with improving process safety effectiveness is to fix any discrepancy between process safety activity needs and the organization's existing capacity to deliver the desired work product(s) of that activity in a timely fashion. The capacity to deliver is a function of the demand rate, the design of the process safety system, and the resources provided to perform the system activity. Situations can exist in which fewer resources are available than are needed to deliver the peak demand for process safety activity results within the required timeframe. To account for such situations, flexibility must to be designed into the process safety system to accommodate demand surges yet not waste resources when they are not required.

What are some sources of information and ways to determine or measure process safety activity demand and resources for a specific situation? The actual demand for and resources available for a process safety activity can readily be determined if accounting systems are established to track process safety activity and to account for resources spent in producing the process safety work product. However, such detailed accounting systems are rare, particularly systems that account for time that facility personnel spend on process safety activities. A facility can normally document how often a process safety activity is performed more accurately than it can the level of resources that was applied. In other words, determining the number or frequency of hazard reviews, training classes, or emergency response drills is simpler than accounting for the number of staff-hours spent reviewing change requests. Estimating resource requirements for some low frequency, high demand activities would present few difficulties. For example, if a team of three people spends approximately two weeks each year reviewing operating procedures for the entire unit, the full-time equivalent staff resource required for that activity is approximately 0.12 (3 persons times two weeks divided by 52 weeks per year). Most process safety activities require very little time to execute, but occur at a relatively high rate. If an accurate estimate of the resource demand rate for these activities is required, the facility needs to keep track of the time spent on the activity on a weekly or monthly basis to get a realistic accounting of the resources used.

Although facilities rarely set up an accounting system to explicitly track this information, a sufficient understanding of process safety activity demand and needed resources can be developed using information sources such as:

- The request dates or completion dates for the process safety activity.
- The number of personnel primarily devoted to specified process safety activities.
- Action tracking system data showing time logged in the system for each specific process safety activity.

In addition, the precision, accuracy, or completeness of this understanding of demand/capacity and resources will also be affected by where the process is in its life cycle. For example, during the preliminary design stage, little may be known about the time that will be needed to conduct preventive maintenance activities. Conversely, many existing facilities maintain historical data on hours charged to each preventive maintenance work order and can provide a very accurate accounting of the resource demand for the work order or a group of similar work orders.

2.2.3   Process Safety Culture Associated with the Company or Facility

Why is process safety culture important to the design of process safety management activities? Leadership, commitment, and employee attitudes at all levels of an organization have a significant impact on the quality of process safety activities. A sound process safety culture that embraces a questioning attitude is more likely to thoroughly examine potential process safety issues. Such beliefs as, "It's not my job," "I only do what I'm told to do," "We'll take shortcuts if necessary to get the job done," or "This process safety stuff is costing too much money" are indicative of a weak process safety culture.

Workers with such attitudes are less likely to use due diligence during the conduct of necessary, but sometimes tedious, process safety activities. A facility with a weak process safety culture often uses the absence of an incident, rather than risk understanding, as the basis for rejecting efforts to improve, or for discarding safe practices that are deemed burdensome.

Process safety culture at a facility tends to be something that a company cannot "manufacture" or easily measure and control; rather, it is the result of everything that has happened, or failed to happen, in the collective memory of the workforce. Thus, the quality of management leadership and commitment that exists at a facility will drive (or limit) the process safety culture. This fundamental character tendency called process safety culture determines what the workforce actually does about process safety when no one is looking, and it sometimes differs from what management or workers say should be done. A well designed process safety management system must be supported by a sufficient process safety culture to deliver safe, sustainable facility operation. In fact, management systems that are not aligned with a facility's culture generally fail over the long term.

Thus, the state of the company or facility process safety culture can greatly influence the level of command and control, as well as the level of performance-based expectations, in process safety activities and guidance. More advanced, stronger cultures can be successful with performance-based process safety systems; whereas, weaker process safety cultures may need more prescriptive process safety management systems. Both extremes of culture can achieve process safety success; however, the process safety professional must consider culture when designing the process safety management system.

What are some sources of information and ways to measure the culture of a specific company or facility situation? The actual culture of a company or facility can never be precisely known; however, a variety of ways exist for companies to develop a better understanding of management leadership and commitment, as well as employee attitudes, behaviors, and potential actions. Such an understanding can be developed through:

- Analyzing employee safety attitude surveys, including survey results regarding management leadership and commitment.
- Making random observations of work practices to discern attention to safety detail.
- Providing anonymous safety issue reporting mechanisms.
- Analyzing audit results, which can reveal the degree of the care in conduct of process safety activities.
- Analyzing incident root cause trends to identify chronic or systemic issues.

The precision, accuracy, or completeness of this understanding may also depend on where in the project life cycle this cultural insight is needed. For example, during the laboratory development stage, little may be known about the specific work environment in which the process will be operated; indeed, the facility may not yet exist. Thus, considerable uncertainty may exist regarding how employees will react to the intrinsic hazards and risks of the process and how careful they will be in executing process safety activities. When a process is currently operating in a facility, more sources of culture insight, as outlined above, are available to guide process safety activity design or improvement efforts.

2.2.4   Examples of How RBPS Criteria Can Affect Element Work Activity Implementation

To help the reader see how the RBPS criteria can affect process safety management system design and implementation, Table 2.2 takes one work activity from four RBPS elements and, using the implementation options contained in the element chapters, illustrates how the perception of risk can affect the extent of work activity implementation. One element is used from each of the four main RBPS accident prevention pillars:

- *Commit to Process Safety* – Chapter 5, Process Safety Competence
- *Understand Hazards and Risk* – Chapter 9, Hazard Identification and Risk Analysis
- *Manage Risk* – Chapter 15, Management of Change
- *Learn from Experience* – Chapter 22, Management Review and Continuous Improvement

Table 2.2 is followed by a short description of how the demand for resources and process safety culture may further affect the level of detail and effort applied to the implementation of an RBPS work activity.

Each of the RBPS element chapters (Chapters 3 through 22) provides a range of implementation options for work activities that could result from applying these RBPS criteria to a particular element. Also, Chapter 23 focuses entirely on RBPS implementation strategies. Examples of using the RBPS criteria are given for a variety of implementation situations, such as adding a new element to an existing system, correcting a deficient element, or improving an existing element.

TABLE 2.2. Examples of How Risk Affects Implementation of RBPS Work Activities

RBPS Element *Key Practice* Essential Feature Work Activity	Perceived Risk Level of Process Where the RBPS Element Activity Is to Be Implemented		
	Low	Medium	High
5. Process Safety Competency			
5.3.2 Execute Activities that Help Maintain and Enhance Process Safety Competency			
Plan Personnel Transitions			
17. Consider individual and organizational competency in succession planning.	a. Program grooms senior managers.	b. Program grooms senior managers and technical personnel; it requires at least one completed process safety assignment.	d. Program is organization-deep and includes competency maintenance, protection, and improvement.
9. Hazard Identification and Risk Analysis (HIRA)			
9.3.1 Maintain a Dependable Practice			
Integrate HIRA Activities into the Life Cycle of Projects or Processes			
2. Determine when HIRAs should be performed.	a. HIRAs are part of normal design review.	c. An initial HIRA, and periodic updates, are performed.	d. A series of HIRAs specific to each stage of the life cycle are performed.
15. Management of Change (MOC)			
15.3.1 Maintain a Dependable Practice			
Involve Competent Personnel			
5. Provide training on the MOC system.	a. Informal training is provided.	b. MOC practice is broadcast via e-mail one time.	d. MOC initial and refresher training provided to affected personnel.
22. Management Review and Continuous Improvement			
22.3.1 Maintain a Dependable Practice			
Define Roles and Responsibilities			
1. Develop a written policy for management review.	a. General guidance applies to all elements.	b. Detailed guidance addresses specific management review requirements.	c. Detailed guidance addresses specific requirements for each RBPS element.

Consider a facility that is classified as medium risk, but has a weak process safety culture. In that case, work activity 15.3.1.5, *Provide training on the MOC system,* should probably be escalated to include initial and refresher MOC training for all personnel, regardless of the demand rate. On the other hand, if that same facility had a sound process safety culture, the training of people involved in MOCs might be relaxed to informal training at low demand rates or maintained with one-time training, even at high demand rates.

A facility with low hazard processes might be content with general guidance for management reviews, regardless of its demand rate or process safety culture. However, a facility with high hazard processes might only accept general guidance for hazard reviews if the demand rate was low and the culture was strong. With a higher demand rate, more detailed guidance would be required for management reviews, and with a weak or evolving process safety culture, detailed guidance for review of each RBPS element should be provided.

2.3   USING ELEMENT CHAPTERS TO DESIGN AND IMPROVE A PROCESS SAFETY MANAGEMENT SYSTEM

These *Guidelines* provide information and tools to aid process safety professionals in (1) designing and implementing a process safety management system or element, (2) fixing a deficient existing system, or (3) charting a path for continuous improvement of a mature system. Chapters 3 through 22 address each of the 20 RPBS elements and include the same six sections:

X.1 – Element overview

X.2 – Key principles and essential features

X.3 – Possible work activities (and associated implementation options)

X.4 – Examples of ways to improve effectiveness

X.5 – Element metrics

X.6 – Management review

where X is the specific chapter number.

Chapter 23 provides several examples illustrating how to use the RBPS approach to (1) develop and implement an initial process safety management system design, (2) repair a defective process safety management system, or (3) improve an existing system using performance and efficiency enhancements. Chapter 24 includes additional ideas for continual improvement.

These *Guidelines* are structured to facilitate their use by practitioners with a variety of backgrounds and needs. Readers should match their objectives to those defined in Table 2.3 to help decide which sections of these *Guidelines* may be of most use to them after reading Chapters 1 and 2. Note that the "X" in the second column of Table 2.3 refers to Chapter 3 through 22. Thus, the X may represent multiple chapters in these *Guidelines*.

Readers who intend to use these *Guidelines* to design and implement a management system for one or more new RBPS elements, or to use these *Guidelines* to completely revamp one or more of their existing process safety management elements, should to scan Chapter 23 prior to reading the chapter on the RBPS element(s) they intend to implement or revamp. Section 23.4 provides four worked case studies along with some commentary. These case studies help explain how to best use the information provided in each element chapter.

These *Guidelines* may be used in many ways that do not require the reader to implement a completely new RBPS system, implement new RBPS elements, overhaul an existing system or element, or make some other significant change. These uses might include:

- Reviewing the effectiveness ideas presented in the fourth section (X.4) of each element chapter.
- Comparing existing metrics to those listed in the fifth section (X.5) of each element chapter.
- Considering implementation of a management review process.
- Conducting a "self assessment" of the process safety management programs at a facility.
- Providing key personnel with a detailed overview of one or more RBPS elements.
- Upgrading process safety overview training and other communication tools.

TABLE 2.3. Advice on Using these *Guidelines* to Meet Specific User Needs

User Need or Objective	*Key Sections*
Develop or implement an initial process safety management system.	X.1, X.2, 23.2, and 23.4.4
Repair a defective process safety management system.	X.2, X.3, X.5, and 23.4.3
Improve an existing system using performance and efficiency enhancements.	X.4, X.5, X.6, 23.3, and 23.4.1
Improve an existing system by adding one or more new elements.	Chapters 3, 5, 7, 17, 20, 22 (based upon which new element[s] are selected), and Section 23.4.2
General overview of one or more RBPS elements.	X.1 and X.2

Examples provided in the fourth section of each element chapter (X.4) can help readers improve the effectiveness of existing process safety management systems. Ideas are included for (1) improving the efficiency of process safety management programs, in other words, obtaining the same or improved outputs while reducing the inputs and (2) increasing performance, or taking the program to the next level. Most of the ideas were provided by CCPS member companies and describe proven strategies that they have used to improve the effectiveness of their process safety management programs.

Two aspects of the continuous improvement/total quality management model presented at the start of this chapter are metrics and management review. Metrics and management review can be very effective tools for improving any management system, including a process safety management system. In fact, these are two of the eight new elements that are being introduced in these *Guidelines*.

Measuring any aspect of RBPS element performance has an intrinsic valve. It indicates that management is interested and encourages workers to pay attention to the measured parameter. The fifth section of each chapter provides a list of metrics that have been provided by CCPS member companies. In addition, Chapter 20 describes how to develop and implement a management system that supports the metrics element.

Management review complements the other three elements in the *learn from experience* accident prevention pillar. First-party management reviews are a particularly effective complement to periodic audits. Audits are typically conducted by one or more experts who are independent of the activity being audited. Audits can effectively identify gaps in process safety management systems and practices and can sometimes identify cases in which a systematic failure to execute work activities in sufficient detail exists. However, auditors are generally present at a facility for a few days every few years and employees generally know when auditors are present. In some cases, facility employees adjust their practices and behavior when they know of an impending audit or that auditors might be present at the facility. Auditors typically use a sampling technique and can miss known deficiencies unless personnel at the facility point these deficiencies out to the audit team. Moreover, audit teams generally have a fixed scope and are a "snapshot in time"; opportunities for improvement that fall outside of this fixed scope or occur at times other than when auditors are present may not be brought forward. Audits have their place, but, by definition, the information that audits can provide is limited.

Conversely, transparent and more frequent management reviews conducted by the management team at a facility are less limited by scope or protocol, and benefit from accurate knowledge of "how things really happen." Facility managers and other key leaders generally know what is going well and where improvements are needed. They are also in a position to facilitate the necessary improvements. In fact, simply holding periodic management review meetings will likely have the same effect as metrics – it demonstrates commitment and encourages commitment at all levels of the organization. When the facility's management team sets aside time on a regular basis to address process safety issues, and does so in an honest and thoughtful manner, it sends a clear message to all employees that process safety is important.

Each element chapter ends with a section on management review. The topics that are presented are not exhaustive; providing such a list would be impossible. Rather, the last section of each element chapter provides a list of issues for the particular element that apply to a wide range of facilities and should be viewed as examples for readers who establish a management review process at their facility. These lists are supplemented by detailed guidance on the management review element in Chapter 22.

Additional ways to use the information included in these *Guidelines* are presented in Chapter 23.

I. COMMIT TO PROCESS SAFETY

Risk-based process safety (RBPS) is based on four pillars: (1) committing to process safety, (2) understanding hazards and risk, (3) managing risk, and (4) learning from experience. To commit to process safety, facilities should focus on:

- Developing and sustaining a culture that embraces process safety.
- Identifying, understanding, and complying with codes, standards, regulations, and laws.
- Establishing and continually enhancing organizational competence.
- Soliciting input from and consulting with all stakeholders, including employees, contractors, and neighbors.

This pillar is supported by five RBPS elements. The element names, along with the short names used throughout the *RBPS Guidelines*, are:

- Process Safety Culture (*process safety culture*), Chapter 3
- Compliance with Standards (*standards*), Chapter 4
- Process Safety Competency (*competency*), Chapter 5
- Workforce Involvement (*involvement*), Chapter 6
- Stakeholder Outreach (*outreach*), Chapter 7

The management systems for each of these elements should be based on the company's current understanding of the risk associated with the processes with which the workers will interact. In addition, the rate at which personnel, processes, facilities, or products change (placing demands on resources), along with the process safety culture at the facility and within the company, can also influence the scope and flexibility of the management system required to appropriately implement each RBPS element.

Chapter 2 discussed the general application of risk understanding, tempered with knowledge of resource demands and process safety culture at a facility, to the creation, correction, and improvement of process safety management systems. Chapters 3 through 7 describe a range of considerations specific to the development of management systems that support efforts to create and sustain the appropriate level of commitment. Each chapter also includes ideas to (1) improve performance and efficiency, (2) track key metrics, and (3) periodically review results and identify any necessary improvements to the management systems that support commitment to process safety.

3

PROCESS SAFETY CULTURE

On January 28, 1986, the space shuttle Challenger exploded 73 seconds after liftoff from Kennedy Space Center, killing all seven astronauts aboard. A field assembly joint in the right-hand solid rocket booster had failed, leaking hot combustion gases which, in turn, breached the liquid hydrogen vessel in the shuttle's external fuel tank assembly. The associated liquid oxygen vessel failed shortly thereafter, and the resulting catastrophic explosion destroyed the shuttle.

A subsequent investigation by a presidential commission revealed significant weaknesses in NASA's safety culture, which had set the stage for this disaster. These weaknesses included: (1) the tolerance of a situation in which production pressures – in this instance, the emphasis on maintaining an aggressive launch schedule – overshadowed safety concerns, (2) the gradual acceptance of increasing levels of damage to the field joints, as determined from post-launch inspections, as being a normal occurrence, even though this was in violation of design specifications and established safety requirements, (3) a can-do attitude, based upon past successes, that limited NASA's sense of vulnerability, and (4) a hierarchical structure and attitude that limited both the free exchange of information (especially disparate opinions) and the credibility given to the technical experts who were lower in the NASA structure or in contractor organizations.

The loss of the Columbia space shuttle, 17 years later, demonstrated additional flaws in the NASA safety culture, including the failure to learn and apply the painful lessons from the Challenger accident. Many of the safety culture problems identified in the Challenger investigation persisted. See Chapter 19 for additional information on the Columbia accident.

3.1   ELEMENT OVERVIEW

Developing, sustaining, and enhancing the organization's process safety culture is one of five elements in the RBPS pillar of *committing to process safety*. This chapter describes what process safety culture means, what the attributes of a sound culture are, and how organizations might begin to enhance their own culture. Section 3.2 describes the key principles and essential features of a management system for this element. Section 3.3 lists work activities that support these essential features, and presents a range of approaches that might be appropriate for each work activity, depending on perceived risk, resources, and organizational culture. Sections 3.4 through 3.6 include (1) ideas to improve the effectiveness of management systems and specific programs that support this element, (2) metrics that could be used to monitor this element, and (3) issues that may be appropriate for management review.

3.1.1   What Is Process Safety Culture?

Process safety culture has been defined as, "the combination of group values and behaviors that determine the manner in which process safety is managed" (Ref. 3.1). More succinct definitions include, "How we do things around here," "What we expect here," and "How we behave when no one is watching."

A culture develops as a group identifies certain attitudes and behaviors that provide common benefit to its members, in this case, attitudes and behaviors that support the goal of safer process operations. As the group reinforces such attitudes and behaviors, and becomes accustomed to their benefits, these attitudes and behaviors become integrated into the group's value system (Ref. 3.2). In an especially sound culture, deeply held values are reflected in the group's actions, and newcomers are expected to endorse these values in order to remain part of the group.

> While this chapter focuses on *process safety culture*, the concepts described here are broadly applicable to the topic of safety culture in its more general context. Indeed, some would assert that a successful *process safety culture* requires the foundation of a sound conventional safety culture.

The process safety culture of an organization is a significant determinant of how it will approach process risk control issues, and process safety management system failures can often be linked to cultural deficiencies. Accordingly, enlightened organizations are increasingly seeking to identify and address such cultural root causes of process safety performance problems.

Table 3.1 contrasts some examples of how process risk control issues might be addressed in a sound and a weak process safety culture.

3.1.2   Why Is It Important?

Investigations of catastrophic events, such as the Longford gas plant explosion (Ref. 3.3) and the Piper Alpha oil platform disaster (Ref. 3.4), have identified common process safety culture weaknesses that are often factors in other serious incidents.   (The Longford and Piper Alpha accidents are briefly described in Chapter 9, *Hazard Identification and Risk Analysis*, and Chapter 11, *Safe Work Practices*, respectively.)   For example, the following essential features of a sound culture were absent or deficient in both of these accidents:

- Establish and enforce high standards of process safety performance.
- Maintain a sense of vulnerability.
- Ensure open and effective communication.
- Provide timely response to process safety issues and concerns.

These essential features will be defined later in this chapter.

TABLE 3.1.   Culture as a Determinant of Process Risk Control Attitudes and Practices

Weak Culture	Sound Culture
Assigns little value to process safety.	Integrates an imperative for safe operations into the organization's core values.
Has a poorly developed sense of its process safety vulnerabilities.	Has a focus on potential failures that drives it to seek a clear understanding of risk and the means to control it.
Devotes minimal resources to controlling residual risk.	Seeks to provide resources proportional to the perceived risk(s) it seeks to control.
Overlooks the weak signals of safety problems.	Places an emphasis on learning from past experience in order to prevent future problems.
Often accepts and normalizes increasingly poor performance.	Seeks to continuously improve performance.
Relies on management to identify hazards and determine what actions should be taken.	Employees are involved in identifying hazards and deciding how they should be addressed. Employees take action to address hazards without management involvement.

Management systems and their associated policies and procedures depend upon the actions of individuals and groups for their successful implementation. For example, a procedure may properly reflect the desired intent and be adequately detailed in its instructions. However, the successful execution of the procedure requires the actions of properly trained individuals who understand the importance of the underlying intent, who accept their responsibility for the task, and who appreciate that taking a shortcut would be inconsistent with the values of the group.

The values of the group (e.g., corporation, facility, shift team) can help shape the attitudes of the individual, which in turn, play a significant role in determining individual behaviors. A sound culture provides its members with the values necessary for understanding why strict adherence to procedures (one aspect of operating discipline) is the right thing to do.

> *Process safety culture* is directly related to the concept of operating discipline, which is addressed in Chapter 17, Conduct of Operations. Chapter 17 describes *conduct of operations* as "the pursuit of excellence in the performance of every task... [as] ... workers at every level... perform their duties with alertness, due thought, full knowledge, sound judgment, and a proper sense of pride and accountability."

While management systems may be heavily reliant upon procedures, no practical procedure can anticipate and address every situation. The values that underlie the process safety culture help the individual understand, accept, and do what is right when no written rules or procedures are in place to address a particular situation, or when procedures are out of date or inconsistent with the organization's common values and objectives. A sound *process safety culture* would ensure that such disparities are brought to the forefront and resolved.

High levels of operational risk, and circumstances in which the rate or magnitude of challenges to the safety management system are high, necessitate sound cultures to help ensure that the organization accepts, understands, and controls this risk in a continuously improving fashion. Useful models exist in high risk/high performance systems such as nuclear aircraft carriers and submarines, air traffic control systems, and nuclear power plants (Ref. 3.5). Valid parallels exist within the chemical process industries, especially as facilities grow larger, become more tightly integrated, operate under more severe process and environmental conditions, and are more cognizant of the risks to which they expose surrounding facilities and populations.

Readers should not construe, however, that a lower risk facility can accept a weak process safety culture. While a lower risk facility may have fewer rules, standards, or expectations than a higher risk facility, the need to follow the rules that are defined is no different. Although a sound culture can assist in making wise risk management decisions, a weak culture could prompt an organization to use the risk-based process safety approach to justify bad decisions, which could lead to under-scoping or avoiding the required rigor for the implementation of one or more RBPS elements. For this reason, organizations should be particularly careful in designing, implementing, and monitoring RBPS programs where cultural weaknesses are known or suspected to exist.

In an increasingly competitive global business environment, maintaining an organization's fundamental safety commitment at all of its locations becomes increasingly difficult because of challenges such as cost pressures, geographical separation, downsizing, and reorganizations. The strength of the culture will help determine how the organization responds to such challenges.

Ultimately, however, the actions or inactions of the individual can be the limiting factor in determining process safety performance. Creating and sustaining a sound *process safety culture* can be a decisive factor in determining the actions of the individual, and in sustaining the success of the organization in managing process hazards (Ref. 3.6). Consequently, a sound culture is essential to maximizing the results from each RBPS element, as well as ensuring proper implementation of the RBPS concept. For this reason, each chapter in the RBPS *Guidelines* describes the relevance of culture to the subject element.

3.1.3 Where/When Is It Done?

Efforts to nurture and sustain a sound process safety culture must occur everywhere, from the boardroom to the production floor. As discussed in Section 3.1.4, everyone in the organization has a role in this.

Organizations that have determined the need for significant enhancement of their culture should take calculated action, recognizing that this is a long-term effort that will require dedicated resources for as long as the organization exists. Cultural changes cannot be quickly achieved – patience and perseverance are required (Ref. 3.1).

While the behaviors and attitudes associated with a sound culture, once established, should become the norm for members of the organization, staff turnover will continually require the instilling of group values and attitudes in new members of the organization. External factors, such as economic pressures, may have a potentially erosive effect on the culture. Thus, while the intensity of effort required for sustaining a culture may not be as great as that required to effect a step change, a continuing effort will nevertheless be required.

3.1.4   Who Does It?

"The only thing of real importance that leaders do is to create and manage culture..." (Ref. 3.2). The leadership of an organization has the primary responsibility for identifying the need for, and fostering, cultural change and for sustaining a sound culture once it is established. However, similar to the concept of safety as a line responsibility, the responsibility for fostering and maintaining a sound culture cascades down through the organization.

As previously noted, culture is based upon shared values, beliefs, and perceptions that determine what come to be regarded as the norms for the organization. In other words, cultures develop from societal agreements about what constitutes appropriate attitudes and behaviors. If the members of an organization feel deeply about a particular behavior, deviation will not be tolerated, and workers will face strong societal pressures to conform (Ref. 3.2). Each individual in the organization has a role in reinforcing such behavioral norms.

3.1.5   What Is the Anticipated Work Product?

The anticipated work product is a sound culture that (1) incorporates the features discussed in Section 3.2 and (2) maximizes the effectiveness of all other RBPS elements and the overall safety management system.

3.1.6   How Is It Done?

An organization's process safety culture is founded on its underlying values regarding process safety. Because leaders cannot change an organization's values and beliefs through edict, cultural change cannot simply be mandated. Successful cultural change requires that expectations of new attitudes and behaviors be communicated and reinforced, that these new attitudes and behaviors demonstrate successful results, and that the members of the organization recognize and appreciate the resulting successes (Ref. 3.2).

Acceptable behaviors must be modeled at all levels of the organization through leadership by example. The rationale for, and anticipated benefits of, expected behaviors must be made evident to all. Positive reinforcement and accountabilities for expected behaviors must be clear and certain. By consistently reinforcing positive behaviors and linking them to the important benefits they bring, management should be able to gradually shift the values of the organization in a positive direction, advancing the organization from a rule-driven culture to a value-based culture.

Many organizations have successfully established sound process safety cultures. Often, these cultures have been developed in response to, and are reinforced by frequent reference to, significant loss events in the company's past. Those organizations fortunate enough not to have experienced such a seminal event may find it helpful to draw upon the experience of others in their, or similar, industries. Other organizations may take justifiable pride in an exemplary process safety record and may seek to inspire employees to maintain their diligence and efforts in order to preserve that record (while seeking to avoid the complacency that past successes may inspire; see *Maintain a Sense of Vulnerability* in Section 3.2.2).

Section 3.3 details more specific activities involved in establishing and nurturing a sound process safety culture.

3.2   KEY PRINCIPLES AND ESSENTIAL FEATURES

This section describes the key principles and essential features of a process safety culture management element. The primary need for culture management will exist initially in immature cultures. As the culture matures, the need for culture management should lessen and be replaced by the need for careful monitoring and guidance by the organization's leadership. A perceived, continuing need to drive the culture should be seen as a strong indication that the culture is still evolving and that a sufficiently mature and sound culture has not yet been achieved.

The following key principles should be addressed when developing, evaluating, or improving any management system for the *process safety culture* element:

- Maintain a dependable practice.
- Develop and implement a sound culture.
- Monitor and guide the culture.

The essential features for each principle are further described in Sections 3.2.1 through 3.2.3. Section 3.3 describes work activities that typically underpin the essential features related to these principles. Facility management should evaluate the risks and potential benefits that may be achieved as a result of improvements in this element. Based on this evaluation, the facility should develop a management system, or upgrade an existing management system, to (1) address some or all of the essential features and (2) execute some or all of the work activities, depending on perceived risk and/or process hazards that it identifies.

3.2.1   Maintain a Dependable Practice

In the context of this book, maintaining a dependable practice means ensuring that the practice is implemented consistently over time. For some elements, this might entail a written procedure as an essential feature of the element. With respect to the *process safety culture* element, the following four essential features will help achieve and maintain a sound process safety culture. The state of each, as it might exist in a sound culture, is described.

Establish process safety as a core value. The organization frequently reinforces the value it places on process safety; consequently, a deeply ingrained sense of that value exists at all levels of the organization. In exceptionally sound cultures, this value is promoted to an ethical imperative. All personnel have an awareness of their responsibilities to self, co-workers, company, and society with respect to their performance. Each person accepts accountability for personal actions as well as the responsibility for counseling against the potentially inappropriate actions of others; therefore, a strong individual and group intolerance for violations of performance norms exists.

Provide strong leadership. Managers are educated in the basic concepts and dynamics of culture and are aware of their roles in fostering a sound process safety culture. Visible, active, and consistent support for process safety programs and objectives exists at all levels of management within the organization. Visionary and inspiring managers are committed to doing what is right, and they demonstrate their values through their communications, actions, priorities, and provision of resources. Performance reviews of leaders and promotions into leadership positions address the individual's commitment to process safety performance improvement. The concept of process safety as a line responsibility is carried down through all levels of the organization.

Establish and enforce high standards of performance. High standards of process safety performance are established and reinforced, for both groups and individuals. When changing circumstances warrant, standards can be carefully modified. However, normalization of deviance, in other words, a gradual erosion of standards as a result of increased tolerance of nonconformances, is never accepted. The organization allows zero tolerance for willful violations of process safety standards, rules, or procedures.

Document the process safety culture emphasis and approach. Culture cannot be designed or manufactured; rather, it is the output of thousands of individual management and employee actions (and inactions) that create the foundation for future individual attitudes and behaviors. Nonetheless, an organization should record, for its own institutional memory, the salient features of any significant, complex activity that the organization wants to be consistently effective over the lifetime of the organization. Thus, a facility should document key principles or activities that support or maintain its process safety culture. This might involve simply recording basic process safety tenets, such as in a policy or mission statement. Or, a facility may choose to formalize its culture-nurturing efforts by documenting, in the form of a simple written management practice, some specific culture evaluation, monitoring, and learning activities that are expected to be carried out by a specified person or group on a periodic basis.

3.2.2 Develop and Implement a Sound Culture

The attitudes and behaviors that an organization accepts as valid and subsequently incorporates into its culture are those that have been demonstrated to successfully deal with the challenges faced by the organization (Ref. 3.2). The following essential features will help an organization manage its process safety challenges:

Maintain a sense of vulnerability. The organization maintains a high awareness of process hazards and their potential consequences and is constantly vigilant for indications of system weaknesses that might foreshadow more significant safety events. The organization strives to avoid the complacency that might be stimulated by past performance and a good safety record. Where uncertainty exists, the organization places the burden of proof on determining that an activity or condition is low risk before it is allowed, rather than requiring employees to prove that it is high risk in order to prevent it.

Empower individuals to successfully fulfill their safety responsibilities. The organization provides clear delegation of, and accountability for, safety-related responsibilities. Accordingly, employees are provided the necessary authority and resources to allow success in their assigned roles. Personnel accept and fulfill their individual process safety responsibilities, and management expects and encourages the sharing of process safety concerns by all members of the organization.

Defer to expertise. The organization places a high value on the training and development of individuals and groups. The authority for key process safety decisions naturally migrates to the proper people based upon their knowledge and expertise, rather than their rank or position. Competent authorities have a substantive role in the deliberation of process safety issues. The organization maintains a sufficient level of expertise required for safe operations.

Ensure open and effective communications. Healthy communications channels exist both vertically and horizontally within the organization. Vertical communications are two way – managers listen as well as speak. Horizontal communications ensure that all workers have the information needed for safe operations. Communications channels are monitored for their effectiveness, and necessary repairs are implemented. Redundant or non-traditional communications channels exist where necessary to provide adequate communications. The organization emphasizes promptly observing and reporting non-standard conditions to permit the timely detection of weak signals that might foretell safety issues (Ref. 3.5).

Establish a questioning/learning environment. The organization strives to enhance risk awareness and understanding as a means to continuous improvement in process safety performance. Enhancements are implemented in various ways, including: (1) performing appropriate and timely risk assessments, (2) promptly and thoroughly investigating incidents, (3) looking beyond the facility or company for applicable lessons, and (4) sharing and applying lessons learned throughout the organization, as appropriate. The organization recognizes that catastrophic events typically have complex causes; consequently, overly simple solutions are avoided when addressing process safety issues. For example, the response to a release caused by pipe corrosion does not stop with replacing the pipe. Was the corrosion rate excessive? If so, why? Does this suggest a processing problem that must be addressed? Was the inspection frequency appropriate for the known corrosion rate? If not, why?

Foster mutual trust. Employees trust managers to do the right thing in support of process safety. Managers trust employees to shoulder their share of responsibility for performance and to report potential problems and concerns promptly. Peers trust the motivations and behaviors of peers. Employees have confidence that a just system exists in which honest errors can be reported without fear of reprisals. Organizational performance, communications, and behaviors are such that the community can trust the facility, and the facility can be confident of a continued license to operate. However, even though mutual trust exists, people are willing to accept others evaluating or checking their actions related to critical tasks/activities that control process safety risks.

Provide timely response to process safety issues and concerns. The organization recognizes that only a brief period often exists between the recognition of a problem and suffering the consequences of the problem. Priority is placed on the timely communication and response to lessons learned from incident investigations, audits, risk assessments, and so forth. Mismatches between practices and procedures (or standards) are resolved in a timely manner to prevent normalization of deviance. The organization emphasizes the timely reporting and resolution of employee concerns.

3.2.3 Monitor and Guide the Culture

Provide continuous monitoring of performance. The organization maintains a healthy attention to its performance. Relevant, clear metrics addressing both leading and lagging indicators are tracked, trended, and responded to. The organization closely and frequently monitors conditions within process operations, management systems, and interpersonal issues that could have a bearing on process safety performance. (Ref. 3.5)

3.3   POSSIBLE WORK ACTIVITIES

The RBPS approach suggests that the degree of rigor designed into each work activity should be tailored to risk, tempered by resource considerations, and tuned to the facility's culture. Thus, the degree of rigor that should be applied

to a particular work activity will vary for each facility, and likely will vary between units or process areas at a facility. Therefore, to develop a risk-based process safety management system, readers should perform the following steps:

1. Assess the risks at the facility, investigate the balance between the resource load for RBPS activities and available resources, and examine the facility's culture. This process is described in more detail in Section 2.2.
2. Estimate the potential benefits that may be achieved by addressing each of the key principles for this RBPS element. These principles are listed in Section 3.2.
3. Based on the results from steps 1 and 2, decide which essential features described in Sections 3.2.1 through 3.2.3 are necessary to properly manage risk.
4. For each essential feature that will be implemented, determine how it will be implemented and select the corresponding work activities described in this section. Note that this list of work activities cannot be comprehensive for all industries; readers will likely need to add work activities or modify some of the work activities listed in this section.
5. For each work activity that will be implemented, determine the level of rigor that will be required. Each work activity in this section is followed by two to five implementation options that describe an increasing degree of rigor. Note that work activities listed in this section are labeled with a number; implementation options are labeled with a letter.

Note: Regulatory requirements may specify that process safety management systems include certain features or work activities, or that a minimum level of detail be designed into specific work activities. Thus, the design and implementation of process safety management systems should be based on regulatory requirements as well as the guidance provided in this book.

3.3.1 Maintain a Dependable Practice

Establish Process Safety as a Core Value

1. Clearly state the importance of process safety in a high level vision statement and in supporting policy documents. Provide a clear vision of what process safety means to the organization and what success would look like.

   a. A vision statement and supporting policy documents exist, but may not be updated in keeping with internal and external drivers, such as evolving stakeholder expectations for improved process safety performance.

   b. A clear, concise vision statement and supporting policy documents exist and are periodically reviewed and updated as circumstances warrant.

   c. Management has engaged the organization by seeking input and involvement in developing and updating the process safety vision and supporting documentation.

2. Ensure that process safety receives emphasis properly balanced against other organizational objectives, such as production, quality, and so forth.

   a. The organization's expectations with respect to relative priorities are variable.

   b. Process safety often, but not always, receives emphasis and resources on par with production, quality, and other initiatives.

   c. Process safety consistently receives emphasis and resources equal to or greater than production, quality, and other initiatives.

3. Speak frequently and consistently to all levels of the organization about the importance of process safety.

   a. The importance of process safety is occasionally communicated within management levels of the organization.

   b. The importance of process safety is frequently communicated within management and supervision levels of the organization.

   c. The importance of process safety is frequently communicated within all levels of the organization.

Provide Strong Leadership

4. Educate managers in process safety culture, vision, expectations, roles, responsibilities, and standards.

   a. Culture, vision, expectations, standards, roles, and responsibilities are discussed with new managers.

   b. A formal training program on process safety culture has been implemented for all new and current managers.

   c. A formal training program on process safety culture with periodic refresher training has been implemented and is updated as needed.

5. Demonstrate personal values, priorities, and concerns for process safety through what is asked about, measured, commented on, praised, or criticized.

    a. Management behaves in a fashion consistent with strong personal values, priorities, and concerns for process safety.

    b. Management seeks opportunities to proactively demonstrate their personal values, priorities, and concerns for process safety.

6. Require that responsibility and accountability for process safety leadership be shared at all levels of the organization.

    a. Responsibilities and accountabilities are established only for middle managers and supervisors.

    b. Responsibilities and accountabilities are established only for all managerial and supervisory levels.

    c. Responsibilities and accountabilities are established for everyone.

Establish and Enforce High Standards of Performance

7. Implement a process for defining process safety goals.

    a. Process safety goals are established in an informal fashion.

    b. A formal system exists for establishing and periodically reviewing process safety goals for the organization.

    c. Employees perform a meaningful role in helping establish and review process safety goals for the organization.

8. Ensure that employees know what is expected of them by effectively communicating the process safety policies, goals, and plans for achieving the desired process safety performance.

    a. Process safety performance expectations are shared with employees in an ad hoc fashion.

    b. A formal communications system exists for sharing information on process safety policies, goals, and plans to achieve the desired process safety performance (for example, through written program documentation or in employee job descriptions and training).

    c. Item (b) and the effectiveness of the communications system is monitored to ensure that this information is reaching all facility personnel.

9. Establish responsibilities and reinforce accountabilities for process safety roles.

    a. Responsibilities and accountabilities for process safety roles are addressed in an ad hoc fashion.

    b. Responsibilities and accountabilities for process safety roles have been established and are informally reinforced.

    c. Responsibilities and accountabilities for process safety roles have been established and are periodically reviewed and updated as warranted. Management response to acceptable and unacceptable performance of process safety responsibilities is timely, consistent, and fair.

10. Implement a policy of zero tolerance for willful violations of process safety policies, procedures, and rules. Delineate those attitudes and behaviors that the organization will not tolerate under any circumstances.

    a. Process safety policies, procedures, and rules are enforced, but sometimes inconsistently.

    b. The consequences of willful violations of safety policies, procedures, and rules have been established and are actively enforced.

11. Ensure that process safety performance rewards and corrective actions are consistently applied.

    a. Rewards and corrective actions are implemented in an informal fashion.

    b. A formal system is established for implementing performance rewards and corrective actions.

    c. Employees or their representatives have an appropriate role in determining performance rewards and corrective actions.

12. Determine and address the causes of significant or persistent noncompliances and failures to fulfill process safety program deliverables.

    a. The causes of noncompliances are investigated in an informal fashion.

    b. Formal root cause analysis is conducted of significant or persistent noncompliances with the intent of identifying root causes and preventing such noncompliances in the future.

Document the Process Safety Culture Emphasis and Approach

13. Document how the organization approaches evaluating and nurturing process safety culture.

    a. Process safety policies or mission statements are documented and communicated to the workforce; these include process safety culture themes.

    b. A detailed written program exists listing the individual responsible for maintaining process safety culture policies, culture monitoring means, and plans for improvement and correction.

3.3.2 Develop and Implement a Sound Culture

Maintain a Sense of Vulnerability

14. Ensure that all staff are adequately educated on, and appreciate, the hazards of operations (see Chapter 9, Hazard Identification and Risk Assessment).

15. Ensure that all staff are adequately educated on, and appreciate, the consequences of deviating from established safe operating practices and conditions (see Chapter 10, Operating Procedures).

16. Ensure that lessons learned from investigations of incidents and near misses, audits, and hazard assessments are broadly, frequently, and effectively shared (see *Establish a Questioning/Learning Environment*).

    a. Lessons learned are distributed to the immediately affected work group.

    b. Lessons learned are widely shared throughout the organization.

    c. The organization (1) seeks to gather and share lessons learned from diverse sources, including from other organizations, locations, or companies, and (2) has an effective means to validate that appropriate action has been taken.

17. Monitor for, and combat, organizational overconfidence that can be stimulated by past good performance.

    a. Managers and supervisors implement informal approaches to maintaining a sense of vulnerability within the organization.

    b. A formal effort has been implemented to maintain a sense of vulnerability within the organization through efforts such as: (1) effectively sharing lessons learned from recent incident and near miss investigations, both from within and outside the organization, (2) periodically refreshing memories of past significant events within the company and industry, (3) effectively sharing lessons learned from hazard assessments, and (4) providing periodic refresher training on hazards of operations.

    c. Item (b) and management reviews and surveys are used to determine if attitudes are softening with respect to the needs for continued process safety diligence. Hazard assessment and investigation reports are audited to determine if teams are appropriately focusing on the potential for what could happen.

Empower Individuals to Successfully Fulfill Their Process Safety Responsibilities

18. Continually reinforce that all employees have responsibilities to themselves, their co-workers, the company, and the community (see *Establish and Enforce High Standards of Performance*).

    a.   Such responsibilities are reinforced in an informal fashion.

    b.   Employees are given formal training on these issues.

    c.   Managers and supervisors, through training, written and oral communications, meetings, performance reviews, and so forth, make a conscientious effort to ingrain these concepts within the culture.

19. Provide employees with the resources necessary to achieve their process safety responsibilities.

    a.   Resources are provided for the more critical initiatives.

    b.   Process safety initiatives are properly resourced, or alternative approaches to suitably achieve objectives with available resources are identified.

    c.   Resource requirements in support of process safety initiatives are given explicit consideration in the budgeting process and quality resources are assigned to the initiative.

20. Give employees the necessary authority and support commensurate with their process safety responsibilities.

    a.   Individual authority is implicitly associated with the process safety responsibilities.

    b.   Authorities are explicitly addressed, for example in job descriptions that outline process safety responsibilities.

    c.   Management provides support when necessary to reinforce an individual's authority.

Defer to Expertise

21. Identify and provide the proper mix and amount of expertise for safely operating the facility, using a formal process to assess the significance of changes (see Chapter 15, Management of Change and Chapter 17, Conduct of Operations).

    a.   Staffing issues are addressed in an informal fashion with a view to maintaining the proper mix.

    b.   Staffing issues, including changes in staffing, are administered under a personnel skill development program, and are assessed against explicit specifications for the mix and amount of expertise for operating the facility.

22. Establish a high value for process safety knowledge and expertise.

    a.   Staff members, on their own initiative, can pursue advancement opportunities that will enhance their work-related knowledge and expertise; they are nominally supported in these efforts.

    b. Staff members are encouraged to pursue advancement opportunities that will enhance their work-related knowledge and expertise, and are generally supported.

    c. Priority is placed on personal development. Formal development plans are required and funded by the organization.

23. Seek out expertise when making critical decisions, but reinforce expectations that experts have a responsibility to speak out on their own.

    a. Expert opinion may be solicited, but in an informal fashion.

    b. Formal roles are established for expert involvement in certain process safety critical contexts.

    c. Formal networks are established to provide a substantive role for subject matter experts in key process safety-related functions.

Ensure Open and Effective Communications

24. Enable employees to fulfill their responsibilities for sharing concerns by listening to them and acting in response.

    a. Management listens to employee concerns, but response may be slow.

    b. Management promptly responds to relevant employee concerns.

    c. Management has implemented a proactive system for soliciting and responding to employee concerns.

25. Stress the importance of timely communication of information.

    a. Training or counseling is provided on this topic in an ad hoc fashion.

    b. This topic is covered with new employees during basic process orientation and process safety management training.

    c. Managers and supervisors seek additional opportunities to reinforce these concepts, for example through incorporation into incident investigation lessons learned or table top drills.

26. Assess relevant avenues of communication to ensure that messages are properly communicated and acknowledged, and that they are not filtered as they are relayed.

    a. Management relies on anecdotal information to gain confidence that process safety messages are properly communicated.

    b. Management relies on personal conversations with employees to gain confidence that process safety critical messages are getting through.

    c. Management monitors a variety of information sources to gain confidence that process safety critical messages are communicated.

Information sources include meeting minutes, employee surveys, formal suggestion programs, and personal conversations.

27. Establish a system for ensuring that management is accessible to the workforce for reporting potential hazards and providing input on operational safety management policy, issues, and needs.
    a. Management relies on ad hoc mechanisms for receiving such information.
    b. Multiple means exist for the workforce to report potential hazards and to provide input to management on operational safety management policy, issues, and needs.
    c. Supplemental, secure means exist for anonymous reporting by staff members who may otherwise fear repercussions from reporting.

Establish a Questioning/Learning Environment

28. Emphasize maintaining vigilance to the off-standard or abnormal (see also Section 17.2.4).
    a. No special training or counseling is provided on this topic, but the concepts are informally communicated.
    b. This topic is covered with new employees during basic process orientation and process safety management training.
    c. Managers and supervisors actively seek additional opportunities to reinforce these concepts, for example, through incorporation into incident investigation lessons learned or table top drills.

Foster Mutual Trust

29. Develop a disciplinary system which has clear criteria for acceptable and unacceptable behaviors, and which distinguishes situations involving willful misconduct from human errors prompted by system causes. Apply such criteria firmly and fairly.
    a. The organization has an unwritten disciplinary system that addresses these issues.
    b. The organization has a written disciplinary system and applies it consistently.
    c. The organization has a written disciplinary system and continuously monitors its application. The organization actively seeks to address system root causes and fairly identifies those situations in which human error can be attributed to such causes. The organization has developed a "no fault" policy for such instances and has communicated it effectively to employees.

Provide Timely Response to Process Safety Issues and Concerns

30. Ensure that risk analyses, audits, management reviews, and so forth are conducted in accordance with credible, established schedules (see Chapters 9, Hazard Identification and Risk Analysis; 21, Auditing; and 22, Management Review and Continuous Improvement).

31. Ensure timely resolution of process safety concerns, issues, suggestions, and recommendations, whatever the source.
    a. Process safety concerns, issues, suggestions, and recommendations are generally resolved, with resolution monitored in an ad hoc fashion.
    b. Formal tracking systems have been established for process safety recommendations and suggestions, and strict accountabilities are imposed for their timely resolution.

3.3.3 Monitor and Guide the Culture

Provide Continuous Monitoring of Performance

32. Implement and track a diverse set of metrics that encompasses a balanced mix of leading and lagging indicators. Trend results over time, and respond to patterns (see Chapter 20, Measurement and Metrics).
    a. The organization places primary emphasis on measuring lagging indicators.
    b. The organization monitors a mix of leading and lagging indicators that are directly relevant to process safety culture.
    c. The organization closely monitors leading and lagging performance trends and promptly responds to indications of weakening process safety culture.

33. Maintain standards of performance with respect to timely, forthright reporting of performance statistics.
    a. Reporting and followup occurs in an informal fashion.
    b. Formal responsibilities and accountabilities have been established for timely, forthright reporting of performance statistics.
    c. Item (b) and management aggressively follows up on missing or late reports.

34. Conduct periodic reviews of the organization's process safety culture.
    a. Informal assessments are conducted by management.
    b. Periodic reviews are conducted by nonfacility, company personnel.
    c. Protocols exist for special reviews by independent third parties when circumstances warrant.

35. Strive to identify and correct cultural issues that underlie failures to adequately fulfill process safety responsibilities (e.g., why does the organization tolerate this?).

a. Investigations of failures to adequately fulfill safety responsibilities are limited to the identification of the immediate causes.

b. Investigations of failures to adequately fulfill process safety responsibilities address the identification of the management system causes.

c. Investigations of failures to adequately fulfill process safety responsibilities are extended to identify and address any underlying cultural causes.

36. Implement an effective management review system (see Chapter 22, Management Review and Continuous Improvement).

3.4   EXAMPLES OF WAYS TO IMPROVE EFFECTIVENESS

This section provides specific examples of industry tested methods for improving the effectiveness of work activities related to the process safety culture element. The examples are sorted by the key principles that were first introduced in Section 3.2. The examples fall into two categories:

1. Methods for improving the performance of activities that support this element.
2. Methods for improving the efficiency of activities that support this element by maintaining the necessary level of performance using fewer resources.

These examples were obtained from the results of industry practice surveys, workshops, and CCPS member-company input. Readers desiring to improve their management systems and work activities related to this element should examine these ideas, evaluate current management system and work activity performance and efficiency, and then select and implement enhancements using the risk-based principles described in Section 2.1.

3.4.1   Maintain a Dependable Practice

Provide safety culture training. The concepts underlying the role and importance of culture in enhancing the process safety performance of the organization are not particularly complex. However, the dynamics involved may not be part of the typical manager's or supervisor's thought processes as they address the daily conduct of their jobs. Organizations should provide training to company and facility management on leadership principles, organizational effectiveness, and the role of culture in determining process safety performance. More general awareness training could be provided to all company and facility employees on the importance of culture and on established company process safety values, objectives, and goals.

Management visibility and support. Include in each manager's job description explicit responsibilities that support process safety culture initiatives, including frequent contact with the work force. Provide certain accountabilities for successfully carrying out these responsibilities.

Change leadership, if required, to change the culture. Company or facility leadership must be prepared to replace managers who resist the changes necessary to effect needed cultural changes. In extreme cases, this could involve changing the entire top leadership at a facility.

Provide true, palpable, balanced accountabilities for process safety performance. Reinforcement for performance, both positive and negative, should be tangible and certain, at all levels in the organization. Expectations regarding attitudes, behaviors, and performance should be included in written job descriptions at all levels of the organization and should be addressed in periodic performance reviews.

- Implement a progressive discipline program to address failures to comply with process safety responsibilities.
- Provide true consequences to managers, not only for their own actions, but also for the organizations they manage, for which they set and should be reinforcing standards of performance. For example, many companies will impose severe career consequences on the manager in charge of a facility at which a fatality has occurred.
- Require that the supervisors and/or managers responsible for a unit or organization in which a serious incident occurred personally present the results of the incident investigation to management.
- Ensure that accountabilities follow managers to new locations or assignments. In other words, managers should understand that they will be held accountable for what they have done and have chosen not to do, even after they move to their next assignment.
- Provide performance-based incentives for superior efforts and achievements in support of process safety initiatives. Avoid situations that offer financial incentives for achieving production forecasts, but have no counterbalancing incentives for achieving process safety expectations. Be alert, however, to the potential that such incentives can induce some individuals to attempt to beat the system. Implement controls to ensure the quality and integrity of the actions and the statistics upon which such incentives are based.
- Implement a process safety suggestion program with appropriate incentives for providing ideas that truly enhance process safety.

Extend the process safety culture emphasis to the entire company. Extending the emphasis on process safety culture throughout all sectors of the company (e.g., R&D, engineering, corporate staff functions, marketing/sales, transportation) will help reinforce and validate the value that the company places on this issue. This will also help ensure continuity as managers move from one functional group to another; process safety concepts and behaviors will not have to be relearned.

Conduct periodic leadership workshops. Conduct periodic workshops for facility and corporate leaders that address the relationship between leadership or organizational failures and process safety incidents. Use case studies, such as the Challenger accident or other significant industry events, to illustrate important lessons learned. The CCPS process safety culture awareness tool could be used for an initial workshop (Ref. 3.7).

Include frequent, brief articles in newsletters for key management and technical contacts. Issue a short newsletter monthly, highlighting key cultural issues and successes. Include tips and lessons aimed at enhancing managerial, supervisory, and technical staff members' awareness and abilities to address such issues. Such bulletins help keep safety on the mind of business and facility leaders.

3.4.2   Develop and Implement a Sound Culture

Involve all levels of the organization in planning and evaluating the performance of process safety initiatives. True employee participation in developing, implementing, and monitoring initiatives to determine their effectiveness will enhance the sense of ownership held by employees, which should increase their level of engagement in such initiatives.

Increase frequency of communications during periods of high need. The frequency of process safety-related communications should be increased during periods of increased need, for example, during reorganizations and other times when the organization is experiencing significant changes, after a significant safety event, or after an unfavorable audit. When people lack access to information they feel they need, they find ways, productive or nonproductive, of filling that gap. Management should pre-empt speculation with actual information.

Distribute newsletters containing culture supportive articles to the general population of the company. Issuing periodic publications, such as the CCPS Process Safety Beacon, to the entire company helps maintain a sense of vulnerability to process hazards.

Integrate culture enhancements into existing corporate initiatives. Take concepts such as loss of vulnerability or normalization of deviance and fold these themes into existing operational excellence policies and programs. This helps avoid the perception that culture initiatives are a new program that is being layered on top of other such initiatives.

Use stand downs to draw attention to acute or chronic issues. In some circumstances, communication of messages can be made more efficient, and more emphatic, through the use of a stand down, wherein normal routines are interrupted so that all staff members can participate in group meetings addressing particular process safety concerns. Such stand downs illustrate the fact that management feels strongly enough about the issues being discussed to forgo revenue generating activities (i.e., management is placing process safety ahead of production). Stand downs also provide an opportunity for higher management to address large numbers of personnel and speak personally about their support of process safety initiatives and their concerns about the circumstances that may have prompted the stand down. Such stand downs are often used in response to a single serious event or a trend of deteriorating performance.

3.4.3 Monitor and Guide the Culture

Conduct periodic audits and reviews. Many companies find that conducting periodic employee surveys addressing attitudes and perceptions toward safety-related topics is very effective in monitoring the health of the organization's process safety culture. Monitoring can be done more frequently for high risk operations or in areas where the culture is recognized to need improvement. Survey results should be trended over time, especially during periods when particular efforts are being implemented to effect changes in the culture. Management should also determine the appropriate degree of resolution for such studies. For example, is a facility-wide survey satisfactory, or does a need exist to be able to discern different attitude patterns in the various facility departments?

Use independent auditors or reviewers. To ensure objectivity, process safety culture evaluators should have a suitable degree of independence from the organization being studied. This may be achieved by using knowledgeable personnel from an unaligned facility within the company, or by employing experts from outside the company.

Perform an annual executive review of process safety culture. Include an evaluation of process safety culture issues and performance in annual reviews during which company executives discuss safety and business performance with facility management.

Search for deeper root causes of process safety performance problems and be alert for weak signals of culture problems. Chronic performance problems can be identified by incidents, audit results, management reviews, equipment failures, and so forth. These situations should be investigated in sufficient depth to identify any underlying cultural issues contributing to the poor performance. For example, facility managers might be aware of staff members circumventing the *MOC* system, yet take no steps to correct their actions. Management must determine what issues exist with respect to facility culture that foster the attitude that policies and procedures can be violated without repercussion.

3.5   ELEMENT METRICS

Chapter 20 describes how metrics can be used to improve performance and when they may be appropriate. This section includes several examples of metrics that could be used to monitor the health of the *process safety culture* element, sorted by the key principles that were first introduced in Section 3.2.

In addition to identifying high value metrics, readers will need to determine how to best measure each metric they choose to track. In some cases, an ordinal number provides the needed information, for example, total number of workers. Other cases, such as average years of service, require that two or more attributes be indexed to provide meaningful information. Still other metrics may need to be tracked as a rate, as in the case of employee turnover. Sometimes, the rate of change may be of greatest interest. Since every situation is different, the reader will need to determine how to track and present data in a manner that best serves the need to monitor the health of RBPS management systems at their facility.

3.5.1   Maintain a Dependable Practice

- *Frequency with which upper managers visit the worksite, or percentage of the scheduled visits that actually take place.* A low value may indicate that upper management places little value on employee input and on personally motivating strong process safety performance.
- *Percentage of managers and supervisors trained on the importance of, and approaches to create and reinforce, a sound process safety culture.* A high value is indicative of the value that the organization places on implementing positive process safety culture change.
- *Percentage of meetings that address process safety and include active participation by a member of upper management.* An imbalance in emphasis may be indicative of a management attitude that process safety is less important.
- *Percentage of employees receiving either rewards or corrective actions related to the quality of their fulfillment of process safety responsibilities.* This metric would indicate the amount of accountability or reinforcement provided for performance. Do these percentages appear appropriate considering current performance statistics?
- *Performance metrics for other RBPS elements.* Does trending of these metrics indicate a deterioration of performance over time that could signify a pattern of normalization of deviance (e.g., increasing delinquencies of relief valve tests, failure to meet risk analysis schedules, failure to meet refresher training schedules)?

- *Relative frequency and emphasis of process safety-related topics and other topics such as cost, quality, and production in management communications.* An imbalance in emphasis may imply that management does not feel that process safety is important.

3.5.2 Develop and Implement a Sound Culture

- *Number of open recommendations (from risk analyses, incident investigations, audits, safety suggestions).* A large number of open recommendations, or an upward trend, may indicate a lack of responsiveness to process safety issues.
- *Number of near misses and incidents reported each month.* Is the rate high enough to indicate a healthy reporting system, without being so high as to indicate serious weaknesses in process safety management systems?
- *Typical, and maximum, durations for completing an incident investigation and issuing the report.* Long durations may indicate a lack of awareness of the importance of promptly addressing process safety issues.
- *Percentage of near misses and incidents identified as being caused by unsafe acts or shortcuts.* A high value may indicate a tendency in the organization to take inappropriate risks.
- *Number of meetings addressing process safety that are conducted per year.* Too few meetings may indicate a lack of process safety emphasis on the part of management.
- *Percentage of the required attendance achieved for meetings addressing process safety.* Low attendance may reflect a low level of employee engagement in process safety.
- *Frequency with which relevant process safety statistics are shared with the organization.* A low value may indicate that management does not adequately appreciate the value of informing the workforce of the organization's process safety performance.
- *Average response time to the resolution of a process safety suggestion.* Slow management responses to suggestions may provide a disincentive to employee participation.
- *Number of process safety suggestions reported each month.* A low value may reflect a low level of employee engagement in improving process safety, or a perception that employee participation offers a low return on the investment in effort.
- *Percentage of employees participating in the process safety suggestion program each month.* A low value may reflect a low level of employee engagement in improving process safety, or a perception that employee participation offers a low return on the investment in effort.

- *Manager attendance at management review meetings.* Poor attendance may indicate a low interest in process safety performance, or in communicating management expectations.

3.5.3   Monitor and Guide the Culture

- *Frequency with which relevant process safety metrics are prepared and shared with leadership.* A low value may indicate a lack of interest in, or a lack of awareness of the importance of, such information on the part of the leadership team.
- *Results of periodic employee attitude or perception surveys.* Do surveys reflect discernable trends, either upward or downward? What do these indicate about the health of the organization's process safety culture?

3.6   MANAGEMENT REVIEW

The overall design and conduct of management reviews is described in Chapter 22. However, many specific questions/discussion topics exist that management may want to check periodically to ensure that the management system for the *process safety culture* element is working properly. In particular, management must first seek to understand whether the system being reviewed is producing the desired results. If the organization's *process safety culture* is less than satisfactory, or it is not improving as a result of management system changes, then management should identify possible corrective actions and pursue them. Possibly, the organization is not working on the correct activities, or the organization is not doing the necessary activities well. Even if the results are satisfactory, management reviews can help determine if resources are being used wisely – are there tasks that could be done more efficiently or tasks that should not be done at all? Management can combine metrics listed in the previous section with personal observations, direct questioning, audit results, and feedback on topics to help answer these questions. Activities and topics for discussion include the following:

- Evaluate housekeeping practices during tours of the facility. While exemplary housekeeping may not be a guarantee of a sound process safety culture, poor housekeeping practices generally correlate with a poor culture.
- Monitor the content of non-verbal communications channels, for example, posters and video displays intended to provide information to the facility population. What topic is most frequently addressed – process safety or some other topic such as cost reduction initiatives?

- Interview employees to determine if they know their most significant process safety responsibilities, or if they understand their greatest process safety hazards.
- Based upon interviews, records of disciplinary actions, and incident reports, determine whether those who take safety shortcuts are disciplined, even if the shortcut is "successful" (e.g., it increases production, albeit in violation of procedures). Are exceptional examples of prudent behavior rewarded?
- Determine whether the potential for catastrophic consequences is adequately addressed during the conduct of risk analyses as well as when generalizing the lessons learned from incident investigations. Does a review of risk analyses indicate that credible major events are adequately addressed or, in contrast, that such events are commonly discounted as unlikely?
- Evaluate whether high performance standards are being maintained. In preparing for a management review, be alert for investigation reports or other information sources suggesting any patterns of tacit approval by supervisors of violations of standards or procedures, especially if operational advantages exist for doing so.
- Determine whether process safety-related decision-making responsibilities are clearly delegated to knowledgeable parties or, alternatively, whether the input from such individuals is sought by decision makers. Do interviews reveal or confirm that technical and process safety authorities feel they have appropriate input into the decision-making process?
- Compare the concerns expressed by employees during conversations and interviews with the issues that are reaching management through other communications channels. Do potential problems exist that would not normally become known through the normal communications channels?
- Determine if safety messages originating with management are actually reaching employees. Are the messages that reach them sufficiently similar to the original messages to confirm the integrity of the communications channels?
- Determine whether the process safety program includes a continuous improvement theme. Is someone at the facility or in the corporation thinking strategically about process safety management system improvements? Are appropriately challenging modifications periodically made to program goals?
- Determine whether an appropriate balance exists between the time management allots for reviewing process safety performance metrics versus the time spent reviewing production-related metrics.
- Determine, based upon interviews and reviews of investigation reports, whether the organization treats incident investigations as a search for the truth or a search for the guilty.

- Evaluate the frequency and duration of management visits to the worksite. Do such visits provide an adequate opportunity for workers to discuss process safety topics with management?
- Spend time directly observing work activities to determine whether work practices conform to applicable process safety requirements.

Perhaps the most effective means for gaining a near real time perspective on the process safety culture is through frequent and substantive contacts with employees. Such contacts allow management to (1) test the communications channels (Are employees hearing what management is saying? Is management hearing what employees are saying?) and (2) directly sample employee attitudes, assuming a sufficiently trusting environment has been established. Such contact promotes the discussion of problem areas and allows management personnel to demonstrate their value for the safety program.

3.7   REFERENCES

3.1   Jones, David, "Turning the Titanic – Three Case Histories in Cultural Change," CCPS International Conference and Workshop, Toronto, 2001.

3.2   Schein, E.H., *Organizational Culture and Leadership*, 3rd Ed., San Francisco: Jossey-Bass, 2004.

3.3   Hopkins, Andrew, *Lessons From Longford*, CCH Australia Limited, 2000.

3.4   UK Department of Energy, *The Public Inquiry Into the Piper Alpha Disaster* (2 Vol.), London: HMSO, ISBN 0101113102, 1990.

3.5   Weick, Karl E. and Sutcliffe, Kathleen M., *Managing the Unexpected*, Jossey-Bass, 2001.

3.6   Jones, D. and Kadri, S., "Nurturing a Strong Process Safety Culture," 20th Annual CCPS International Conference, AIChE Spring Meeting, Atlanta, GA, April 2005.

3.7   Center for Chemical Process Safety, "Building Process Safety Culture: Tools to Enhance Process Safety Performance," http://www.aiche.org/CCPS/PSCulture.aspx.

Additional reading

Frank, W. L., "Essential Elements of a Sound Safety Culture," AIChE, Process Plant Safety Symposium, Atlanta, GA, April 2005.

UK Health and Safety Executive, *Safety Culture: A Review of the Literature*, HSL/2002/25, 2002.

Columbia Accident Investigation Board Report, U.S. Government Printing Office, 2003.

4

COMPLIANCE WITH STANDARDS

On February 20, 2003, an explosion and fire damaged a manufacturing facility in Corbin, Kentucky, fatally injuring seven workers. The facility produced fiberglass insulation for the automotive industry. Investigators found that the explosion was fueled by resin dust that had accumulated in a production area, which was likely ignited by flames from a malfunctioning oven. The resin involved was a phenolic binder used in producing fiberglass mats. The investigation also determined that the company was largely unaware of the hazards of dust explosions and recommended prevention practices and technology available for preventing dust explosions. These technologies are listed in numerous industry standards and guidelines (Ref. 4.1).

4.1   ELEMENT OVERVIEW

Identifying and addressing relevant process safety standards, codes, regulations, and laws over the life of a process is one of the five elements in the RBPS pillar of *committing to process safety.* This chapter describes a process for maintaining adherence to applicable standards, codes, regulations, and laws (*standards*), the attributes of a *standards* system, and the steps an organization might take to implement the *standards* element. Section 4.2 describes the key principles and essential features of a management system for this element. Section 4.3 lists work activities that support these essential features and presents a range of approaches that might be appropriate for each work activity, depending on perceived risk, resources, and organizational culture. Sections 4.4 through 4.6 include (1) ideas for improving the effectiveness of management systems and specific programs that support this element, (2) metrics that could be used to monitor this element, and (3) management review issues that may be appropriate.

4.1.1   What Is It?

Standards is a system to identify, develop, acquire, evaluate, disseminate, and provide access to applicable standards, codes, regulations, and laws that affect process safety. The *standards* system addresses both internal and external standards; national and international codes and standards; and local, state, and federal regulations and laws. The system makes this information easily and quickly accessible to potential users. The *standards* system interacts in some fashion with every RBPS management system element.

4.1.2   Why Is It Important?

Knowledge of and conformance to *standards* helps a company (1) operate and maintain a safe facility, (2) consistently implement process safety practices, and (3) minimize legal liability. Changes in *standards* may occur at irregular intervals or on a fixed schedule, and the *standards* system must keep up with such changes so the company can adjust its compliance activities. The *standards* system also forms the basis for the standards of care used in an audit program to determine management system conformance.

4.1.3   Where/When Is It Done?

To promote consistent interpretation, implementation, and efficiency, the initial identification of and ongoing monitoring of changes in *standards* is frequently done at a company level. On the other hand, *standards* activities are also performed at each facility, where the staff are more familiar with state and local laws and regulations. *Standards* activities should begin early in the process life cycle to ensure that process designs meet applicable codes and standards from the outset, rather than having to make expensive changes later. The *MOC* element should address changes initiated by the *standards* system, and the *auditing* element monitors the compliance actions that must be taken.

4.1.4   Who Does It?

Generally, identification of applicable standards and codes is done by someone with a technical background who has a need for such information, such as an equipment or facility designer. Determination of applicable regulations and laws is typically done by someone knowledgeable with federal, state, and local agencies that adopt such provisions. Frequently, this person works closely with the company legal department to ensure accurate interpretations.

4.1.5   What Is the Anticipated Work Product?

The main products of a *standards* system are an accurate, complete, up-to-date, and accessible set of documents, data, and information. Categories of information include internal company guidelines, consensus codes and standards, applicable regulations, and laws. Ancillary products include company guidance documents that may be created to help ensure efficient, consistent conformance to *standards*. Outputs of the *standards* element are used to facilitate the performance of other elements. For example, determining the relevant external codes and standards to which a company must adhere will help form the scope of the *auditing* element.

4.1.6 How Is It Done?

If the *standards* element work is done at a company level, then the responsible party keeps a list of all applicable requirements and copies of all such updated documents. This information is typically communicated to division- and facility-level personnel responsible for local compliance activities.

Facilities that exhibit a high demand rate for maintaining compliance with frequently changing *standards* may need greater specificity in the *standards* procedure and larger allocation of personnel resources to fulfill the defined roles and responsibilities. Lower demand situations can allow facilities to operate a *standards* protocol with greater flexibility – possibly with a single person providing the advisory service at a divisional or corporate level for multiple facilities. Facilities with strong process safety cultures generally will have more performance-based *standards* procedures, allowing trained employees to use good judgment in managing compliance. Facilities with an immature or evolving process safety culture may require more prescriptive *standards* procedures, more frequent auditing, and greater command and control management system features to ensure good *standards* implementation discipline.

4.2   KEY PRINCIPLES AND ESSENTIAL FEATURES

Safe operation and maintenance of facilities that manufacture, store, or otherwise use hazardous chemicals requires robust process safety management systems. The primary objective of the *standards* element is to ensure that a facility remains in conformance with applicable standards, codes, regulations, and laws, including voluntary ones adopted by the company over the life of the facility. Long-term conformance to such standards of care helps ensure that the facility is operated in a safe and legal fashion.

A *standards* system establishes a formal process for maintaining institutional awareness about these obligations. Other elements, including *auditing* and *management review*, use the standard of care established by the *standards* element to determine whether the company is meeting its process safety obligations.

Table 4.1 provides some examples of the types of process safety obligations addressed by this element. *Note: This alphabetical list contains examples and does not represent a complete list.*

The following key principles should be addressed when developing, evaluating, or improving any system for the *standards* element:

- Maintain a dependable practice.
- Conduct compliance work activities.
- Follow through on decisions, actions, and use of compliance results.

The essential features for each principle are further described in Sections 4.2.1 through 4.2.3. Section 4.3 describes work activities that typically underpin the essential features related to these principles. Facility management should evaluate the risks and potential benefits that may be achieved as a result of improvements in this element. Based on this evaluation, management should develop a management system, or upgrade an existing management system, to address some or all of the essential features. However, none of these work activities will be effective if the accompanying *culture* element does not embrace the use of reliable management systems. Even the best *standards* activities, without the right culture, will not produce sustainable conformance to standards, codes, regulations, and laws over the life of a process.

In the case of the *standards* element, risk takes on a two-fold meaning: (1) the risk of experiencing an incident and (2) the risk of experiencing an adverse regulatory outcome as a result of a noncompliance issue. In either case, higher risk situations usually dictate a greater need for formality and thoroughness in the implementation of a *standards* protocol. For example, a higher risk facility may develop a detailed written program that (1) identifies applicable standards, (2) specifies exactly how the facility will comply with each standard, and (3) how it will stay abreast of updates to *standards*. Companies having lower risk situations may appropriately decide to adopt a general policy about keeping abreast of applicable codes and standards.

4.2.1 Maintain a Dependable Practice

When a company identifies or defines an activity to be undertaken, that company likely wants the activity to be performed correctly and consistently over the life of the facility. The following essential features help ensure that process safety management activities are executed dependably across a facility involving a variety or people and situations.

TABLE 4.1. Examples and Sources of Process Safety Related Standards, Codes, Regulations, and Laws

Voluntary Industry Standards
• American Petroleum Institute Recommended Practices (Ref. 4.2)
• American Chemistry Council Responsible Care* Management System and RC 14001 (Ref. 4.3)
• ISO 14001 — Environmental Management System (Ref. 4.4)
• OHSAS 18001 — International Occupational Health and Safety Management System (Ref. 4.5)
• Organization for Economic Cooperation and Development — Guiding Principles on Chemical Accident Prevention, Preparedness, and Response, 2003 (Ref. 4.6)

Consensus Codes
• American National Standards Institute (Ref. 4.7)
• American Petroleum Institute (Ref. 4.2)
• American Society of Mechanical Engineers (Ref. 4.8)
• The Chlorine Institute (Ref. 4.9)
• The Instrumentation, Systems, and Automation Society/International Electrotechnical Commission (Ref. 4.10)
• National Fire Protection Association (Ref. 4.11)

U.S. Federal, State, and Local Laws and Regulations
• U.S. OSHA — Process Safety Management standard (29 CFR 1910.119) (Ref. 4.12)
• U.S. Occupational Safety and Health Act — General Duty Clause, Section 5(a)(1) (Ref. 4.13)
• U.S. EPA — Risk Management Program regulation (40 CFR 68) (Ref. 4.14)
• Clean Air Act — General Duty Requirements, Section 112(r)(1) (Ref. 4.15)
• California Risk Management and Prevention Program (Ref. 4.16)
• New Jersey Toxic Catastrophe Prevention Act (Ref. 4.17)
• Contra Costa County Industrial Safety Ordinance (Ref. 4.18)
• Delaware Extremely Hazardous Substances Risk Management Act (Ref. 4.19)
• Nevada Chemical Accident Prevention Program (Ref. 4.20)

International Laws and Regulations
• Australian National Standard for the Control of Major Hazard Facilities (Ref. 4.21)
• Canadian Environmental Protection Agency — Environmental Emergency Planning, CEPA, 1999 (section 200) (Ref. 4.22)
• European Commission Seveso II Directive (Ref. 4.23)
• Korean OSHA PSM standard (Ref. 4.24)
• Malaysia — Department of Occupational Safety and Health (DOSH) Ministry of Human Resources Malaysia, Section 16 of Act 514 (Ref. 4.25)
• Mexican Integral Security and Environmental Management System (SIASPA) (Ref. 4.26)
• United Kingdom, Health and Safety Executive COMAH Regulations (Ref. 4.27)

Ensure consistent implementation of the standards system. For consistent implementation, the *standards* activities should be documented to an appropriate level of detail in a procedure or a written program addressing the general management system aspects discussed in Section 1.4. The scope of what a *standards* system is supposed to address should be well defined to help ensure that all process safety obligations are addressed and all significant sources of potential changes to those obligations are monitored. Specific mechanisms for monitoring the status of significant process safety obligations – particularly existing and anticipated regulations – should be considered and defined to an appropriate level of detail. The *standards* element should have someone assigned to monitor the overall activity for this element and available to respond to interpretation questions/issues involving special situations and regulatory anomalies. In addition, a competent authority should be available who can grant a dispensation or waiver from an internal standard. Any such departures should be the exception and not the rule, and their technical basis should be documented.

Identify when standards compliance is needed. Companies should determine the applicability of potentially relevant process safety-related obligations, the standards of care invoked by these obligations, and interpretations that could affect work activity implementation at facilities. Companies should also monitor compliance deadlines and changes to these obligations. Knowledge of the compliance deadlines is critical for timely incorporation of external process safety obligations into RBPS activities. Companies should also determine the applicability and deadlines for internal or voluntary standards, such as adopting new risk tolerance criteria.

Involve competent personnel. Even within a detailed *standards* program, performance will only be as good as the people who are involved in conducting the *standards* practices. Key personnel should have a basic awareness of the *standards* system and its significant process safety obligations. Those individuals who are assigned specific *standards* practice duties will require more detailed training.

Ensure that standards compliance practices remain effective. Once a *standards* system is in place, periodic monitoring, maintenance, and corrective action may be needed to keep it operating at peak performance and efficiency. To ensure effectiveness during routine operation, companies should consider implementing a few *standards* system performance and efficiency metrics; however, data collection and analysis will be minimal because evaluating the real-time status of the *standards* system will rarely be necessary. Periodic audits are sufficient to monitor the performance of most *standards* system work activities, and they can spotlight whether the *standards* system is providing adequate information to process safety personnel.

4.2.2 Conduct Compliance Work Activities

The actual work required to maintain compliance to standards, codes, regulations, and laws is conducted in the other RBPS elements. The *standards* system provides a communication mechanism for informing management and personnel about the company's obligations and compliance status. In addition, the *standards* element is the focal point for monitoring changes to obligations and the potential impact of those changes on the company.

Provide appropriate inputs to standards activities. Facilities or individual users define their own requirements for monitoring interpretations, anticipated changes, and new initiatives related to the *standards* element. Companies should provide adequate access to information on standards, codes, and updates. Some obligations may require that key personnel maintain qualifications or competency in a specific process safety activity or discipline.

Conduct compliance assurance activities. On an ongoing basis, the *standards* system should identify all current and proposed obligations and make that information available to all potentially affected personnel. Summaries of these obligations may be provided, and change notices may be broadcast to affected personnel as they occur. Based on changes to existing obligations, or on potential new obligations, companies should determine the applicability to and compliance impact on facility operations. These changes may give rise to additional regulatory risk. *Standards* element personnel should provide information to management, regulatory affairs personnel, and industry groups regarding company concerns about process safety obligation changes.   They should also forecast necessary activities to maintain compliance with process safety obligations to ensure adequate resources.

Determine compliance status periodically as required and provide a status report to management. The *standards* system should coordinate with the *auditing* and *management review* elements to ensure that the appropriate compliance scrutiny is given to applicable operations.   The status of compliance should be communicated regularly to management, particularly if noncompliance issues arise. Some companies have independent compliance audits performed, while others use combined purpose audits. Whichever approach is used, the *standards* element personnel should request the *auditing* element to conduct compliance assurance reviews in advance of established deadlines. *Standards* element personnel may need to participate in such audits as necessary to ensure that the appropriate standards of care are applied to subject process safety management systems and activities.   Once these compliance audits or reviews are performed, the *standards* element is the mechanism for providing feedback to management concerning the compliance status.

Review the applicability of standards as new information or changes arise. The *standards* system should facilitate the company's regulatory affairs personnel in assessing potential new obligations, or changes to existing obligations, and help the facility determine the potential applicability and compliance impact to facility operations.   The *standards* system should provide information to facility management, regulatory affairs personnel, and industry groups regarding company concerns about process safety obligation changes.

4.2.3   Follow Through on Decisions, Actions, and Use of Compliance Results

The results of compliance status evaluations may dictate action by the company. If compliance is achieved, then no action is typically needed beyond possible notification of compliance to outside parties, as required. When compliance is not achieved, then management is informed, and the *standards* element participates in activities to regain compliance. The *standards* element is the archive for all compliance records.

Update compliance documents and reports as needed. Some obligations are met simply by executing process safety work activities in a prescribed manner meeting the specified standard of care. Other obligations carry notification or reporting requirements. As compliance status changes, *standards* element personnel may need to update compliance status reports or other documents. In addition, the company should update whatever awareness-level communication and training items they use to convey changes in process safety obligations to personnel.

Communicate conformance or submit compliance assurance records to the appropriate external entity. Notify the appropriate *standards* organization and submit necessary compliance reports to applicable organizations and government agencies. Communicate with applicable organizations and government agencies regarding compliance status, as required.

Maintain element work records. The main objective of the *standards* element is to provide information to other RBPS elements, allowing them to conduct their work activities in a manner that ensures compliance with all process safety obligations. Facilities should decide what types of records related to the *standards* element that they want to keep, including records that help demonstrate regulatory compliance. Based on that decision, management should develop a retention policy. Simple copies of regulatory submittals may be all that some facilities need.

4.3   POSSIBLE WORK ACTIVITIES

The RBPS approach suggests that the degree of rigor designed into each work activity should be tailored to risk, tempered by resource considerations, and tuned to the facility's culture. Thus, the degree of rigor that should be applied to a particular work activity will vary for each facility, and likely will vary between units or process areas within a single facility. Therefore, to develop a risk-based process safety management system, readers should perform the following steps:

1. Assess the risks at the facility, investigate the balance between the resource load for RBPS activities and available resources, and examine the facility's culture. This process is described in more detail in Section 2.2.
2. Estimate the potential benefits that may be achieved by addressing each of the key principles for this RBPS element. These principles are listed in Section 4.2.

3. Based on the results from steps 1 and 2, decide which essential features described in Sections 4.2.1 through 4.2.3 are necessary to properly manage risk.

4. For each essential feature that will be implemented, determine how it will be implemented and select the corresponding work activities described in this section. Note that this list of work activities cannot be comprehensive for all industries; readers will likely need to add work activities or modify some of the work activities listed in this section.

5. For each work activity that will be implemented, determine the level of rigor that will be required. Each work activity in this section is followed by two to five implementation options that describe an increasing degree of rigor. Note that work activities listed in this section are labeled with a number; implementation options are labeled with a letter.

Note: Regulatory requirements may specify that process safety management systems include certain features or work activities, or that a minimum level of detail be designed into specific work activities. Thus, the design and implementation of process safety management systems should be based on regulatory requirements as well as the guidance provided in this book.

4.3.1 Maintain a Dependable Practice

Ensure Consistent Implementation of the Standards System

1. Develop a written program that identifies all process safety obligations.
   a. An informal procedure exists.
   b. A simple written policy addresses compliance with major legal/regulatory requirements.
   c. A written program addresses all regulatory requirements.
   d. A detailed written program addresses all process safety-related obligations.

2. Establish a *standards* element owner.
   a. An ad hoc *standards* element owner is chosen.
   b. A part-time *standards* element owner, either local or offsite, is designated.
   c. The facility assigns a single, full-time *standards* element owner.
   d. Multiple *standards* element owners are designated across the facility according to the source of the obligation.

3. Define the roles and responsibilities for personnel assigned to perform activities to help ensure compliance with applicable process safety-related standards.

    a.   The *standards* element is everyone's responsibility.

    b.   A single worker is assigned responsibility for all *standards* in each facility area.

    c.   All *standards* element roles/responsibilities are assigned to job functions/departments.

Identify When Standards Compliance Is Needed

4.   Identify all sources and applicability of process safety obligations, voluntary and required.

    a.   Major regulatory obligations are identified.

    b.   All process safety-related obligations are defined.

    c.   Obligations associated with recognized and generally accepted good engineering practices are identified for each specific unit/facility.

5.   Determine the schedule for compliance with all process safety obligations.

    a.   Compliance deadlines for major regulations and laws are generally known.

    b.   Compliance deadlines for all major process safety-related regulatory/legal obligations are maintained on an action schedule.

    c.   Compliance deadlines for all process safety-related required and voluntary obligations are maintained on an action schedule.

6.   Develop the appropriate risk tolerance criteria or guidance for use in risk-based decision making situations (see also Chapter 9).

    a.   Informal risk tolerance criteria/guidance is used.

    b.   Risk tolerance criteria/guidance is defined for a few situations on an ad hoc basis.

    c.   Risk tolerance criteria/guidance is defined for some situations using industry standards.

    d.   Risk tolerance criteria/guidance is defined for many situations using company-specific information.

Involve Competent Personnel

7.   Define the technical and regulatory knowledge/skills needed for compliance.

    a.   Experience requirements are defined for some compliance tasks.

    b.   General compliance disciplines are defined for major regulations and laws.

    c.   Compliance qualifications and knowledge/skills are defined for all major regulatory/legal obligations associated with *standards*-related activities.

    d.   Compliance qualifications and knowledge/skills are defined for all major *standards*-related activities, including voluntary initiatives.

8.   Provide a competent technical and regulatory authority.

      a.   Several personnel informally serve in the role of *standards* expert.

      b.   Several personnel are formally designated as *standards* experts.

      c.   A single competent authority is designated to manage all compliance issues.

9.   Provide initial and refresher awareness training on relevant standards to the workforce.

      a.   Informal training is provided.

      b.   *Standards* practices are broadcast once (e.g., through e-mail) to everyone.

      c.   Initial *standards* element awareness training is provided once to affected personnel.

      d.   Initial and refresher *standards* element awareness training is provided to affected personnel.

Ensure that Standards Compliance Practices Remain Effective

10.   Create a list of all relevant process safety-related *standards*.

      a.   The *standards* list is kept by each facility, either electronically or as a hard copy.

      b.   The *standards* list is kept by the *standards* element coordinator and is available to all affected personnel.

      c.   In addition to item (b), the *standards* list is made accessible via a web-based system.

11.   Create an activity record for all *standards* compliance assurance activities.

      a.   Informal, ad hoc records are kept.

      b.   *Standards* records are kept by the corporate regulatory compliance manager.

      c.   *Standards* records are kept electronically at each facility by the *standards* element coordinator.

      d.   *Standards* records are kept by the *standards* element coordinator on the web and are accessible to all affected employees.

12.   Collect and evaluate performance/efficiency data on *standards* compliance activities.

      a.   Informal activity data are collected.

      b.   Basic regulatory activity data are collected and evaluated.

      c.   *Standards* element performance indicators are collected and evaluated annually.

      d.   *Standards* performance and efficiency indicators are collected and evaluated regularly.

13.   Provide input to improvement activities, such as metrics, management reviews, and audits.

      a.   Inputs are provided on an informal, as-needed basis.

    b.   Basic *standards* suggestions are provided to management prior to audits.

    c.   *Standards* compliance indicators are collected annually.

    d.   *Standards* performance and efficiency indicators are regularly discussed at periodic management reviews.

4.3.2   Conduct Compliance Work Activities

Provide Appropriate Inputs to Standards Activities

14. Solicit needs from within the company concerning *standards*, changes to or interpretation of *standards*, or other new initiatives.

    a.   Informal communication occurs regarding *standards*.

    b.   Communication occurs when a potential obligation requires facility input.

    c.   Key individuals know who to contact within the company, and an informal network exits.

    d.   A formal network or system is in place for users to identify their needs related to *standards*.

Conduct Compliance Assurance Activities

15. Coordinate with company legal counsel regarding regulatory and legal compliance issues.

    a.   *Standards* issues are informally monitored by regulatory and legal personnel.

    b.   *Standards* issues are regularly coordinated between facility and legal/regulatory departments.

    c.   *Standards* issues are discussed regularly with all affected parties and communicated as needed to affected personnel.

16. Communicate compliance obligations to all appropriate personnel.

    a.   Ad hoc communication occurs whenever a need is identified.

    b.   Informal communication/training on obligations occurs for some personnel.

    c.   A formal system exists for informing/training operating personnel on obligations.

    d.   A formal system exists for informing/training personnel on obligations for all potentially affected personnel.

17. Provide access to *standards* materials for all personnel who need them.

    a.   Personnel gain access on their own initiative.

    b.   A *standards* archive is compiled in a central location, and personnel are informed once about its existence.

    c.   A *standards* archive is compiled in a central location, and personnel are informed regularly about its existence.

      d. A *standards* archive is compiled electronically, and personnel are informed regularly about its existence.

18. Participate in or coordinate with *auditing* element activities.

      a. *Auditing* personnel are provided summary information on the *standards* archive.

      b. *Auditing* personnel are provided access to the *standards* archive.

      c. Item (b), and *auditing* personnel are trained on company decisions regarding applicable *standards*.

19. Monitor changes to *standards* via appropriate means.

      a. Personnel informally monitor sources of *standards* as needed.

      b. *Standards* sources are informally monitored by a single individual.

      c. *Standards* sources are regularly monitored by assigned personnel.

      d. *Standards* sources and changes are identified and communicated as needed to affected personnel.

20. Maintain a compliance schedule for relevant *standards* sources.

      a. A variety of informal compliance schedules exist for some obligations.

      b. A facility authority keeps an informal compliance schedule.

      c. A variety of formal compliance schedules exist.

      d. A single, unified compliance schedule exists for all known obligations.

21. Maintain compliance with each process safety-related obligation.

      a. Compliance is monitored informally by interested parties for a single important obligation.

      b. Compliance with some obligations is formally monitored by interested parties.

      c. Compliance with all important obligations is formally monitored by appropriate personnel.

      d. Compliance with all obligations is regularly monitored by appropriate personnel.

Determine Compliance Status Periodically as Required and Provide a Status Report to Management

22. Request the *auditing* element to conduct compliance assurance reviews in advance of established deadlines.

      a. Informal requests to assess compliance with obligations are made on an ad hoc basis.

      b. A formal schedule exists for requesting compliance reviews.

      c. Item (b), and *standards* element personnel regularly request compliance reviews.

23. Participate in audits as necessary to ensure that the *auditing* personnel use the appropriate standards of care for the subject process safety obligations.

    a. *Standards* element personnel participate occasionally in compliance reviews.

    b. *Standards* element personnel participate regularly in major compliance reviews.

    c. *Standards* element personnel participate regularly in all compliance reviews.

24. Provide feedback to management concerning the compliance status.

    a. Compliance conformance feedback is provided on an ad hoc basis.

    b. Compliance conformance feedback is provided on a regular basis for a few regulations.

    c. Compliance conformance feedback is provided on a regular basis for all regulations.

    d. Compliance conformance feedback on all obligations is tracked and communicated broadly on a regular basis.

Review the Applicability of Standards as New Information or Changes Arise

25. Based on potential new obligations or changes to existing obligations, determine the potential applicability and compliance impact to facility operations.

    a. An informal evaluation process exists to evaluate changes to obligations.

    b. Obligations are reviewed based on perceived changes.

    c. Personnel evaluate changes in applicability for major existing and new obligations.

    d. Personnel evaluate all compliance obligations based on need.

26. Provide information to management, regulatory affairs personnel, and industry groups regarding company concerns about changes to process safety obligation.

    a. An informal communication process exists.

    b. New applicability or obligations are communicated for major situations.

    c. New applicability or obligations are communicated routinely as a formal part of the company governance process.

    d. Item (c), and new applicability or obligations are communicated via the company network and are tracked until completion.

4.3.3 Follow Through on Decisions, Actions, and Use of Compliance Results

Update Compliance Documents and Reports as Needed

27. Modify compliance documents as needed based on changes in status.

    a. New applicability or obligations documents are updated on an informal basis.

      b. Compliance guidance is occasionally updated for major obligations.

      c. Compliance guidance is routinely updated for major obligations using a formal process.

      d. Compliance guidance is regularly updated for all obligations using a formal process.

28. Modify awareness-level communication and training items to convey changes in process safety obligations to personnel.

      a. An informal means exists to communicate some obligations on an ad hoc basis.

      b. An informal rollout exists to communicate changes to major obligations.

      c. A formal rollout communication exists for changes to most obligations.

      d. Refresher training is provided for changes to all obligations on a routine basis.

Communicate Conformance or Submit Compliance Assurance Records to the Appropriate External Entity

29. Submit necessary compliance reports to applicable organizations and government agencies.

      a. Individual process owners or facility personnel communicate compliance status as needed.

      b. Corporate managers oversee all compliance reporting obligations.

      c. The facility regulatory compliance manager ensures that all required reports are made.

      d. Items (b) and (c) occur.

Maintain Element Work Records

30. Maintain records concerning compliance activities.

      a. Informal records are kept by issue/process owners.

      b. Formal compliance obligation records for major regulations are kept by single person.

      c. Formal compliance obligation records for all regulations are kept by single person.

      d. Item (c), and an electronic or web-based system for *standards* obligations records is maintained.

4.4   EXAMPLES OF WAYS TO IMPROVE EFFECTIVENESS

This section provides specific examples of industry tested methods for improving the effectiveness of work activities related to the *standards* element. The examples are sorted by the key principles that were first introduced in Section 4.2. The examples fall into two categories:

1. Methods for improving the performance of activities that support this element.
2. Methods for improving the efficiency of activities that support this element by maintaining the necessary level of performance using fewer resources.

These examples were obtained from the results of industry practice surveys, workshops, and CCPS member-company input. Readers desiring to improve their management systems and work activities related to this element should examine these ideas, evaluate current management system and work activity performance and efficiency, and then select and implement enhancements using the risk-based principles described in Section 2.1.

4.4.1 Maintain a Dependable Practice

Develop a specific standards review procedure for each obligation. Companies may be subject to many *standards* obligations and may have opportunities to assume voluntary compliance targets. Before such commitments are made, a thorough review of the benefits, costs, and compliance schedule should be performed to determine if the proposed obligation and schedule is the best compliance approach.

4.4.2 Conduct Compliance Work Activities

Specify a minimum level of detail expected from standards reviews for each obligation. This element should provide the standard of care for the *auditing* element to use in conducting audits. Concurrently, companies should define the expectations for the level of detail of reviews to avoid overworking the compliance assessment process.

Provide detailed compliance guidance to personnel to ensure understanding. Some regulations and standards are complex and require special understanding of technical or performance-based language and requirements. Detailed guidelines can help ensure that company personnel aim at the correct compliance targets.

Maintain access to a newsletter service to keep up to date on changes. Various organizations provide hard copy or electronic regulatory and industry news services. Subscribing to one of these is an efficient way of keeping abreast of new initiatives and helps ensure that a company is informed of impending obligations.

Provide training to appropriate personnel on relevant standards. To comply with *standards*, company employees and contractors must understand them. Proper training can provide the knowledge necessary for those individuals involved in process safety to understand the *standards* and know how best to achieve compliance.

Participate in the standards setting process. Companies that are active in industry groups that monitor regulatory activities can preview upcoming regulations and may even be able to positively influence their outcome. Industry groups that create *standards* can significantly affect company facilities. Participating in pertinent groups and committees may help ensure that the activity considers a company's technical position before finalizing its requirements, thus reducing the burden on the company.

4.4.3 Follow Through on Decisions, Actions, and Use of Compliance Results

Review regulatory coverage. Periodic review of regulatory coverage will help ensure that facilities remain aware of new or revised interpretations and help minimize the likelihood of receiving regulatory citations.

4.5 ELEMENT METRICS

Chapter 20 describes how metrics can be used to improve performance and when they may be appropriate. This section includes several examples of metrics that could be used to monitor the health of the *standards* element, sorted by the key principles that were first introduced in Section 4.2.

In addition to identifying high value metrics, readers will need to determine how to best measure each metric they choose to track. In some cases, an ordinal number provides the needed information, for example, total number of workers. Other cases require that two or more attributes be indexed to provide meaningful information, such as averaging employees' length of service to obtain a unit's average years of experience per employee. Still other metrics, such as employee turnover, may need to be tracked as a rate. Sometimes, the rate of change may be of greatest interest. Since every situation is different, the reader will need to determine how to track and present data to most effectively monitor the health of RBPS management systems at their facility.

4.5.1 Maintain a Dependable Practice

- *The number of new sources of standards identified and adopted during the past year.* Identifying no new sources over an extended period may indicate an ineffective program.
- *The number of people trained on standards activities.* A high or increasing number may indicate an active program.

- *The average amount of calendar time taken for standards reviews.* A high or increasing number may indicate a need for efficiency improvements. On the other hand, coupled with a greater than average number of reviews or complex reviews, the time required may be reasonable.

4.5.2 Conduct Compliance Work Activities

- *The number of existing standards revised per year.* A low number over an extended period may indicate a lapse in *standards* review activities.
- *The number of standards organizations meetings attended per year.* A low or decreasing number may indicate reduced involvement of company personnel in *standards* outreach and monitoring activities.
- *The number of audits in which standards element personnel participated.* A low or zero number may indicate a lapse in *standards* interfacing activities with the *auditing* element.
- *The number of identified standards applicability changes.* A low number may indicate a lapse in program activity; a high number may foreshadow new compliance burdens that management should be aware of for resource budgeting purposes.

4.5.3 Follow Through on Decisions, Actions, and Use of Compliance Results

- *The number of compliance violations per year.* A high number may indicate a breakdown in the *standards* interfaces with facility personnel or the *auditing* program for regulatory compliance.
- *The number of non-conformances to non-regulatory standards per audit.* A high number may indicate a breakdown in the *standards* interfaces with facility personnel or the *auditing* program for voluntary or internal standards.
- *The average amount of calendar time between standards system review completion and closeout of all action items.* A large period may indicate an inefficient program or one in which follow-up activities are lapsing.

4.6 MANAGEMENT REVIEW

The overall design and conduct of management reviews is described in Chapter 22. However, many specific questions/discussion topics exist that management may want to check periodically to ensure that the management system for the *standards* element is working properly. In particular, management must first seek to understand whether the system being reviewed is producing the desired results. If the organization's level of conformance to accepted standards, codes, regulations, and laws is less than satisfactory, or it is not improving as a result of management system changes, then management should identify possible corrective actions and pursue them. Possibly, the organization is not working on the correct activities, or the organization is not doing the necessary activities well. Even if the results are satisfactory, management reviews can help determine if resources are being used wisely – are there tasks that could be done more efficiently or tasks that should not be done at all? Management can combine metrics and indicators listed in the previous section with personal observations, direct questioning, audit results, and feedback on various topics to help answer these questions. Activities and topics for discussion include the following:

- Review feedback from federal, state, and local regulators.
- Determine whether company standards and guidelines are up to date.
- Review the schedule for reviewing and revising company standards and guidelines.
- Determine whether compliance implementation plans are being achieved.
- Verify that the appropriate personnel are participating in regulatory and industry monitoring activities.
- Review the attendance of company personnel at recent industry meetings.
- Review the status of the latest rollout/implementation plan with a company standard.
- Determine whether action items dealing with company standards/guidelines development are overdue.
- Determine whether any regulatory citations have been received.
- Verify the last time that regulatory interpretation guidance was updated.
- Review resource utilization for staff groups addressing compliance issues.

In addition, metrics are updated and the *standards* element owner normally makes a special effort to understand the reasons for any trends or anomalies in the metrics. Finally, if any major near-term *standards* activities to address known regulatory issues are planned, a briefing is often prepared for the management review committee.

The results of a management review of *standards* activities should demonstrate that leadership at the facility is aware of and values *standards*. Management review offers an opportunity for facility personnel to ask whether sufficient time and resources are being applied to efforts to maintain regulatory, industry, and internal standards compliance.

Management review meetings help provide focus to efforts (1) to implement planned *standards* activities, (2) to participate in regulatory advocacy efforts and industry groups, and (3) to develop internal guidelines. Such meetings also provide an opportunity to identify ways to improve the efficiency of work activities that support this element. In addition, an effective management review process educates the entire leadership team on the importance of *standards* and the role the *standards* program can play in helping to maintain the compliance status and image of the facility in the eyes of the industry, community, and government.

4.7   REFERENCES

4.1   *Combustible Dust Fire and Explosions at CTA Acoustics, Inc. Corbin, Kentucky February 20, 2003*; U.S. Chemical Safety and Hazard Investigation Board, February 15, 2005. http://www.csb.gov/index.cfm?folder=completed_investigations&page=info&INV_ID=35

4.2   American Petroleum Institute, 1220 L Street, NW, Washington, DC 20005. www.api.org

4.3   American Chemistry Council 1300 Wilson Blvd., Arlington, VA 22209. www.americanchemistry.com

4.4   ISO 14001 – Environmental Management System, International Organization for Standardization (ISO), Geneva, Switzerland. www.iso.org/iso/en/iso9000-14000/index.html

4.5   OHSAS 18001 – International Occupational Health and Safety Management System. www.ohsas-18001-occupational-health-and-safety.com/

4.6   Organization for Economic Cooperation and Development – Guiding Principles on Chemical Accident Prevention, Preparedness, and Response, 2nd edition, 2003, Organisation for Economic Co-Operation and Development, Paris, 2003. www2.oecd.org/ guidingprinciples/index.asp

4.7   American National Standards Institute, 25 West 43rd Street, New York, NY 10036. www.ansi.org

4.8   American Society of Mechanical Engineers, Three Park Avenue, New York, NY 10016. www.asme.org

4.9   The Chlorine Institute, 1300 Wilson Blvd., Arlington, VA 22209, www.chlorineinstitute.org

4.10   The Instrumentation, Systems, and Automation Society/International Electrotechnical Commission, 67 Alexander Drive, Research Triangle Park, NC 27709. www.isa.org

4.11   National Fire Protection Association, 1 Batterymarch Park, Quincy, Massachusetts, 02169. www.nfpa.org

4.12   Process Safety Management of Highly Hazardous Chemicals (29 CFR 1910.119), *U.S. Occupational Safety and Health Administration, May 1992. www.osha.gov*

4.13   Section 5(a)(1) – General Duty Clause, Occupational Safety and Health Act of 1970, Public Law 91-596, 29 USC 654, December 29, 1970. www.osha.gov

4.14   Accidental Release Prevention Requirements: Risk Management Programs Under Clean Air Act Section 112(r)(7), 40 CFR 68, U.S. Environmental Protection Agency, June 20, 1996 Fed. Reg. Vol. 61[31667-31730]. www.epa.gov

4.15   Clean Air Act Section 112(r)(1) – Prevention of Accidental Releases – Purpose and general duty, Public Law No. 101-549, November 1990 www.epa.gov

4.16   California Accidental Release Program (CalARP) Regulation, CCR Title 19, Division 2 – Office of Emergency Services, Chapter 4.5, June 28, 2004. www.oes.ca.gov

4.17   Toxic Catastrophe Prevention Act (TCPA), New Jersey Department of Environmental Protection Bureau of Chemical Release Information and Prevention, N.J.A.C. 7:31 Consolidated Rule Document, April 17, 2006. www.nj.gov/dep

4.18   Contra Costa County Industrial Safety Ordinance. www.co.contra-costa.ca.us/

4.19   Extremely Hazardous Substances Risk Management Act, Regulation 1201, Accidental Release Prevention Regulation, Delaware Department of Natural Resources and Environmental Control, March 11, 2006. www.dnrec.delaware.gov/

4.20   Chemical Accident Prevention Program (CAPP), Nevada Division of Environmental Protection, NRS 459.380, February 15, 2005. http://ndep.nv.gov/bapc/capp/

4.21   Australian National Standard for the Control of Major Hazard Facilities, NOHSC: 1014, 2002. www.docep.wa.gov.au/

4.22   Environmental Emergency Regulations (SOR/2003-307), Environment Canada. www.ec.gc.ca/CEPARegistry/regulations/ detailReg.cfm?intReg=70

4.23   Control of Major-Accident Hazards Involving Dangerous Substances, European Directive Seveso II (96/82/EC). http://europa.eu.int/comm/environment/seveso/

4.24   Korean OSHA PSM standard, Industrial Safety and Health Act – Article 20, Preparation of Safety and Health Management Regulations. Korean Ministry of Environment – Framework Plan on Hazardous Chemicals Management, 2001-2005. www.kosha.net/jsp/board/viewlist.jsp?cf=29099&x=19565&no=3

4.25   Malaysia – Department of Occupational Safety and Health (DOSH) Ministry of Human Resources Malaysia, Section 16 of Act 514. http://dosh.mohr.gov.my/

4.26   Mexican Integral Security and Environmental Management System (SIASPA), 1998. www.pepsonline.org/Publications/pemex.pdf

4.27   Control of Major Accident Hazards Regulations (COMAH), United Kingdom Health & Safety Executive, 1999 and 2005. www.hse.gov.uk/comah/

5

PROCESS SAFETY COMPETENCY

On April 21, 1995, a blender containing a mixture of sodium hydrosulfite, aluminum powder, potassium carbonate, and benzaldehyde exploded and triggered a major fire at a specialty chemical plant in Lodi, New Jersey. Five employees were killed and many more were injured. Most of the plant was destroyed as a result of the fire, and other nearby businesses were destroyed or significantly damaged. Property damage at the plant was estimated at $20M. An EPA/OSHA joint chemical accident investigation team (JCAIT) determined that the immediate cause of the explosion was the inadvertent introduction of water and heat into water-reactive materials during the mixing operation. During an emergency operation to offload the blender contents, the material ignited and a deflagration occurred. The JCAIT identified the following root causes and contributing factors of the accident:

- An inadequate process hazards analysis was conducted, and appropriate preventive actions were not taken.
- Standard operating procedures and training were inadequate.
- The decision to re-enter the plant and offload the blender was based on inadequate information.
- The blender used was inappropriate for the materials blended.
- Communications between the plant and the company that provided the blending technology was inadequate.
- The training of fire brigade members and emergency responders was inadequate.

This incident illustrates how weaknesses in process safety competency can result in poor understanding of hazards at all levels of the organization, and ultimately in poor decision making. For example, the joint investigation report issued by EPA and OSHA (Ref. 5.1) specifically addressed:

- Failure to properly identify and analyze hazards during the process hazard analysis.
- Less than adequate operator training.
- Less than adequate information available to guide emergency response decisions.

5.1   ELEMENT OVERVIEW

Developing, sustaining, and enhancing the organization's process safety competency is one of five elements in the RBPS pillar of *committing to process safety*. This chapter describes what process safety competency means, the attributes of process safety competency, and how organizations might enhance their own competency. Section 5.2 describes the key principles and essential features of a management system for this element. Section 5.3 lists work activities that support these essential features, and presents a range of approaches that might be appropriate for each work activity, depending on perceived risk, resources, and organizational culture. Sections 5.4 through 5.6 include (1) ideas to improve the effectiveness of management systems and specific programs that support this element, (2) metrics that could be used to monitor this element, and (3) issues that may be appropriate for management review.

5.1.1   What Is It?

Developing and maintaining process safety competency encompasses three interrelated actions: (1) continuously improving knowledge and competency, (2) ensuring that appropriate information is available to people who need it, and (3) consistently applying what has been learned.

The learning aspect includes efforts to develop, discover, or otherwise enhance knowledge. It ranges from narrowly defined tasks that develop new information based on a specific request, such as, conducting experiments that provide data needed by hazard identification and risk analysis teams, to wide ranging efforts to maintain and advance the knowledge base of the entire organization or even a sector of the chemical industry. The learning aspect also includes structured means to retain people-based knowledge, including succession planning.

Process safety competency is closely related to the *knowledge* and *training* elements of the RBPS system. While the competency element often generates new information, the *knowledge* element provides the means to catalog and store information so that it can be retrieved on request. The *competency* element focuses primarily on organizational learning, whereas the *training* element addresses efforts to develop and maintain the competence of each individual worker.

The *competency* element involves increasing the body of knowledge and, when applicable, pushing newly acquired knowledge out to appropriate parts of the organization, sometimes independently of any request. Most important, this element supports the application of this body of process knowledge to situations that help manage risk and improve plant performance.

5.1.2 Why Is It Important?

Although catastrophic process safety incidents are relatively rare, the losses associated with the incidents can be devastating. Learning must be proactive, and lessons must not be forgotten. Business seems to be changing at an ever increasing rate. Acquisitions, divestitures, reorganizations, and resignations of key individuals make it more difficult than ever to maintain competency simply by relying on the knowledge in people's heads. Simultaneously, the growth in information technology enables businesses to efficiently control, expand, and manage a storehouse of information and access this information from anywhere at any time. However, only competent people can transform information into knowledge. Knowledge management, not information management, helps organizations understand and manage risk and remain competitive.

5.1.3 Where/When Is It Done?

Developing and maintaining competency is done almost everywhere in an organization. Organizational competency is enhanced wherever observations and analysis collide to produce discovery, and it benefits from the more mundane tasks of searching literature and attending technical meetings. In high performing organizations, any time is appropriate for learning; maintaining and improving organizational competency is a continuous process.

5.1.4 Who Does It?

The *competency* element involves a wide variety of personnel. In many cases, a single engineer or technical specialist is assigned to spearhead the efforts of a team charged with gathering and maintaining the knowledge relevant to a particular technology or process. However, the value of this information is limited unless it is put to use. Hence, the information must somehow be captured and applied throughout the organization, often via collaborative projects and improvement efforts involving corporate gurus, supervisors, process engineers, operators, maintenance personnel, and facility management.

5.1.5   What Is the Anticipated Work Product?

The main product of the *competency* element is an understanding and interpretation of knowledge that helps the organization make better decisions and increases the likelihood that individuals who are faced with an abnormal situation will take the proper action. This differs from the *knowledge* element described in Chapter 8. The *knowledge* element is mainly a collection of data and information in written form. Although the *competency* element also produces written records, its main output to the other RBPS elements is understanding. Information provided by the *knowledge* element and understanding provided by the *competency* element underpin almost every other RBPS element.

The *competency* element interfaces with many other RBPS elements. For example, it directly complements the *training* element, which primarily focuses on individual competence. It also links to the *standards* element, particularly in the area of sharing knowledge with external organizations and influencing industry-specific standards and recommended practices.

Normally, the most tangible work product produced by the *competency* element is a technology manual. Even though much of the information collected involves written documentation, transcribing what experienced employees intuitively know or feel is not always possible or feasible. Thus, another important work product is a means to effectively manage personnel transitions.

Collecting information is often relatively inexpensive – it includes activities such as attending meetings, reading papers, supporting collective projects (e.g., the CCPS Design Institute for Physical Property Research), or participating in industry-wide technical committees. Conversely, improving process understanding can require a long-term research effort. Regardless, a work product that consists solely of a pile of technical papers that are routed to managers, supervisors, and technical personnel at operating facilities is likely to provide little benefit. To maximize return on investment, the information must be evaluated, made relevant to operating units, and stored in a format that will support learning, remembering, and when appropriate, action.

5.1.6   How Is It Done?

Unlike some of the other RBPS elements, no simple answer exists as to "how" *competency* is "done." The single most important factor is a commitment by senior management to support efforts to learn, and to share new information and insights among units at a facility, with sister facilities within the company, and potentially with other companies. Once the commitment is in place, opportunities to learn and interact with others abound. Some information will have to be passed along through mentorship and collaboration; both of these activities typically require active management support to ensure success. A closely related activity is succession planning, which is an intentional activity that helps ensure that key positions are staffed with individuals who possess specific knowledge and experience.

5.2 KEY PRINCIPLES AND ESSENTIAL FEATURES

Safe operation and maintenance of facilities that manufacture, store, or otherwise use hazardous chemicals requires process safety competency. That is, facilities should implement management systems to (1) help the organization proactively identify learning needs that are critical to process safety, (2) support efforts to learn or obtain the critical knowledge, (3) maintain knowledge in a manner that helps promote risk-informed decision making, and (4) share the information with other facilities (including, in some cases, competitors). In addition, competent organizations are constantly in search of opportunities to learn from others, or from their own experience.

The following key principles should be addressed when developing, evaluating, or improving any management system for the *competency* element:

- Maintain a dependable practice.
- Execute activities that help maintain and enhance process safety competency.
- Evaluate and share results.
- Adjust plans.

The essential features for each principle are further described in Sections 5.2.1 through 5.2.4. Section 5.3 describes work activities that typically underpin the essential features related to these principles. Facility management should evaluate the risks and potential benefits that may be achieved as a result of improvements in this element. Based on this evaluation, the facility should develop a management system, or upgrade an existing management system, to (1) address some or all of the essential features and (2) execute some or all of the work activities, depending on perceived risk and/or identified process hazards. However, these steps will be ineffective if the accompanying *culture* element does not embrace the use of reliable management systems. Even the best management system, without the right culture, will not produce a competent organization.

5.2.1 Maintain a Dependable Practice

Almost all companies profess to be learning organizations that aspire to a high degree of competency. However, those that are successful in this pursuit intentionally foster learning by establishing objectives and making plans to achieve the objectives. Normally, one or more of several conditions are necessary for an organization to invest in process safety competency:

- A business case describes the expected benefits and the level of resources that must be invested to achieve those benefits.
- The organization inherently values technology and places particular value on enhancing its process safety competency.
- The organization believes that decisions should be based on knowledge that is supported by facts, and any significant improvement in the body of knowledge will lead to better decisions, thereby reducing risk and improving performance.

Without a strong business case or cultural driver, written policies and procedures aimed at increasing competency within the organization can quickly vanish when the business encounters a downturn.

Establish objectives. Establishing specific goals that are jointly supported by the technical and manufacturing organizations (and valued by the business) helps ensure that the effort provides meaningful results. Conversely, if the only measure of success is driving the occurrence of infrequent loss events even lower, management will find it difficult to determine if the effort is paying off. Hence, sustaining support for these activities will also be difficult. If objectives are established that are measurable and can be tied to overall business performance, sustaining the resources needed to execute the necessary work activities becomes easier.

Appoint a champion. If maintaining and enhancing process safety competency is everyone's responsibility, the likelihood is low that it will be a high priority for any single individual. To address this potential problem, overall responsibility for this element is often assigned to a senior manager in the technology; environmental, safety, and health; or engineering group. In other cases, this role is assigned to a widely respected senior engineer. Appointing a corporate-wide champion helps ensure consistent practice and facilitate exchange of information between facilities.

Identify corollary benefits. In most cases, efforts to better understand and manage risk will produce other more tangible benefits. For example, the same loss of institutional knowledge that can lead to a repeated process safety incident likely leads to even more frequent losses associated with lower yield and output, higher maintenance costs, increased quality defects and customer complaints, and so forth. In fact, adequately maintaining process safety competency would be difficult without also maintaining and enhancing overall process knowledge, which should benefit the company in areas other than safety. Closely linking efforts to maintain and enhance process safety knowledge with efforts to maintain and enhance competency throughout the organization should result in direct financial benefits.

Develop a learning plan.. Once objectives and ownership are established, determine the resources that will be needed to achieve the objectives, keeping one eye on the expected value of the results compared to the required resources/investment.   David Garvin characterizes a learning plan as something that:

- Incorporates the results of uncertainty cataloging, which is proactively asking, "What else might we need to know and what benefit might this information provide?"
- Spells out assumptions about each uncertainty.
- Presents approaches for testing each assumption and resolving each critical uncertainty through experimentation and learning.
- Prioritizes the assumption-testing tasks and defines a path for moving forward as quickly and inexpensively as possible.
- Serves as a log of efforts to maintain competency and a reference for future learning plans (Ref. 5.2).

In addition to identifying learning objectives, success depends on appointing an owner, translating objectives into specific action statements, identifying corollary benefits, and establishing a budget.

Promote a learning organization.   A learning organization promotes activities that help create, acquire, interpret, transfer, and retain knowledge. Learning organizations are adept at translating new knowledge into new ways of behaving, and managing the learning process so that it is focused and purposeful. Learning occurs by design and in pursuit of clearly defined needs.

David Garvin identifies the following six characteristics of a learning organization:

- **Recognizes and accepts differences**, and supports discussion and evaluation of divergent opinions and data.
- **Provides timely feedback** and flexibility in the means used to conduct work activities. Current methods are evaluated if a new, and potentially better, approach is introduced.
- **Stimulates new ideas** to promote a step change in risk understanding and operational performance. Small incremental changes can play a significant role in developing and maintaining competency and managing risk, but the incremental approach should not preclude new ways of thinking and acting.
- **Maintains an external focus**. Ideas or approaches developed outside the organization are not automatically discounted, but are evaluated to determine if they fit the facility's objectives. Similarly, when studying previous incidents within the industry, a learning organization understands that "it can happen here" and takes steps to improve safety.

- **Tolerates errors and mistakes**, but learns from them. Obviously, mistakes that can have catastrophic consequences are not tolerable, but failure to encourage innovation, analyze errors, and promote further learning stifles improvement. Do not demand perfection.
- **Establishes and periodically updates the learning plan** to help focus efforts to increase competence and revalidate the perceived benefits and expected costs of learning activities.

Garvin goes on to identify five "learning disabilities" that are often encountered in organizations. Signs of any of these learning disabilities should sound an alarm. They are:

- **Blind spots** – narrow focus, poor assumptions, disruptive technologies
- **Filtering** – ignoring or downplaying information that doesn't fit into the existing paradigm
- **Lack of information sharing** – ineffective sharing, information hoarding
- **Flawed interpretation** – poor logic, emotional bias, hindsight
- **Inaction** – inability or unwillingness to act

5.2.2 Execute Activities that Help Maintain and Enhance Process Safety Competency

Owners, budgets, plans, and objectives alone are normally insufficient to bring about positive change. These need to be transformed into actions that improve competency.

Appoint technology stewards. Appoint a person to be responsible for maintaining the collective knowledge regarding each process, including process safety-related knowledge. This role often involves coordinating work done by others; one single person is unlikely to have the range of skills necessary to address the different types of knowledge and experience needed. Normally, this is a part-time assignment for a senior engineer or technologist who has been closely involved with the process and its technology for many years.

In some cases, this concept is extended to broad technology areas that are critical to maintaining and enhancing competency. For example, some companies appoint a corporate steward in the area of materials science/corrosion. Many other technology areas exist that companies often consider critical and continuously work to develop expertise. In some instances, companies work to advance the state-of-the-art throughout the industry via jointly funded initiatives, for example, supporting professional societies and associations that issue codes, standards, and recommended practices.

Document knowledge. One means to sustain and share knowledge is to document information in a technology manual. Technology manuals often reference internal design standards or external codes/standards, which helps to document the design team's thought process. Understanding the designer's intent is critical to evaluating changes to the design. In many cases, the technology manuals are now completely electronic, sometimes with hyperlinks to related documents.

Documenting process-specific knowledge is increasingly important during this era of divestitures, acquisitions, and other rapid transitions. A divestiture can terminate access to process knowledge or expert opinion with the stroke of a pen. Likewise, a diligent buyer will evaluate all aspects of a pending acquisition, including the knowledge and expertise that goes along with the physical assets, and may choose to devalue an asset that is not supported by well documented knowledge and sound expertise.

Ensure that information is accessible. To be useful, information gathered as part of the work activities for this element needs to be accessible. One significant difference between information gathered as part of this element and corresponding information gathered as part of the *knowledge* element (see Chapter 8) is that the data stored in the process safety files requires a much higher degree of structure. For example, if a process engineer has a question regarding the sizing of the emergency pressure relief system, the engineer likely knows (1) that the data exist and (2) where the data are located within the files maintained as part of the *knowledge* element. Conversely, if an engineer is charged with evaluating the suitability of a room that is being considered for a shelter in place location, the engineer is less likely to know that a technical paper describing a sound technical approach for selecting shelter in place locations exists in the proceedings of the 2001 CCPS conference, and that the book is available in the central engineering library at the corporate office. To be accessible, information must be cataloged or indexed.

Most government agencies have information posted on their internet sites and a search engine can be used to locate relevant documents. The same can be done by organizations by placing information on internal networks and using the indexing capabilities of a search engine to identify the relevant documents.

Provide structure. Even if information is well communicated when it is developed or discovered, need for this information may not be immediate. Thus, a good method for communicating new information is important but not sufficient to ensure that information is available when needed.

Historically, information was managed via a stable workforce. Most people spent their entire career with a single company, often in the same business unit or facility. Companies also employed librarians or records clerks to help maintain and manage information. In the future, this role will likely be replaced with very efficient and extremely smart software tools to conduct searches. The immediate challenge is how to manage information in this time of rapid transition. One obvious means is to provide a structured catalog listing a standard set of technical issues/content with links to relevant documents. A more advanced and user friendly (but maintenance intensive) method is to provide embedded links that allow users to quickly jump to related documents. Such links can be provided with either pointers or electronic hyperlinks. Assuming the related documents also contain links, the user is more likely to locate relevant information by exploring this web of information than by accessing a document via a single pointer (e.g., akin to a single reference in a central card file at a library).

Push knowledge to appropriate personnel. Developing and maintaining a network of process safety professionals provides a means to "push" new information throughout the corporation. Organizations that maintain a broad understanding at all levels of what can go wrong, how bad it might be, how likely it may be, and what can or should be done to manage risk are likely to manage risk more effectively than organizations in which people operate on autopilot, or worse yet, are unaware of risk. However, to be most effective, communications must be targeted, potentially useful, and technically sound. Seminars that provide an opportunity for learning, networking, and developing consensus on how new knowledge should be applied can help provide the needed clarity and context to new information. Mass communication tools such as e-mail are efficient, but can quickly become ineffective if overused. In addition, these tools provide limited opportunity to provide context and improve clarity.

Apply knowledge. While efforts to stockpile useful information and promote learning should be applauded, they hold little meaning without proper application. As part of any learning event or investment to improve knowledge, understanding, or organizational competence, facilities should also identify any subsequent steps that might be necessary. Try to establish plans that will deliver real value and that can be audited. Doing so will help the organization better manage risk and improve operations, while simultaneously making the case for future investments in similar activities.

Update information. Keeping information up to date can be even more difficult than providing structure to facilitate searches. This task is relatively straightforward with discreet and well-defined documents such as operating procedures or engineering drawings. However, loosely structured technical information, some of which is published by third parties, is more difficult to update. For example, if a unit changes the catalyst it uses for a particular reaction, the new material safety data sheet will likely be filed, recipes will be updated, procedures and training materials will be updated, and the change will be authorized and recorded via the *management of change* element. However, notations pertaining to the catalyst used at the facility are sometimes hand written in the margins of the product handling guides included in the technology manual from the catalyst supplier. Such facility-specific information is likely to be overlooked and not be updated when the catalyst change is authorized.

Promote person-to-person contact. Workers must apply rules, skill, and knowledge to successfully execute their daily activities. Rules can be read and skills can be practiced, but knowledge and understanding, which can be critical to recognizing hazards and managing risk, normally requires interaction with subject matter experts. This interaction provides an opportunity for others to (1) observe how the expert processes information and (2) better understand what issues the expert identifies as critical, as well as how and why the expert identified those issues that were most important for managing risk.

Plan personnel transitions. Companies often face the need to maintain intellectual capital through transition planning. This involves identifying critical positions within the organization, particularly positions that require specific skills or experience. Once these positions are identified, management proceeds to (1) determine who in the organization may be able to replace the incumbent upon retirement, resignation, or other transition (sometimes designating more than one possible replacement) and (2) train those who have been identified to fill the vacancy with as little disruption as possible. This training process may require months or years, which is typically much longer than the time normally available when the incumbent opts for a new position. Failing to plan for personnel transitions not only increases risk, it results in loss of significant value when knowledgeable workers move on to other opportunities.

Solicit knowledge from external sources. A single individual or single organization rarely has a monopoly on knowledge or possesses all of the answers related to a particular operation. Even if a facility has no direct counterpart, management systems and approaches used to manage risk can be benchmarked against those of other organizations, both within the company and within the industry.

As Cole and Klein point out in their paper titled, *How to be a Safety Star*, it is simply not possible for any one person, or even any one organization, to have all the answers to all the questions (Ref. 5.3). And yet a wide range of risks must be managed. Cole and Klein propose that star safety performers have access to a broad group of people with expertise in a variety of areas, which are often connected to other networks of their own.

The AIChE, along with many sectors of the chemical industry, has a long standing tradition of promoting cross-learning, particularly in the area of process safety. Examples include (1) the annual Ammonia Safety Symposium sponsored by the AIChE, (2) the Chlorine Institute, and (3) the many chemical producer associations including companies that produce or use chemicals such as acrylates, methacrylates, cyanides, ammonia, and ethylene. In addition, some companies actively promote activities to compare their process safety management systems to parallel systems in the nuclear or aviation industries to try to stimulate new thinking.

5.2.3   Evaluate and Share Results

Good management systems have a plan-do-check-act feature. In some cases, the steps are obvious and difficult to miss. If a facility undertakes a project to expand the output of a unit by 20%, the obvious "check" step, operate at the increased rate, will be integrated into the project. However, this model is often not applied to softer work activities that are part of the *competency* element; an organization may continue to provide resources to an activity simply because it always has. Companies that periodically check the value derived from activities that are part of the *competency* element are more likely to maintain the vitality of these activities.

Evaluate the utility of existing efforts. Evaluate the benefits that are realized from various work activities that make up the *competency* element, and periodically assess whether these activities are (1) meeting the objectives that were established or (2) providing other benefits that justify the resource demand.

Solicit needs from operating units. Work activities are often executed by process engineers or process safety professionals who support a business unit, a number of facilities, or a number of operating units at a large facility. Close alignment between the *competency* element objectives and the operating units helps ensure success of *competency* element work activities, and also helps strengthen the relationship between process safety professionals and operating teams.

5.2.4   Adjust Plans

Periodically (e.g., annually) review the status of efforts to promote process safety competency. With one eye looking toward what is currently working well and the other focused on upcoming challenges, revise the plans to more closely align the activities with the perceived needs. This management review activity is addressed in Section 5.6.

5.3   POSSIBLE WORK ACTIVITIES

The RBPS approach suggests that the degree of rigor designed into each work activity should be tailored to risk, tempered by resource considerations, and tuned to the facility's culture. Thus, the degree of rigor that should be applied to a particular work activity will vary for each facility, and likely will vary between units or process areas within a facility. Therefore, to develop a risk-based process safety management system, readers should perform the following steps:

1. Assess the risks at the facility, investigate the balance between the resource load for RBPS activities and available resources, and examine the facility's culture. This process is described in more detail in Section 2.2.
2. Estimate the potential benefits that may be achieved by addressing each of the key principles for this RBPS element. These principles are listed in Section 5.2.
3. Based on the results from steps 1 and 2, decide which essential features described in Sections 5.2.1 through 5.2.4 are necessary to properly manage risk.
4. For each essential feature that will be implemented, determine how it will be implemented and select the corresponding work activities described in this section. Note that this list of work activities cannot be comprehensive for all industries; readers will likely need to add work activities or modify some of the work activities listed in this section.
5. For each work activity that will be implemented, determine the level of rigor that will be required. Each work activity in this section is followed by two to five implementation options that describe an increasing degree of rigor. Note that work activities listed in this section are labeled with a number; implementation options are labeled with a letter.

Note: Regulatory requirements may specify that process safety management systems include certain features or work activities, or that a minimum level of detail be designed into specific work activities. Thus, the design and implementation of process safety management systems should be based on regulatory requirements as well as the guidance provided in this book.

5.3.1 Maintain a Dependable Practice

Establish Objectives

1. Develop a set of measurable objectives for maintaining and enhancing process safety competency.
   a. General goals are identified, but they are not translated into measurable objectives for individual personnel.
   b. Some objectives are documented in key individuals' annual performance plans.

   c.   Objectives are established for one or more departments; the
        objectives, along with periodic updates on progress toward
        achieving each objective, are widely published.

Appoint a Champion

2.   Assign responsibility for championing efforts to maintain and enhance
     process safety competency.

   a.   Process safety competency is generally limited to ensuring
        compliance with regulations and industry standards, and
        responsibility for these activities is clearly defined.

   b.   Responsibility for maintaining process safety competency is
        specifically included in the job description of the process safety
        management coordinator or process safety management engineer.

   c.   Responsibility for maintaining process safety competency for each
        element is included in the job description of appropriate subject
        matter experts within the organization, for example, a senior
        reliability engineer is assigned to assess changes in inspection
        practices and incorporate new technologies when appropriate.

   d.   Responsibility for maintaining process safety competency is
        championed by a senior manager who is supported by a formal
        network representing a broad range of functions within the
        company. Activities are seamlessly integrated with efforts to
        maintain and enhance the organization's overall competency in
        core areas. Management periodically reviews progress toward
        objectives that have been established. Where appropriate,
        individual contributions are also included in performance
        objectives and reviewed periodically as part of the company's
        performance management process.

Identify Corollary Benefits

3.   Link efforts to maintain and enhance process safety knowledge to
     initiatives that support near-term business objectives.

   a.   Benefits are primarily cast in terms of reductions in incident or near
        miss rates or increased regulatory compliance; no overt link exists
        between process safety competency and other aspects of the
        business.

   b.   Some linkage exists, but the efforts are primarily driven by efforts
        to provide corollary benefits. For example, a study is conducted by
        the reliability group to determine if increasing the test interval for
        relief valves will not exceed risk tolerance criteria that have been
        established.

    c.   The organization takes a holistic approach to maintaining and enhancing technical competency; initiatives to enhance process safety competency are part of an overall strategy to achieve business goals.

Develop a Learning Plan

4.   Based on an agreed set of objectives, identify and fund activities that are likely to support progress toward organizational objectives that promote competency.

    a.   Formal objectives that promote competency are in place, but budgets are not tied to initiatives to achieve these objectives.

    b.   Objectives exist and the required technical resources are funded, but no funding has been allocated to operating facilities to implement new initiatives that support the objectives.

    c.   Objectives have been established to support both development and implementation of new initiatives that support the objectives.

    d.   Specific activities that support the objectives are identified, funded, and staffed.

5.   Establish a longer term (e.g., 3- to 5-year) plan for competency work activities.

    a.   Some objectives exist and certain managers have a mental vision for long-term improvement, but no written strategic plan is in place and no budget has been allocated for activities that support the objectives.

    b.   A plan exists, and activities that underpin the plan are either supported based on expected return on investment or funded on an ad hoc basis.

    c.   A written plan to promote process safety competency is included in the strategic plan for the business unit. Budgets have been established to support both development and implementation of new initiatives that support the plan.

    d.   In addition to item (c), key personnel are assigned to the tasks that support the long-term plan, with anticipated outcomes identified.

Promote a Learning Organization

6.   Establish organizational objectives and promote activities that help create, acquire, interpret, transfer, and retain knowledge.

    a.   Management supports activities undertaken by individuals to improve competency through learning, particularly if the individuals involved make a significant contribution in terms of personal time/resources.

b. The organization recognizes and values activities that promote learning and liberally supports opportunities identified by individuals within the organization, but it has not developed a management ˙strategy for improving overall organizational competency through learning.

c. The organization is keenly aware of the link between learning and competency, strives to identify opportunities to improve competency through learning, evaluates the likely benefit that might be realized, and on that basis, develops and funds a plan to promote learning in a targeted manner.

5.3.2 Execute Activities that Help Maintain and Enhance Process Safety Competency

Appoint Technology Stewards

7. Appoint a technology steward for each type of process operated by the company.

a. Manufacturing or engineering managers at each facility are responsible for maintaining and enhancing technical knowledge for the processes at their facility.

b. Senior engineers or scientists are appointed to maintain knowledge of technical areas of competency, such as emergency pressure relief, catalysts, and so forth. However, no corresponding formal system exists to appoint technology stewards (i.e., process experts).

c. A technology steward is assigned to each type of process, mostly to help with process troubleshooting and specific improvement projects; information is largely maintained in this person's memory and/or personal files.

d. A technology steward is assigned to each type of process and is supported by a formal network of key individuals who represent a range of functions within the organization. Improvement goals are established based on work products developed by the technology steward and this network of key individuals.

e. In addition to item (d), a technology steward is assigned to proactively monitor potential changes to standards that are generally relevant to process safety and specifically relevant to processes operated by the company or facility. (This work activity often overlaps with work activities for the *standards* element listed under the key principle titled "Conduct Standards Compliance Work Activities.")

8. Appoint a technology steward for broader technology areas that are critical to the company, such as corrosion, inspection of fixed equipment, predictive maintenance methods for rotating equipment, dust explosion hazards, and so forth.

    a. Most efforts for this activity are initiated by senior engineers who want to expand their personal competence in a specific area.

    b. Technical managers routinely review the company's competency, including process safety competency, and appoint technology stewards or adjust existing appointments as needed.

Document Knowledge

9. Create a technology manual that documents the history of the process as well as knowledge that is critical to maintaining process safety competency.

    a. Process knowledge is primarily maintained in the facility's engineering files; process history is often limited to knowledge maintained by long-term employees. Written historical information is limited to such items as incident investigation records, change logs, equipment inspection records, and the like, and some of these records are purged periodically.

    b. Copies of all significant reports and engineering documents related to a process are maintained in a designated location, such as a specific file drawer, that is maintained by the technology steward.

    c. A formal system exists to capture certain documents; the documents are then indexed or filed in a retrievable manner, and are made available upon request.

    d. The technology steward is actively involved in compiling process information, including information related to process safety. This person maintains a controlled register of information that is available to all affected personnel/organizations (e.g., production, maintenance, engineering, research and development) and allocates time to periodically search for new information from sources both within and outside of the company and update the register accordingly.

    e. The technology steward is tasked with documenting, in a retrievable manner, the basis for past design, operational, and maintenance decisions. For example, a notation could be added to a procedure stating, "Step 4.3 of the procedure was inserted because of information we received from the valve manufacturer about…"; or use of a specific component could be addressed with, "We specifically use a NAMCO valve here with a pneumatic actuator because plant X had a motor-operated valve that failed, resulting in...".

10. Control changes to the technology manual.

    a. No formal manual or central compilation of information exists; however, many documents are maintained in personal files and technology stewards attempt to maintain up-to-date records.

    b. Access to files is limited, either by physical locks or by protected electronic file folders, but a number of people do have access. No specific individual has been assigned as a "gatekeeper" responsible for validating information that is added to the manual.

    c. An informal system exists to control changes to the technology manual; for example, general practice is to consult with the technology steward prior to making changes to the technology manual.

    d. The technology manual is included in the scope of the facility's formal document control system and a process for reviewing and approving changes to the technology manual is established that includes review/approval by the technology steward who "owns" the manual.

Ensure that Information is Accessible

11. Document what information is available in a manner that facilitates searches.

    a. Information is consolidated in a single location, and a register of files/documents exists.

    b. A register has been developed; most information is stored on computer networks that can be accessed from anywhere within the company.

    c. Most information is stored on computer networks that can be accessed from anywhere within the company. In addition, a register of documents and a means to search for key words or phrases is provided. Personnel have a clear expectation that technical information will be accessible when it is needed and that the information will be shared.

Provide Structure

12. Provide a means to quickly locate technical information, facilitate maintenance of existing information, and file new information in a logical manner.

    a. Information is collected and filed by the most convenient datum; for example, process history data and incident reports are filed by date, information on reactivity hazards is kept in the engineering files concerning sizing of emergency pressure relief systems.

b. A standard data structure is provided for use throughout the company, related documents are listed in accordance with this structure, and technical personnel who routinely add or revise documents are familiar with, and help maintain, the structure.

c. Related documents include active links or cross references, and these links are routinely maintained and updated.

d. In addition to item (c), a data librarian is designated to manage the information/data.

Push Knowledge to Appropriate Personnel

13. Proactively push safety-critical information to potentially affected facilities and key personnel; do not depend on others to discover a need to know about an issue.

a. The content of initial and refresher training programs for production operators and supervisors is periodically compared to information in the technology manual and updated as needed.

b. Newly generated information is sent to persons responsible for updating procedures and training manuals; however, the communication is one-way unless the recipient asks for clarification.

c. Initial and refresher training is provided to technical support personnel to ensure that these personnel are aware of information contained in the technology manual as well as how the information is structured. New information is transmitted to all affected personnel in a timely and targeted manner.

d. Item (c) is performed in a manner that facilitates discussion of the issues and how they apply to the facility. For example, a new standard for sizing relief valves would be accompanied by a seminar at which engineers could question subject matter experts.

Apply Knowledge

14. Develop and periodically review plans to apply process knowledge in an effort to better manage risk, improve operational performance, or achieve some other important organizational objective.

a. Opportunities to improve knowledge are generally supported, but staff members have no general expectation that specific plans will be developed to apply new knowledge.

b. Management goes out of its way to recognize instances in which new knowledge or learning events produce tangible benefits.

c. All levels of the organization expect that new knowledge will be applied; following each significant learning event, written action plans are generated and implemented.

d. In addition to item (c), management regularly examines the benefits that are derived from the application of knowledge and prioritizes future activities based partly on documented benefits from previous efforts.

Update Information

15. In addition to controlling changes, update information when it becomes obsolete or is affected by a change to the process.

a. The technology steward annually reviews copies of change logs to determine what changes should be made to the technology manual.

b. The technology steward updates the information if the unit (1) believes the change is significant enough to affect the manual and (2) provides details of the change to the technology steward.

c. The technology steward is notified of all changes. In addition to any duties associated with authorizing the change, the technology steward determines if the technology manual should be updated. Actions needed to maintain an up-to-date technology manual are integral to the *management of change* element.

d. In addition to modifying the technology manual based on changes made or new information generated at a facility, the technology steward modifies the technology manual based on occurrences outside the facility, such as incidents or new technology developments.

Promote Person-to-Person Contact

16. Allocate time for the technology steward to be present at operating units to gain a first-hand understanding of how each unit is operating and to identify opportunities for improvement.

a. Maintaining and enhancing process safety competency is largely the responsibility of the operations or engineering group at each facility.

b. Specific "experts" are identified and consulted when the facility's staff (1) cannot resolve a process-related problem or (2) identify the cause(s) for a particularly significant incident.

c. Interaction between "experts" and production units is routine, most often occurring during plant trials or commissioning of new equipment.

d. The technology steward and facility personnel interact routinely.

Plan Personnel Transitions

17. Consider individual and organizational competency in succession planning.
    a. A succession planning program is in place to groom senior managers, and most individuals in the program serve in at least one assignment that exposes them to process safety practices.
    b. A succession planning program is in place to groom senior manufacturing managers and technical personnel. The program includes a formal requirement to serve in at least one assignment in which incumbents are exposed to process safety practices.
    c. A succession planning program is in place that reaches throughout the organization, extending to positions such as production and maintenance supervisors, operators, craftspersons, and maintenance planners.
    d. A succession planning program is in place as stated in item (c), and the stated objectives for the program include (1) maintaining the organization's competency and critical knowledge through transitions and (2) enhancing process safety competency over time.

18. Extend succession planning efforts to technical and staff functions, including process safety professionals.
    a. Rather than developing internal skills, the company recruits external job candidates when needed to fill positions that require specific knowledge, assuming no qualified internal candidate is identified/available.
    b. A program is in place to expose individuals to process safety principles with the intent of developing a baseline level of competence throughout the technical organization, resulting in a number of candidates who are qualified for a position that opens up in the process safety field.
    c. The company's succession planning program extends to key technical positions, including process safety professionals.
    d. A succession planning program is in place as stated in item (c), and the stated objectives for the program include (1) maintaining the organization's competency and critical knowledge through transitions and (2) enhancing competency over time.

Solicit Knowledge from External Sources

19. Participate in industry associations and other networks that provide insight to how process safety is managed at other companies.
    a. Attendance at technical meetings and exchanges is inconsistent, but management supports these types of activities unless business performance is extremely poor.

b. Participation in technical meetings and exchanges is encouraged.

c. Certain employees are encouraged to take leadership roles in technical or trade associations so that the company (1) can share information with other industry members by publishing what it perceives as best practices or by participating in the development of codes, standards, or guidelines and (2) stays abreast of what other companies in the industry are doing.

5.3.3 Evaluate and Share Results

Evaluate the Utility of Existing Efforts

20. Periodically compare the objectives that were established in the overall competency improvement plan to the benefits that have been derived from work activities that support the *competency* element.

a. No formal plan or objectives exist, but a management review process is in place to periodically examine the benefits derived from work activities.

b. Some objectives are documented in key individual's annual performance plans and progress is reviewed on an individual basis as part of the annual performance appraisal process.

c. The status of ongoing efforts to maintain and enhance process safety competency is a standing agenda item at periodic management meetings.

d. A formal management review process is used to determine what measurable benefits have been achieved and to compare the actual benefits to the goals that were established when the plans were developed.

Solicit Needs from Operating Units

21. When identifying objectives and evaluating the utility of work activities for the *competency* element, query personnel at the operating unit level to determine what needs remain unmet from their perspective.

a. Process safety professionals consider the perceived needs of operating units when determining whether or not to pursue or continue specific activities that help maintain or improve process safety competency.

b. Technical organizations present ideas at project review meetings, and senior manufacturing managers determine which ideas warrant funding.

c. Process safety professionals and other technical personnel jointly work with operating units to identify needs, understand the potential benefits associated with meeting the needs, and promote new initiatives, or continuation of existing initiatives, based on understanding of risk and how the plans may affect risk.

d. Based on the results of the work described in item (c), senior manufacturing and business managers champion specific initiatives or activities.

5.3.4 Adjust Plans

22. Based on a periodic review with senior management and key personnel from operating facilities, adjust plans or the resources provided to various plans/activities.

a. Management supports efforts to maintain and enhance process safety competency but a tendency remains to cut resources for all *competency* element work activities when the business in underperforming. Conversely, management tends to increase funding when the business is performing well without careful scrutiny of the expected benefits.

b. The process safety function is staffed and funded at a fixed level (possibly adjusted for inflation), and work activities that support the *competency* element compete for a share of this fixed budget.

c. A formal process is in place to periodically evaluate and adjust priorities and resources; resources are adjusted for work activities that support the *competency* element in a logical and transparent manner.

5.4 EXAMPLES OF WAYS TO IMPROVE EFFECTIVENESS

This section provides specific examples of industry tested methods for improving the effectiveness of work activities related to the *competency* element. The examples are sorted by the key principles that were first introduced in Section 5.2. The examples fall into two categories:

1. Methods for improving the performance of activities that support this element.
2. Methods for improving the efficiency of activities that support this element by maintaining the necessary level of performance using fewer resources.

These examples were obtained from the results of industry practice surveys, workshops, and CCPS member-company input. Readers desiring to improve their management systems and work activities related to this element should examine these ideas, evaluate current management system and work activity performance and efficiency, and then select and implement enhancements using the risk-based principles described in Section 2.1.

5.4.1 Maintain a Dependable Practice

Appoint a corporate owner for each element of the process safety program. These owners facilitate sharing and cross learning and are often distributed throughout the company. Element owners who support a single facility can easily become the company's guru for activities related to a particular element. In most cases, these corporate element owners establish a network of element owners at each facility, and these networks review, prioritize, and work on issues of common interest.

Designate technology stewards for the technical aspects of process safety management. These stewards are responsible for maintaining the organization's knowledge related to specific process safety issues. Many organizations charter standing work groups or assign responsibility to a senior engineer to maintain the organization's collective competence in specialized areas, such as chemical reactivity, emergency relief vent sizing (particularly when the scenario may involve runaway reactions), flammability and fire protection, and so forth.

Designate someone to monitor other companies in the same industry, industry sponsored organizations, and technical literature. This person is responsible for gathering lessons learned or other technical information that may be pertinent to processes operated by the company. If this duty is not formally assigned to a specific individual, (1) many valuable lessons may go unnoticed or (2) the potential exists for more than one person to be performing this task, which can be inefficient. (This activity is often integrated into the duties of the technology steward.)

Designate one or more individuals in the organization to monitor incidents, new technical developments, and other information. The list of possible information sources includes electronic publications such as CCPS's *Process Safety Beacon* and *Process Safety Progress*, the Chemical Safety Board's news alerts, and many similar services. However, if monitoring these sources of new information is considered everyone's responsibility, it becomes likely that either (1) nobody will feel personally responsible for completing this task and critical information may go unnoticed or (2) staff time will be poorly spent because multiple engineers will be monitoring, researching, and trying to address the same issues. Assigning this activity to an appropriate person should help minimize the chances of missing an important issue while simultaneously not significantly impacting overall productivity.

Establish unit-specific training modules for new unit managers and supervisors. All too often, supervisors and unit managers are assigned to a new position with little or no formal training on the process, the hazards of the process, or the systems that are critical to maintaining safety. The technology manual described in Section 5.2.2 provides an opportunity to document this information in a clear and concise form. Regardless of the communication tool, newly assigned managers and supervisors must quickly become aware of the hazards of the process and the controls that are critical to managing the risk associated with those hazards.

Establish a process safety curriculum for new engineers. Rarely does a new engineering graduate have an appreciation for the variety of management systems and technologies that underpin process safety. New engineers cannot be expected to simply absorb this range of information, and the consequences of not learning it range from economic losses associated with process or safety system redesigns to human losses resulting from catastrophic accidents. Some companies maintain a formal job rotation schedule for new hires. For example, new hires might spend 6 months in each of four functional areas to acquire a more complete view of the organization. Including the process safety function in the rotation for some fraction of the engineers will provide the organization with staff members who have a good appreciation for process safety.

Hold periodic technical seminars on subjects related to process safety. In addition to a structured curriculum for new engineers, periodically invite experts to conduct seminars and training on subjects related to process safety that are of broad interest throughout the organization.

Benchmark your process safety programs with other facilities or companies. No facility has a monopoly on process safety competence; all are able to learn something from others. Many companies establish a structured program of benchmarking, and others take it to the next step and benchmark with similar facilities operated by other companies within their industry. Obviously, companies are keen to protect trade secret information from competitors and benchmarking is generally limited. However, organizations can benefit from (1) having representatives of nearby facilities meet to discuss aspects of a specific process safety work activity, such as evaluating the performance of local contractors, or (2) inviting companies in the same industry to share details of activities or systems that have proven to be effective, for example, some specific aspect of the *management of change* element.

5.4.2 Execute Activities that Help Maintain and Enhance Process Safety Competency

Designate technology stewards. These individuals are responsible for collecting, organizing, and maintaining all technical information related to a specific type of unit within the corporation.

Establish centers of excellence. Each center is responsible for various aspects of process safety, and some centers may be virtual instead of physical (i.e., the people are not co-located, but the organization operates seamlessly). This approach can be particularly effective for issues that are common to most facilities yet require specialized technical understanding. For example, companies often designate an individual or group to be the company's center of excellence for design of emergency pressure relief systems, prevention of dust explosions, calorimetry, mitigation of static electricity hazards, and fire protection. In many cases, this same concept is applied to advanced risk analysis, using a specially trained and highly experienced expert or central group to lead quantitative risk studies. If a company chooses to establish centers of excellence, the existence of the centers should be widely publicized.

Continually invest in research and development efforts in process safety. When justified, invest in computer models, analytical methods, or other tools that can increase the efficiency of process safety professionals or the effectiveness of activities related to process safety. In his book, *The 7 Habits of Highly Effective People*, Steven Covey writes about the importance of "sharpening the saw" (Ref. 5.4). Using worn out tools, or the wrong tool, at best adversely affects productivity and can lead to a poor quality work product. In the area of process safety, effective tools can significantly improve performance; for example, an atmospheric release model that can quickly and accurately predict the downwind effects on members of the public can provide valuable information to emergency responders.

Maintain a detailed intranet website that provides process safety management tools and information to all employees at all facilities. This site could contain information on current and historical performance, annual objectives, key resources, experts for all RBPS elements, training programs, external links to other databases, incident trends, tools, checklists for hazard identification and risk analyses, operational readiness reviews, technical information that supports the *asset integrity* element, and any other aspects deemed necessary. The content should address each RBPS element, have an assigned owner/content steward, and be periodically reviewed for accuracy and currency.

Sample diverse training programs, but be consistent once a decision is made. Organizations often develop or improve competency via training, and much of the training needed to improve competence is offered by external companies. External training can be quite expensive. Consider pilot testing a training program prior to making a large investment. This will allow the organization to evaluate the utility of methods presented in the respective programs, assess the overall applicability of the program, and identify areas in which the training must be customized or modified to be consistent with internal organizational practices. However, once a training program is selected, rules should be established to ensure that personnel attend only approved training programs to increase consistency in the approach taken to process safety or related issues within the organization.

Incorporate lessons from previous incidents in training and similar activities. Many companies will send out an e-mail or similar notification listing the lessons learned from each incident, but the communication is sent only once, and it is normally not timely. For example, lessons learned from incidents involving maintenance shutdowns should be reviewed prior to shutdown rather than when the information is initially collected or published.

Previous incidents can also add context to training to help emphasize important aspects of a procedure. For example, a brief description of how failure to follow a certain step in a procedure led to an accident helps focus trainees' attention. This also helps convert rote memorization of a work rule to an understanding of the basis for the rule.

Many facilities require supervisors or senior staff members to conduct a daily or weekly safety talk. Providing information on lessons learned to the personnel assigned to lead the talks will help ensure that the lessons are not forgotten.

Ensure that the training program addresses topics important to process safety competence. To help ensure competence in operational activities, companies often spend considerable time and effort developing initial and refresher training curricula for operators and crafts personnel. Developing formal training curricula (in addition to experience and skill requirements) for positions such as process safety coordinator, process hazard analysis team member or team leader, and incident investigation team member or team leader is likely to provide similar benefits.

Hold frequent process safety talks. Process safety incidents are normally rare events. One way to maintain competence is to periodically talk about what might go wrong, emphasizing the important safeguards that exist at the facility to help prevent or mitigate the accident scenario. Consider emphasizing these safety talks as a two-way interaction. For example, lay out the facts related to an incident, and ask the group members to use their knowledge and experience to (1) postulate what might have happened and (2) propose some engineering or administrative controls that might have failed during the accident sequence. For maximum effect, pose questions that the group can discuss throughout the shift and then reconvene the next day to talk about the group's conclusions and to compare them to what actually happened if the safety talk was based on a real-world incident. This notion can be extended to challenge the group to answer process-specific questions such as, "Why do we always check the pH of the blend tank before we transfer the next lot of material to the tank," particularly if the hazards associated with high (or low) pH are not obvious or well understood.

Keep company legends alive. In addition to establishing a process safety curriculum and holding frequent process safety talks, find ways to help the organization remember the specific process safety lessons it has learned. For example, have each facility develop a list of short vignettes and share them between all of the facilities in the company. Using these stories as "tool box" or "5-minute" safety talks at the start of each shift will (1) reinforce safety in employees' minds, (2) help the company retain its memory, (3) provide relevant subjects for safety talks, and (4) alleviate the burden on supervisors to come up with a safety topic each day.

5.4.3 Evaluate and Share Results

When possible, express objectives in measurable units. Objectives for some aspects of this element will have discrete measures, but in many cases, soft measures must suffice. For example, determining if the training curriculum for new process engineers addresses process safety is simple. More difficult is determining if the curriculum *adequately* addresses process safety. More difficult yet is measuring the return on investment for this activity. However, some work activities that make up this element, such as appointing a technology steward and creating a technology manual, should contribute to improvements in well-established metrics such as output, yield, and availability. Whenever a link can be established between one of these traditional metrics and a process safety competency work activity, expect positive results in the metric and carefully examine the effectiveness of the supporting RBPS work activity if performance deteriorates for no obvious reason.

5.4.4 Adjust Plans

Include enhancement of competency in the business's strategic plan. Strategic plans are carefully reviewed by senior management and periodically updated. If the business believes that maintaining process safety competency is vital, it will likely want to plan for improvement, and periodically adjust its plans to achieve those results just as it does for other important aspects of the business.

5.5   ELEMENT METRICS

Chapter 20 describes how metrics can be used to improve performance and when they may be appropriate. This section includes several examples of metrics that could be used to monitor the health of the *competency* element, sorted by the key principles that were first introduced in Section 5.2.

In addition to identifying high value metrics, readers will need to determine how to best measure each metric they choose to track. In some cases, an ordinal number provides the needed information, for example, total number of workers. Other cases, such as average years of experience, require that two or more attributes be indexed to provide meaningful information. Still other metrics may need to be tracked as a rate, as in the case of employee turnover. Sometimes, the rate of change may be of greatest interest. Since every situation is different, the reader will need to determine how to track and present data to most efficiently monitor the health of RBPS management systems at their facility.

5.5.1 Maintain a Dependable Practice

- *Comparison of actual to budgeted spending for activities associated with execution of the learning plan.* The intensity of training and related learning activities can become more closely related to business performance than perceived need. Significant deviations from the spending plan (either up or down) should be examined to determine if (1) increases are truly delivering value and (2) decreases are not diminishing the organization's process safety competency.
- *Presence of objectives related to enhancing process safety competence in each manager's, supervisor's, and technical staff member's personal performance plans, and in the trend over time.* The built-in infrastructure offered by annual performance plans (or personal objectives) will help ensure that (1) enhancement of process safety competency is integrated into systems that help develop human resources and (2) the specific tasks that help promote gains in competency at the individual level are reviewed and adjusted on a periodic basis.
- *Opinion surveys regarding the effectiveness of programs to promote learning.* Evaluating employee survey results is one method for determining if investments made in promoting a learning organization are paying off, and if the utility of the investment is changing over time.

5.5.2 Execute Activities that Help Maintain and Enhance Process Safety Competency

- *Frequency with which incidents recur because the organization has allowed safeguards that were implemented as a result of a previous incident to lapse.* This metric is a direct measure of process safety competency, but it may suffer from significant lag.
- *Number (or percent) of technology steward positions that are currently staffed.* Unstaffed positions are an indication that the facility or business unit does not value maintaining process safety competency.
- *Staff-hours (per year) devoted by the technology steward to face-to-face contact with operating units.* This metric will indicate the level of effort devoted to maintaining and enhancing process (and process safety) competency; a downward trend over time may indicate a lapse in management support.
- *Frequency with which incident investigation teams determine that the basic physical or chemical phenomenon that caused an incident was not known within the organization.* This is a direct measure of process safety competency; however, because of the low frequency of process

safety incidents (including near miss events), significant lag may exist between a decline in competency and a noticeable change in this metric.

- *Use of the technology manual ("hits" if it is implemented in a web-based fashion).* A steady upward trend indicates that facilities find the technology manual useful; conversely, a steady downward trend may indicate that the information is hard to locate, out of date, or otherwise not relevant.

- *Average response time for questions posed to the technology steward or center of excellence.* An upward trend in response time may indicate insufficient resources, with the potential for complete failure of this system as a result of a loss in interest at operating facilities and/or less than adequate resources to maintain competency or provide current and accurate information.

- *Staff-hours (per year) devoted by the technology steward to supporting troubleshooting efforts for/with each operating unit.* A steady downward trend (or steep decline) may indicate that facilities do not value this resource, that the resource is not effective, or that management does not support efforts to spread knowledge between facilities.

- *Number of questions posed to the center of excellence per unit time.* A steady upward trend may indicate that facilities find this to be an effective resource, or it may indicate a loss of competence at operating facilities.

- *Ratio of process changes to updates to the technology manual.* A decrease in this metric may be an early warning that the information in the technology manual is not being maintained current and accurate.

- *Number of facilities or business units within the company that maintain up-to-date succession plans.* This metric is a direct measure of management's commitment to maintaining competence in key positions. Note that these plans are typically updated periodically, rather than continuously, and personnel can reasonably expect that the plan will be updated as scheduled.

- *Pareto analysis of the topics discussed with the technology steward or with centers of excellence, or researched using the technology manual (assuming it is web-based).* The fraction of questions that address a particular subject should be considered when planning for future efforts to invest in competency and/or developing the learning plan for future years.

- *Opinion surveys on how technical information is stored and the efficacy of searches.* Surveys may provide a reliable and easy to execute method for evaluating the effectiveness of information storage and retrieval mechanisms

5.5.3 Evaluate and Share Results

- *Opinion surveys about the organization's competence, including opinions on the trend over time.* Although achieving consistency in polling is likely difficult, particularly over time, a decline in the organization's perception of its competence should be closely examined.
- *Opinion surveys on the usefulness of each center of excellence.* Normally, when companies establish centers of excellence, some experimentation takes place regarding which technology areas to emphasize and how to best orchestrate communication and other interaction between operating facilities and the center of excellence. In addition, the relevance of a particular technology area may change with time. Thus, opinion surveys can help correct existing issues and guide future plans.

5.5.4 Adjust Plans

- *Surveys on the usefulness of attending technical meetings.* Companies often invest considerable resources in participating in technical organizations, including attending technical meetings. Periodic surveys or other metrics to measure the value derived from these efforts will help companies adjust their investment strategy to better fit perceived needs.

5.6   MANAGEMENT REVIEW

The overall design and conduct of management reviews is described in Chapter 22. However, there are many specific questions/discussion topics that management may want to check periodically to ensure that the management system for the *competency* element is working properly. In particular, management must first seek to understand whether the system being reviewed is producing the desired results. If the organization's level of *competency* is less than satisfactory, or it is not improving as a result of management system changes, then management should identify possible corrective actions and pursue them. Possibly, the organization is not working on the correct activities, or the organization is not doing the necessary activities well. Even if the results are satisfactory, management reviews can help determine if resources are being used wisely: are there tasks that could be done more efficiently or tasks that should not be done at all? Management can combine metrics and indicators listed in the previous section with personal observations, direct questioning, audit results, and feedback on various topics to help answer such questions. Activities and topics for discussion include the following:

- Review the objectives that have been established for the *competency* element. Identify any important needs that should be added and terminate or refocus any existing efforts that are no longer pertinent. Also, review progress toward achieving each remaining objective.
- Review the learning plan to determine if it is helping to improve the organization's competency.
- Discuss Garvin's "learning disabilities" (blind spots, filtering, lack of information sharing, flawed interpretation, and inaction) and critically examine the organization's behavior pattern with regard to learning.
- Review conformance with plans to help promote organization learning and competence. Discuss changes that occurred between plan approval and execution. Do these changes reflect an effort to continuously improve, or do they reflect a lack of management support for this element?
- Review the benefits that can be attributed to efforts to expand the organization's performance. What efforts in this area directly led to changes that help better manage risk or improve performance?
- Review the list of technology stewards. Query supported facilities to determine if technology stewards are providing support as needed. (This management review item may be more appropriate as a corporate function.)
- Review a representative sample of technology manuals to determine if they are (1) up to date, (2) accessible, and (3) user friendly, in other words, the information can be readily located.
- If the company uses networks of corporate-led teams to drive improvements in its process safety management programs (presumably with a representative from each facility), survey some of the personnel assigned to these teams to determine if they believe that their participation is a good use of time. Can they point to tangible benefits that the effort delivered to their facility? Can they identify instances in which their input helped another facility within the company? Do they support continued participation on this team? If not, provide this information to the corporate sponsor for the initiative and request specific improvements or reduce the support provided to the initiative.
- Review succession plans for key personnel, including
  o The management team at each facility.
  o Technology stewards.
  o Key members of each center of excellence.
  o Process safety professionals and RBPS element owners.
  o Key technical staff.
- Review training and development plans for personnel who have been identified as candidates for key positions and take action to restart any plans that have stalled.

- Examine the process for determining and maintaining the expertise required to safely operate the facility. Does such a process exist and is it being followed?
- Discuss any known upcoming changes in key staff positions or senior technical positions, and confirm that appropriate transition plans are in place.

Regardless of the questions that are asked, the management review should try to evaluate the organization's depth on several levels. Is the depth of knowledge sufficient? Does understanding extend beyond operational experience? When things go wrong, is it likely that key personnel will have the depth of understanding of design intent required to make the right decisions? Also, is there sufficient depth in key personnel? If the entire organization depends on a single individual to field all of the really tough questions, what will happen if that person suddenly resigns, or wins the lottery and elects to retire early? Management reviews provide an opportunity for the organization to honestly assess its depth and take action to address any concerns. Proactively addressing this issue is vitally important; loss of competency at best leads to loss of a valuable asset that requires significant time and money to restore, and at worst can represent a crack in the RBPS pillar that mandates a commitment to process safety.

5.7 REFERENCES

5.1 Joint Chemical Accident Investigation Team investigation report no. EPA 550-R-97-002, U.S. Environmental Protection Agency, October 1997.

5.2 Garvin, David, *Learning in Action*, Harvard Business School Press, 2000.

5.3 Cole, Bruce C. and Klein, James A., *How to be a Safety Star*, American Society of Safety Engineers Professional Development Conference, June 25, 2003.

5.4 Covey, Stephen R., *The 7 Habits of Highly Effective People*, Simon and Schuster, New York, 1989.

6

WORKFORCE INVOLVEMENT

One of the more humbling, and formative, experiences in a process engineer's career is to describe in detail how some aspect of a process or piece of equipment operates – only to have a seasoned operator respond, "That's all well and good, but let me tell you what *really* happens on a Sunday at 3:00 a.m." The young engineer learns then that operators can provide a valuable source of process knowledge and expertise that complements books, computer models, and theory.

Organizations often have a similar opportunity to tap more deeply into the knowledge and expertise of workers. Operators and maintenance personnel may have a deeper and more authentic understanding of a particular problem, and its solution, than does the engineer. The engineer may have a more profound understanding of potential safety and business risks of a certain course of action than does the facility manager. The contract maintenance employee who happens to notice a sound or a smell that is different from what it was on the first two days on the job may, on that third day, have a more important role to play in the safety of the facility than any 30-year veteran company employee.

6.1   ELEMENT OVERVIEW

Promoting the active involvement of personnel at all levels of the organization is one of five elements in the RBPS pillar of *committing to process safety*. This chapter addresses the diversity of roles that workers can fulfill in support of process safety management system development, implementation, and enhancement. Section 6.2 describes the key principles and essential features of a management system for this element. Section 6.3 lists work activities that support these essential features, and presents a range of approaches that might be appropriate for each work activity, depending on perceived risk, resources, and organizational culture. Sections 6.4 through 6.6 include (1) ideas for improving the effectiveness of management systems and specific programs that support this element, (2) metrics that could be used to monitor this element, and (3) issues that may be appropriate for management review.

6.1.1 What Is It?

Workers, at all levels and in all positions in an organization, should have roles and responsibilities for enhancing and ensuring the safety of the organization's operations. However, some workers may not be aware of all of their opportunities to contribute. Some organizations may not effectively tap into the full expertise of their workers or, worse, may even discourage workers who seek to contribute through what the organization views as a nontraditional role. *Workforce involvement* provides a system for enabling the active participation of company and contractor workers in the design, development, implementation, and continuous improvement of the RBPS management system.

Effective *workforce involvement* involves developing a written plan of action regarding worker participation, consulting with workers on the development of each element of the RBPS management system, and providing workers and their representatives access to all information developed under the RBPS management system.

Workforce involvement provides for a consultative relationship between management and workers at all levels of the organization. This element is not intended to create a system whereby any worker or group can dictate the content of the RBPS management system; however, for *workforce involvement* to succeed, management must provide due and fair consideration of the input provided by workers.

Unless a reason exists for distinguishing between company employees and contract workers, the term **worker**, as subsequently used in this chapter, should be interpreted as referring to both groups.

6.1.2 Why Is It Important?

Those workers directly involved in operating and maintaining the process are most exposed to the hazards of the process. The *workforce involvement* element provides an equitable mechanism for workers to be directly involved in protecting their own welfare. Furthermore, these workers are potentially the most knowledgeable people with respect to the day-to-day details of operating the process and maintaining the equipment and facilities, and may be the sole source for some types of knowledge gained through their unique experiences. *Workforce involvement* provides management a formalized mechanism for tapping into this valuable expertise.

Workforce involvement also ensures that mechanisms exist for workers to access the information they need to perform their jobs, including fulfilling their roles in support of the implementation of the RBPS management system.

Workforce involvement either directly implements or helps reinforce a number of the essential features of a sound process safety culture, as outlined in Chapter 3. For example:

- **Individual empowerment.** *Workforce involvement* provides explicit roles, responsibilities, and authorities for workers in the planning, implementation, and improvement of the RBPS management system.
- **Deference to expertise.** *Workforce involvement* provides a mechanism for workers to share their expertise in the operation and maintenance of the process. By having a role in helping define their training needs, workers can also help direct the enhancement of their expertise.
- **Open and effective communications.** *Workforce involvement* provides various mechanisms for workers and managers to communicate.
- **Mutual trust.** By enhancing dialogue and interaction between workers and management on process safety issues, *workforce involvement* provides opportunities for fostering mutual trust within the organization.
- **Responsiveness.** The manner and timeliness of management response to worker suggestions will be a primary determinant of the degree of success that can be achieved by the *workforce involvement* program. Slow, no, or superficial response can both cause the loss of time-critical opportunities to respond to problems and serve as a disincentive to future worker participation.

6.1.3 Where/When Is It Done?

By its nature, *workforce involvement* is associated with virtually every process safety activity, whenever and wherever it occurs. Thus *workforce involvement* should begin during the design of the RBPS management system and continue on through its implementation and continuous improvement.

Certain *workforce involvement* activities may be scheduled periodically; for example, a periodic operator opinion survey regarding the adequacy of the refresher training program. In addition, a mechanism permitting continuous input or feedback from workers should be provided as part of the *workforce involvement* program.

Management's responsibility for providing access to information developed under the RBPS management system provides a continuing role for management as such information is developed and modified (see Chapter 8, Process Knowledge Management).

6.1.4   Who Does It?

Management, with worker involvement, establishes the procedures and systems that constitute the *workforce involvement* element, which, in turn, describe the process for identifying opportunities for the workforce to be engaged in the development and implementation of each RBPS element. The *workforce involvement* procedures also establish a responsibility for workers to make suggestions for the development, implementation, and improvement of the RBPS management system and, for management to respond to such suggestions.

Management ensures that required information is available to workers under the *workforce involvement* program.

Finally, workers must support the *workforce involvement* element through their active participation (involvement) in process safety management system activities.

This chapter addresses *workforce involvement* as it pertains to only two groups – the company and the workers. More complex dynamics may exist at a unionized facility at which the union organization and its contract work rules will likely affect *workforce involvement* activities.

6.1.5   What Is the Anticipated Work Product?

The output of the *workforce involvement* element is a RBPS management system that:

- Defines the roles and responsibilities of all involved.
- Meets the current needs of all constituencies.
- Is maintained in an evergreen condition through both management attention and ongoing worker input.
- Provides workers the information necessary to understand the hazards to which they may be exposed, and to support their roles in RBPS management system implementation.

The *workforce involvement* element yields suggestions for the implementation and improvement of any RBPS element. Such suggestions are directed to the respective element custodians for consideration, response, and potential implementation.

Another product of this element should be the active engagement of workers at all levels of the organization, and a correspondingly greater sense of worker ownership of, and commitment to, the successful implementation of the RBPS management system.

6.1.6 How Is It Done?

Implementation of the *workforce involvement* element normally involves (1) identification of the more common roles that workers can and should play in the implementation of the RBPS management system and (2) the establishment of mechanisms to facilitate this participation. Additional opportunities for worker involvement may be identified and formalized as the *workforce involvement* program evolves. The resulting working environment should be flexible enough to allow capitalizing on unexpected *workforce involvement* opportunities as they arise.

Table 6.1 (adapted from Ref. 6.1) lists general areas of activity in which *workforce involvement* could be encouraged, as identified by the United Kingdom (UK) Health and Safety Executive (HSE).

One worker role would be to provide commentary on, and suggestions related to, the design, development, implementation, and continuous improvement of the RBPS management system. Mechanisms for soliciting and submitting suggestions should be established, as well as protocols for evaluating these suggestions and providing management response.

TABLE 6.1. UK HSE Workforce Involvement Suggestions

Policy	Workforce involvement in development or review of policy statement.
Organizing	
Control	Giving employees specific health and safety responsibilities.
Communication	Employees are involved in delivering health and safety messages.
Competence	Employees are involved in design and delivery of training.
Cooperation	Structure of safety committees. Suggestion schemes.
Planning	
Objectives/plans	Employees are involved in setting health and safety plans/objectives.
Risk assessments	Employees participate in risk assessments.
Procurement	Employees are involved in the procurement of equipment, materials, etc.
Design	Employees help design new ways of working.
Problem solving	Employees are involved in problem solving.
Operation of risk control systems	Employees are involved in planning risk control systems.
Measurement	
Active monitoring	Employees assist in carrying out inspections, observations, etc.
Reactive monitoring	Employees participate in accident and near miss investigations and hazard spotting.
Audit and Review	Employees participate in audits of the efficiency, effectiveness, and reliability of the health and safety system and in systematic reviews of performance, based on data from monitoring and audits.

In addition, information required to be shared with workers must be identified and mechanisms established for providing this information. Employers have the right to require that workers seeking access to trade secret information sign confidentiality agreements before being given access to such information.

Written policies, protocols, or procedures may be appropriate for documenting and controlling some of the activities mentioned above (see Section 6.2.1).

6.2   KEY PRINCIPLES AND ESSENTIAL FEATURES

Safe operation and maintenance of facilities that manufacture, store, or otherwise use hazardous chemicals requires the active involvement of workers who (1) are aware of the hazards in the workplace, (2) understand the engineered controls and management systems provided to address those hazards, and (3) accept and strive to fulfill their roles and responsibilities in support of providing a safe work environment. The following key principles should be addressed when developing, evaluating, or improving any management system for the *workforce involvement* element.

- Maintain a dependable practice.
- Conduct work activities.
- Monitor the system for effectiveness.
- Actively promote the *workforce involvement* program.

The essential features for each principle are further described in Sections 6.2.1 through 6.2.4. Section 6.3 describes work activities that typically underpin the essential features related to these principles. Facility management should evaluate the risks and potential benefits that may be achieved as a result of improvements in this element. Based on this evaluation, management should develop a management system, or upgrade an existing management system, to (1) address some or all of the essential features and (2) execute some or all of the work activities, depending on perceived risk and/or process hazards that it identifies. However, these steps will be ineffective if the accompanying *process safety culture* element does not embrace the use of reliable management systems. Even the best management system, without the right culture, will not produce an actively involved work force.

6.2.1   Maintain a Dependable Practice

When a company identifies or defines an activity to be undertaken, that company likely wants the activity to be performed correctly and consistently over the life of the facility. For the *workforce involvement* practice to be executed dependably across a company or facility involving a variety of people and situations, the following essential features should be considered.

Ensure consistent implementation. For consistent implementation, the *workforce involvement* program should be documented to an appropriate level of detail, addressing the general management system aspects discussed in Section 1.4. At union facilities, some aspects of the *workforce involvement* program may also be reflected in the work rules negotiated between the union and the company.

Each RBPS element will offer several opportunities for worker involvement in the design, development, implementation, and continuous improvement of the element. The more common activities should be identified and documented in element-specific program documentation. Protocols for addressing worker suggestions should be established (e.g., how a proposed procedure modification, or a suggested change in a maintenance frequency, is handled). Additional, more general program documentation may be required for the *workforce involvement* element itself.

While follow-up responsibility for suggestions made regarding particular elements will rest with the corresponding element owner, organizations may find it beneficial to assign one individual to monitor the implementation of the *workforce involvement* element (i.e., to function as the *workforce involvement* element owner).

Involve competent personnel. All facility personnel should have a basic awareness of the *workforce involvement* program to enable them to interact with it and contribute to it. All workers must understand their personal responsibility for actively participating in the design, development, implementation, and continuous improvement of the RBPS management system. Newly assigned personnel should be oriented on the *workforce involvement* program as part of their initial process safety management system training. Periodic reminders within workforce safety and information meetings and toolbox meetings should help maintain awareness of the program and serve to solicit input.

Managers responsible for the various elements must understand their roles and responsibilities with respect to (1) soliciting and accepting worker participation and (2) providing open-minded and timely responses to suggestions from workers.

6.2.2   Conduct Work Activities

Provide appropriate inputs. Inputs to the *workforce involvement* program include the suggestions from, and active participation of, workers in the design, development, implementation, and continuous improvement of the RBPS management system. Written program documentation should, at a minimum, identify opportunities for worker participation that are required by corporate or regulatory requirements, if such exist.

Apply appropriate work processes and create element work products. *Workforce involvement* work practices and products will be specific to the various elements. For example, for the *operating procedures* element, one work process might involve (1) an operator submitting a request for a procedure modification, (2) review of the suggestion by the procedure coordinator, who decides whether the revision should be adopted, adopted in principle, or rejected, (3) response to the individual who made the suggestion, explaining the basis for the action taken, (4) review of the proposed action with other employee representatives, and (5) necessary action, if any, to implement the decision made in step (2). The resulting work product(s) could be the revised procedure (if appropriate) and the records of the submitted suggestion and its resolution.

All of the steps outlined above for the work practice are important; however, it is essential that the decision-making process in step (2) be open-minded and transparent, and that appropriate feedback be given in step (3), especially if the suggestion is rejected. For the *workforce involvement* program to thrive, workers must see a positive return on their investment of time and effort devoted to participation. This does not mean that suggestions cannot be critically reviewed and, when appropriate, rejected.

6.2.3 Monitor the System for Effectiveness

Ensure that the workforce involvement practices remain effective. Once the *workforce involvement* program is in place, periodic monitoring, maintenance, and corrective action will be needed to keep it operating at peak performance and efficiency. A carefully selected set of relevant metrics should be identified for monitoring the role of *workforce involvement* in enhancing the effectiveness of the associated RBPS elements (see Section 6.5).

Evaluations of RBPS element performance problems should include consideration of the amount and relevance of *workforce involvement* in the implementation of the element. For example, an audit finding stating that operating procedures are generally out of date would present the opportunity for greater involvement by operators in identifying and addressing procedure deficiencies. Alternatives for addressing this gap could include a more formalized operating procedure revision program or a team-based approach in which operators work collectively to review and update procedures on a periodic basis. Workers should be involved in selecting which alternative is chosen.

Section 6.6 suggests a number of management review activities that could be used for monitoring *workforce involvement* effectiveness.

6.2.4 Actively Promote the Workforce Involvement Program

Stimulate workforce participation. By its very nature, the *workforce involvement* program cannot achieve its intended goals without active worker participation. Initiatives may be required, especially initially, to stimulate such participation. This may be particularly true for organizations (1) lacking a tradition of seeking or accepting worker input on such matters or (2) whose past management support for safety programs has been weak.

Adopt new workforce participation opportunities. A list of tasks included in the program documentation is unlikely to comprehensively address all opportunities for worker participation in the design, development, implementation, and improvement of the RBPS management system. As the RBPS management system develops, and as the organization's process safety culture matures, new opportunities for worker involvement may be created or otherwise become apparent. The *workforce involvement* program should be sufficiently flexible to embrace such opportunities as they are identified.

Publicize the success of the workforce involvement program. Sharing the results from the implementation of the *workforce involvement* program should help stimulate worker interest in participation. Demonstrating the positive benefits yielded by the program should illustrate both a return on the investment of effort made by workforce participants and the receptivity of management to the involvement of workers in the design, development, implementation, and continuous improvement of the RBPS management system.

6.3 POSSIBLE WORK ACTIVITIES

The RBPS approach suggests that the degree of rigor designed into each work activity should be tailored to risk, tempered by resource considerations, and tuned to the facility's culture. Thus, the degree of rigor that should be applied to a particular work activity will vary for each facility, and likely will vary between units or process areas at a facility. Therefore, to develop a risk-based process safety management system, readers should perform the following steps:

1. Assess the risks at the facility, investigate the balance between the resource load for RBPS activities and available resources, and examine the facility's culture. This process is described in more detail in Section 2.2.
2. Estimate the potential benefits that may be achieved by addressing each of the key principles for this RBPS element. These principles are listed in Section 6.2.

3. Based on the results from steps 1 and 2, decide which essential features described in Sections 6.2.1 through 6.2.4 are necessary to properly manage risk.

4. For each essential feature that will be implemented, determine how it will be implemented and select the corresponding work activities described in this section. Note that this list of work activities cannot be comprehensive for all industries; readers will likely need to add work activities or modify some of the work activities listed in this section.

5. For each work activity that will be implemented, determine the level of rigor that will be required. Each work activity in this section is followed by two to five implementation options that describe an increasing degree of rigor. Note that work activities listed in this section are labeled with a number; implementation options are labeled with a letter.

Note: Regulatory requirements may specify that process safety management systems include certain features or work activities, or that a minimum level of detail be designed into specific work activities. Thus, the design and implementation of process safety management systems should be based on regulatory requirements as well as the guidance provided in this book.

6.3.1   Maintain a Dependable Practice

Ensure Consistent Implementation

1. Develop written program documentation for managing the overall *workforce involvement* element and for administering *workforce involvement* activities within the various RBPS elements. Such documentation should (1) describe for each RBPS element relevant worker activities for participating in the design, development, implementation, and continuous improvement of the RBPS management system, (2) identify those activities for which worker involvement is mandatory, for example, activities that address specific corporate or regulatory requirements, and describe mechanisms for obtaining this input, and (3) include specific roles and responsibilities for RBPS element owners, and others in management, in regard to workforce involvement issues.
   a. Written procedures address only explicit regulatory requirements.
   b. Written procedures have been prepared to address a diverse range of *workforce involvement* opportunities.

2. Establish an owner for the *workforce involvement* element to monitor and ensure its effectiveness on a routine basis.

a. Individual RBPS element owners are expected to monitor *workforce involvement* within their elements.

b. A facility *workforce involvement* owner has been designated; this person monitors the element on a time-available basis.

c. A facility *workforce involvement* owner has been designated and is provided the time and resources to regularly monitor, advocate, and strengthen the element.

Involve Competent Personnel

3. Provide awareness training on the *workforce involvement* element to all workers.

a. Information on the *workforce involvement* element is shared in an informal fashion, such as through e-mail.

b. Initial *workforce involvement* awareness training is provided once, perhaps as part of new hire orientation.

c. Periodic refresher training is provided for *workforce involvement* awareness training.

4. Provide detailed training to all affected workers who are assigned specific roles within the RBPS management system.

a. Initial *workforce involvement* detailed training is provided once to key personnel.

b. Periodic refresher training for *workforce involvement* detailed training is provided to key personnel.

6.3.2 Conduct Work Activities

Provide Appropriate Inputs

5. Provide systems for scheduling and facilitating the active involvement of workers in the implementation of the various RBPS elements; for example, inviting workers to participate on risk analysis teams.

a. Formal systems have been established for commonly anticipated activities.

b. Formal systems have been established for a diverse range of activities in all RBPS elements.

6. Provide systems for workers to provide their input on issues related to the design, development, implementation, and continuous improvement of the RBPS management system. For example, establish a formal system for suggesting operating procedure revisions or for soliciting feedback at information and toolbox meetings.

a. Informal systems have been established.

b. Formal systems have been established for some activities.

c. Formal systems have been established for all commonly anticipated activities.

Apply Appropriate Work Processes and Create Element Work Products

7. Establish protocols for management to follow when considering and responding to worker suggestions. Such protocols help ensure proper and timely management consideration and resolution of worker suggestions related to the RBPS management system.
    a. Responses are provided, but response time may be slow.
    b. Timely consideration and response to suggestions is a policy requirement.
    c. Response time is monitored and corrective actions are implemented, as required, when response times are excessive.
8. Ensure that workers who make suggestions under the *workforce involvement* program receive appropriate feedback, including an explanation of the manner and rationale for resolving their suggestions.
    a. Informal feedback is provided.
    b. Feedback is formally documented and tracked.
    c. Records of worker participation are maintained, and exceptional levels of participation are commended.
9. Provide a mechanism for resolving disputes, such as a worker dispute of a management decision to reject a suggestion.
    a. Disputes are resolved though informal mechanisms.
    b. A formal program exists with management arbitrating all disputes.
    c. A panel with both management and worker representation arbitrates all disputes.
10. Ensure timely implementation of recommendations accepted under the *workforce involvement* program.
    a. Implementation timeliness is addressed in an ad hoc fashion.
    b. Timely implementation is a policy requirement.
    c. Implementation time is monitored and corrective actions are taken, as required, when response times are excessive.

6.3.3 Monitor the System for Effectiveness

Ensure that Workforce Involvement Practices Remain Effective

11. Develop key metrics for monitoring the performance and effectiveness of the *workforce involvement* element (see Section 6.5).
    a. A few, predominately lagging, indicators are tracked.
    b. A broader set of leading and lagging indicators have been established and are tracked intermittently.
    c. A comprehensive set of leading and lagging indicators have been established and are tracked regularly.
12. Provide timely resolution of, and improvement opportunities for, any *workforce involvement* performance problems identified by the metrics.
    a. Resolution timeliness is addressed in an ad hoc fashion.
    b. Timely resolution is a policy requirement.

    c.   Implementation time is monitored and corrective actions are taken, as required, when response times are excessive.

6.3.4   Actively Promote the Workforce Involvement Program

Stimulate Participation in the Workforce Involvement Program

13. Encourage formal and informal activities that enhance *workforce involvement*, for example, safety suggestion programs, safety lotteries, other incentives for participation, job observation programs, safety councils and focus teams, and management by walking around.

    a.   Occasional, ad hoc efforts exist to stimulate *workforce involvement*.

    b.   Organized efforts exist for some of the more significant *workforce involvement* initiatives.

    c.   Diverse efforts exist to stimulate a variety of *workforce involvement* initiatives. The effectiveness of these efforts is monitored, and they are modified as required.

14. Implement controls to ensure the integration of contract workers into the *workforce involvement* program (contractor organizations, especially transient or short term contractors, may be especially difficult to engage and actively involve in the process).

    a.   Contract workers participate in the *workforce involvement* program on an ad hoc basis.

    b.   Contract workers most directly exposed to process hazards regularly participate in the *workforce involvement* program.

    c.   Participation in the *workforce involvement* program has been integrated throughout the contractor ranks.

Adopt New Opportunities for Participation in the Workforce Involvement Program

15. Modify the written procedures for managing *workforce involvement* to include any new opportunities that should be formalized.

    a.   New *workforce involvement* opportunities are pursued on an ad hoc basis.

    b.   New *workforce involvement* opportunities, when identified, are integrated into the written procedures.

    c.   The facility proactively seeks to identify new *workforce involvement* opportunities and integrate them into the written procedures.

Publicize the Success of the Workforce Involvement Program

16. Actively publicize the workforce involvement program, its goals, the progress of its implementation, and notable successes achieved.

a. Publicity for the *workforce involvement* program is provided in an ad hoc fashion.
b. The *workforce involvement* program is publicized when more notable successes are achieved.
c. The facility aggressively publicizes the *workforce involvement* program and its successes.

6.4   EXAMPLES OF WAYS TO IMPROVE EFFECTIVENESS

This section provides specific examples of industry tested methods for improving the effectiveness of work activities related to the *workforce involvement* element. The examples are sorted by the key principles that were first introduced in Section 6.2. The examples fall into two categories:

1. Methods for improving the performance of activities that support this element.
2. Methods for improving the efficiency of activities that support this element by maintaining the necessary level of performance using fewer resources.

These examples were obtained from the results of industry practice surveys, workshops, and CCPS member-company input. Readers desiring to improve their management systems and work activities related to this element should examine these ideas, evaluate current management system and work activity performance and efficiency, and then select and implement enhancements using the risk-based principles described in Section 2.1.

Additional examples are provided in Reference 6.1, which can be downloaded from the UK Health and Safety Executive website.

6.4.1   Maintain a Dependable Practice

- *Develop the basic framework or standards for the workforce participation plan at the corporate level.* Facility efforts can focus on developing detailed implementation plans and assigning of roles and responsibilities to meet facility-specific needs.
- *Define, subject to risk-based considerations, classes of issues that can be resolved by workers without management involvement.* Examples of such issues might include editorial and other minor changes to operating procedures. Establish worker-implemented management of change controls for these issues.
- *Provide for substantive workforce participation in the creation or revision of safety policies and procedures, and the establishment of*

safety goals and plans. People tend to find more relevance in the rules and objectives that they helped establish (Ref. 6.1).

- *Create positions for safety champions, staffed by workforce volunteers, to serve in an advisory and mentoring role.* Provide additional training to safety champions, addressing topics such as hazard recognition, job safety evaluations, and behavioral-based safety. Use the safety champions to assist in new employee orientation, communication of safety-related messages, conduct of safety meetings, and so forth (Ref. 6.1).

- *Provide training on hazard identification and basic risk assessment principles to all operators and maintenance personnel.* A basic understanding of process hazards and their evaluation should help workers understand the importance of maintaining and enhancing the RBPS management system.

6.4.2 Conduct Work Activities

- *Institute a worker job safety observation program.* Programs that provide direct workforce involvement, such as using work observation cards or forming a peer review and counsel/challenge system, may be a stepping stone to engaging workers more broadly in the RBPS management system.

- *Implement a suggestion submission and response program independent of any particular RBPS element.* Individual elements may have unique processes for making suggestions (e.g., an operating procedure suggestion program). A more general suggestion program may be warranted for the balance of the RBPS elements.

- *Include personnel from all levels of the organization in a regularly scheduled program of field safety and housekeeping inspections.* Emphasize that these inspection teams should include, and not necessarily be led by, upper facility management and their direct reports. (See Chapter 17, Conduct of Operations.)

- *Implement a program of informal what-if exercises, such as table top drills, as part of the process safety training program.* These programs should be developed and conducted with significant worker involvement. (See Chapter 14, Training and Performance Assurance.)

- *Conduct an annual process technology or process safety school developed and taught with significant workforce involvement.* This program should serve as both an enhanced learning experience for the workforce developers and instructors and a mechanism for getting more of the workforce involved in nontraditional roles.

- *Institute a formal mentoring program in which senior, experienced workers assist in the development of less experienced personnel.* A program such as this should help preserve the organizational memory not captured by formal training programs. Appropriate oversight may be warranted (1) to ensure that organizational memory does not propagate bad habits and misconceptions and (2) to capture relevant lessons learned that could be added to formal training programs.

- *Assign experienced operators and maintenance personnel to project design teams.* This allows workers to share their practical knowledge of real-life operability and maintenance issues with the project team.

- *Conduct periodic offsite meetings, during which workers from all levels collaboratively identify potential opportunities for system improvements.* Such meetings (1) help focus participants on the task at hand by freeing them from worksite distractions and (2) emphasize that the organization values workforce involvement sufficiently to incur the added costs of such forums.

- *Use a web-based electronic survey to collect feedback from manufacturing and research facilities.* Such surveys provide another avenue for identifying issues, tools, and solutions that can be further leveraged to improve RBPS management system quality. These surveys may reach portions of the workforce, such as researchers and engineers, who are less likely to participate in other, more traditional suggestion programs. (See Chapter 5, Process Safety Competency.)

- *When communicating safety messages or safety policies, include a convenient way for the reader to provide feedback (e.g., a tear-off sheet for paper communications).* Feedback is more likely to be provided if it is easier for the worker to do so (Ref. 6.1).

- *Form functional teams for relevant RBPS elements, with worker representation from all levels.* For example, a central management of change (MOC) team might review the more significant change requests to ensure consistent treatment across the facility and to ensure that no critical MOC controls are missed. The organization may elect to establish a Process Safety Steering Committee with workforce membership that meets regularly, reviews all RBPS elements on a regular basis, and reviews and approves all changes to RBPS elements.

- *Reassign selected workers from their normal duties and dedicate them to accomplishing a key RBPS task.* Temporarily relieving workers of their normal job responsibilities, such as placing an operator on day shift to assist in rewriting procedures or developing a new RBPS management system, ensures that they have time to focus on the task and reinforces the organization's value for the *workforce involvement* program.

- *Budget time into work schedules to allow workforce members to fulfill formalized workforce involvement activities.* If fully relieving workers of their routine responsibilities (as discussed above) is not feasible, appropriate allowances must be made to permit workers to satisfy their *workforce involvement* responsibilities (Ref. 6.1).

- *Establish and adhere to schedules for senior staff to spend time in work areas.* Direct contact and discussions between managers and workers enables workers to provide feedback on RBPS management system performance issues and to seek information from staff. It is helpful if workers can anticipate these opportunities, and count on them occurring.
- *Use a quality circle approach to addressing RBPS management system problems.* Involving diverse worker teams to address problems and propose solutions may enhance the quality and productivity of the process.
- *Establish inter-facility networks to address common issues.* Such systems could allow each facility to solicit input from, or otherwise tap into the expertise and perspective of, workers at other company facilities who are potentially seeking to solve the same problems.
- *Strive to motivate a broad range of participation.* The same workers may perennially volunteer to fulfill the same *workforce involvement* roles. Without discouraging reliable contributors, strive to motivate others to accept their fair share of the responsibility so the company benefits from the widest possible range of inputs.

6.4.3 Monitor the System for Effectiveness

- *Involve the workforce in identifying suitable RBPS metrics and in monitoring and communicating this information to management.* Workers should become more attuned to improving RBPS management system performance if they have a substantive role in monitoring that performance (Ref. 6.1).
- *Maintain auditable records documenting workforce involvement activities.* Such records may be mandatory in situations for which *workforce involvement* activities are dictated by corporate or regulatory requirements. Maintain records of other nonmandatory activities in sufficient detail to allow management to determine whether the *workforce involvement* program is encouraging a level of activity appropriate to stimulate the improvement of the RBPS management system.
- *Conduct periodic surveys to monitor worker attitudes and to solicit inputs.* Such surveys may provide leading indicators, which may provide early warning of a change in level of workforce engagement and involvement.

6.4.4 Actively Promote the Workforce Involvement Program

- *Create success stories to stimulate interest in the workforce involvement program.* Identify needed RBPS improvements and empower worker teams to address these issues. Provide the resources and autonomy necessary for the teams to succeed, then publicize the successes when they occur. Emphasize that management is receptive to, and awaiting, worker suggestions for the next round of problems to be addressed.

- *Institute a program that provides recognition or awards for safe behaviors to be given by peers.* Peer recognition can be a significant motivator. However, more tangible (e.g., financial) incentives should be focused on levels of performance that go beyond simply performing the day-to-day job as anticipated.
- *Conduct small, informal meetings with the facility manager and just a few workers.* Use these meetings to discuss safety issues, gather worker feedback, and stress *workforce involvement.* Schedule an ongoing series of meetings so that all workers will have an opportunity to participate.
- *Make use of electronic bulletin boards to complement face-to-face meetings.* Post meeting minutes, other communications, improvement requests and resolutions, preliminary incident reports, *workforce involvement* successes, and so forth for all workers to read.

6.5 ELEMENT METRICS

Chapter 20 describes how metrics can be used to improve performance and when they may be appropriate. This section includes several examples of metrics that could be used to monitor the health of the *workforce involvement* element, sorted by the key principles that were first introduced in Section 6.2.

In addition to identifying high value metrics, readers will need to determine how to best measure each metric they choose to track. In some cases, an ordinal number provides the needed information, for example, total number of workers. Other cases, such as average years of experience, require that two or more attributes be indexed to provide meaningful information. Still other metrics may need to be tracked as a rate, as in the case of employee turnover. Sometimes, the rate of change may be of greatest interest. Since every situation is different, the reader will need to determine how to track and present data to most efficiently monitor the health of RBPS management systems at their facility.

6.5.1 Maintain a Dependable Practice

- *Percentage of workers trained on workforce involvement and their responsibilities.* A low value may indicate that the organization provides insufficient emphasis and priority on workforce involvement.
- *Percentage of managers trained on workforce involvement and their responsibilities.* A low value may indicate that the organization provides insufficient emphasis and priority on workforce involvement.

6.5.2 Conduct Work Activities

- *Percentage of workers who have participated in key defined workforce involvement activities, such as submitting a suggestion, serving on a risk analysis team, or participating in an investigation, over the last 12 months.* A low value may indicate either a lack of awareness on the part of workers, or a lack of emphasis on workforce involvement within the organization.

- *Rate of submittal of worker suggestions, and changes in rate with time.* A low rate may reflect a low level of employee engagement in improving process safety, or that employees perceive a low return on investment for their participation. A change in the rate, up or down, may indicate increasing or decreasing employee engagement or may simply be a result of changes in the rate at which modifications are being made to management systems.

6.5.3 Monitor the System for Effectiveness

- *Number of suggestions that have not been evaluated (no decision made to accept or reject) and average/maximum delinquency.* A high backlog of unresolved suggestions, or long delinquency periods, may indicate that the organization provides insufficient emphasis and priority on *workforce involvement.*

- *Number of accepted suggestions that have not been implemented and average/maximum delinquency.* A high backlog of accepted suggestions that have not been implemented, or long delinquency periods, may indicate that the organization provides insufficient emphasis and priority on *workforce involvement.* Further inquiry may be required to determine if this problem is unique to the *workforce involvement* element or whether the organization is generally slow in implementing recommendations from other RBPS elements.

- *Percentage of suggestions accepted.* A low value could reflect either (1) low management receptivity to suggestions or (2) suggestions that are either of poor quality or are not consistent with the focus of the suggestion program (indicating a possible need for additional worker training).

6.5.4 Actively Promote the Workforce Involvement Program

- *Results of worker attitude surveys with respect to acceptance of process safety responsibilities.* Surveys showing a high acceptance of process safety responsibilities may indicate successful implementation of the *workforce involvement* element.

6.6   MANAGEMENT REVIEW

The overall design and conduct of management reviews is described in Chapter 22. However, many specific questions/discussion topics exist that management may want to check periodically to ensure that the management system for the *workforce involvement* element is working properly. In particular, management must first seek to understand whether the system being reviewed is producing the desired results. If the organization's level of *workforce involvement* is less than satisfactory, or it is not improving as a result of management system changes, then management should identify possible corrective actions and pursue them. Possibly, the organization is not working on the correct activities, or the organization is not doing the necessary activities well. Even if the results are satisfactory, management reviews can help determine if resources are being used wisely: are there tasks that could be done more efficiently or tasks that should not be done at all? Management can combine metrics listed in the previous section with personal observations, direct questioning, audit results, and feedback on various topics to help answer such questions. Activities and topics for discussion include the following:

- Determine whether *workforce involvement* program documentation is periodically updated as new involvement opportunities are identified and developed.
- Evaluate what housekeeping practices indicate with regard to worker acceptance of safety responsibilities. Poor housekeeping practices may indicate a lack of worker engagement that may extend into the *workforce involvement* element.
- Identify which RBPS elements are most frequently and least frequently addressed by worker suggestions. Are the less frequently mentioned elements truly so robust, or are they out of sight and out of mind, perhaps warranting increased management emphasis?
- Discuss process safety responsibilities with workers. Do workers have a prevalent sense of ownership for process safety programs, or is enhancement of the RBPS management system considered someone else's responsibility?
- Review a sampling of safety observation cards to gain a sense of the scope and quality of suggestions. For example, do these cards often recommend that someone address issues that the submitter could have resolved personally? If so, this may indicate a low sense of awareness of personal responsibility for safety performance.
- Determine which workers are most commonly involved in RBPS management system activities. Are activities consistently staffed by the same small group of reliable contributors, or is there broader involvement across the organization?

- Discuss participation in RBPS management system activities with workers to identify their motivation. Do they participate because it is obligatory, or because they see value in doing so?
- Review the content of communications that address RBPS management system implementation. Do these communications provide sufficient focus on *workforce involvement* success stories?
- Determine from discussion with workers whether they believe that management is sufficiently responsive to worker suggestions for improving RBPS management system implementation. Do workers feel that their suggestions are given appropriate consideration? Are accepted suggestions implemented in a timely fashion?
- Based upon records reviews and discussions with contractor personnel, determine if contractors are effectively being integrated into the *workforce involvement* program.

A RBPS management system will be more effective if workers share in the ownership of the program. The *workforce involvement* element provides mechanisms for workers to have a substantive, tangible, and important role in guiding and enhancing the programs that help ensure their safety.

6.7 REFERENCES

6.1   UK Health and Safety Executive, Employee Involvement in Health and Safety: Some Examples of Good Practice, WPS/00/03, 2001, www.hse.gov.uk/research/hsl_pdf/2001/employ-i.pdf.

Additional reading

UK Health and Safety Executive, Workforce Participation in the Management of Occupational Health and Safety, HSL/2005/09, 2004, http://www.hse.gov.uk/RESEARCH/hsl_pdf/2005/hsl0509.pdf.

7

STAKEHOLDER OUTREACH

The stakeholder outreach (*outreach*) element has not been a formal part of traditional process safety management systems. However, two situations have sparked interest and activity in stakeholder outreach. First, following the 1984 accident in Bhopal, India, companies belonging to the Canadian Chemical Producers Association, followed by members of the American Chemistry Council (ACC, formerly the Chemical Manufacturers Association), created a mandatory initiative called Responsible Care®. One of the foundational ACC Responsible Care management practices was called Community Awareness and Emergency Response (CAER) (Ref. 7.1). Many facilities established Community Advisory Panels (CAPs) as a result of CAER (Ref. 7.2).

Later, in the late 1990s, as a result of regulations from the U.S. Environmental Protection Agency (EPA), many facilities handling hazardous materials that could affect the public were required to undertake risk communication efforts with their communities to comply with the EPA's Risk Management Plan (RMP) rules (Ref. 7.3). In highly populated areas having a large industrial base, many companies collaborated by having regional RMP communication events. In some cases, multi-year programs were created to (1) engage stakeholder groups, (2) identify their concerns and needs, (3) develop communication/outreach plans, (4) conduct planned activities, and (5) follow up on the results. One of the largest such activities was conducted in the Houston, Texas, area where more than 120 facilities coordinated their RMP outreach activities over a 4-year period. Those activities helped nurture relationships with communities, regulators, local emergency response agencies, and nongovernmental community groups.

> The following is an example of how stakeholder relations can affect process safety and plant operations. In 1998, a specialty gas repackaging firm serving the semiconductor industry in California applied for a permit to expand its facility. Despite the opposition of a national environmental group, the company's long-term effective outreach program had fostered overwhelming local support, and the permit was granted.

7.1   ELEMENT OVERVIEW

Having good relationships with appropriate stakeholders over the life of a facility is one of the five elements in the RBPS pillar of *committing to process safety*. This chapter describes a process for identifying, engaging, and maintaining good relationships with appropriate external stakeholder groups (*outreach*); the attributes of an *outreach* system; and the steps an organization might take to implement *outreach*. Section 7.2 describes the key principles and essential features of a management system for this element. Section 7.3 lists work activities that support these essential features and presents a range of approaches that might be appropriate for each work activity, depending on perceived risk, resources, and organizational culture. Sections 7.4 through 7.6 include (1) ideas for improving the effectiveness of management systems and specific programs that support this element, (2) metrics that could be used to monitor this element, and (3) management review issues that may be appropriate for *outreach*.

7.1.1   What Is It?

Stakeholder outreach (*outreach*) is a process for (1) seeking out individuals or organizations that can be or believe they can be affected by company operations and engaging them in a dialogue about process safety, (2) establishing a relationship with community organizations, other companies and professional groups, and local, state, and federal authorities, and (3) providing accurate information about the company and facility's products, processes, plans, hazards, and risks. This process ensures that management makes relevant process safety information available to a variety of organizations. This element also encourages the sharing of relevant information and lessons learned with similar facilities within the company and with other companies in the industry group. Finally, the *outreach* element promotes involvement of the facility in the local community and facilitates communication of information and facility activities that could affect the community.

7.1.2   Why Is It Important?

Sharing information with industry peers will promote better process safety for everyone. Sharing information in proactive ways with community and government stakeholders builds trust and commitment. Trust supports a facility's license to operate, and regulators may be more willing to work with the facility to resolve issues when they arise. The more the public understands the facility's process safety systems, the more confident they will feel that the company takes every reasonable precaution to protect public safety and the environment. By promoting openness and responsiveness, an effective *outreach* program will increase all stakeholders' confidence in the company.

7.1.3   Where/When Is It Done?

Information sharing and relationship building occurs (1) during employees' community activities, such as speaking to school groups, service clubs, and senior centers, (2) during planned events, such as industry conferences and community meetings, and (3) when episodic events such as incidents and regulatory inspections occur. Events involving the subject facility, other company facilities, or other facilities in the same industry can place the spotlight upon a company. In addition, many companies routinely sponsor volunteer community activities, such as school activities, neighborhood cleanups, beautification efforts, historic preservation, "greenspace" improvements, and plant tours that also have a desirable relationship-nurturing effect.

7.1.4   Who Does It?

Any person who is in a position to convey information or an impression of the company is a potential communicator. Key personnel include not only those having management responsibility, but also people who are in routine contact with the community, such as phone operators, receptionists, or security guards. Communications training for key *outreach* personnel should prepare them for a wide scope of outreach activities ranging from planned events to a press conference in the aftermath of an incident. Keep in mind, however, that appropriate communication and legal personnel should be consulted in the design and execution of the *outreach* program in order to protect confidential business information.

Facilities with solid process safety cultures will generally have more performance-based *outreach* procedures, allowing trained employees to use good judgment in communicating with stakeholder groups. Facilities with an evolving or uncertain process safety culture may require more prescriptive communication procedures, more frequent training, and greater command and control management system features to ensure good public risk communication discipline.

7.1.5   What Is the Anticipated Work Product?

The main products of an *outreach* system are successful communications with relevant stakeholders that build trust and good will. A reservoir of favorable stakeholder opinion can be an invaluable resource when adverse events or conflicts occur. Ancillary products include communication plans, messages, tools, brochures, training materials, and records. Outputs of the *outreach* element can also be used to facilitate the performance of other elements. For example, understanding the community concerns and participating in local emergency planning committee activities can help provide inputs to the ongoing improvement of the facility emergency response plan developed as part of the *emergency* element.

7.1.6   How Is It Done?

Companies train key personnel to interact with important stakeholder groups during planned events and provide resources for all employees to use in their everyday encounters with the public. Crisis communication and outreach training is provided to senior management to help deal with episodic events.

Higher risk situations usually dictate a greater need for formality and thoroughness in the implementation of the *outreach* element. Conversely, companies having lower risk situations may appropriately decide to pursue *outreach* activities in a less rigorous fashion. In the case of the *outreach* element, risk takes on a two-fold meaning: (1) the risk of experiencing an incident and (2) the risk of experiencing an adverse stakeholder reaction as a result of a process safety issue at the facility or other facilities within the company or industry. A higher risk situation may demand a more formal risk communication program that provides detailed information to stakeholders and keeps them updated. In a lower risk situation, a general community outreach policy, via informal practices by trained key employees, may be sufficient.

7.2   KEY PRINCIPLES AND ESSENTIAL FEATURES

Safe operation and maintenance of facilities that manufacture, store, or otherwise use hazardous chemicals requires robust process safety management systems. The primary objective of the *outreach* element is to ensure that a facility retains a good relationship with stakeholders, and establishes and maintains connections with all relevant industry, government, and public groups over the life of the facility. These relationships facilitate the transfer of information, foster mutual understanding of each group's views, and promote long term interaction and mutual trust.

The following key principles should be addressed when developing, evaluating, or improving any system for the *outreach* element:

- Maintain a dependable practice.
- Identify communication and outreach needs.
- Conduct communication/outreach activities.
- Follow through on commitments and actions.

The essential features for each principle are further described in Sections 7.2.1 through 7.2.4. Section 7.3 describes work activities that typically underpin the essential features related to these principles. Facility management should evaluate the risks and potential benefits that may be achieved as a result of improvements in this element. Based on this evaluation, management should develop a management system, or upgrade an existing management system, to address some or all of the essential features. However, none of these work activities will be effective if the accompanying *culture* element does not embrace the use of reliable management systems. Even the best *outreach* activities, without the right culture, will not produce good stakeholder relations over the life of a facility.

7.2.1 Maintain a Dependable Practice

When a company identifies or defines an activity to be undertaken, that company likely wants the activity to be performed correctly and consistently over the life of that process and other similar processes. In order for *outreach* activities to be executed dependably across a company involving a variety or people and situations, the following essential features should be considered:

Ensure consistent implementation. For consistent implementation, the *outreach* element should have someone assigned to monitor communications and respond to special questions, issues, or emergencies. The scope of the *outreach* system should contain (1) strategies for handling all types of outreach/communications situations and (2) methods for addressing changes in *outreach* needs.

Involve competent personnel. As with all RBPS elements, performance will only be as good as the people who are involved in conducting the *outreach* activities. All personnel in the company should have a basic awareness of community outreach activities so that they will know how to properly support these initiatives. However, not everyone will need to know all the details about interactions within the industry or with regulators. Key personnel assigned specific communication/outreach roles will need specialized training, including crisis communications and outreach training.

Keep practices effective. Once an *outreach* system is in place, periodic monitoring, maintenance, and corrective action will be needed to keep it operating at peak performance and efficiency. Section 7.5 provides examples of performance and efficiency metrics that may be useful for monitoring the scope, frequency, and effectiveness of *outreach* activities.

To help ensure outreach effectiveness, companies should consider implementing metrics by setting up data collection and analysis systems to efficiently acquire and evaluate the real-time status of the *outreach* system. Technical and communication professionals should consider designing *outreach* audit activities that help ensure the long-term effectiveness of the *outreach* element. Management review activities involving *outreach* are discussed in Section 7.6. Typically these reviews and more formal audits will reveal gaps in communication/outreach practice execution and identify ineffective activities for which corrective action is needed.

7.2.2 Identify Communication and Outreach Needs

Effective communication, outreach, and relationship building cannot happen unless key stakeholders are identified, specific audiences are targeted, perspectives are understood, company messages and themes are developed, and delivery venues are planned (Ref. 7.4). The level of effort and rigor should be based upon risk and perceived stakeholder needs. A low need situation is exemplified by a facility with minimal potential offsite impacts that generally has a good relationship with neighbors. An example of a high need situation is a facility with a worst-case scenario that could affect people many miles away having a poor safety and environmental record. Companies/facilities should implement effective means of determining the level of need, and on that basis, identify the types of information and messages that should be communicated for each stakeholder group. For an *outreach* system to address all potentially significant situations involving stakeholder needs and concerns, the following essential features should be considered:

Identify relevant stakeholders. Many sources of information can be used to identify who the relevant stakeholder groups are for a facility. Feedback from community groups and community advisory panels can help determine interested parties. Employees and contractors, through *workforce involvement* activities, can also help identify interested groups.

Define appropriate scope. Potential topics and limits on public communications should be established and understood so that outreach efforts can focus on the issues that are of importance to stakeholders while protecting business confidential information. Concentrating on the most important and relevant topics will help meet stakeholder needs in an efficient manner. Sometimes one issue will dominate the community discussion, such as environmental concerns; however, that does not mean that a company should ignore other important issues, such as the risk of an accident or information about a facility's process safety and emergency response programs.

The *outreach* system should prioritize its activities to maximize the impact given its opportunities to reach stakeholders and its resources. Crisis communication plans should anticipate situations in which the company may need to rapidly increase its outreach/communication efforts and engage external resources.

7.2.3 Conduct Communication/Outreach Activities

Once stakeholder groups are identified and key audiences are targeted, company messages and facility information can be blended to develop outreach/communication tools (Ref. 7.4). The organization should create plans for supporting regular interactions with industry and community groups and regulators. In order for companies/facilities to adopt and implement appropriate communication and outreach strategies, the following essential features should be considered:

Identify appropriate communication pathways. A myriad of opportunities exist for interaction between company representatives and stakeholders. Normally, resources are limited and companies will have to select which events they believe offer the best opportunities to reach the target audiences.

Develop appropriate communication tools. Once communication/ outreach targets are identified, management can determine the message(s) they want to convey, choose the most appropriate media to employ, and select the best design for the communication products. Tools developed for outreach communications should be actively used in appropriate venues, which range from live presentations to interactive web sites. As products are used in the outreach/communication events, they should be evaluated to determine their effectiveness.

Share appropriate information. Companies should define a process for screening information from their routine activities to determine if it may be of stakeholder interest. Once discovered, designated individuals should deliver important information via established communication pathways.

Maintain external relationships. Efforts should be made to stay in routine contact with important stakeholder groups. Companies should identify regularly scheduled stakeholder meetings and establish a routine attendance pattern so that company involvement is seen as a normal activity, not one that only occurs when a problem exists. In addition, lessons learned from significant encounters should be captured and broadly communicated to key personnel.

7.2.4 Follow Through on Commitments and Actions

If during an outreach/communication activity, a company representative commits to provide additional information, those promises should be tracked and diligently completed (Ref. 7.4). The following essential features should be considered:

- *Followup commitments to stakeholders and receive feedback.* One of the best ways to develop trust is to consistently deliver on promises. Companies should be careful with the commitments they make and be even more careful to follow through on such commitments. A tracking system should be established to monitor and track the *outreach* commitments to completion.
- *Share stakeholder concerns with management.* Once feedback is received from stakeholders, these concerns should be shared with management for appropriate response and to ensure that management stays connected to the concerns and issues of the community.
- *Document outreach encounters.* Documenting these activities helps ensure institutional memory. These encounters can frequently provide new insights into stakeholder perceptions and issues. They should be distilled and evaluated, and provided to management for integration into future *outreach* plans or revision of existing *outreach* plans.

7.3 POSSIBLE WORK ACTIVITIES

The RBPS approach suggests that the degree of rigor designed into each work activity should be tailored to risk, tempered by resource considerations, and tuned to the facility's culture. Thus, the degree of rigor that should be applied to a particular work activity will vary for each facility, and likely will vary between units or process areas within a single facility. Therefore, to develop a risk-based process safety management system, readers should perform the following steps:

1. Assess the risks at the facility, investigate the balance between the resource load for RBPS activities and available resources, and examine the facility's culture. This process is described in more detail in Section 2.2.
2. Estimate the potential benefits that may be achieved by addressing each of the key principles for this RBPS element. These principles are listed in Section 7.2.
3. Based on the results from steps 1 and 2, decide which essential features described in Sections 7.2.1 through 7.2.4 are necessary to properly manage risk.
4. For each essential feature that will be implemented, determine how it will be implemented and select the corresponding work activities described in this section. Note that this list of work activities cannot be comprehensive for all industries; readers will likely need to add work activities or modify some of the work activities listed in this section.

5. For each work activity that will be implemented, determine the level of rigor that will be required. Each work activity in this section is followed by two to five implementation options that describe an increasing degree of rigor. Note that work activities listed in this section are labeled with a number; implementation options are labeled with a letter.

> *Note: Regulatory requirements may specify that process safety management systems include certain features or work activities, or that a minimum level of detail be designed into specific work activities. Thus, the design and implementation of process safety management systems should be based on regulatory requirements as well as the guidance provided in this book.*

7.3.1 Maintain a Dependable Practice

Ensure Consistent Implementation

1. Establish a program to guide external communication, information sharing, and *outreach* activities.
    a. An informal practice is in place.
    b. A simple written procedure covers all communications.
    c. A detailed written program describes *outreach* activities.
    d. A detailed written program exists for different communication protocols for each stakeholder group.
2. Assign an owner of the *outreach* system to routinely monitor its effectiveness.
    a. A part-time owner, local or offsite, is assigned.
    b. Multiple owners are responsible for different facets of the *outreach* element.
    c. A single, full-time owner is designated for the *outreach* element.
3. Define the scope of the *outreach* system so that types of information to be shared and sources of changes are monitored.
    a. The scope is informally defined.
    b. The types of information and circumstances for sharing are generally defined.
    c. Multiple information types and circumstances for sharing are specifically defined.
    d. Multiple information types and circumstances for sharing are defined with context-sensitive examples.

Involve Competent Personnel

4. Define the roles and responsibilities of personnel involved with information sharing and outreach.

    a.   Roles and responsibilities are informally accepted by ad hoc participants.

    b.   Communication/outreach is the facility manager's responsibility.

    c.   A communication/outreach role is specifically defined at the facility.

    d.   All communication and outreach roles/responsibilities are specified.

5.   Provide awareness training and refresher training on the *outreach* system to all employees.

    a.   Informal training is provided.

    b.   *Outreach* practice is broadcast once (e.g., through e-mail) to everyone.

    c.   Initial *outreach* awareness training is provided to the involved personnel.

    d.   Initial and refresher *outreach* awareness training is provided to the involved personnel.

6.   Provide training to all affected management employees who are assigned specific roles within the *outreach* system.

    a.   Informal communications/outreach training is provided.

    b.   Generic communications training is broadcast once (e.g., through e-mail) to everyone.

    c.   Detailed communications/outreach training is provided once to key personnel.

    d.   Detailed initial/refresher training is provided to key personnel. Training includes workshops involving role playing exercises using interviewers with significant experience in broadcast journalism.

Keep Practices Effective

7.   Keep a lessons learned summary of all communications/outreach encounters to aid in day-to-day management of the *outreach* process.

    a.   An ad hoc summary is kept in minutes of plant staff meetings.

    b.   A summary is kept in the *outreach* coordinator's office.

    c.   Detailed notes on each encounter are kept by the *outreach* coordinator and are accessible to key personnel.

    d.   Electronic or web-based summaries are kept and are accessible to all involved personnel.

8.   Establish and collect data on *outreach* performance indicators and efficiency indicators.

    a.   Ad hoc data are collected.

    b.   Basic community attitude information is occasionally collected.

    c.   Stakeholder attitude indicators are collected annually.

    d.   Comprehensive stakeholder surveys are performed and analyzed regularly.

7.3.2 Identify Communication and Outreach Needs

Identify Relevant Stakeholders

9. Define the relevant audience groups from among the following: employees, contractors, regulatory authorities, community groups, other facilities in the company, and other companies within industry.
    a. Community stakeholders are identified based on minimum regulatory requirements.
    b. Stakeholders are identified by a few management personnel.
    c. A comprehensive list of stakeholders is developed and maintained from internal and external sources.
10. Develop plans for engaging each stakeholder group.
    a. Ad hoc plans are developed for the primary stakeholder groups.
    b. Formal plans are developed for the primary stakeholder groups.
    c. An informal *outreach* strategy is developed for all affected stakeholder groups.
    d. A comprehensive multi-stakeholder strategy is developed and maintained.

Define Appropriate Scope

11. Identify specific categories of information that are to be made available to each stakeholder group.
    a. An ad hoc list is developed for the primary stakeholder groups.
    b. A general list is developed for interested stakeholder groups.
    c. A formal, customized list is developed and maintained for each stakeholder group.
12. Identify regular opportunities for information sharing/outreach.
    a. An informal schedule is developed for the primary stakeholder groups.
    b. A formal schedule is developed for interested stakeholder groups.
    c. In addition to item (b), the schedule is regularly monitored and updated.
13. Develop a crisis communication plan for episodic events.
    a. A plan is developed for primary stakeholder groups.
    b. A formal plan is developed for affected stakeholder groups.
    c. Formal plans are developed and maintained for all identified stakeholder groups for multiple types of crisis communication events.

7.3.3 Conduct Communication/Outreach Activities

Identify Appropriate Communication Pathways

14. Determine the communication/outreach pathways and methods to be used with each identified stakeholder group.

a.   Communication methods are ad hoc.
b.   Formal communication pathways are identified for the primary stakeholder groups.
c.   Formal communication pathways are determined for all identified stakeholder groups.

Develop Appropriate Communication Tools

15. Develop appropriate communication tools.
    a.   Ad hoc communication tools are used.
    b.   General communication tools are developed for each communication channel (e.g., small meetings, community gatherings, conferences).
    c.   In addition to item (b), the tools are customized for each stakeholder group.
16. Use lessons learned from communications/outreach encounters to revise/update communication tools.
    a.   Communication tools are normally updated only if a participant reports an error or a problem with the content.
    b.   The person who distributes communication tools is expected to confirm that they are up to date.
    c.   A formal management system exists for updating all communication tools.
    d.   In addition to item (c), lessons learned from previous *outreach* efforts are used to revise/update communication tools for all identified stakeholder groups.
17. Identify and develop new communications tools as needed.
    a.   New communication tools are created only when needs are identified.
    b.   A formal management system exists for identifying new communication needs.
    c.   A formal approach exists for identifying new communication needs for each stakeholder group.

Share Appropriate Information

18. Participate in information sharing/outreach activities.
    a.   *Outreach* events are attended on an ad hoc basis.
    b.   A representative regularly attends *outreach* events.
    c.   Key employees attend all planned *outreach* events.
    d.   In addition to item (b), a formal approach is taken for selecting appropriate personnel to attend planned *outreach* events for each stakeholder group.

19. Implement crisis communications plans when unplanned communication "opportunities" occur, for example, facility incidents, company/industry incidents.
    a. The crisis communication plan is only activated for facility events.
    b. The crisis communication plan is activated for any news media inquiries.
    c. In addition to item (b), information is proactively provided to the news media.
20. Turn positive accomplishments and events, such as meeting safety goals or being recognized by a regulator for good performance, into communication and *outreach* opportunities.
    a. Major positive events are communicated when they occur.
    b. Positive events are communicated on an ad hoc basis.
    c. All positive events are communicated on a regular basis.
21. Participate in value adding, industry-specific organizations to share and gather information from peer companies.
    a. Participate in informal, unplanned *outreach* experience sharing activities.
    b. Participate in *outreach* experience sharing activities regularly.
    c. A formal approach is used for having appropriate personnel participate in all planned *outreach* experience sharing events.
    d. In addition to item (c), co-sponsor *outreach* experience sharing events.

Maintain External Relationships

22. Monitor each stakeholder group to determine whether its information needs are being met.
    a. Needs/concerns of the primary stakeholder groups are checked on an ad hoc basis.
    b. Needs/concerns of the primary stakeholder groups are regularly monitored.
    c. Needs/concerns of the all stakeholder groups are regularly monitored.
23. Participate in industry groups to ensure that the company is involved in current industry *outreach* efforts.
    a. Participation occurs on an ad hoc basis.
    b. Formal participation in *outreach* efforts occurs regularly.
    c. In addition to item (b), co-sponsor and plan industry events.

7.3.4 Follow Through on Commitments and Actions

Follow-up Commitments to Stakeholders and Receive Feedback

24. Maintain a communications action item follow-up log to ensure that requests for information are met in a timely fashion.

   a. An ad hoc log is kept.
   b. A formal log is kept for requests from the primary stakeholder groups.
   c. A formal log is kept for requests from the primary stakeholder groups, and the response date is recorded.
   d. A formal log is kept for all identified stakeholder group requests, response actions are tracked to completion, and the log is archived for future reference.
25. Solicit feedback from stakeholders following individual encounters to determine if their concerns and issues were addressed.
   a. Feedback is obtained on an ad hoc basis.
   b. Informal feedback is solicited from the primary stakeholder groups.
   c. Formal feedback is solicited from all stakeholder groups.
   d. In addition to item (c), key contacts are called to discuss their thoughts on any feedback.

Share Stakeholder Concerns with Management

26. Provide periodic briefings to key personnel regarding stakeholder involvement activities, attitudes, and concerns.
   a. Ad hoc briefings are provided.
   b. Generic information is broadcast (e.g., through e-mail) to everyone annually.
   c. Detailed briefings are provided to key personnel after major events.
   d. Detailed refresher briefings are provided to key personnel regularly.
27. Determine near-term company plans that may affect stakeholders and begin planning activities to solicit input, identify possible concerns, and resolve issues.
   a. Activities are developed for the primary stakeholder groups.
   b. Activities are developed for all stakeholder groups.
   c. Item (b) is executed before formal plans are announced to all stakeholder groups.

Document Outreach Encounters

28. Keep records on the activities involved in conducting the information sharing.
   a. Individual records are kept.
   b. The *outreach* activities of key personnel are documented.
   c. Comprehensive stakeholder interaction information is documented from all *outreach* activities.
29. Keep records of lessons learned to guide future encounters.
   a. Some records are collected on an ad hoc basis.

b. Stakeholder interaction lessons learned are documented and shared among key personnel.
c. Comprehensive stakeholder interaction lessons learned are documented from all *outreach* activities and shared among all participants.

7.4 EXAMPLES OF WAYS TO IMPROVE EFFECTIVENESS

This section provides specific examples of industry tested methods for improving the effectiveness of work activities related to the *outreach* element. The examples are sorted by the key principles that were first introduced in Section 7.2. The examples fall into two categories:

1. Methods for improving the performance of activities that support this element.
2. Methods for improving the efficiency of activities that support this element by maintaining the necessary level of performance using fewer resources.

These examples were obtained from the results of industry practice surveys, workshops, and CCPS member-company input. Readers desiring to improve their management systems and work activities related to this element should examine these ideas, evaluate current management system and work activity performance and efficiency, and then select and implement enhancements using the risk-based principles described in Section 2.1.

7.4.1 Maintain a Dependable Practice

- *Periodically review the number, frequency, and distribution across stakeholder groups of the various communication and outreach activities undertaken.* A facility may have a variety of stakeholders with which it wishes to maintain relationships. Sometimes in the bustle of company life, a facility may fail to maintain contact with a key stakeholder group. Formally charting outreach activities for each group can help a company maintain a spectrum of relationships and avoid losing track of one or more important stakeholder groups.
- *Keep track of hours spent to determine if any inefficiencies are evident.* Any significant *outreach* efforts inevitably require staff time. Keeping track of time spent by type of activity will help highlight when too much time is being spent on certain activities or groups.
- *Periodically review the number of people who attend professional conferences and industry group meetings.* Maintaining contact with other companies is an important source of information and outreach to

industry. Monitoring the number and activity level for conferences and industry gatherings can alert management to under- and over-commitment of resources to any particular event.

- *Provide advanced communication training to key personnel.* Placing untrained personnel in the media spotlight is certain to generate poor *outreach* results. Providing realistic training on public and media communications is critical to ensuring *outreach* success.

- *Involve senior/executive level management in strategic outreach activities.* Stakeholders want interaction with credible company representatives who can speak with authority and deliver on promises. Generally, stakeholders want executive commitments regarding long range plans or major initiatives. However, facility managers and other personnel normally have more specific knowledge and credibility when discussing facility-specific activities or precautions. Thus, companies should try to match the message with the most effective messenger.

- *Conduct outreach refresher training concerning troublesome issues.* Some companies may have chronic *outreach* issues with the community. Providing issue-specific training will help key personnel make the best use of their *outreach* opportunities in critical situations.

7.4.2 Identify Communication and Outreach Needs

- *Reduce effort and resources spent on outreach events that no longer attract targeted stakeholders or generate significant stakeholder response.* If targeted outreach events are no longer reaching the intended audiences, or if stakeholder concern appears to have ebbed, reducing involvement/effort is a prudent step in stewarding outreach resources. On the other hand, sometimes being active in a minimal way helps preserve a connection with a group that may be useful in the future.

- *Survey stakeholder groups to determine important issues/concerns and focus plans on the major issues.* Facility management must understand what issues are of greatest concern to a target audience. Surveys of stakeholder groups, formal focus groups, or more informal monitoring of community concerns such as reading newspaper articles or letters to the editor can help ensure that you are directing *outreach* efforts toward areas that can have the greatest impact.

7.4.3 Conduct Communication/Outreach Activities

- *Create customizable templates for communication/outreach tools.* Develop key message templates, print media templates, brochures, and other standard items for ready use when needed, rather than having to craft a situation-specific piece each time a need arises.

- *For the most frequent and time consuming outreach activities, provide checklists to aid outreach specialists.* To ensure that all outreach activities are completed, develop a checklist for typical recurring events.
- *Reduce the number of reviews needed to approve press releases and communication pieces.* Design a review process that includes only the necessary departments and people.

7.4.4   Follow Through on Commitments and Actions

- *Integrate commitment tracking systems.* Many elements within the RBPS system generate commitments with deadlines. Integrating *outreach* commitments with those of other elements, such as *risk*, *incidents*, and *audits*, will help ensure that all commitments are tracked and completed in a timely fashion.

7.5   ELEMENT METRICS

Chapter 20 describes how metrics can be used to improve performance and when they may be appropriate. This section includes several examples of metrics that could be used to monitor the health of the *outreach* element, sorted by the key principles that were first introduced in Section 7.2.

In addition to identifying high value metrics, readers will need to determine how to best measure each metric they choose to track. In some cases, an ordinal number provides the needed information, for example, total number of media stories on a particular issue. Other cases require that two or more attributes be indexed to provide meaningful information, such as average attendance at community advisory panel (CAP) meetings. Still other metrics, such as the number of negative or adverse letters to the editor of the local newspaper per quarter, may need to be tracked as a rate. Sometimes, the rate of change may be of greatest interest. Since every situation is different, the reader will need to determine how to track and present data in a manner that most effectively monitors the health of RBPS management systems at their facility.

7.5.1   Maintain a Dependable Practice

- *Number of CAP members that choose to stay involved.* A decline may indicate that members are less concerned about the risk posed by the facility or believe that their participation does not positively impact facility operations.

- *The number of complaints received by the facility.* Any notable increase in the rate of complaints should be investigated to determine if a chronic adverse situation has developed. Reductions may indicate either a lack of interest by community members or improved performance.
- *Community attitude survey results.* A well-designed survey can provide valid insight to the effectiveness of *outreach* element activities and help the facility focus its *outreach* activities to maximize the return on the resources it invests in this area.
- *Percentage of prepared "key messages" issued that appear in media coverage.* If press releases and other "key messages" are not being aired or are being aired less frequently, management should contact the media outlet(s) to uncover any underlying issues.
- *Positive statements about the company made by regulators in public forums.* Positive statements coming from parties that are impartial and competent are important for establishing credibility, and a decreasing number may indicate a deteriorating relationship.
- *Reduction in the number of activist group complaints/demonstrations made against the company.* Fewer negative reactions may mean that the company's outreach activities are effective.
- *Requests granted by regulators.* Positive responses to requests for permits or regulatory relief may mean that facility performance and outreach efforts are improving and the regulators are rewarding a company's proactive efforts for continued improvement.
- *The cost associated with regulatory citations.* An increase in costs may indicate that the company performance or relationship with regulators is worsening.
- *The cost incurred for communication/outreach training.* An unplanned increase may indicate a reduction in efficiency or a required response to adverse events. Conversely, a significant decrease may indicate that training efforts have lapsed.

7.5.2 Identify Communication and Outreach Needs

- *Number of information requests from the community.* An increase may indicate greater community interest resulting from *outreach* efforts or increase in external issues that the company should address in its *outreach* planning. A reduction could mean the company *outreach* efforts are successfully meeting community information needs or that the community is becoming apathetic.
- *The annual number of inquiries from regulators.* A decrease may indicate that regulators are comfortable with company process safety efforts and that *outreach* efforts toward the regulatory agency have been successful.

- *The number of new stakeholders identified.* A decrease may indicate a greater need for *outreach* planning and resources; an increase may indicate an effective program.
- *The number of new/revised communications plans.* New/revised plans may indicate increased *outreach* activity, or greater *outreach* need.

7.5.3 Conduct Communication/Outreach Activities

- *Number of CAP meetings and attendance rate.* A high or increasing number indicates greater community interest and a need for additional *outreach* planning.
- *Number of community members attending planned outreach functions such as plant tours and open houses.* A high or increasing number indicates greater community interest and opportunity for the *outreach* program; a low number or decreasing rate may indicate either decreasing community interest or that company *outreach* activities are satisfying community needs for interaction and information.
- *Number of key personnel that have received initial or refresher communications, outreach, or crisis management training.* Declines indicate that the outreach program is in jeopardy.
- *Number of industry group meetings at which company presenters shared significant lessons learned.* Declines may indicate decreasing outreach activities to industry groups.
- *Number of press briefings on the company.* A stable or increasing number means company communication plans are active.
- *The number of outreach activities per month/year.* An on-target number indicates that company activities are proceeding; and a lower number may mean that company *outreach* efforts are lax; an above-target number may mean an increase in *outreach* needs.
- *The number of outreach meetings held.* A high or increasing number indicates greater community interest and need for *outreach* activities and resources.
- *The amount of time spent preparing for and conducting CAP meetings.* A high or increasing amount may signal a need for improved efficiency, or it may indicate that more difficult communication issues are being addressed.
- *The cost incurred for attendance at industry group meetings.* A high or increasing number may indicate more active company efforts or a need to improve efficiency.

7.5.4 Follow Through on Commitments and Actions

- *Number of commitments made to the community versus the number of commitments completed.* A high ratio means the company needs to do a better job in responding to promises made to the community.
- *Length of time to respond to community inquiries.* A long or increasing time may indicate the need for greater *outreach* resources or increased response efficiency.
- *The cost of responding to requests for information.* An increase may indicate a need to increase *outreach* resources or improve efficiency.
- *The number of management review meetings held to discuss outreach issues.* A low number means management interest needs renewal.

7.6 MANAGEMENT REVIEW

The overall design and conduct of management reviews is described in Chapter 22. However, many specific questions/discussion topics exist that management may want to check periodically to ensure that the management system for the *outreach* element is working properly. In particular, management must first seek to understand whether the system being reviewed is producing the desired results. If stakeholder attitudes about a facility are not positive or are not improving as a result of management system changes, then management should identify possible corrective actions and pursue them. Possibly, the organization is not working on the correct activities, or the organization is not doing the necessary activities well. Even if the results are satisfactory, management reviews can help determine if resources are being used wisely – are there tasks that could be done more efficiently or tasks that should not be done at all? Management can combine metrics and indicators listed in the previous section with personal observations, direct questioning, audit results, and feedback on various topics to help answer these questions. Activities and topics for discussion include the following:

- Review feedback from all stakeholder groups to determine (1) if any indications of a change in interest for certain topics are present and (2) if stakeholders appear to be receiving the intended messages.
- Examine the request rate and the responsiveness to requests for information from community groups. If the request rate is changing, determine if scaling back on certain efforts as a result of a decline in interest would be appropriate, or if additional effort is needed to try to re-engage stakeholders. If the responsiveness appears to be declining, examine why this has occurred and if additional resources or a change in focus is needed.

- Review any feedback from local emergency planners or the fire department. This information should be conveyed to the *emergency* element for resolution, but any change in the rate or tone of feedback may indicate a need for additional *outreach* activities.
- Note the number of people attending a plant open house event. Was it successful? In the future, should it be promoted more extensively or should personal invitations be extended to certain stakeholder groups?
- Review comments made by attendees of *outreach* activities. Are messages being received as intended? If not, why not?
- Evaluate what employees hear from their friends and community organizations. Consider performing employee surveys. Is the facility considered a good neighbor? If not, can *outreach* activities help improve the facility's reputation?
- Review the rate, apparent severity, and trend of recent citizen complaints. Are citizens complaining more often or about smaller issues? Is one person or a small activist group the source of most complaints? If so, could this lead to bigger issues for the facility, such as increased oversight by local authorities?
- Review the web site hit rate. Does management understand the reason for any abrupt changes?
- Evaluate the resource utilization for communication issues. Does the resource utilization trend mirror plans and objectives for this element? If not, should the plans be revised?

When metrics are updated, the *outreach* element owner should make a special effort to understand the reasons for any trends or anomalies. In addition, if any near-term major communication/*outreach* activities to address known public relations issues and community concerns are planned, those should all be included in a briefing for the management review committee.

In addition, an effective management review process reminds the entire leadership team of the importance of *outreach* and the role it can play in helping to secure the status and image of the facility in the eyes of the industry, community, and government. Often the management reviewers are key individuals in *outreach* efforts. If deficiencies are found, the reviewers should ask themselves whether they have been devoting the necessary time and energy to *outreach* activities.

7.7 REFERENCES

7.1   *Community Awareness and Emergency Response Guide (CAER)*, American Chemistry Council, Arlington, VA, 1995.

7.2   *Community Advisory Panel Guide*, American Chemistry Council, Arlington, VA, 1996.

7.3   The Chemical Safety Information, Site Security, and Fuels Regulatory Relief Act, Public Law (PL) 106-40, August 5, 1999.   http://yosemite.epa.gov/ oswer/ceppoweb.nsf/ content/csissfrra.htm

7.4   *RMP Communication Manual,* American Chemistry Council, Arlington, VA, 1999.

II.   UNDERSTAND HAZARDS AND RISK

Risk-based process safety (RBPS) is based on four pillars: (1) committing to process safety, (2) understanding hazards and risk, (3) managing risk, and (4) learning from experience. To understand hazards and risk, facilities should focus on:

- Collecting, documenting, and maintaining process safety knowledge.
- Conducting hazard identification and risk analysis studies.

This pillar is supported by two RBPS elements. The element names, along with the short names used throughout the *RBPS Guidelines*, are:

- Process Knowledge Management (*knowledge*), Chapter 8
- Hazard Identification and Risk Analysis (*risk*), Chapter 9

The management systems for each of these elements should be based on the company's current understanding of the risk associated with the processes with which the workers will interact. In addition, the rate at which personnel, processes, facilities, or products change (placing demands on resources), along with the process safety culture at the facility and within the company, can also influence the scope and flexibility of the management system required to appropriately implement each RBPS element.

Chapter 2 discussed the general application of risk understanding, tempered with knowledge of resource demands and process safety culture at a facility, to the creation, correction, and improvement of process safety management systems. Chapters 8 and 9 describe a range of considerations specific to the development of management systems that support efforts to understand hazards and risks. Each chapter also includes ideas to (1) improve performance and efficiency, (2) track key metrics, and (3) periodically review results and identify any necessary improvements to the management systems that support commitment to process safety.

8

PROCESS KNOWLEDGE

MANAGEMENT

On February 19, 1999, an explosion in Lehigh County, Pennsylvania, resulted in 5 fatalities, 14 injuries, and damage to several nearby buildings. The explosion occurred during manufacture of the facility's first production-scale lot of hydroxylamine. According to the Chemical Safety and Hazard Investigation Board's (CSB's) investigation report (Ref. 8.1):

> *The incident demonstrates the need for effective process safety management and engineering throughout the development, design, construction, and startup of a hazardous chemical production process . . . [D]eficiencies in "process knowledge and documentation" and "process safety reviews for capital projects" significantly contributed to the incident.*

The CSB report goes on to say:

> *[T]he development, understanding, and application of process safety information during process design was inadequate for managing the explosive decomposition hazard of hydroxylamine. During pilot-plant operation, management became aware of the fire and explosion hazards of hydroxylamine concentrations in excess of 70-wt percent, as documented in the MSDS. This knowledge was not adequately translated into the process design, operating procedures, mitigative measures, or precautionary instructions for process*

> operators. *[The] hydroxylamine production process, as designed, concentrated hydroxylamine in a liquid solution to a level in excess of 85 wt-percent. This concentration is significantly higher than the MSDS referenced 70 percent concentration at which an explosive hazard exists.*
>
> The CSB investigators concluded that lack of process knowledge and failure to properly apply the available process knowledge directly led to the explosion.

8.1 ELEMENT OVERVIEW

Developing, documenting, and maintaining process knowledge is one of two elements in the RBPS pillar of **understanding hazards and risk**. Section 8.2 describes the key principles and essential features of a management system for this element. Section 8.3 lists work activities that support these essential features, and presents a range of approaches that might be appropriate for each work activity, depending on perceived risk, resources, and organizational culture. Sections 8.4 through 8.6 include (1) ideas for improving the effectiveness of management systems and specific programs that support this element, (2) metrics that could be used to monitor this element, and (3) issues that may be appropriate for management review.

8.1.1 What Is It?

The *knowledge* element primarily focuses on information that can easily be recorded in documents, such as (1) written technical documents and specifications, (2) engineering drawings and calculations, (3) specifications for design, fabrication, and installation of process equipment, and (4) other written documents such as material safety data sheets (MSDSs). Throughout this chapter, the term *process knowledge* will be used to refer to this collection of information. The *knowledge* element involves work activities associated with compiling, cataloging, and making available a specific set of data that is normally recorded in paper or electronic format. However, *knowledge* implies understanding, not simply compiling data. In that respect, the *competency* element (see Chapter 5) complements the *knowledge* element in that it helps ensure that users can properly interpret and understand the information that is collected as part of this element.

The *knowledge* and *competency* elements are closely linked in other ways as well. The *competency* element involves work activities that (1) promote personal and organizational learning and (2) help ensure that the organization retains critical information in its collective memory. Technology manuals and other written documents produced as part of the *competency* element are often stored and distributed via the *knowledge* element's management system. A key distinguishing feature of information developed as part of the *knowledge* element is that it generally has a high degree of structure, which applies to all processes. In addition, the primary objective of the knowledge element is to maintain accurate, complete, and understandable information that can be accessed on demand. Conversely, a technology manual compiled as part of the *competency* element often includes historical information; it is also less structured and is much more likely to include sections that are "under development" based on ongoing projects within the company's research and technology functions.

Information collected and maintained by the *knowledge* element tends to answer questions starting with "What," for example, "What is the area and overall heat transfer coefficient for this heat exchanger?" Whereas the *competency* element helps process knowledge users answer questions starting with "Why", such as, "Why is this heat exchanger ten times larger than it needs to be for routine service?" The *knowledge* element includes work activities to ensure that the information is (1) kept current and accurate, (2) stored in a manner that facilitates retrieval, and (3) accessible to employees who need it to perform their process safety-related duties.

8.1.2   Why Is It Important?

Risk understanding depends on accurate process knowledge. Thus, this element underpins the entire concept of risk-based process safety management; the RBPS methodology cannot be efficiently applied without an understanding of risk. Process knowledge also supports other RBPS elements. For example, the *procedures, training, asset integrity, management of change*, and *incidents* elements all draw on information that is collected and maintained as part of the *knowledge* element.

8.1.3   Where/When Is It Done?

Development and documentation of process knowledge starts early and continues throughout the life cycle of the process. For example, early laboratory efforts to develop new materials, characterize these materials, and evaluate the synthesis route (including the potential for runaway reaction or other inherent hazards) normally become part of the process knowledge. Efforts continue through the design, hazard review, construction, commissioning, and operational phases of the life cycle. Many facilities place special emphasis on reviewing process knowledge for accuracy and thoroughness immediately prior to conducting a risk analysis or management of change review. Knowledge of special hazards often becomes critical to safe mothballing, decommissioning, and demolition of process units. Knowledge is typically developed and maintained at a number of physical locations, but, in general, process knowledge should always be available to key personnel at operating facilities.

8.1.4 Who Does It?

Knowledge grows and evolves throughout the life cycle of the process and thus is the responsibility of a number of organizations. Early in the life cycle of a process, knowledge is normally developed by the central research, development, and engineering groups. Detailed design work is often performed by external engineering firms, and much of the information developed during the detailed engineering and construction phases of the life cycle is developed externally. Around the time of plant commissioning and startup, responsibility for maintaining and expanding knowledge typically shifts to the facility at which the unit is located. In other cases, the knowledge is maintained by a group external to the facility, such as central engineering or the engineering contractor that designed the unit. Chemical hazard information is developed mainly by suppliers or corporate research and provided to the facility. For example, much of the hazard information on raw materials is documented in MSDSs and product- or chemical-specific guidelines published by the company that manufactures the material.

8.1.5 What Is the Anticipated Work Product?

The main product of the *knowledge* element is (1) accurate, complete, and up-to-date information to support process safety activities at each life cycle phase and (2) a system to store and retrieve this information. Knowledge normally includes:

- Written technical documents and specifications.
- Process design basis.
- Equipment design basis.
- Engineering drawings and calculations.
- Specifications for design, fabrication, and installation of process equipment.
- Other documents such as chemical hazards information, MSDSs, etc.

Another work product is a listing of relevant engineering guidelines or standards that are described in detail in Chapter 4. Documenting the engineering guidelines or standards helps ensure process safety by (1) establishing agreed-to minimum specifications/requirements and (2) identifying what equipment may be affected by changes in guidelines or standards.

These work products define technical content and work activities for several other RBPS elements. For example, the maximum allowable working pressures and temperatures for vessels and other equipment often directly define safe operating limits in the *procedures* element. The list of safety systems that is maintained by the *knowledge* element likewise directly corresponds to the description of safety systems in the operating procedures and is used as input for efforts to identify equipment that should be included in the scope of the *asset integrity* program. The strong link between the *knowledge* and *competency* elements is more fully described in Section 5.1.5. Finally, the work products of this element are a vital input to the *risk* element; understanding risk is central to the entire concept of risk-based process safety management.

8.1.6   How Is It Done?

Process knowledge is typically collected and cataloged as (1) hard copy documents stored in file cabinets or libraries and (2) electronic files or databases maintained on computer networks. Although the more technically advanced database approach should simplify searches and improve accessibility, neither approach will work well unless the facility staff (1) is trained on how to access the information, (2) understands the content, and (3) makes a commitment to keep the information current and accurate.

8.2   KEY PRINCIPLES AND ESSENTIAL FEATURES

Safe operation and maintenance of facilities that manufacture, store, or otherwise use hazardous chemicals requires that hazards be identified, and in most cases, that the risk associated with the hazards be evaluated. Accurate and complete process knowledge is vital to identifying hazards and evaluating risk. Knowledge of design intent, operating methods, equipment ratings, and process hazards underpins any effort to apply the risk-based methods described in this book. In addition, the *knowledge* element serves as a central data repository that provides critical information for many other RBPS elements, including *procedures*, *training*, and *asset integrity*.

The policy governing the *knowledge* element will typically state what information must be collected for all facilities/units and what information should be collected based on hazards, for example, the presence of a highly hazardous chemical in greater than a stated quantity. Also included is information that should be collected, based on special hazards, to support hazard identification, risk analysis, and other RBPS activities during specific phases of the life cycle of the process.

The following key principles should be addressed when developing, evaluating, or improving any management system for the *knowledge* element:

- Maintain a dependable practice.
- Catalog process knowledge in a manner that facilitates retrieval.
- Protect and update process knowledge.
- Use process knowledge.

The essential features for each principle are further described in Sections 8.2.1 through 8.2.4. Section 8.3 describes work activities that typically underpin the essential features related to these principles. Facility management should evaluate the risks and potential benefits that may be achieved as a result of improvements in this element. Based on this evaluation, management should develop a system, or upgrade an existing management system, to address some or all of the essential features, and execute some or all of the work activities, depending on perceived risk and/or process hazards that it identifies. However, these steps will be ineffective if the accompanying *culture* element does not embrace the use of reliable management systems. Even the best management system, without the right culture, will not reliably provide the information that is critical to efforts to (1) understand hazards and manage risk and (2) support other RBPS elements that contribute to effective risk management.

8.2.1 Maintain a Dependable Practice

Accurate and complete process knowledge is required to thoroughly identify process hazards and analyze risk. The RBPS approach cannot be applied without an understanding of hazards and risk, which in turn depends on the *knowledge* element.

Designing and building a modern, efficient, and safe process unit requires a significant investment. In addition to this initial investment, almost all units are modified over time to increase throughput and/or efficiency. The information required to design, construct, and optimize a unit represents a significant, and valuable, corporate asset. Because process knowledge provides the foundation for long-term viability and continued success of the business, a management system should be established to protect and promote the use of this information. Establishing a dependable practice to collect, maintain, and protect a company's process knowledge helps protect an important asset which simply makes good business sense. The management system should include the essential features outlined in the following paragraphs.

Ensure consistent implementation. A written policy will help define the scope, describe different roles, and ensure consistent implementation of the management system. This policy will also assign specific responsibilities for compiling and maintaining each part of the process knowledge. A written policy is especially important if responsibility for this element is shared among several organizations, which is often the most efficient method for managing process knowledge. For example, the facility's safety department may be responsible for maintaining current and accurate MSDSs, while a senior process engineer may also be responsible for maintaining knowledge of the process technology. In some cases, more advanced portions of the knowledge are developed and maintained by a "technology steward" who may also be responsible for understanding and documenting this information for all of the company's facilities that operate a particular type of unit (Chapter 5 describes this concept in more detail). Although this distribution of responsibilities can provide significant benefits, it can lead to gaps in process knowledge or failures to maintain process knowledge if roles and responsibilities are not clearly defined. Finally, the policy should address record retention requirements to (1) protect current and accurate process knowledge from disposal simply because it has not changed in a number of years and (2) implement the company's policy regarding retention of superseded documents.

In short, the policy governing the knowledge element should address the:

- Types of information that will be maintained by the organization.
- Name of the person or department responsible for maintaining each part of the information stored within the *knowledge* element.
- Location(s) of *knowledge* element documents needed by employees/ contractors to perform their work.
- Events that typically lead to the creation or update of each type of document/work product stored within the *knowledge* element.
- Archival of out-of-date process knowledge and periodic purging of unneeded information.
- Specific regulatory requirements.
- Document retention requirements.

Define the scope. The scope of the *knowledge* element contains several parts. First, a facility needs to decide if the RBPS management systems will be applied throughout the facility or only to certain units. Since the RBPS approach requires knowledge of hazards and risk, this element will apply unless the facility elects to use some other approach. For example, a standards-based approach to process safety is often more appropriate for utility systems that serve office buildings. However, standards-based approaches, as well as other well-established approaches to managing process safety, require certain process knowledge. A facility also needs to determine what types of information should be included.

To support efforts to analyze risk, and to effectively manage the information required to support ongoing operations, facilities typically choose to collect the necessary information, such as that listed in Table 8.1, or they determine that specific items are not applicable to a particular process (Ref. 8.2).

TABLE 8.1. Typical Types of Process Knowledge

Chemical Hazards Information

- Toxicity information
- Permissible exposure limits
- Physical data
- Reactivity data
- Corrosivity data
- Thermal and chemical stability data
- Hazardous effects of inadvertent mixing of typical contaminants (e.g., air, water) with different materials contained in process streams and utility systems
- Thermodynamic data
- Calorimetric data
- Special hazards
  - o Shock sensitivity
  - o Pyrophoric properties
  - o Chemical stabilizers, including effects of purification (removal of a stabilizer or other chemical species)
- Maximum deflagration or detonation pressure and flame speed
- Industrial hygiene data

Process Technology Information

- A simplified process flow diagram, or block flow diagram for very simple processes
- Process chemistry, including laboratory notebooks that provide information discovered during the early stages of product or process development
- Hazards related to credible undesired chemical reactions (e.g., production of byproduct dioxin when processing chlorinated organic compounds)
- Material and energy balances
- Maximum intended inventory
- Safe upper and lower limits for process parameters such as temperatures, pressures, flows, or compositions, and the likely consequences of deviations from safe limits
- Adiabatic reaction temperature and the corresponding system pressure, based on both intended and worst credible case material composition
- Separation equipment design information and design basis (e.g., minimum reflux ratio required to maintain safe operation)
- A description of control system logic in narrative format and/or simple figures
- Maps and/or tables showing zones/distances of concern for overpressurization or toxic exposure hazards based on consequence analysis

Process Equipment Information

- Materials of construction
- Piping and instrumentation diagrams (P&IDs)
- Electrical classification diagrams
- Relief system design basis and calculations, including any flare system
- Ventilation system design basis and calculations
- Listing of design codes and standards applicable to the process
- Safety systems (e.g. interlocks, detection, or suppression systems)
- Mechanical data/design basis sheets for process equipment
- Shop fabrication drawings
- Piping specifications
- Isometric drawings
- Control system logic diagrams, loop sheets, and interlock tables
- Instrument data, including a register or database of key parameters for field instruments, alarms, interlocks, and so forth
- Electrical data, including one-line diagrams, a motor database, and grounding/bonding drawings
- Facility data, including plot plans that document the location of underground utility and process piping, structural drawings and structural analysis, design and design basis information for fixed fire protection systems, and information on heat/blast loads and fire/blast walls
- Design basis and analysis for fixed or dedicated hoists
- Location of safety showers/eye wash stations, fire extinguishers, and other safety equipment
- Portable multi-unit equipment

Finally, a facility needs to determine what requirements apply over the life cycle of a process. Early in the life of a process, hazard information is needed to support various hazard identification activities that are part of the *risk* element. At later stages of development, more detailed risk analysis activities require more detailed engineering data. Thus, the policy governing the *knowledge* element should stage the requirements so that they match what is needed for the *risk* element without placing an undue burden on the process development team to produce data or information that is not yet needed and is often likely to change prior to when it *is* needed.

Note that process knowledge requirements related to specific process equipment should be based on (1) the type of equipment and (2) the failure modes and consequences of failure for the equipment, in other words, the risk. The process knowledge that might be documented for a specific type of shell and tube heat exchanger varies little from one exchanger to the next, yet the information needed to properly manage risk strongly depends on its service. For example, the *knowledge* requirements for a heat exchanger that must maintain cooling to prevent a runaway reaction will be much greater than for a heat exchanger used to cool hydraulic fluid, assuming that no safety consequences are associated with loss of function of the hydraulic system. Thus, the decision regarding what information to collect and maintain for a given piece of equipment strongly depends on risk.

Some facilities choose to use the management system for their *knowledge* element as a "one stop shop" for all process related information. Thus, information generated by or needed by each process safety element resides in a single management system. This information ranges from original equipment manufacturer's operation and maintenance manuals needed for the *asset integrity* element to a complete log of incidents and copies of incident investigation reports for the *incidents* element.

A facility also needs to determine if different requirements should be developed for various units or areas, and if so, what information should be included in the scope of the *knowledge* element for each unit or area. For example, a facility may choose to document electrical one-line diagrams in all process/manufacturing areas to facilitate maintenance and troubleshooting of electrical equipment, but not require similar information for a finished product warehouse that stores dry goods.

The required depth of understanding or knowledge also depends on whether the process is unique, well understood but operated in relatively few locations, or composed entirely of off-the-shelf equipment. For example, a facility that converts ammonia to ammonium nitrate will need greater understanding of chemical reactivity hazards than a facility that uses ammonia as a refrigerant.

Finally, facilities need to realistically compare the information that is currently available for each process to the list of process knowledge in Table

8.1. Newer and/or highly regulated facilities will likely find that they have most, if not all, of the information that applies to their facility. (Some of the knowledge listed in Table 8.1, such as special hazards, will not apply to most facilities.) Using the risk-based approach, the scope of the knowledge needed to manage risk at a facility that receives a wide range of nonhazardous chemicals; blends them with chemically stable flammable liquids, such as alcohols or ketones, to produce a consumer product; and packages these products for sale to consumers is substantially less than the scope of the knowledge needed by the facility that converts basic hydrocarbons into the flammable, but otherwise stable, solvents used by the blender.

In some cases, older facilities may simply have lost much of their knowledge, particularly knowledge related to equipment and technology. This can be a significant problem for a facility where the process knowledge was developed and maintained by corporate engineering until the facility was subsequently sold to another company. Readers who find themselves in this situation may wonder what process knowledge to assemble first and/or how much is really needed.

The foundation of the risk-based approach is to formulate process safety management systems on the basis of risk; and understanding risk depends on having sufficient knowledge of the process. Bypassing the *knowledge* element and jumping directly to the *risk* element would be imprudent. Even if risk is well understood, reliably applying several other RBPS elements with little or no documented process knowledge is very difficult. For example, effectively applying the *management of change* element would be difficult if the question, "Does it meet the design specification?" cannot be answered because no specifications or other information on design basis are available. Although the cost to create (or recreate) a comprehensive set of process knowledge can be prohibitive, facilities that do not have this information will struggle to develop sound RBPS management systems for many of the elements described in this book.

Therefore, the only clear answer regarding how to proceed with applying RBPS in the absence of complete process knowledge is, "very carefully." Regulations may require that facilities maintain parts of the knowledge listed in Table 8.1. In those cases, the decision whether to comply with the regulatory requirement is a legal/business decision as well as a risk-based decision. The RBPS method offers no guidance on legal decisions other than companies should, at a minimum, comply with laws and regulations. In the absence of regulatory requirements, companies may apply the following "filters" when prioritizing what information to develop:

- Hazards, rather than risk, are usually a good first screening tool. If no flammable materials or combustible dusts are present at the facility, electrical classification diagrams are likely to be low priority.

- Time and expense to obtain information are also good initial screening tools. Chemical hazards information is often available at no cost from suppliers. Some of the process technology information can simply be written down by a knowledgeable chemist or engineer.
- Information needed to support the *risk* element should be located toward the top of the priority list. If a facility decides to use the hazard and operability method as its primary risk analysis tool, it will (at a minimum) need most of the chemical hazard and process technology information listed in Table 8.1. It will also need some of the equipment information listed, for example, a complete listing of safety systems, and up-to-date piping and instrumentation diagrams (P&IDs), including information on control logic if it is not shown on the P&IDs. Many of the items listed in the process equipment information section will not be needed by the PHA team, including isometric drawings and design basis data for hoists, although team members may be assigned to field check items related to these data elements during the course of the meetings.
- Information needed to support the balance of the "manage risk" elements (see Chapters 10 through 18) should be given higher priority. For example, piping specifications help prevent changes that can undermine process safety, and some isometric drawings are normally needed to support piping inspections that are part of the *asset integrity* element.

Thoroughly document chemical reactivity and incompatibility hazards. The property that makes many chemicals useful – that they will react to form other materials – can also make them very hazardous. Teams tasked with identifying hazards and analyzing risk are primary consumers of process knowledge. If the process being analyzed involves reactive chemicals, the team will need information on reactivity hazards.

MSDSs often provide some information on reactivity and incompatibility hazards. However, this information is sometimes insufficient or hard to interpret. Furthermore, MSDSs are unlikely to answer questions such as, "What happens if there is no flow of cooling water to the condenser?", "What happens if the operator fails to add the solvent to the reactor prior to adding the catalyst?", or "What if the agitator stops turning?" Without clear, understandable information on chemical reactivity hazards, teams may (1)-miss hazards or (2) recommend expensive systems to prevent or mitigate reactivity hazards even though the process is inherently safe. Both outcomes are undesirable. The CCPS book titled *Essential Practices for Managing Chemical Reactivity Hazards* is an excellent resource for additional information related to chemical reactivity hazards (Ref. 8.3).

Assign responsibilities to competent personnel. Competent personnel are critical to maintaining and expanding knowledge. Through advances in technology, the amount of information that can be generated or collected has greatly increased, and a wide array of information is available to anyone with a computer and a connection to the internet. This has led to a tendency to relegate collection and maintenance of information to persons with limited technical skills or operational experience. For example, many companies assign a clerk to keep MSDS files current (either in hard copy or electronic form). Although this is not an inherently bad idea, a technically competent person must determine what has changed when a new/updated MSDS is received and the implication of the changes on the safety and health of workers. Since the rate that people can assimilate information has not increased at the same rate that information can be generated, it is more important than ever to assign "experts" to review newly acquired knowledge so that they can quickly identify and assess the impact of any changes.

8.2.2   Catalog Process Knowledge in a Manner that Facilitates Retrieval

Information that cannot be efficiently accessed becomes clutter. Too often, key data such as design bases, manufacturer's drawings/data reports, specifications, and other process knowledge are thrown away because the documents are not well organized or cataloged. A random mixture of current and out-of-date information in the same storage area is generally worse than clutter. In extreme cases, it is a trap set to catch an unsuspecting user of the out-of-date information.

Make information available and provide structure. See the essential features described in the competency element (Section 5.2.2).

Protect knowledge from inadvertent loss. Many facilities now store much or all of their process knowledge on network servers. This provides many benefits, including the ability to store large volumes of information inexpensively, access the information from nearly anywhere, produce backup copies, protect the material in read-only files, and sort the material in a logical structure. Some facilities insert hyperlinks into the documents to provide better interconnectivity between related documents.

One drawback to this approach is that certain documents may need to be available at all times, but in the event of a natural disaster, they may not be available at all. Facilities that use the computerized approach should ensure that current versions of critical documents, such as MSDSs, are always available. This normally requires keeping an up-to-date paper copy of critical documents somewhere at the facility or maintaining copies on a stand-alone computer with an emergency power source.

Process knowledge should be easy to locate. This is generally less of an issue at facilities that have central engineering groups; one typical function of central engineering is to maintain files on the equipment and the process. However, central engineering groups have been abolished at many facilities, leading to rapid erosion of the formal filing system once responsibility for maintaining the information is dispersed throughout the organization. Once lost, process knowledge can be very expensive to locate or recreate.

Store calculations, design data, and similar information in central files. Too often, data, calculations, and similar "backup" information are stored in the file cabinet of the person who developed a specific part of the process knowledge, while the results of the activity are stored in the facility's "official" process knowledge database. This greatly reduces the volume of information that is stored centrally, and can make it easier to find the results. However, the detailed information will likely be lost over time and may have to be recreated in the future. For example, consider a process that involves an exothermic chemical reaction in an inert solvent with the intent of removing heat by boiling the solvent, condensing it, and returning the condensate to the reactor. The original process hazard analysis team likely wanted to know the capacity of the condenser relative to the expected heat load, and the design engineer may have determined that the condenser is rated for eight times the expected heat load. Years later, another hazard analysis team might be charged with evaluating the hazards of a proposed increase in the ratio of reactants to solvent (to increase throughput). If the original design calculations are stored with the process knowledge, the change in the margin of safety that will result from the proposed change will be much easier to estimate. If the calculations cannot be located, they will need to be redone, which introduces an additional opportunity for error as well as causes extra work for the process engineering group. Note that this is an excellent opportunity to use hyperlinks effectively, because the results of the calculations can be stored in summary documents and hyperlinks can be provided to the document that contains the detailed calculations.

Document information in a user-friendly manner. Documents should be developed with the end user(s) in mind. A consistent data structure is normally beneficial. Developing and enforcing standards for symbols, terminology, analysis methods, and software helps reduce the cost of training personnel as they move to a new position within the organization. Using consistent notation on P&IDs is particularly important as they are frequently used by the *risk* element when evaluating hazards and during work activities associated with many RBPS elements that help manage risk. In addition to using consistent symbols, some facilities have developed a standard nomenclature for tag numbers applied to equipment. For example, the pneumatically actuated ball valve on the cooling water line to heat exchanger HE-1234 might be numbered FV-1234-CS and the flow control valve in the cooling water return line might be numbered FCV-1234-CR. Consistently using a standard format, and a standard set of prefixes to designate the type of equipment and suffixes to designate the service helps reduce human error. *Warning . . . this idea should probably not be applied to existing processes because changing tag numbers will create many error-likely situations and be a relatively expensive endeavor.* (Changing equipment tag numbers would affect all P&IDs, field labeling, the computerized maintenance management system, control room displays, equipment files, etc.)

A standard set of tables that summarize information or point users to where they can locate specific data are very useful and are normally inexpensive to create. For example, tables that summarize the results of calculations for the emergency pressure relief systems design bases can allow the user to quickly determine which cases were considered by designers, the results for each case, and the capacity of the system. This is a much more efficient way of storing information than simply providing the raw calculations. For example, users would not have to wade through pages of calculations to determine if a particular cause of overpressure that is of interest to a hazard analysis team was considered by the emergency relief system design team.

If regulatory requirements, corporate standards, or local procedures exist that require certain information that does not apply to a particular process, clearly document that fact to prevent someone from spending considerable effort looking for information that does not exist.

8.2.3   Protect and Update Process Knowledge

Control or limit access to out-of-date documents. Users should be able to easily distinguish between current and out-of-date information. In many cases, this can be accomplished simply by making only current copies of information available and have users print copies of documents when needed. For example, P&IDs may be stored in a manner that everyone at the facility can view and print the most current version, with edit privileges and access to previous revisions limited to the draftsman and one or two other individuals. Many facilities choose to implement robust document control systems, including automatic or procedural controls to assign an expiration date to all documents when they are printed. Other controls include physically isolating out-of-date printed documents and electronic files in areas that can be accessed only by a small number of employees (who clearly understand that this information is out of date and is being kept for archival purposes only).

In some cases, certain sections of documents may remain current while other sections have been superseded. For example, some of the emergency relief system sizing calculations may not be affected by a change in the analysis method used by the company, while other calculations that are affected must be redone. In this case, it is very important to mark out-of-date sections of otherwise current documents and provide a pointer to where the current information can be found.

Another common cause of document control failures is the practice of creating a "process safety information file room" for the primary purpose of protecting the information. This is often done is response to an audit finding that part of all of the process knowledge has been misplaced. To populate the files, the facility assigns someone to locate and copy documents currently found in files maintained by the engineering, maintenance, production, and safety departments. Thus, a set of parallel files are developed to protect against loss of information. However, a file room that is created solely for this purpose often results in (1) significant effort to continuously synchronize the files or (2) failure to keep the documents in the file room current and accurate, which degrades their value.

Ensure accuracy. The *knowledge* element provides a static "snapshot" of a facility at a given time. However, facilities are rarely static. Changes are frequently made to improve operations, or as a result of requirements that are externally imposed on the facility. Failure to actively identify these sources of change and update the process knowledge will quickly lead to unreliable and inaccurate process knowledge. If the process knowledge is not updated as changes are made, a facility may eventually have to do a complete technical review and update of all its process knowledge, which can be very expensive.

Some facilities do not have the resources to maintain all of the electronic files associated with its process knowledge. This is particularly true for drawings, as many facilities no longer have an engineering department or a drafting group. These facilities often depend on a corporate group or a contractor to update drawings and other documents that require special software and/or expertise. In that case, it is important to establish a robust management system for which one set of hard copies of the drawings is the official set, and these drawings are "red-lined" or otherwise updated concurrent with changes. These "redlines" are periodically copied and sent out for revision. If this method is used, it is important to (1) clearly mark any other copies of these documents as unofficial or out of date (e.g., drawing files on the computer network), (2) train all potentially affected workers and reinforce the management system with a written procedure, (3) update files with revised drawings or other documents when they are returned, and (4) establish a management system to capture changes that occur between the time that the documents are sent out for revision and the time they are returned (e.g., making these changes in a different color and transferring them to the revised drawings when they are returned).

Process knowledge may contain some incorrect information as a result of human errors. Over time, errors should be discovered and corrected. Therefore, one might assume that the accuracy and completeness of process knowledge continually increases. However, external changes, such as a vendor changing the specifications of a replacement part, often occur outside the company's *knowledge* management system and are therefore not updated in the records. Efforts to maintain current and accurate process knowledge go beyond simply filing documents. Periodic reviews of process knowledge should be built into the management system. For example:

- Chemical hazard information can be reviewed periodically by safety professionals and as part of hazard analyses or other activities associated with the *risk* element.
- Process technology information can be reviewed at the same time that corresponding operating procedures are reviewed to (1) help the reviewer determine if operating procedures have not been updated to reflect important changes in the corresponding parts of the process knowledge and (2) identify process knowledge that needs to be updated.
- P&IDs and other critical equipment information can be field-verified prior to the periodic revalidation of a unit's process hazard analysis.
- A sampling of equipment information can be reviewed to judge the overall accuracy of the equipment information prior to periodic management reviews of this element.

Protect against inadvertent change. Many facilities allow corrections to process knowledge with minimal review and approval. Insisting on a complex hazard review process prior to correcting an error on a P&ID would likely be a deterrent to correcting information when errors are discovered. However, a "P&ID error" could potentially be the result of an inadvertent or unauthorized change to the process. Correcting the P&ID casts the change in stone with little or no review! Therefore, changes to process knowledge should be controlled and authorized by competent personnel who should make a determination that a "correction" does not permanently authorize an inadvertent/unreviewed change to the process.

Protect against physical (or electronic) removal or misfiling. Documents that contain critical process knowledge are sometimes lost or physically misplaced. One common cause is that users physically remove a document from the central file and fail to return it. Documents maintained in personal files are often lost when the individual moves to another position. The organization responsible for the document repository can be dissolved, leaving the documents vulnerable to inevitable house cleaning, which is often conducted by new tenants of the physical space where the files were stored rather than persons with expertise in the subject matter. Even if documents are stored in well-indexed central files, they can be lost as a result of a reorganization or divestiture. Therefore, controls should be implemented to physically protect master or official copies of the documents that collectively make up the process knowledge.

With the advent of copiers that can rapidly scan documents to electronic format and computer networks that provide inexpensive storage and robust procedures for backing up data, many facilities are moving their process knowledge to the computer network. This type of system offers many advantages, but care should be taken to protect against inadvertent loss of data. Since many of the backup actions occur each night, the system is capable of eventually overwriting good documents or data with corrupted data. Facilities that choose to exclusively use electronic files should carefully evaluate the potential for data corruption and loss and implement appropriate safeguards to protect against the loss of data. For example, data could be backed up to an archive section of the network (or to archive disks) after each periodic review. Facilities should also ensure that they maintain the hardware necessary to read removable electronic storage media and software capable of opening data files.

Support efforts to properly manage change. The *management of change* element includes a key principle regarding recognizing changes (see Chapter 15). In addition, this section of Chapter 8 includes a brief discussion concerning the acceptance of unreviewed/unapproved changes as efforts to update process knowledge. The *knowledge* element should not drive the management of change process. Rather, the *management of change* element should include a means to (1) notify the custodian of the process knowledge of proposed changes, (2) provide basic data, drawings, and other information to persons assigned to design, review, and approve the change, and (3) provide either (a) a set of updated process knowledge or (b) a list of changes that need to be made to the process knowledge. The *knowledge* element should include a means to confirm that the process knowledge has been properly updated when the change is implemented. For example, the knowledge element should insist that "as built" drawings be produced or that the "approved for construction" drawings be field verified after a change is made.

Stewards of process knowledge can support the *management of change* element in two specific ways. First, stewards should always inquire about changes to process knowledge without an approved change request. Although updates may occasionally be needed without change authorization, it is unusual. A questioning attitude on the part of persons assigned to maintain the process knowledge will help prevent inadvertent/unauthorized changes. Second, if the *knowledge* element serves as the central repository of technical and engineering information, it may help identify situations in which two or more changes are authorized simultaneously that may interact with each other. For example, a process engineer may be implementing a small change at the same time that the central engineering group is working on a major change that mostly affects another unit, but ties into the unit where the small change is being made. Providing a system to check out drawings and other information for revision helps ensure that the two parties are aware of related changes that are currently being proposed.

8.2.4 Use Process Knowledge

Process knowledge provides little value if it sits dormant in a file. This knowledge will be underused if any of the following conditions exist:

- Documents are not accessible.
- Information cannot be readily located within documents.
- Personnel have low confidence that the process knowledge is current and accurate.
- Personnel are unaware of how to access process knowledge.

The first three conditions were addressed in key principles 8.2.1 through 8.2.3. Companies make a substantial investment to generate, gather, and organize process knowledge. To make this investment pay dividends, personnel must be trained on what process knowledge is available, why it is important, how to access the knowledge, what to do if a change is made that could affect the information, how to update the information when an approved change is made, and the policy governing document control.

Ensure awareness. Facilities should inform new employees of the location and content of process knowledge. The level of detail provided to new employees should correspond to the new employee's job responsibilities and the degree to which the employee is likely to need to access the process knowledge. Work activities to ensure awareness should be integrated into the training/orientation for new employees who routinely access the process knowledge. Rarely does a new engineering graduate have an appreciation for the gamut of management systems and technology that underpin process safety. The consequences of expecting new engineers to simply glean this information through unstructured work experience can range from economic losses associated with recreating process knowledge to catastrophic accidents. Training should emphasize what knowledge is available, how to access the knowledge, and how to update the knowledge.

Ensure that process knowledge remains useful. Access and awareness may not be sufficient. Facilities should also ensure that information is current, accurate, and complete. Completeness should not be measured solely in terms of a predefined checklist; any assessment of gaps in process knowledge should include knowledge that is needed to understand risk through the life cycle of the process.

8.3   POSSIBLE WORK ACTIVITIES

The RBPS approach suggests that the degree of rigor designed into each work activity should be tailored to risk, tempered by resource considerations, and tuned to the facility's culture. Thus, the degree of rigor that should be applied

to a particular work activity will vary for each facility, and likely will vary between units or process areas at a facility. Therefore, to develop a risk-based process safety management system, readers should perform the following steps:

1. Assess the risks at the facility, investigate the balance between the resource load for RBPS activities and available resources, and examine the facility's culture. This process is described in more detail in Section 2.2.
2. Estimate the potential benefits that may be achieved by addressing each of the key principles for this RBPS element. These principles are listed in Section 8.2.
3. Based on the results from steps 1 and 2, decide which essential features described in Sections 8.2.1 through 8.2.4 are necessary to properly manage risk.
4. For each essential feature that will be implemented, determine how it will be implemented and select the corresponding work activities described in this section. Note that this list of work activities cannot be comprehensive for all industries; readers will likely need to add work activities or modify some of the work activities listed in this section.
5. For each work activity that will be implemented, determine the level of rigor that will be required. Each work activity in this section is followed by two to five implementation options that describe an increasing degree of rigor. Note that work activities listed in this section are labeled with a number; implementation options are labeled with a letter.

Note: Regulatory requirements may specify that process safety management systems include certain features or work activities, or that a minimum level of detail be designed into specific work activities. Thus, the design and implementation of process safety management systems should be based on regulatory requirements as well as the guidance provided in this book.

8.3.1 Maintain a Dependable Practice

Ensure Consistent Implementation

1. Create a written policy governing the *knowledge* element.
   a. Process knowledge is compiled and maintained on an ad hoc basis.
   b. Responsibility for compiling and maintaining process knowledge is included in the job description of the PSM coordinator or PSM engineer.

    c.   Responsibility for compiling and maintaining process knowledge is integrated into well established facility practices, and responsibilities are widely understood.

    d.   A formal written policy exists for governing the *knowledge* element that addresses all of the items listed in the description of this essential feature in Section 8.2.1. The practice of compiling and maintaining process knowledge is integrated into well established facility practices.

Define the Scope

2.   Specify in the written policy the scope of the knowledge element, including the various types of information and documentation that should be created/compiled for each unit at the facility.

    a.   Process knowledge is compiled and maintained on an ad hoc basis.

    b.   The written policy describes what process knowledge is required for certain designated areas at the facility.

    c.   The written policy clearly describes what process knowledge is required for each unit or process area.

    d.   The written policy clearly describes what process knowledge is required for each unit or process area, along with standards regarding how the information will be compiled and stored.

    e.   In addition to item (d), the written policy includes checklists or forms that indicate the type of information required for common types of equipment. These forms are tailored based on risk, for example, the requirements for a compressor that is used to supply hydrocarbons to high-pressure process equipment are much more rigorous than for compressors that are used to supply utility air.

3.   Compile chemical hazard information, process technology information, and process equipment information listed in Table 8.1.

    a.   Process knowledge is compiled and maintained on an ad hoc basis.

    b.   Certain types of process knowledge are compiled, maintained, and well indexed (e.g., MSDSs); however, most process knowledge is compiled and maintained on an ad hoc basis.

    c.   Most of the basic process knowledge is complied and maintained, but no infrastructure has been developed to support collection of knowledge that is not specifically required by regulatory or insurance company requirements.

    d.   A comprehensive set of process knowledge is compiled and maintained, and a consistent structure is used for all process areas; the level of detail varies based on hazards and perceived risk.

4. Include in the written policy governing the *knowledge* element what process knowledge is needed as a function of the life cycle of a process.
    a. Process knowledge is compiled and maintained on an ad hoc basis.
    b. The policy governing the *knowledge* element only addresses process knowledge requirements during the plant-scale design and operating phases of the life cycle.
    c. The policy governing the knowledge element addresses process knowledge requirements during all phases of the life cycle.

Thoroughly Document Chemical Reactivity and Incompatibility Hazards
5. Include in the written policy governing the *knowledge* element a specific standard for documentation of chemical reactive hazards.
    a. Documentation of chemical reactivity hazards is generally limited to MSDSs.
    b. A written policy requires that units maintain MSDSs for all chemicals present at the unit, and that data be recorded using a specific form/matrix to summarize the hazards of mixing for chemicals that are normally present at the unit (including utility streams, water, air, and any common contaminants).
    c. A written policy clearly describes what hazards must be addressed, and references tools that help users evaluate special hazards such as self reactivity, potential for a runaway reaction, shock sensitivity, potential for spontaneous combustion or a dust cloud explosion, alternate chemical reactions that present a special hazard, and other hazards that are related to physical attributes such as particle size.
    d. In addition to item (c), the written policy requires that chemical reactivity hazards be evaluated using appropriate laboratory methods.

Assign Responsibilities to Competent Personnel
6. Ensure that competent personnel are responsible for (1) maintaining current and accurate process knowledge and (2) reviewing changes to the process knowledge.
    a. Process knowledge is compiled and maintained by a designated senior engineer. Most files are stored in this person's office. Changes initiated external to the facility are normally not reviewed to determine how they might impact the facility; for example, the facility has no process to review updates to MSDSs or a supplier's product safety guideline.
    b. Responsibility for compiling and maintaining process knowledge is well defined, but this work activity is generally limited to keeping files up to date. Changes initiated external to the facility are normally not reviewed.

c. Responsibility for compiling and maintaining process knowledge is well defined. A process safety professional or competent engineer reviews changes or updates to process knowledge, but does not generally inform unit personnel of changes.

d. In addition to item (c), updates to process knowledge are routinely summarized and distributed to appropriate personnel at the facility.

8.3.2 Catalog Process Knowledge in a Manner that Facilitates Retrieval

Make Information Available and Provide Structure

See the corresponding work activities for the *competency* element (Section 5.3.2).

Protect Knowledge from Inadvertent Loss

7. File documents by equipment or type of information rather than by capital project number, change authorization number, date, project leader's name, an so forth.

a. Process knowledge is maintained in project files in a central file room and/or in personal files by the engineer assigned to install the equipment. The project files are sorted by project number, resulting in files being maintained in chronological order.

b. Process knowledge is maintained in project files in a central file room, sorted by unit.

c. Process knowledge is stored by equipment tag number, and obsolete information is clearly marked, removed to an archive file, or discarded.

8. Seek to eliminate parallel copies of process knowledge, particularly if they are not sanctioned as part of the *knowledge* element.

a. Process knowledge is maintained in a central file room and sorted by unit, but some members of the organization tend to make personal copies.

b. Process knowledge is stored in a central location and the organization's practice is to use this central source rather than information in project or personal files.

Store Calculations, Design Data, and Similar Information in Central Files

9. Retain the calculations and data that support the process knowledge.

a. Process knowledge and supporting data/calculations are maintained in personal files indefinitely.

b. Process knowledge includes the results of calculations or key data, and the name of the person who has the background information that supports these results.

    c.   All calculations and data that support process knowledge are kept in the central files.

    d.   Process knowledge is maintained in summary form, and any relevant data or calculations are referenced (by hyperlink or file number) in the tables that summarize process knowledge.

Document Information in a User-Friendly Manner

10.   Develop standards for analysis methods, symbols, terminology, and software.

    a.   Although no standard exists within the company, external standards are used. For example, P&ID symbols for some projects are based on a standard provided by the design contractor, while P&ID symbols for other projects are based on the Instrumentation, Systems, and Automation Society's (ISA's) recommended practice.

    b.   An unwritten standard practice is in place for consistent use of symbols, terminology, and software; however, the person who develops the needed information is free to choose an appropriate method and to deviate from the standard practices.

    c.   The company has standards for software and references standards for methods, symbols, and terminology that are available in the literature. However, these references may be unavailable at operating facilities; users are allowed relatively wide discretion to deviate from the standards.

    d.   The company has standards for software and references standards for methods, symbols, and terminology that are available in the literature. The referenced documents are generally available to operating facilities. Deviation from standards is strongly discouraged.

    e.   The company has developed standards for software, methods, symbols, and terminology that are based on well established practices in the industry. The requirements are well described in the standards. Examples are provided as needed to increase understanding, and the standards are sometimes supplemented by reference documents that are available to each facility.

11.   Use tables or other highly structured methods to summarize results and/or point to the location in the facility's files where process knowledge can be found.

    a.   Process knowledge is compiled and maintained on an ad hoc basis.

    b.   Process knowledge is maintained in a designated file room; however, the files are maintained in a structured, but inefficient order (e.g., chronological) and are not cross referenced.

     c.   Each unit has established a register of process knowledge, including the location where it can be found.

     d.   A standard structure is used for storing process knowledge by all units at the facility or facilities within the business. This structure is widely known and well documented.

12. If a company policy or regulation requires a particular type of process knowledge that does not apply to a specific unit or process, clearly document that it does not apply and why it does not apply.

     a.   A standard file structure has been established for process knowledge, but files are simply empty if the knowledge is deemed to be not applicable or not important.

     b.   A standard file structure has been established for process knowledge. A note is placed in the files if the knowledge is not applicable, explaining the basis for the decision to not collect and maintain this information.

8.3.3   Protect and Update the Process Knowledge

Control or Limit Access to Out-of-Date Documents

13. Implement a means to control/limit access to out-of-date process knowledge.

     a.   Process knowledge is compiled and maintained on an ad hoc basis, with no system to control this information.

     b.   Process knowledge is maintained in central files; users are expected to locate the current version of documents.

     c.   Any out-of-date process knowledge is clearly marked.

     d.   A register is maintained of all process knowledge, including revision numbers and issue dates. Users are instructed to compare revision numbers or issue dates to this register before using a document.

     e.   Out-of-date process knowledge is archived in a separate, clearly marked location, or out-of-date sections of documents are clearly marked to prevent inadvertent use.

Ensure Accuracy

14. Periodically review the accuracy of process knowledge.

     a.   Process knowledge is generally reviewed before use. An expectation exists among users that they must first confirm the accuracy of engineering drawings or other facility-specific technical information before using it.

     b.   The responsibility for maintaining and periodically reviewing process knowledge is formally included in job descriptions. An expectation exists within the organization that process knowledge (1) will be checked periodically and (2) is current and accurate.

    c. In addition to the activities listed under item (b), a formal system is in place to periodically conduct a thorough assessment of process knowledge to help ensure that it remains complete and accurate.

15. Periodically assess the adequacy of process knowledge, and gather new information when needed or request new information from the *competency* element.

    a. Efforts are made to update and/or expand process knowledge prior to conducting hazard analyses and risk assessments, and in conjunction with other activities that depend heavily on process knowledge. Any errors or omissions are corrected at the time that they are discovered.

    b. The *competency* and *knowledge* elements are closely linked; new information discovered as part of the competency element activities is transferred to the knowledge element on an ongoing basis.

Protect Against Inadvertent Change

16. Assign persons with proper knowledge and experience to review and approve corrections or changes to process knowledge.

    a. No review or approval is needed for a process engineer or designer to update the process knowledge. When certain process knowledge, such as P&IDs, is field verified, conditions in the field are generally believed to be correct. Thus, the process knowledge is updated to reflect the current field configuration with minimal review.

    b. Differences between process knowledge and equipment in the field are generally brought to the attention of a designated process engineer who determines which is correct, and on that basis appropriate changes are made.

    c. Differences between process knowledge and equipment in the field are generally brought to the attention of one or more senior technical persons with recognized expertise in the affected area for resolution.

    d. Differences between process knowledge and equipment in the field are brought to the attention of one or more senior technical persons who determine which is correct based on an understanding of the designer's intent. A pattern of significant discrepancies is investigated as a chronic near miss incident.

Protect Against Physical (or Electronic) Removal or Misfiling

17. Survey the organization periodically to determine if different departments have established redundant file systems for knowledge. If systems are discovered that are not specifically for archival purposes or are not fully described in the written policy governing the *knowledge* element, take steps to address the situation.

    a. Some persons or organizations maintain files that parallel the official process knowledge files, and this is generally tolerated by management as long as these files are dated and clearly marked as copies.

    b. Parallel file systems are discouraged.

    c. The discovery of a sanctioned, parallel, and uncontrolled file system is treated as a near miss incident in terms of a management system failure.

18. Maintain a protected archive of the process knowledge at a separate facility and update the archive file on a regular basis.

    a. A second copy of all of the process knowledge is stored at another location (normally the business unit engineering office), and these files are synchronized periodically.

    b. Process knowledge is mostly stored in computer files; backup tapes or disks are made occasionally and are stored at a safe offsite location. Backup copies of information that are not stored in computer files are also maintained at the same offsite location. Updates are made to both sets of files simultaneously.

    c. In addition to item (b), care is taken to ensure that hardware and supporting software remains available to read data archived on electronic media. When necessary, data files are transferred to newer media or saved in a new format.

Support Efforts to Properly Manage Change

19. Ensure that designers and other persons assigned to compile and maintain information understand the definition of a "change" and the scope of the *management of change* element. As part of the routine work flow for updating process knowledge, have these individuals confirm that any changes have been authorized.

    a. Engineers and other personnel normally verify process knowledge before using it.

    b. The *management of change* element includes specific steps to determine if process knowledge needs to be updated. Requests to update information are processed quickly.

    c. In addition to item (b) above, the *knowledge* element includes a step to confirm that the change has been authorized.

20. Provide a means to ensure the fidelity of process knowledge and prevent unauthorized changes that would corrupt the information.
    a. Process knowledge is filed in central files with administrative controls, such as sign-out cards, but unlimited access is allowed. Facility personnel are trained to not add or change any information unless they believe that the existing information is incomplete or incorrect.
    b. Process knowledge is filed in central files with controlled access. Some administrative controls exist to prevent users from adding or changing information without some form of authorization. Sign-out cards are used whenever files are removed from the area where process knowledge is stored.
    c. Process knowledge is largely stored in read-only files on the computer network and is typically viewed on a computer terminal or printed on demand.
21. Implement a means to check out copies of process knowledge for revision.
    a. Process knowledge is stored in a protected area and controlled by a single person; this person should recognize when two or more persons are working to revise the same process or system simultaneously.
    b. A paper-based sign-out card system should alert personnel if multiple persons are working on changes related to the same part of the facility; however, this system works only if original documents are borrowed, not if they are copied and returned to the library.
    c. A change request number is issued early in the change authorization process and process knowledge that may be affected by the change is flagged. Requestors are warned that a change request is under way whenever they request a flagged document.

(See Section 15.3.2 for work activities associated with recognizing changes.)

8.3.4 Use Process Knowledge

Ensure Awareness
22. Determine if employees who need to use process knowledge thoroughly understand how to use it and how to interpret data it contains. If gaps exist, provide suitable training.

a. New engineers and other employees are informed of the existence of significant parts of the process knowledge during orientation training or via a formal mentoring system.

b. A module exists in initial training to introduce new engineers and other technical personnel to process knowledge and the *knowledge* element. The training also explains how this knowledge should be maintained and used.

c. In addition to a formal training module, audits are planned periodically to measure the organization's competence regarding the existence and proper use of process knowledge. For example, operators are asked to locate and interpret information on MSDSs, engineers are asked to locate and interpret drawings, and maintenance personnel are asked to locate and use records such as equipment repair history.

Ensure that Process Knowledge Remains Useful

23. Assess whether the information is adequate to meet the needs of the *risk* element and other RBPS elements at each point in the unit's life cycle.

a. Periodic assessments or audits primarily compare process knowledge files to a checklist that specifies minimum requirements.

b. Periodic audits help determine if the process knowledge needed for each unit or phase of operation is (1) adequate, (2) available, (3) user friendly, and (4) actually used.

c. The management review process for this element examines whether the process knowledge is sufficient for all modes of operation and life cycle stages; gaps are identified and addressed independent of the *audits* element.

8.4   EXAMPLES OF WAYS TO IMPROVE EFFECTIVENESS

This section gives specific examples of industry tested ways to improve the effectiveness of work activities related to the *knowledge* element. The examples are sorted by the key principles that were first introduced in Section 8.2. The examples fall into two categories:

1. Methods for improving the performance of activities that support this element.

2. Methods for improving the efficiency of activities that support this element by maintaining the necessary level of performance using fewer resources.

These examples were obtained from the results of industry practice surveys, workshops, and CCPS member-company input. Readers desiring to improve their management systems and work activities related to this element should examine these ideas, evaluate current management system and work activity performance and efficiency, and then select and implement enhancements using the risk-based principles described in Section 2.1.

8.4.1 Maintain a Dependable Practice

Develop and widely distribute a simple index that clarifies accountability for maintaining process knowledge. Two common reasons that facilities fail to maintain accurate process knowledge is that this responsibility is either (1) formally assigned to a single person or (2) the collective responsibility of the entire organization. These two ends of the spectrum each have drawbacks. While assigning the task to maintain process knowledge to a single person results in clear accountability, it effectively creates an overhead position. In addition, no single individual is likely to be able to evaluate the completeness, accuracy, or quality of all of the information, particularly at larger, multi-unit facilities.

At the other extreme, facilities often list the various parts of process knowledge in their overall policy, and then leave it up to the entire organization to ensure that this information is collected and maintained. When responsibility for collecting and maintaining process knowledge is delegated to the entire organization in this manner, the process knowledge becomes vulnerable, particularly if the organization undergoes significant restructuring. In addition, this decentralized mode of information storage makes it difficult to know where the official process knowledge resides, which can result in low efficiency or potentially risky consequences such as (1) recreating information, (2) maintaining information is personal or local files that will likely become out of date, or (3) making decisions based on incomplete or inaccurate information.

A useful tool for controlling process knowledge, when using a distributed approach, is a simple 3-column table. The first column lists the various parts of the process knowledge, the second column contains the location of official copies of the information, and the third column lists the name or job title of the person who owns the information and is responsible for ensuring that the information is current, accurate, and sufficiently protected.

Develop a centralized infrastructure. Many companies provide a centralized infrastructure, such as databases residing on computer networks or intranet-based systems, to facilitate the distribution and maintenance of process knowledge. Without this infrastructure, each department tends to develop its own system, which increases the overall cost for maintaining process knowledge, increases training costs, and may not survive organizational realignments.

Use central technical groups to develop and maintain process knowledge. Some companies use central engineering or technology groups to develop and maintain process knowledge. This can be particularly efficient when multiple smaller facilities are operating similar or identical processes. Also, process knowledge is normally developed by central groups or engineering firms for larger capital projects and delivered to the facility (or in some cases, to this central engineering group) upon completion of the project.

Document safe conditions/features that are incorporated into the design of the process. Process designers often specify operating conditions based on operating in a safe, stable manner. For example, the designers of a process that involves a reaction between two chemicals in an inert solvent may specify a minimum mass ratio for solvent to reactants based on a loss of cooling failure. Over time, the demand for the product may grow while production remains limited by the step requiring separation of the solvent from the product. To increase throughput, the production group may eventually agree to reduce the ratio of solvent to reactants. The production group may reason that, since the reactor had operated without incident for ten years and the change would be less than 20%, the reduced ratio would not cause any problems. In fact, reducing the recycle ratio by 20% may not introduce a hazard, but the next 20% reduction might. More important, a runaway reaction may not occur until a power failure occurs that extends for several hours on an occasion when the operators are unable to start the diesel-powered emergency cooling water pumps. This scenario may be unlikely in any given year, but is very credible over the life of the facility. If the facility had known the technical basis for the minimum recycle ratio, management likely would not have authorized this change. Therefore, when safe operating conditions are specified, the reasoning behind those decisions should also be documented to prevent future changes resulting in unsafe conditions.

Extend the process knowledge to encompass information needed for efficient operation. At some facilities, government regulations and corporate audits create an incentive to (1) limit the scope of process knowledge and (2) maintain a "pristine" central set of files structured so they directly reflect the regulatory requirements or corporate standards. Maintaining a second set of files that are not readily accessible and cover only selected equipment is expensive. The value provided of having pristine files does not normally justify this expense.

Make the process knowledge management element the hub for all process knowledge. Modern databases and networks provide the opportunity to easily link information together. Data such as (1) incident reports, (2) lists of equipment included in the *asset integrity* program, and (3) the register of approved (and rejected) changes are not part of the process knowledge. However, there is no reason that the data index cannot be expanded to include all information and knowledge related to process safety. Thus, the *knowledge* element can serve as the starting point for persons seeking information, either with maps that direct users to other documents, an up-to-date index, or live "hot links" to other documents containing relevant information.

8.4.2 Catalog Process Knowledge in a Manner that Facilitates Retrieval

Implement a centralized database for process knowledge. Some companies route all of their process knowledge to a central group that administers a database program used to maintain the information. Read-only access is provided as appropriate. Since database design and management is not normally a core competency of technical groups at operating facilities, splitting the functions of generating process knowledge and storing the knowledge can lead to increased efficiency. A corollary benefit is that this system helps provide a high degree of structure to the process knowledge.

Develop a standard template or set of forms for recording process knowledge. Similar to a database, standard forms help ensure that the information is complete, in that empty cells on the forms indicate that information is missing. In addition, standard forms help provide structure and facilitate future transition to an electronic database.

Create partially completed forms or databases for process knowledge. Some companies operate nearly identical processes at many locations around the world. One very efficient way to create a comprehensive set of process knowledge is to establish a standard set of forms or a database that is partially filled out by a central corporate group, completed by facility personnel, and maintained at the facility. These forms can be electronic spreadsheets that produce some of the results, such as heat and material balances based on user input for flow rates. This system reduces redundant effort, helps reduce the likelihood that information will be missing or incorrect, and helps clarify who in the organization is responsible for developing and maintaining each part of the process knowledge. Also, if the process knowledge is maintained in a highly structured database, changes to the information developed by the corporate group can be made very efficiently.

Include specifications for delivery of process knowledge in the scope of work for new capital projects. Many companies use outside engineering firms for design and construction of new facilities or major changes to existing facilities. Unless the intended format of the process knowledge is included in the scope of work, the documents provided by the engineering firm will most likely not conform to the facility's structure for process knowledge, or the documents may not address all of the facility's requirements.

Provide context for process knowledge. Without context or definition, data contained in the *knowledge* element can become a collection of miscellaneous facts and figures. Process knowledge is not solely for the benefit of process safety professionals; do not assume that users will understand all of the jargon, or that they will have immediate access to a definition or an explanation of the jargon.

Reduce demand on engineering resources for troubleshooting, plant turnarounds, and minor capital projects. Without current and accurate process knowledge, efforts to troubleshoot process problems, plan major overhauls, or implement minor capital projects frequently start with extensive field tracing of piping and equipment. One facility that kept careful track of the level of effort spent on its annual turnarounds noted that the engineering effort required to plan for and support turnarounds decreased by 75% between 1992 and 1998. This reduction reportedly resulted from an increase in the accessibility and accuracy of information, including process knowledge, equipment maintenance plans, and formal procedures for preparing equipment for maintenance.

Create company- or facility-specific safety data sheets. By design, MSDSs are not specific to any user's facility, and they contain a significant amount of information that is not applicable to operators, maintenance personnel, and others who use them on a day-to-day basis. To reduce confusion, some companies extract data from suppliers' MSDSs and augment this with facility-specific standards to create a safety data sheet for each hazardous process chemical. For example, an MSDS may recommend use of nitrile rubber gloves, whereas a facility-specific safety data sheet may specify "blue rubber gloves" or a similar term that is commonly used by operators and maintenance personnel at the facility. In a similar manner, some companies develop quick guides for information contained in MSDSs that is routinely needed by operators and maintenance personnel (e.g., PPE requirements), emergency responders (e.g., hazard ratings, specific material incompatibilities, PPE requirements), and design/engineering teams (e.g., physical properties).

8.4.3   Protect and Update Process Knowledge

Treat mismanagement of process knowledge as a near miss incident. Companies spend significant resources to develop process knowledge and other technical information. However, if facilities do not maintain this information, it quickly erodes and becomes of little use. Once this occurs, the company often spends quite a bit of money to update or recreate the process knowledge. A lack of current and accurate process knowledge can also prevent a risk analysis team from identifying and properly analyzing an accident scenario, possibly resulting in unacceptable risk or an incident that should have been prevented. Either way, process knowledge is a significant capital asset that should be protected, and loss of process knowledge should be treated as any other incident involving property loss or damage.

Electronically link the knowledge and management of change elements. Some facilities use software to route change requests for approval and to record the results of these requests. These software packages often include check boxes to specify which documents are affected by the change and must be updated. Linking a "check" in each of these boxes to an electronic communication to the owner of the corresponding information (1) increases the likelihood that appropriate people will be notified of the need to update information and (2) reduces the time required to determine who needs to be notified. As with any system, communication should be two-way, and the person with overall responsibility for the change should ensure that affected documentation is actually updated, not simply that a request to update documentation was sent to respective document owners.

8.4.4   Use the Process Knowledge

Fully integrate maintenance of process knowledge into day-to-day activities. Assign responsibility for maintaining process knowledge to departments that are most likely to derive value from the information. For example, if maintenance department personnel must routinely access technical information contained in original equipment manufacturers' maintenance manuals, consider assigning ownership for this part of the process knowledge to the maintenance department rather than to the engineering or safety department. This approach requires that the responsibility for maintaining process knowledge be widely understood throughout the organization.

8.5   ELEMENT METRICS

Chapter 20 describes how metrics can be used to improve performance and when they may be appropriate. This section includes several examples of metrics that could be used to monitor the health of the *knowledge* element, sorted by the key principles that were first introduced in Section 8.2.

In addition to identifying high value metrics, readers will need to determine how to best measure each metric they choose to track. In some cases, an ordinal number provides the needed information, for example, total number of workers. Other cases, such as average years of experience, require that two or more attributes be indexed to provide meaningful information. Still other metrics may need to be tracked as a rate, as in the case of employee turnover. Sometimes, the rate of change may be of greatest interest. Since every situation is different, the reader will need to determine how to track and present data to most efficiently monitor the health of RBPS management systems at their facility.

8.5.1   Maintain a Dependable Practice

- *Number of incident investigations that include an element of discovery.* Incident investigation teams sometimes discover basic hazard information that was (1) documented in the process knowledge but not widely known at the unit level or (2) widely reported in the literature but not documented in the process knowledge or process technology manual (see Chapter 5 for a description of a technology manual). This metric should decrease with time if the *knowledge* element is effective.

- *Number of PHA team recommendations that include an indication of less than adequate process knowledge where the information <u>was not</u> available.* Recommendations that start with "Determine if (a certain hazard exists) and ..." or "Evaluate the capability of (a safety system) and ..." may indicate that important information was not included in the knowledge element.
- *The number or percent of blank records in the process knowledge database.* This is one measure of the completeness of the process knowledge. Completeness can easily be tracked if a database application is used to store process knowledge. Note that this metric does not help ensure that information is accurate, nor does it necessarily provide a risk-based means to compare the state of the process knowledge between units or facilities because some "blank records" will be much more risk significant than others.

8.5.2   Catalog Process Knowledge in a Manner that Facilitates Retrieval

- *Number of PHA team recommendations that include an indication of less than adequate process knowledge where the information actually <u>was</u> available.* This metric could be confused with the previously proposed metric related to process knowledge that was <u>not</u> available. Recommendations that start with "Determine if (a certain hazard exists) and ..." or "Evaluate the capability of (a safety system) and ..." could indicate that important information does not exist. They could also indicate that process knowledge is difficult to locate. Either situation is undesirable, but the corrective actions to address them are likely very different.
- *Results of periodic surveys to determine if users of process knowledge believe it is accessible.* Although opinion surveys can be somewhat subjective, the data are relatively easy to gather and reflect users' beliefs about the state of the *knowledge* element.
- *Number of instances in which maintenance planners or purchasing agents cannot locate specifications or similar data.* One function of process knowledge is to provide information needed to help ensure that maintenance materials and repair parts conform to specifications. Having to wait for engineers to develop or find this information can be costly in both time and money.

- *If process knowledge is web-based, the number or percent of dead links.* Posting process knowledge on the company's intranet has many advantages. Likewise, linking related information can be very helpful. However, these links must be maintained and updated to maintain their usefulness (e.g., changes in target files may disable a link).

8.5.3 Protect and Update Process Knowledge

- *Accuracy of process knowledge during periodic reviews.* Results from existing work activities, such as field verification of P&IDs, that are normally completed prior to a risk assessment activity can also provide insight into how the accuracy of the process knowledge is changing over time.
- *Number of times during audits or assessments that process knowledge (or duplicate copies of reports, etc.) must be retrieved from personal files.* This metric is a clear indicator of whether the management system that stores and protects process knowledge is working – the higher the number, the less efficient the system. However, to be useful as a metric that would allow for comparisons over time (or between facilities), a standardized statistical sampling method would need to be developed.
- *Number of change requests initiated to "correct" process knowledge.* This is one simple way to determine if the accuracy of process knowledge appears to be changing with time.
- *Engineering staff time spent recreating process knowledge.* Recreating process knowledge is often time consuming and expensive. If a substantial and unexpected level of resources is applied to this task, the effectiveness of the *management of change* element and/or the link between the *management of change* and *knowledge* elements should be closely examined.
- *Backlog of change requests related to completing updates to process knowledge.* An unexplained increase in this metric may indicate that insufficient resources are available to update the process knowledge. Conversely, a natural cycle may exist as a result of the way that the business is structured; for example, if almost all changes and modifications occur during brief turnarounds, the backlog might temporarily increase concurrent with the turnaround.
- *Results of random checks of process knowledge files after change requests are closed.* Facilities with rigorous management of change systems will include a step to audit some (or maybe all) closed change requests to determine if all actions, including updating process knowledge, were completed. The results of these audits, possibly on a normalized basis such as errors per change authorization, provide

insight to the link between the *management of change* and *knowledge* elements.

- *Average number of days required to have a drawing revised.* This is a particularly important metric for facilities where the expectation for finding current and accurate information on the computer network is high.

- *Ratio of approved change requests (involving equipment changes) to updates to P&IDs.* A statistically significant change in this metric indicates that some part of the management system has changed or that some other factor that has affected either the *management of change* or *knowledge* elements.

- *Results of periodic opinion surveys to determine if users of process knowledge believe that it is current and accurate.* Although opinion surveys can be somewhat subjective, the data are relatively easy to gather and reflect users' beliefs about the state of the *knowledge* element.

- *Results of random checks of MSDS files to determine if they are complete, current, and accurate.* These data can be gathered quickly and consistently across a number of facilities. Given its high profile, deficiencies are strong indicators of weaknesses in the entire *knowledge* element.

8.5.4   Use the Process Knowledge

- *Frequency that process knowledge is accessed.* If the process knowledge is stored such that usage data can be easily collected, this metric may provide some insight to its usefulness. However, the rate of access may be strongly influenced by turnarounds or other infrequent events.

8.6   MANAGEMENT REVIEW

The overall design and conduct of management reviews is described in Chapter 22. However, many specific questions/discussion topics exist that management may want to check periodically to ensure that the management system for the *knowledge* element is working properly. In particular, management must first seek to understand whether the system being reviewed is producing the desired results. If the organization's documentation of process knowledge is less than satisfactory, or it is not improving as a result of management system changes, then management should identify possible corrective actions and pursue them. Possibly, the organization is not working on the correct activities, or the organization is not doing the necessary activities well. Even if the results are satisfactory, management reviews can help determine if resources are being used wisely – are there tasks that could be done more efficiently or tasks that should not be done at all? Management can combine metrics listed in the previous section with personal observations, direct questioning, audit results, and feedback on various topics to help answer such questions. Activities and topics for discussion include the following:

- Review the facility's policy for the knowledge element to ensure that roles and responsibilities have been assigned. Is a particular person, by name or job title, responsible for maintaining each aspect of the process knowledge?
- Sample some of the data repositories, such as drawing files, MSDS files, or equipment records, to determine if the data exist and appear to be current and accurate.
- Review audit results to determine if any significant gaps were identified. Confirm that (1) these gaps have been closed, (2) the root cause for the gaps is understood, (3) other gaps that resulted from the same root cause have been identified and closed, and (4) changes have been made to the management system to eliminate the root cause(s) of the situation.
- If a policy is in place to review process knowledge prior to a risk analysis, develop a "process knowledge report card" for the as-found condition (i.e., number and type of corrections required). Is overall accuracy acceptable? Do any particular units show very good results? Can the practices applied in these units be extended to the entire facility?
- Interview relatively new employees to determine if they can navigate their way through the process knowledge that pertains to their job duties. For example, operators should be able to locate information in MSDSs, process engineers should be able to locate information involving any aspect of the process knowledge, and so forth. Also, determine if this information can be located in a timely manner.
- Review the means used to protect process knowledge from loss or inadvertent change. Is it protected against loss caused by a single event, such as structural fire, flood, or hard disk or other electronic media failure?
- Review how archived (out-of-date) documents are stored. Can an archived document or drawing possibly be confused with the current version of the same document?
- Survey the organization to determine if any effort exists to maintain part of the process knowledge or closely related information in an unofficial system. If multiple, asynchronous systems are discovered, determine why they exist (e.g., personnel lack confidence in the fidelity of the primary data repository) and what should be done address the concern.

- If any relevant projects are under way at the time of the review, such as conversion from a paper-based system to an electronic storage system, determine if the project is going as planned. If not, identify possible roadblocks that could be cleared by management.
- Determine how much engineering effort is being spent recreating lost or misplaced information.
- Determine how much effort has been applied to developing new process knowledge.
- Determine how much effort has been applied to updating existing process knowledge.

The management review for this element can take a variety of paths, but all paths should provide insight to three questions:

- Is the information needed to properly manage risk and safely operate the facility well documented and widely understood?
- Is this information accurate?
- Is it likely that facility personnel can locate current and accurate information that collectively makes up the process knowledge in a timely manner?

A negative answer to one or more of these questions should be investigated. Management reviews should provide an opportunity to identify important gaps in the management system for this and all RBPS elements and to chart a course to promptly and effectively address those gaps.

8.7   REFERENCES

8.1   *The Explosion at Concept Sciences: Hazards of Hydroxylamine*, Chemical Safety and Hazard Investigation Board Report 1999-13-C-PA, Washington, DC, March 2002.

8.2   Center for Chemical Process Safety, *Guidelines for Process Safety Documentation*, American Institute of Chemical Engineers, New York, 1995.

8.3   Center for Chemical Process Safety, *Essential Practices for Managing Chemical Reactivity Hazards*, American Institute of Chemical Engineers, New York, 2003.

Additional reading

Kletz, Trevor, *Case Histories of Process Plant Disasters*, 4th edition, Elsevier Science, Burlington, Mass., 1999.

Kletz, Trevor, *Lessons from Disaster: How Organizations Have No Memory and Accidents Recur*, Gulf Publishing Company, Houston, Texas, 1993.

Lees, Frank P., *Loss Prevention in the Process Industries, Hazard Identification, Assessment, and Control*, 2nd edition, Butterworth-Heinemann, Oxford, England, 1996.

Perry, Robert H. and Green, Don W., *Perry's Chemical Engineers' Handbook*, 7th edition, McGraw-Hill, New York, 1997.

9

HAZARD IDENTIFICATION AND RISK ANALYSIS

In 1998, a major explosion and fire occurred at the Longford gas-processing facility in Victoria, Australia (Ref. 9.1). Two employees were killed and eight others were injured. The incident caused the destruction of one gas separation plant and the shutdown of two others at the facility. This disrupted gas supplies across the state, resulting in 250,000 workers being sent home as factories and businesses were forced to shut down. The incident was caused by workers attempting to recover from a process upset that had embrittled the metal of a heat exchanger. If properly conducted, a risk study would have identified the potential for a loss of lean oil to create dangerously low temperatures in the process equipment. A risk study had been planned three years prior to the accident, but had not been done. This incident illustrates why the management system should ensure that risk studies are performed in a timely manner, why both normal and upset operating situations should be considered, and why the results of risk studies should be communicated to the workers.

9.1   ELEMENT OVERVIEW

A thorough Hazard Identification and Risk Analysis, or *risk*, system is the core element in the RBPS pillar of ***understanding hazards and risk***. This chapter describes the meaning of risk for RBPS purposes, the attributes of a risk system, and the steps an organization might take to implement a robust program for identifying hazards and analyzing risk. Section 9.2 describes the key principles and essential features of a management system for this element. Section 9.3 lists work activities that support these essential features, and presents a range of approaches that might be appropriate for each work activity, depending on perceived risk, resources, and organizational culture. Sections 9.4 through 9.6 include (1) ideas for improving the effectiveness of management systems and specific programs that support this element, (2) metrics that could be used to monitor this element, and (3) issues that may be appropriate for management review.

9.1.1   What Is It?

Hazard Identification and Risk Analysis (HIRA) is a collective term that encompasses all activities involved in identifying hazards and evaluating risk at facilities, throughout their life cycle, to make certain that risks to employees, the public, or the environment are consistently controlled within the organization's risk tolerance.   These studies typically address three main risk questions to a level of detail commensurate with analysis objectives, life cycle stage, available information, and resources. The three main risk questions are:

- Hazard – What can go wrong?
- Consequences – How bad could it be?
- Likelihood – How often might it happen?

When answering these questions, the objective is to perform only the level of analysis necessary to reach a decision, because insufficient analysis may lead to poor decisions and excessive analysis wastes resources. A suite of tools is available to accommodate varying analysis needs: (1) tools for simple hazard identification or qualitative risk analysis include hazard and operability analysis (HAZOP), what-if/checklist analysis, and failure modes and effects analysis (FMEA), (2) tools for simple risk analysis include failure modes, effects, and criticality analysis (FMECA) and layer of protection analysis (LOPA), and (3) tools for detailed quantitative risk analysis include fault trees and event trees (Refs. 9.2, 9.3, 9.4). For example, some companies may judge the mere existence of an explosion hazard to be an unacceptable risk, regardless of its likelihood. Others may be willing to tolerate an explosion risk if proper codes and standards are followed. Still others may be unwilling to accept an explosion risk unless it can be shown that the expected frequency of explosions is less than 10^{-6}/y.

HIRA encompasses the entire spectrum of risk analyses, from qualitative to quantitative.   A process hazard analysis (PHA) is a HIRA that meets specific regulatory requirements in the U.S.   Figure 9.1 illustrates the increasing rigor of risk analyses possible as the scope of the study becomes more focused on specific accident scenarios. Note that as risk studies become more focused and detailed, the cost per scenario analyzed increases, but the overall cost may decrease if only a few representative or bounding scenarios are analyzed.

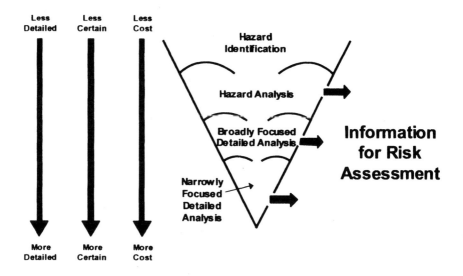

FIGURE 9.1. Levels of Hazard Evaluation and Risk Assessment

9.1.2 Why Is It Important?

To manage risk, hazards must first be identified, and then the risks should be evaluated and determined to be tolerable or not. The earlier in the life cycle that effective risk analysis is performed, the more cost effective the future safe operation of the process or activity is likely to be. The risk understanding developed from these studies forms the basis for establishing most of the other process safety management activities undertaken by the facility. An incorrect perception of risk at any point could lead to either inefficient use of limited resources or unknowing acceptance of risks exceeding the true tolerance of the company or the community.

9.1.3 Where/When Is It Done?

HIRA reviews may be performed at any stage in a project's life cycle – conceptual design, preliminary design, detailed design, construction, ongoing operation, decommissioning, or demolition. In general, the earlier that a hazard is identified (e.g., during conceptual design), the more cost-effectively it can be eliminated or managed. Studies performed during the early design stages are typically done at corporate or engineering offices. Studies performed once a process is near startup, during operation, or before decommissioning are typically done in a plant environment.

9.1.4   Who Does It?

A HIRA study is typically performed by a team of qualified experts on the process, the materials, and the work activities. Personnel who have formal training on risk analysis methods usually lead these teams, applying the selected analysis technique(s) with subject matter experts from engineering, operations, maintenance, and other disciplines as needed.

A simple early-in-life hazard identification study may be performed by a single expert; however, a multi-disciplined team typically conducts more hazardous or complex process risk studies, especially during later life cycle stages. Involving operating and maintenance personnel early in the review process will help identify hazards when they can be eliminated or controlled most cost effectively. When the study is complete, management must then decide whether to implement any recommended risk reduction measures to achieve its risk goals.

9.1.5   What Is the Anticipated Work Product?

The main process safety products of a *risk* system are: (1) guidelines for planning and conducting studies, (2) documented understanding of the risks of the process or activity, (3) documented risk tolerance criteria, (4) possible risk control measures, resolutions, and implemented actions, (5) documented understanding of the residual risks after control measures are taken, and (6) completed risk analysis reports. Other work products may include recommendations for improving asset integrity, procedures, and training as well as up-to-date action item tracking lists and risk communication materials. The scope of HIRAs is sometimes broadened to include operability issues, so the work products may also include recommendations to (1) improve quality and yield, (2) reduce equipment damage, and/or (3) reduce unplanned downtime. The results of risk studies are normally kept for the life of the process and are communicated to those who may be affected. Outputs of the *risk* element can also be used to facilitate the performance of other elements. For example, identifying potential accidents will help define scenarios the *emergency* element must address, and understanding the existing risks may enable the *management of change* element to identify the risks of a change.

9.1.6   How Is It Done?

At each stage in the project life cycle, a review team questions process experts about possible hazards and judges the risk of any hazards that are identified. Several common methods exist for questioning a design, ranging from simple qualitative checklists to complex quantitative fault tree analyses (Refs. 9.2, 9.3, 9.4). The results of the review process are typically documented in a worksheet form, illustrated in Figure 9.2, which varies in detail, depending on the stage of the project and the evaluation method used. Risk studies on operating processes are typically updated or revalidated on a regular basis.

Hazard	Causes	Consequences	Safeguards	Recommendations
1. Fire				
2. Spill				

FIGURE 9.2. Typical Qualitative Risk Analysis Documentation Form

9.2 KEY PRINCIPLES AND ESSENTIAL FEATURES

Safe operation and maintenance of facilities that manufacture, store, or otherwise use hazardous chemicals requires an effective system to identify hazards and judge whether the risks associated with the hazards are adequately controlled. The *risk* element establishes a formal, documented review process for all new and existing processes throughout their life cycle and focuses on the materials, the equipment, and their location with respect to people, sensitive environments, and other vulnerable assets. The results of these reviews, summarizing the risks associated with the process, should be communicated to all potentially affected workers and stakeholders.

The following key principles should be addressed when developing, evaluating, or improving any management system for *risk*:

- Maintain a dependable practice.
- Identify hazards and evaluate risks.
- Assess risks and make risk-based decisions.
- Follow through on assessment results.

The essential features for each principle are further described in Sections 9.2.1 through 9.2.4. Section 9.3 describes work activities that typically underpin the essential features related to these principles. Facility management should evaluate the risks and potential benefits that may be achieved as a result of improvements in this element. Based on this evaluation, the facility should develop a management system, or upgrade an existing management system, to address some or all of the essential features, and execute some or all of the work activities, depending on perceived risk and/or process hazards that it identifies. However, these steps will be ineffective if the accompanying *culture* element does not embrace the identification of hazards and control of risks.

9.2.1   Maintain a Dependable Practice

When a company identifies or defines an activity to be undertaken, that company likely wants the activity to be performed correctly and consistently over the life of that process and other similar processes. For the risk management system to be executed dependably across a facility involving a variety of people and situations, the following essential features should be considered:

Document the intended risk management system. For consistent implementation, the *risk* program should be documented with the appropriate level of detail in a procedure or a written program addressing the general management system aspects discussed in Section 1.4. The written objectives of the HIRA program should clearly state the benefits to the company and express the benefits in terms that demonstrate to management and employees of the value of the activities.

Integrate HIRA activities into the life cycle of projects or processes. A *risk* system should identify the points in a project's life cycle when risk analyses should be performed. The scope of each study should be well defined to help ensure that all types of processes and activities are considered. Procedures for reviews at specific points in the life cycle should be defined to an appropriate level of detail, along with guidance for selecting an appropriate analysis methodology.

Clearly define the analytical scope of HIRAs and assure adequate coverage. A *risk* system should address all of the types of process risks that management wants to control. For example, some companies may consider the possibility of a steam burn to be a process hazard, while others may exclude it because they do not consider steam to be a hazardous chemical within the scope of the *risk* system. Companies should define what types of processes and materials are subject to HIRAs, and what types of consequences are of interest, for example, worker fatality, public evacuation, or process shutdown.

In addition, the system should address when HIRAs should be performed. Some companies may require a series of analyses over a project's life cycle, while others are content with an engineering design review. Each required study should have a defined objective, such as identifying inherently safer design options, that is consistent with the process safety information available for the review. When a prior study is being updated or revalidated, any changes in the analytical scope must be explicitly addressed. Evolved systems should consider documenting the technical basis for any exclusion from the *risk* program.

Determine the physical scope of the risk system. Based on the types of process risks that management wants to control, management should identify the physical areas to which the HIRA practice applies. Some companies may choose to include everything within a facility, while others may include only those processes containing more than a threshold amount of a specific list of chemicals. The physical scope should then be subdivided into manageable units, areas, or activities subject to HIRAs. The objective is to have clearly defined boundaries so that specific pieces of equipment, especially shared equipment such as a flare that might be connected to several units, are clearly identified. Ideally, each unit should have a well defined set of inputs, outputs, and interfaces with other units. In addition, the work activities included in the scope should be defined. For example, one batch reactor might have to be analyzed for each separate product line it produces. Hazards associated with startup, shutdown, maintenance, and other nonroutine activities should also be included in the scope definition.

Involve competent personnel. Even within an explicit *risk* program, performance will only be as good as the people who are involved in the HIRAs. In addition to their subject matter expertise, people who are assigned specific HIRA duties may need more detailed training as team leaders, scribes, or participants. The *risk* program should have someone assigned to (1) manage HIRA activities, (2) schedule HIRAs and revalidations as required, (3) coach HIRA leaders, (4) track resolution of HIRA recommendations, and (5) advise management on risk, as requested.

Make consistent risk judgments. Management should communicate its risk tolerance to the risk assessment teams so they can make risk judgments. This guidance may be in a simple, qualitative form, such as an instruction that the facility must conform to recognized and generally accepted good engineering practices, or that risks must be reduced as low as reasonably practicable (ALARP). Management may fund the development of corporate process safety standards and require adherence to them as well. However, neither of these approaches provides guidance for all situations. Thus, some organizations develop a risk matrix, such as the one shown in Figure 9.3, to communicate their tolerance for any scenario falling within a specific range of severity and likelihood. These categories may be defined either qualitatively or quantitatively. Beyond that, some companies choose (or regulators impose) some absolute risk criterion that events of this magnitude shall not exceed a specified likelihood, for example, the likelihood of a worker fatality at the facility shall not exceed 10^{-4} events/y.

In addition to specific risk tolerance criteria, the company should also specify its preference for the types of risk control measures employed. For example, a company may prefer inherently safer design alternatives to those requiring risk controls, it may prefer passive controls to active controls, or it may prefer engineered controls to administrative controls.

Risk	Serious danger in immediate area	Serious danger inside battery limits	Serious danger site wide	Serious danger offsite
More than once per year	Action required unless risk ALARP	Action required at first opportunity	Immediate action required	Immediate action required
Once every few years	Action required unless risk ALARP	Action required unless risk ALARP	Action required at first opportunity	Immediate action required
Once in the facility's lifetime	No action required	Action required unless risk ALARP	Action required unless risk ALARP	Action required at first opportunity
Not expected in the facility's lifetime	No action required	No action required	Action required unless risk ALARP	Action required unless risk ALARP

FIGURE 9.3. Example Risk Matrix

Verify that HIRA practices remain effective. Once a *risk* system is in place, periodic monitoring, maintenance, and improvement activities may be needed to maintain peak performance and efficiency. To ensure effectiveness, companies should verify that HIRAs are being periodically updated and revalidated to reflect changes that have been made to the facility. Process safety personnel should use audits as part of longer term verification of management practices. Typically, these audits will reveal gaps in HIRA execution requiring corrective action. They also often identify erosion of key technical skills necessary to sustain HIRA practices.

9.2.2 Identify Hazards and Evaluate Risks

Once the analytical and physical scope is defined, hazards can be identified. Many techniques for identifying hazards exist, but in all cases, the basic question, "What could go wrong?" must be answered and compared against the consequences of interest. The risks associated with those hazards that have the potential to exceed the threshold consequences of interest should then be determined, considering the severity of consequences, likelihood of causes, and effectiveness of any safeguards. In order for facilities to adopt and implement appropriate review protocols for relevant hazard types, the following essential features should be considered:

Gather and use appropriate data to identify hazards and evaluate risks. Thorough review of possible hazards must be based on accurate process knowledge. (See Chapter 8, Process Knowledge Management.) Facilities should consider establishing a list of data that must be available to the HIRA team. This list of input information should be documented in the written *risk* program, in the review procedure, or in another manner. This list of inputs will vary based upon when in the life cycle the HIRA is performed and the perceived risk associated with the subject processes. This list will also vary based upon whether a qualitative or quantitative analysis is being performed.

Select appropriate HIRA methods. An appropriate methodology should be selected for each anticipated type of risk analysis. In simple, low hazard situations, only one basic review process may be necessary. In more complicated, high-hazard situations, multiple review processes may be defined using differing types of review protocols. Each review protocol includes the number and disciplines of reviewers, and it specifies the order of the review (series, parallel, or team-based reviews). The decision on the design of these review protocols should consider the following issues:

- List of information available for review.
- Expertise needed for review.
- Acceptable methods for identifying hazards and evaluating risks.
- Type, accuracy, precision, and certainty of results necessary for risk-based decision making.

Ensure that HIRA reviewers have appropriate expertise. Normally, the HIRA leader should be independent of the process being analyzed. The leader should not have an ego investment in the design of a new process or in the ongoing operation of an existing process. Beyond that, reviewers should have diverse backgrounds commensurate with the process type and perceived risks in order to facilitate a thorough hazard review. Appropriate review personnel may be specified by name, job function (e.g., operations, process engineer, process chemist, maintenance), or experience and training (e.g., mechanic vs. reliability engineer, field operator vs. control room operator). Complex processes may require special skills that are developed through years of experience and formal training. If those skills are not available on site, assistance from other facilities, vendors, or consultants may be required.

Perform risk activities to the appropriate level of technical rigor commensurate with the life cycle stage and the available process information. HIRAs will likely be conducted using a variety of techniques, depending upon the process' life cycle stage, available information, and evaluation needs. Regardless of which method is selected and applied, certain technical issues should be addressed in the HIRA. These issues are also a function of life cycle stage and the purpose of the study. In addition, a company may wish to provide guidance on the technique and appropriate level of detail for each stage of the process life cycle.

Prepare a thorough risk assessment report. The minimum documentation of any HIRA should describe the scope of the analysis, the hazards and risks that were identified, and any recommendations for further analysis or risk reduction. The report should not only inform the reader of the risks, as they are understood at the time of the study, but also provide the basis for subsequent reviews or revalidations at later stages in the life cycle. The report should provide the rationale for any recommendations so that management can determine whether proposed resolutions provide the desired risk reduction. In some situations, the risk assessment report must also follow a format or include specific topics imposed by regulatory authorities.

9.2.3   Assess Risks and Make Risk-based Decisions

Once hazards have been identified and the risks associated with them have been analyzed, the acceptability of the risk must be judged. Some companies may judge a risk acceptable if the system conforms to a minimum standard, such as a regulation or code, while other companies may require that risks be reduced as low as reasonably practicable. Some companies may judge the risk to be unacceptable under any circumstances and require that the process be relocated or abandoned unless an inherently safer alternative can be found (Ref. 9.5). The paragraphs below discuss some essential features that companies should consider when adopting and implementing risk judgment protocols.

Apply the risk tolerance criteria. Management must communicate its expectation as to how the risk tolerance criteria will be applied in each risk analysis over the life cycle of the project. If a review identifies only hazards, then the risk tolerance criteria must be applied at that level. For example, Technology A has the potential for offsite effects and Technology B does not. If management (or the community) will not tolerate any offsite risk, then Technology B must be chosen regardless of the advantages that Technology A may offer.

Later in the life cycle, the risk tolerance criteria can be applied to each scenario identified. For example, when evaluating accident scenarios related to offloading, a team may determine that a hose may burst and cause a large material release, but means exist to isolate the release and mitigate the consequences. On a qualitative basis, the unloading system may conform to applicable codes and be judged acceptable. However, companies with more stringent criteria may require a LOPA of the scenario to determine if it meets their risk tolerance, or they may require a detailed quantitative risk analysis.

In all cases, the risk tolerance criteria should guide the risk analysis team in deciding when a recommendation is required, when a recommendation is optional, and when a recommendation would be superfluous. The management guidance should also acknowledge that some considerations might override the normal risk tolerance criteria. For example, if a design does not meet a code or policy requirement, corrective action is required and risk judgment is irrelevant. Or if a tolerable risk can be further reduced at minimal cost, the desire to reduce risk as low as reasonably achievable may warrant consideration of the recommendation.

Select appropriate risk control measures. Management should develop a system to ensure that the most appropriate risk control measures are selected, considering its risk tolerance criteria. In particular, engineering design tradeoffs should consider the true life cycle costs (and not just the capital costs) of active and administrative controls versus inherently safer and passive controls. Otherwise, facilities tend to select controls with lower initial capital costs that may have much higher life cycle costs because of their operating and maintenance expenses.

Another difficulty is that the implementation of any one recommendation to reduce risk may also decrease (or increase) the risk in other areas. Thus, the priority of other proposed risk controls may decrease (or increase) accordingly. If all of the recommendations from any one risk analysis are adopted with no consideration of these interactive effects, resources may be misdirected to unnecessary controls for some risks while the controls for other risks are inadequate.

9.2.4 Follow Through on the Assessment Results

Management should formally resolve each recommendation made by risk analysis teams, either by implementing the proposed change, implementing an alternative risk reduction measure, or by accepting the risk "as is" and documenting the rationale for rejecting the recommendation. Recommendations from early reviews may simply be incorporated into the next stage of design. But if the process is in operation, any corrective action must follow the *management of change* procedure. Sometimes action on a recommendation is deferred because it addresses a lower risk issue; nevertheless, these items should also be carefully tracked and resolved as soon as possible. To ensure that approved HIRA results are properly followed-up, the following essential features should be considered:

Communicate important results to management. Once a HIRA is completed, the study results should be communicated to management.

Obviously, management should review all of the recommendations for further risk reduction and decide whether to follow the team's advice. But management should also review the risks that the team judged tolerable with existing controls. In particular, some organizations require that a summary of accident scenarios with potentially catastrophic consequences be summarized for management and that the responsible manager(s) formally acknowledge their acceptance of the risk. This helps ensure that managers understand the importance of the controls necessary to mitigate those risks.

Document the residual risk. Residual risks are those that remain after risks have been reduced to levels as low as reasonably practicable. Implicitly, residual risks are deemed tolerable risks if the process is operated. Thus, the risk assessment provides a baseline from which future risk judgments can be made. For example, a change in public attitudes may force a company to reconsider its risk tolerance. If the residual risks are clearly documented, then it is a straightforward task to compare the existing risks to any new risk targets. Also, if changes are proposed, the impact of the change on residual risk can be evaluated.

Resolve recommendations and track completion of actions. All recommendations resulting from risk assessments should be resolved in a timely manner. Each recommendation should be assigned to an appropriate person, with a deadline for initial evaluation. That evaluation should consider the costs and benefits of the recommendation, its complexity, and its difficulty of implementation. If the recommendation is modified or rejected, the rationale should be documented. Specific action plans should be developed for accepted recommendations. (Note that resolution of a single recommendation may require multiple action plans, and a single action plan may address multiple recommendations.) The action plan should establish responsibilities and implementation deadlines. The management system should track the status of all actions until they are resolved, and should periodically audit the system to ensure compliance.

Resolution of recommendations resulting from early risk assessments can be evaluated during subsequent reviews. However, corrective actions implemented on a unit in operation are subject to the *management of change* system. Any special or temporary measures necessary to control the risk until a permanent control can be implemented should also be managed as changes.

Communicate results internally. Facilities should have a process for (1) deciding who needs to be informed of the risks, (2) defining the content and extent of the communication, in other words, whether simple awareness is acceptable or more detailed training is required, (3) determining the means for making sure communication takes place in a timely fashion, and (4) defining the means for ensuring that the risks were properly communicated and understood to the extent needed by the worker.

Communicate results externally. Some of the risk information may be of significance to the community or other external organizations. In operational life stages, summarizing the risk results and preparing them for communication to the public, local emergency management authorities, or regulators might be appropriate. Such communications should be carefully designed, taking effective risk communication practices and legal considerations into account as discussed in Chapter 7, Stakeholder Outreach.

Maintain risk assessment records. A critically important work product of a *risk* system is a properly documented and reviewed HIRA. However, if records of the HIRA review process are not kept, updating, revalidating, or auditing the HIRA system is very difficult. Companies should decide what types of HIRA records they want to keep and develop a retention policy. HIRA records are increasingly being kept electronically using typical word processing, spreadsheet, or database applications, or using software specifically designed for performing and documenting qualitative and quantitative risk assessments. However, this is appropriate only if the ability to read older electronic files will be preserved for the life of the process. Paper-based records systems may still be appropriate for simple systems operated at low demand rates and may be the simplest way to facilitate access to the records.

9.3 POSSIBLE WORK ACTIVITIES

The RBPS approach suggests that the degree of rigor designed into each work activity should be tailored to risk, tempered by resource considerations, and tuned to the facility's culture. Thus, the degree of rigor that should be applied to a particular work activity will vary for each facility, and likely will vary between units or process areas at a facility. Therefore, to develop a risk-based process safety management system, readers should perform the following steps:

1. Assess the risks at the facility, investigate the balance between the resource load for RBPS activities and available resources, and examine the facility's culture. This process is described in more detail in Section 2.2.
2. Estimate the potential benefits that may be achieved by addressing each of the key principles for this RBPS element. These principles are listed in Section 9.2.
3. Based on the results from steps 1 and 2, decide which essential features described in Sections 9.2.1 through 9.2.4 are necessary to properly manage risk.
4. For each essential feature that will be implemented, determine how it will be implemented and select the corresponding work activities

described in this section. Note that this list of work activities cannot be comprehensive for all industries; readers will likely need to add work activities or modify some of the work activities listed in this section.

5. For each work activity that will be implemented, determine the level of rigor that will be required. Each work activity in this section is followed by two to five implementation options that describe an increasing degree of rigor. Note that work activities listed in this section are labeled with a number; implementation options are labeled with a letter.

Note: Regulatory requirements may specify that process safety management systems include certain features or work activities, or that a minimum level of detail be designed into specific work activities. Thus, the design and implementation of process safety management systems should be based on regulatory requirements as well as the guidance provided in this book.

9.3.1   Maintain a Dependable Practice

Document the Intended Risk Management System

1. Establish and implement formal procedures to manage risk.
    a. A general policy describes risk management.
    b. Item (a), with general guidance that applies to all processes.
    c. Item (a), with detailed guidance that addresses how *risk* activities should be performed at each point in the life cycle and periodically revalidated.
    d. Item (c), with an automated, electronic work flow system for *risk*.

Integrate HIRA Activities into the Life Cycle of Projects or Processes

2. Determine when HIRAs should be performed:

    - Laboratory-scale development
    - Pilot-scale or semi-works operation
    - Conceptual design
    - Before selecting a plant site
    - Before ordering long-lead equipment
    - Detailed design
    - Before construction
    - Before startup
    - During operation
    - Before shutdown
    - Before demolition

    a. HIRAs are part of normal design review.
    b. HIRAs are performed before startup or acquisition.
    c. An initial HIRA, and periodic updates, are performed.
    d. A series of HIRAs specific to each stage of the life cycle are performed.

3. Specify the allowable time before the risk assessment of an operating process is revalidated in the *risk* system.
   a. Revalidation is required only for major changes.
   b. Time limit is specified for periodic revalidation, even if no changes are made.
   c. Item (b), and the analysis must be redone if specified criteria are met.

Clearly Define the Analytical Scope of HIRAs and Ensure Adequate Coverage

4. Determine the types and severity of consequences to be addressed in the program, for example:
   - Worker injuries exceeding a threshold
   - Public injuries exceeding a threshold
   - Environmental damage exceeding a threshold
   - Property damage exceeding a threshold
   - Business interruption losses exceeding a threshold
   - Company reputation damage exceeding a threshold

   a. Only qualitative types of consequences are defined, such as lost time, fire, or toxic release.
   b. Categories of consequences are defined, for example, those resulting in 10 to 30 days downtime.
   c. Quantitative criteria are defined, such as, 3×10^{-5} fatalities per year.
5. Determine the process systems and chemicals to be addressed in the program.
   a. Process systems and chemicals are selected at management's discretion.
   b. Only process systems with inventories exceeding a threshold amount of specific chemicals are included.
   c. Processes involving specific chemicals are included, regardless of quantity.
   d. All process systems, chemicals, and energy sources are included.
6. Determine the minimum objectives of each required analysis.
   a. Broad objectives are defined, such as safety improvements.
   b. Specific objectives can be inferred from consequences of interest.
   c. Item (b), and specific objectives are defined, for example, inherently safer design options, compliance with regulations and standards, or risk below a target value.

Determine the Physical Scope of the Risk System

7. Develop a list of units and activities to which the *risk* system applies.
   a. Physical scope is selected at management's discretion.
   b. Applicable *risk* areas and activities are identified.

    c.   Item (b), and process boundaries are clearly defined.

    d.   HIRAs are required fenceline to fenceline for all activities.

Involve Competent Personnel

8. Define the roles and responsibilities for *risk* activities.

    a.   The project manager or operations manager is responsible for *risk* activities.

    b.   A designated individual is responsible for overseeing the *risk* program and negotiating for resources.

    c.   All roles and responsibilities are assigned to job functions or departments and included in their measures of performance.

9. Assign a job function as the owner of the *risk* system to monitor its effectiveness on a routine basis. This person has the authority to resolve conflicts and implement programmatic corrective actions as needed.

    a.   The *risk* element owner is a temporary or rotating assignment.

    b.   A single, permanent *risk* element owner is designated.

10. Provide awareness training on the *risk* system to all affected employees and contractors.

    a.   *Risk* practices are broadcast once (e.g., through e-mail) to everyone.

    b.   Initial training is provided once to affected workers.

    c.   Initial and refresher training are provided to affected workers.

11. Provide detailed training to all employees and contractors who are assigned specific roles within the *risk* system.

    a.   Detailed initial training is provided once to HIRA leaders.

    b.   Detailed initial training is provided to HIRA leaders, and they must lead HIRAs periodically to maintain their qualification.

    c.   Detailed initial and refresher training are provided to key HIRA personnel – leaders, scribes, and team members.

Make Consistent Risk Judgments

12. Define the basis for judging risks and risk tolerance.

    a.   Management discretion is used.

    b.   Qualitative criteria are used that are consistent with current regulations and standards.

    c.   Category criteria, such as low, medium, and high, are used.

    d.   Quantitative criteria, such as <1 fatality/10,000 years, are used.

13. Address the criteria for selecting risk control measures in the HIRA procedure.

    a.   HIRA procedure specifies risk controls that meet minimum codes and standards.

    b.   Item (a), and it specifies risk controls that meet risk criteria.

    c.   HIRA procedure provides guidance for selecting controls that will reduce risk as low as reasonably practicable.

14. Maintain current risk tolerance criteria.
    a. Risk tolerance is revised based on management discretion or in reaction to an incident.
    b. Loss data are collected regularly, and the risk criteria are reviewed periodically.
    c. Risk criteria is revised proactively, based on losses and stakeholder expectations.

Verify that Risk Practices Remain Effective

15. Keep a running status of all HIRA reviews.
    a. A HIRA log is kept for each process.
    b. A HIRA log is kept by the facility *risk* coordinator and is accessible to all affected personnel.
    c. A HIRA log is electronically accessible, and it automatically issues notifications when studies are due.
16. Establish *risk* performance and efficiency metrics.
    a. Basic *risk* activity data are collected.
    b. *Risk* performance data are collected annually.
    c. *Risk* performance and efficiency data are collected regularly.
17. Provide input to internal audits of *risk* practices based upon the HIRA performance metrics.
    a. Internal reviews are performed, but only qualitative information is provided to help reviewers target HIRA areas.
    b. HIRA performance metrics are provided to internal reviewers periodically.

9.3.2 Identify Hazards and Evaluate Risks

Gather and Use Appropriate Data to Identify Hazards and Evaluate Risks

18. Consider all of the types of information necessary to properly evaluate risks within the scope of the *risk* element. Facilities should consider developing checklists of appropriate sources of input information for reviewers to use.
    a. Teams are instructed to proceed as scheduled with whatever information is available, and to request missing information as needed.
    b. An informal understanding exists among review leaders about the minimum types of information needed to perform a HIRA.
    c. A checklist of minimum data for each type of risk assessment is provided and enforced.

Select Appropriate HIRA Methods

19. Use appropriate HIRA methods.
    a. Reviewers use their discretion and judgment.
    b. A generic safety issues/hazard checklist is provided for use in review of all hazards.
    c. Qualitative guidance is provided for selecting an appropriate technique.
    d. Specific analysis techniques are required under specified circumstances.
20. Specify (1) issues that must be addressed in a review regardless of technique used and (2) quality parameters for review results.
    a. Each reviewer assembles her own list of issues.
    b. A simple, generic list of issues to be considered is provided.
    c. Criteria for which issues must be considered are provided.

Ensure that HIRA Participants Have Appropriate Expertise

21. Specify the qualifications for participating in HIRAs.
    a. Specific job functions are specified.
    b. Participant qualifications are specified, and only qualified personnel are allowed to participate.
22. Provide a description of necessary disciplines needed on a HIRA review team for each type of analysis.
    a. An informal practice exists for senior personnel to be involved.
    b. General disciplines are given without regard to analysis method.
    c. Specific disciplines are provided for each analysis method.
23. Use qualified leaders.
    a. Leaders should have some experience with the risk analysis methods used.
    b. Leadership training on risk analysis methods is provided to key personnel.
    c. Leaders and team members are trained in risk analysis techniques.
24. Train team leaders and participants in the use of risk tolerance criteria.
    a. Understanding of risk tolerance guidelines is inferred from on-the-job experience.
    b. Informal training on risk tolerance guidelines is provided as needed at the beginning of each analysis.
    c. HIRA leaders are formally trained on risk tolerance guidelines.
    d. HIRA leaders and team participants are trained on risk tolerance guidelines.

Perform HIRA Activities to the Appropriate Level of Technical Rigor Commensurate with the Life Cycle Stage and the Available Process Information

25. Conduct HIRAs at an appropriate level of rigor commensurate with the level of detail that is available for the process life cycle stage.
    a. General guidelines on dividing a process are given.
    b. Detailed guidelines on sectioning a process are provided.
    c. Level of resolution guidance is provided for different life cycle stages.
26. Address the appropriate technical risk issues based on the available information at that stage of the process life cycle.
    a. Basic issues are specified.
    b. Issues are specified for operating life stage only.
    c. Issues are specified for each life cycle stage.
27. Address inherent safety issues in an appropriate fashion considering the stage of the process life cycle.
    a. Inherent safety issues are specified, but no plan is established.
    b. Inherent safety issues are specified and addressed for each life cycle stage.

Prepare a Thorough HIRA Report

28. Document the hazards, risks, and recommendations from the analysis in a formal report.
    a. Informal report guidelines are provided.
    b. Formal report writing and reviewing guidelines are provided.
    c. Formal report writing and reviewing guidelines are provided, along with model reports and criteria.
29. Document the risk control measures.
    a. The primary control is compliance with standards.
    b. The report lists specific engineered and administrative controls.
    c. Item (b), and risk controls subject to the asset integrity program are explicitly identified.
    d. Item (c), with safe operating limits and limiting conditions for operation.

9.3.3   Assess Risks and Make Risk-based Decisions

Apply the Risk Tolerance Criteria

30. Address the options that teams have in making decisions in the HIRA procedure.
    a. An informal understanding exists about what options the team has.
    b. A written program exists, but roles and responsibilities are not specified.

c. Teams are given a general list of options through the written program.

d. Teams are given specific lists of options and are accountable for their role in risk management.

Select Appropriate Risk Control Measures

31. Define preferred risk controls.

a. Engineered controls are preferred to administrative controls.

b. Passive controls are preferred over active controls.

c. Inherently safer approaches are preferred.

9.3.4   Follow Through on Assessment Results

Communicate Important Results to Management

32. Inform management of the risks identified and the recommendations for risk reduction.

a. Communication occurs in response to management inquiries.

b. A formal system exists for informing management.

c. A formal system exists for informing management and documenting the resolution of any recommendations or acceptance of the risk.

Document the Residual Risk

33. Document the residual risk for each operation, unit, and facility.

a. An informal system is in place to document some residual risk.

b. A formal system exists to document residual risk.

c. Item (b), and the information is periodically updated.

Resolve Recommendations and Track Completion of Actions

34. Address HIRA recommendations and document their resolution, including specific actions that address the recommendations. Prioritize actions based on risk and benefit.

a. Action items are tracked on an ad hoc basis.

b. A formal system is used to track action items to completion.

c. A formal system is used to prioritize and track action items to completion.

Communicate Results Internally

35. Communicate risks to potentially affected personnel, including contractors.

a. An informal system is in place to inform personnel about risks.

b. A formal system exists to inform operating personnel about risks.

   c.   A formal system exists to inform all potentially affected personnel about risks.

Communicate the Results Externally

36. Inform stakeholders, including emergency management agencies, of the hazards and risks identified, the recommendations for risk reduction, and the residual risks.
   a.   Communication occurs in response to stakeholder requests.
   b.   A formal system initiates written communication with local emergency planning committees (LEPCs).
   c.   A formal system initiates both written communication with LEPCs and outreach to community leaders.

Maintain Risk Assessment Records

37. Archive the HIRA results, along with key materials and information used by reviewers. Preserve the results on diverse media in redundant locations. Retain the HIRA results and key materials and information for a specified period (e.g., 1 year, 5 years, the life of the process) to support other RBPS work activities.
   a.   Minimal HIRA records are kept for a brief period.
   b.   Extensive HIRA records are kept until the next analysis or revalidation.
   c.   Extensive HIRA records are reliably archived for the life of the process or beyond.

9.4   EXAMPLES OF WAYS TO IMPROVE EFFECTIVENESS

This section provides specific examples of industry tested methods for improving the effectiveness of work activities related to the *risk* element. The examples are sorted by the key principles that were first introduced in Section 9.2. The examples fall into two categories:

1. Methods for improving the performance of activities that support this element.
2. Methods for improving the efficiency of activities that support this element by maintaining the necessary level of performance using fewer resources.

These examples were obtained from the results of industry practice surveys, workshops, and CCPS member-company input. Readers desiring to improve their management systems and work activities related to this element should examine these ideas, evaluate current management system and work activity performance and efficiency, and then select and implement enhancements using the risk-based principles described in Section 2.1.

9.4.1 Maintain a Dependable Practice

Use standard documentation forms. A HIRA may be correctly documented in dozens of ways, but this variability often causes inefficiencies when someone else later attempts to revalidate the study or use the results. To reduce such inefficiencies, a company should decide what documentation style it prefers for each HIRA technique and standardize it. For example, some organizations prefer to include a risk ranking for each deviation in a HAZOP study. In that case, the standard HAZOP documentation form should include columns for risk-ranking information, even if the rankings are not always used. This will help the leader and team ensure that they have met all of the study requirements or have a logical reason for any deviation from it. Having standard forms will also help anyone later using the study quickly find the information they seek.

The concept of standardized documentation forms can be expanded to include entire analysis worksheets. For example, typical causes and safeguards might be included on a standard HAZOP worksheet. The leader could then use the information as a brainstorming aid during the meeting and customize the information as appropriate to the specific situation. The template worksheet makes it easy and efficient for study leaders to prepare complete, consistent analyses.

Use standard checklists. Many HIRA studies rely on checklists of questions to identify hazards or ensure that the risks are adequately controlled. However, rather than force each study leader to develop a personal library of checklists, it is far more efficient if the organization develops standard checklists and makes them available for all study leaders. For example, a hazard identification checklist might include questions about falling objects, dust, and chemical fumes. These hazards may not be present, but the discipline of having everyone consider the possibility improves the overall quality of the study. Even though no obvious falling objects might be present, steam discharges from a vent in winter might freeze into a massive icicle that could fall and damage delicate equipment below. Without this specific checklist question about falling objects, the team might not have even considered that possibility.

The *risk* element owner is the logical individual to compile a comprehensive list of questions and post them on the company's intranet. Anyone performing a hazard study would be expected to download the latest checklist and use it. If someone later identified additional questions to improve the checklist, the checklist can be easily updated, and all subsequent studies will have the benefit of that improvement.

Standardizing analysis approaches has similar benefits. For example, a company may want facility siting issues addressed consistently at all of its facilities, or it may want high consequence scenarios evaluated with a LOPA. By posting the standard approach and any associated tools on the intranet, all HIRA leaders can efficiently and consistently use it.

Provide model studies. Many HIRA studies are similar to others that have been performed, but unfortunately, the study leader may be unaware of that work. Sometimes, adapting a previous similar study to address a current situation is more efficient than developing a completely new analysis and report. Thus, compiling a searchable library of past HIRA studies and making them available to all HIRA study leaders would improve efficiency. The *risk* element owner is the logical person to take on this activity. When a study meeting the organization's standards is completed, a read-only version can be posted on the company's intranet. Using keywords or text phrases, anyone can search the database for similar analyses and use the best match as a starting point or benchmark for their own study.

Develop and use HIRA subject matter experts (SMEs). Almost anyone can be trained to provide cardiopulmonary resuscitation, but few have the skill of a professional medical technician. Similarly, anyone can be taught to lead a HIRA, but few will be as effective as a trained SME who leads HIRAs routinely. The ideal number of HIRA leaders for an organization depends on workload – each should have regular opportunities to lead without being overburdened. Having several studies led by one experienced person is far more efficient than having them led by a new person each time. Small facilities with limited HIRA needs can achieve this efficiency by sharing an SME with other facilities or hiring an outside professional as needed.

Consolidate small studies. A significant amount of overhead cost is associated with each HIRA study – preparing for the analysis, assembling a knowledgeable team, and documenting the results. Therefore, analyzing one unit in its entirety would likely be more efficient than dividing it into several smaller studies. However, the efficiency gain is quickly lost if the required calendar time extends to the point that team members must be substituted, or the number of required experts makes the team large and unwieldy.

Train designers in process safety principles. Ideally, HIRA studies should affirm that the designers have identified the hazards and implemented appropriate risk controls, thus no further action is required. In reality, HIRA teams usually find many deficiencies in the design or planned activities. Each finding or recommendation can result in significant expenditures to correct the situation, particularly for issues discovered late in the design cycle. Most designers whose work has undergone HIRA studies learn how to avoid similar issues in future designs. However, proactively teaching process safety principles to designers so that errors can be avoided is much more efficient.

Define and enforce stage gates for projects. Correcting a design flaw or applying for a discharge permit after a unit is constructed is very expensive. Identifying and mitigating risks early in the project life cycle is a far more efficient approach. Stage gates that govern the progress of a design from one life cycle stage to the next should be developed. These may be in the form of checklists to confirm, for example, that (1) the proposed emergency response plan will be capable of handling releases of the proposed chemicals, (2) the facility has (or can get) relevant permits, (3) required HIRAs have been completed, and (4) any issues have been resolved satisfactorily. Table 9.1 lists types of HIRA issues that might be addressed at various stages in the life cycle of a project to help minimize the chance of expensive surprises.

Diversify analysis methods. Each HIRA method has unique strengths and weaknesses. If only one method is used to analyze a high hazard system, some risks may be underestimated or overlooked. In some situations, re-analyzing the system with a different technique may find issues that the original study missed.

Expand the analysis scope to include economic impacts. Organizations seeking to comply with regulations tend to focus their *risk* element on safety and environmental concerns. However, long before regulatory drivers existed, some companies were voluntarily performing HIRAs for the economic benefits, such as less frequent process shutdowns, higher plant availability, and increased yield. Even though expanding the analysis scope to include economic impacts will increase resource requirements, the benefits almost always exceed the added cost.

Use an independent risk study leader. Risk studies are much more effective if led by an independent reviewer because few people are effective critics of their own work. If the same person who designed a system attempts to lead the review, any currently existing risks will likely be underestimated or overlooked. Similarly, if the person responsible for operating an existing system attempts to lead the review, familiar risks are likely to be misjudged.

9.4.2   Identify Hazards and Evaluate Risks

Audit RBPS elements during HIRA preparation. Many of the other RBPS elements (e.g., *knowledge, reliability, management of change*) are routinely evaluated in preparation for a HIRA, and an audit of their effectiveness and completeness can be performed with little or no additional effort. A simple form can be developed with checklist questions, such as

- Were the drawings available?
- Were the drawings accurate?
- Were there undocumented changes?
- Are safeguards being tested as scheduled?

TABLE 9.1. Example Issues that Can Be Addressed at Various Life Cycle Stages

Life Cycle Stage	Technical Issue To Be Addressed
Conceptual Design	Chemical toxicity and reactivity hazards
	Fire/explosion hazards
	Environmental hazards
	Permit limits
	Siting criteria
	Inherent safety
Process Flow Diagram	Layout spacing
	Building requirements
	Electrical classification
	Worst-case scenarios
	Emergency response plans
	Inherent safety
	Permit limits
	Hazardous material inventories
	Qualitative estimate of impacts
Detailed Design	Operating errors
	Single failure points
	Multiple failure points
	Safety integrity levels
	Utility failures
	Process deviations
	Quantitative estimate of frequencies and impacts
Construction	Heavy lifts
	Field changes
	Staffing levels
	Disruption of existing processes
	Disruption of utilities
Post-startup	Operating errors
	Maintenance errors
	Temporary activities
	Safety integrity levels
	Emergency response plans
	Process deviations
Decommissioning	Release of residues
	Reaction of residues
	Disposal of waste
	Disruption of remaining processes

If serious deficiencies are found, they must be corrected before the HIRA can proceed; otherwise, the form can be filed and kept available to support the next audit.

Provide administrative support to document HIRAs. Experienced HIRA leaders and team members are typically highly paid, senior staff. If the leader attempts to simultaneously lead and document the meeting, a significant portion of the team's time can be wasted waiting on the leader to catch up on his notes. In addition, the leader, who typically could be doing technical work, must instead devote time afterwards to producing the analysis report. Selecting an administrative assistant or junior staff member to document the meeting notes and help the leader prepare the HIRA report is far more efficient and economically prudent.

Provide appropriate team composition. Regulations and company standards sometimes impose a minimum requirement on team composition. Beyond that, companies should make a risk-based judgment about the number of necessary team members. For a simple, low hazard system, a team meeting those minimum requirements is adequate; in higher hazard situations, a larger team with more experience and diverse skills may be necessary. However, even when a larger skill set is necessary, having some specialized individuals available "on call" is more efficient than having them sit through hours of review just to answer a few questions.

Include team members from other facilities to foster cross-learning. The effectiveness of HIRAs is a direct function of the team's knowledge and diversity. By including experienced team members from different shifts, from similar units at the facility, and from similar units at other facilities, the HIRA benefits from the diversity of experience. In addition, the team members themselves benefit from the group discussions, which usually reveal facts and experiences of which they were previously unaware, thus improving the quality of risk judgments. This cross-learning also offers substantial benefits when the team members return to their normal job duties with new ideas about how to improve or troubleshoot issues in their own area.

Conduct virtual meetings. Significant cost is often associated with assembling a knowledgeable team for a HIRA study. However, having everyone physically at the same location during the review meetings is not necessary. In many cases, internet or videoconference technology is a more efficient means of involving experts (possibly including the team leader) from other locations without the time and expense of travel. This can be particularly valuable for simple analyses and for small facilities that do not have resident experts who can participate. However, the efficiency gain is quickly lost if the remote participants are distracted by other activities or if the virtual meeting equipment limits their participation, making it difficult to see, hear, discuss, share, and so forth.

Select analysis methods based on hazards and potential consequences.
HIRA tools range from simple checklists to complex consequence models.
Applying complex tools when simple ones would suffice is very inefficient;
similarly, simple tools cannot always adequately analyze complex systems.
To select an appropriate analysis tool, perform an initial screening based on
hazards and potential consequences. If, for example, the screening does not
identify any toxic chemical hazards, or if those hazards exist but the quantities
are so small that the maximum consequences are limited, then a simpler, less
rigorous HIRA method may be sufficient to analyze the risks without wasting
resources over-analyzing a simple system. Conversely, a screening that
identifies severe consequences indicates that more rigorous HIRA methods
may be appropriate.

Limit discussions to analysis scope and threshold severity. HIRA
meetings may open a Pandora's Box of issues; however, the team should
discuss only those issues that fall within the defined scope. For example, if a
study is tasked with finding only process safety issues, allowing lengthy
discussions of economic improvement opportunities, and then documenting
those discussions and recommendations, would be very inefficient. Within the
defined scope, a minimum threshold should be defined for consequences of
interest. For example, if the study scope is personal injury, the threshold
severity might exclude injuries requiring only first-aid treatment. The leader
should manage the team discussions to stay narrowly focused within the
study's scope and threshold severity. Other issues that may arise should be
briefly noted for management consideration or further study, but the
discussion should be terminated at that point.

Automate parts of the analysis. Software tools can significantly enhance
the effectiveness of the HIRA team. Numeric tools have traditionally been
used in quantitative analyses of event frequencies, consequences, and risks.
However, computers can also substantially accelerate qualitative risk analyses.
For example, inexpensive software can generate a HAZOP report template
based on nodes defined by the user. Spreadsheets can automate LOPAs and
improve the consistency of the results. Wherever computers can automate and
standardize portions of HIRA studies, they significantly improve the
effectiveness of the analysts. Automated methods should be standardized
across the organization to enhance efficiency as well as effectiveness.

Review previous incidents. One good way to improve the critical thinking
of a HIRA team is to review previous incidents from the subject (or similar)
processes. In addition to reminding team members of the facts, this helps
preempt the mindset that "bad things can't happen here." The *incidents*
element should be the primary source of information, but reviewing incidents
from other facilities is also helpful. Many incident descriptions are posted on
the internet by government authorities such as OSHA, EPA, CSB, HSE, and so
forth. Other sources of published incident descriptions include: *Process Safety*

Progress by AIChE, the Process Safety Incident Database by CCPS, a number of books authored by Trevor Kletz, and *Safety Digest of Lessons Learned* by the American Petroleum Institute.

9.4.3 Assess Risks and Make Risk-based Decisions

Manage analysis depth based on risk ranking. A simple risk matrix can be used with most HIRA tools to manage the depth of the analysis. When an accident scenario is identified, the leader can direct the team to rank its risk based on existing protections. If the risk falls in a tolerable zone of the risk matrix, then no further action is necessary and the team should move on. Otherwise, the leader should urge the team to consider the scenario in more detail until practical recommendations to reduce the risk are identified.

Compliance with standards instead of independent analyses. In many situations, standard designs that conform to recognized and generally accepted good engineering practices are automatically judged to reduce risk to tolerable levels because these practices are distilled from hundreds of years of operating experience in a wide variety of applications. In those situations, ensuring that the design meets recognized standards is more efficient than performing an independent series of HIRA studies to discover recommendations that are already required by code or by industry practice. These "good engineering" requirements can be reviewed as checklist questions, and the absence of any unique hazards can be confirmed with what-if questions. This knowledge-based approach can usually be performed during normal engineering design reviews, and the final report can consist of the signed-off drawings and a listing of the standards that were applied.

Review all high consequence scenarios. When a high consequence scenario is identified, its risk will either be deemed acceptable or recommendations for risk reduction will be developed. In either case, management would be prudent to subject such scenarios to additional review to either affirm their acceptance of the risk or to ensure company standards for risk reduction are applied consistently across the organization. These reviews can also be used as mentoring sessions to help less experienced leaders improve their skill.

9.4.4 Follow Through on Assessment Results

Maintain current HIRAs. Most organizations require that HIRAs for operating units be revalidated periodically. This typically involves updating the original report to reflect any changes since the last revalidation (Ref. 9.6). Rather than wait several years to update the HIRA, some companies maintain an up-to-date version. When a change occurs, the review team revises, adds, or deletes the affected node(s) as appropriate. This expedites the next revalidation because any changes are already incorporated in the document.

The revalidation team primarily needs to confirm the accuracy of the changes and look for (1) any undiscovered interactions between them or (2) unrecognized effects on the collective risk.

Prioritize corrective actions based on risk. All recommendations from HIRA studies should be resolved as soon as possible, but not all of the corrective actions are equally important. Scoring the corrective actions based on risk is an effective way of allocating scarce resources to the most important issues first. However, deferring a corrective action when resources are available simply because it has a relatively low risk score is not appropriate. Documenting recommendations and their associated risk scores in a table, spreadsheet, or database is an effective way to facilitate the risk-based ranking. This allows anyone to efficiently sort the results to check for internal consistency (similar risks should have similar scores) and easily select the higher risks for corrective action first.

Consolidate and automate corrective action tracking. Nearly every HIRA study will generate recommendations for corrective action. The most efficient way to ensure that these recommendations are resolved in a timely manner is to consolidate all of them into a centralized tracking database. This allows management to track the status of corrective actions and to reprioritize work as necessary to get them resolved. A central database also provides traceability for anyone who needs to know the status or final resolution of a particular corrective action.

9.5   ELEMENT METRICS

Chapter 20 describes how metrics can be used to improve performance and when they may be appropriate. This section includes several examples of metrics that could be used to monitor the health of the *risk* element, sorted by the key principles that were first introduced in Section 9.2.

In addition to identifying high value metrics, readers will need to determine how to best measure each metric they choose to track. In some cases, an ordinal number provides the needed information, for example, total number of workers. Other cases, such as average years of experience, require that two or more attributes be indexed to provide meaningful information. Still other metrics may need to be tracked as a rate, as in the case of employee turnover. Sometimes, the rate of change may be of greatest interest. Since every situation is different, the reader will need to determine how to track and present data to most efficiently monitor the health of RBPS management systems at their facility.

9.5.1   Maintain a Dependable Practice

- *Number of HIRAs of each type that are overdue.* A high number usually indicates resource constraints or a lack of management support for the program. In regulated situations, a high number also highlights areas in which obligations are not being met.

- *Percentage of intended revalidations that require the study to be completely redone.* An abnormally high number may indicate that the original studies were poorly performed or that the *management of change* element has been ineffective.

- *Number of audit findings.* Audits may target specific issues related to company or regulatory standards, such as topics reviewed, report completeness, or results effectiveness. An abnormally high number may indicate that the studies were poorly performed or may reflect deficiencies in the *knowledge* element.

- *Number of qualified HIRA leaders, scribes, and participants.* Unexpected decreases in this number may indicate imminent backlogs resulting from a lack of resources.

- *HIRA resource demand and team efficiency.* HIRA resource demand is simply the number of staff hours charged to HIRA activities (excluding resolution of recommendations), and efficiency divides that total by the number of HIRAs performed, meeting days, nodes analyzed, or recommendations produced. Unexpected declines in this metric may indicate that management is failing to provide the resources necessary to perform the work. Unexpected increases in this metric may indicate a decline in the quality of data inputs to HIRAs, such as incomplete drawings, or an attempt to co-opt HIRAs into normal engineering design reviews. Increases may also indicate a decline in team leadership skills or in the competence of team members.

9.5.2   Identify Hazards and Evaluate Risks

- *Number of HIRAs of each type scheduled.* This metric provides an indication of resource requirements. A downward trend or unexpected drop may indicate that required studies are not being performed.

- *Time required to issue a HIRA report.* Increases in this metric may indicate that team leaders or management reviewers are giving HIRA activities a low priority.

- *Technique used.* Unexplained changes in this metric may indicate a shift to less appropriate techniques.

9.5.3   Assess Risks and Make Risk-based Decisions

- *Percentage of recommendations for administrative controls, active engineered controls, passive engineered controls, and inherently safer alternatives.* This metric indicates whether early life cycle reviews are effective in implementing inherently safer options and whether teams understand company preferences for controlling risk. It also indicates whether managers are attempting to shift the burden of controlling risk from a capital expense to an operations and maintenance expense.

- *Ratio of actual losses to risk tolerance criteria.* Actual losses that consistently exceed the expected losses based on a company's risk tolerance criteria may indicate that the scope of the *risk* program is inadequate or that teams are assuming too much credit for the effectiveness of engineered and administrative controls.

- *Number of recommendations per study or per year.* Theoretically, for a given unit, this number should decline over time. However, integrated over a facility, new issues are always arising from changes in processes, equipment, procedures, personnel, and risk tolerance. Unexpected declines in this number may indicate that teams are not being thorough enough, that the analysis teams are losing competence, or that the analysis teams are accepting higher risk.

- *Number of recommendations per revalidation.* An abnormally high number may indicate that the original study was poorly performed or that the *management of change* element has been ineffective.

9.5.4 Follow Through on Assessment Results

- *Number of recommendations unresolved by their due date.* This metric provides an indication of management's commitment to achieving tolerable risk levels.

- *Percentage of repeat recommendations.* A high number may indicate that lessons learned from previous studies are not being communicated through the organization or that prior corrective actions have been ineffective.

- *Average time corrective actions require for completion.* Increases in this metric often indicate an accumulation of low priority corrective actions that remain scheduled for months or years. In regulated situations, it highlights failure to mitigate risks "as soon as possible."

- *Percentage of recommendations rejected by management.* A high number may indicate that the team failed to understand the company's risk tolerance guidelines or that management is now willing to accept more risk than the risk tolerance guidelines indicate.

- *Management exceptions to risk criteria (accepting higher risk).* A high number may identify managers who do not accept the company's risk tolerance guidelines or that management is now willing to accept more risk than the risk tolerance guidelines indicate.

- *Residual risks.* This metric tracks the number of scenarios that fall into each category of a company's risk matrix after corrective actions are applied. A large number of scenarios falling into categories in which risk reduction is optional may indicate that the teams (or managers) are not striving to make risks as low as reasonably achievable. On a corporate-wide basis, this metric indicates whether actual loss experience is consistent with the risks predicted through HIRA activities.

- *Internal/external risk communications.* Low numbers may indicate that employees are not being kept involved in the *risk* element or that the company is not attempting to maintain a dialogue with the community.

9.6   MANAGEMENT REVIEW

The overall design and conduct of management reviews is described in Chapter 22. However, many specific questions/discussion topics exist that management may want to check periodically to ensure that the management system for the *risk* element is working properly. In particular, management must first seek to understand whether the system being reviewed is producing the desired results. If the organization's *risk* performance is less than satisfactory, or it is not improving as a result of management system changes, then management should identify possible corrective actions and pursue them. Possibly, the organization is not working on the correct activities, or the organization is not doing the necessary activities well. Even if the results are satisfactory, management reviews can help determine if resources are being used wisely – are there tasks that could be done more efficiently or tasks that should not be done at all? Management can combine metrics listed in the previous section with personal observations, direct questioning, audit results, and feedback on various topics. Activities and topics for discussion include the following:

- Review the HIRAs scheduled for current processes or projects to determine whether they are appropriate for their stage in the life cycle.
- Determine whether HIRA methods are being selected based on the risk of units and activities.
- Verify that the scope of the HIRAs is broad enough to capture the full range of risks.
- Verify that the required HIRAs are being completed in a timely manner.
- Investigate any unexpected delays in starting scheduled HIRAs, and determine whether delays were caused by a lack of information, a lack of leadership or expert resources, or poor planning.
- Verify that adequate resources, including appropriate subject matter experts, are available for upcoming HIRAs.
- Verify that the *knowledge* element is up to date for scheduled analyses.

- Review the amount of overtime required for team resources to participate in the HIRAs.
- Considering the pool of qualified leaders and team members, investigate whether only a few individuals are actually participating in the reviews.
- Determine whether the latest versions of applicable resources (e.g., checklists, report templates) are being used.
- Determine whether risk studies incorrectly list failures of protective systems as causes of the upsets they protect against, for example, failure of the firewater system as the cause of a fire.
- Determine whether risk studies are unduly optimistic in claiming credit for engineered controls that are not included in the *asset integrity* program.
- Verify that HIRA forms are being properly completed.
- Determine whether the risk judgments by different teams, areas, and departments are consistent with the risk tolerance guidelines and each other.
- Determine whether recommendations are made for every scenario with an unacceptable risk score and for most scenarios that are marginally tolerable.
- Investigate any long delays between the end of analysis meetings and the issuance of the report.
- Verify that recommendations are being resolved in a timely manner, particularly for simple recommendations, such as adding a caution statement to a procedure.
- Determine whether the resolution activities of any recommendations are overdue or have been repeatedly rescheduled.
- Investigate any audit findings of *risk* program deficiencies and determine whether corrective actions have been implemented.
- Investigate whether incidents have occurred that were not anticipated in HIRAs, or if anticipated, whether the controls were inadequate.
- Investigate any serious incidents in units that were considered low risk.
- Verify that the actual losses, as tracked through the *incidents* element, are consistent with the HIRA predictions and corporate risk tolerance criteria.

Regardless of the questions that are asked, the management review should try to evaluate the organization's depth on several levels. Is the depth of hazard identification sufficient? Does understanding of risk extend beyond operational experience? When things go wrong, is it likely that key personnel will have the risk understanding required to make prudent decisions? Also, is there sufficient depth in key personnel? If the entire organization depends on a single individual to make all of the really tough risk judgments, what will happen if that person suddenly resigns or falls ill? Management reviews provide an opportunity for the organization to honestly assess its depth, and take actions to address any concerns before experiencing a loss event resulting from an insufficient knowledge of the hazards, and before losing the risk understanding that it has worked long and hard to amass. Any weaknesses revealed by the management review should be resolved as described in Chapter 22.

9.7   REFERENCES

9.1   *Report of the Longford Royal Commission,* Government Printer for the State of Victoria, 1999.

9.2   *Guidelines for Hazard Evaluation Procedures (Second Edition with Worked Examples)*, Center for Chemical Process Safety, New York, 1992.

9.3   *Layer of Protection Analysis – Simplified Process Risk Analysis*, Center for Chemical Process Safety, New York, 2001.

9.4   *Guidelines for Chemical Process Quantitative Risk Analysis (Second Edition)*, Center for Chemical Process Safety, New York, 1999.

9.5   *Inherently Safer Chemical Processes: a Life Cycle Approach*, Center for Chemical Process Safety, New York, 1997.

9.6   *Revalidating Process Hazard Analyses*, Center for Chemical Process Safety, New York, 2001.

Additional reading

Safety Digest of Lessons Learned, American Petroleum Institute Publication 758, 1981.

III. MANAGE RISK

Risk-based process safety (RBPS) is based on four pillars: (1) committing to process safety, (2) understanding hazards and risk, (3) managing risk, and (4) learning from experience. To manage risk, facilities should focus on:

- Developing written procedures that (1) describe how to safely start up, operate, and shut down processes, (2) address other applicable operating modes, and (3) provide written instructions that operators can execute when they encounter process upsets/unsafe conditions.
- Implementing an integrated suite of safe work policies, procedures, permits, and practices to control maintenance and other nonroutine work.
- Executing work activities to ensure that equipment is fabricated and installed in accordance with specifications, and that it remains fit for service over its entire life cycle.
- Managing contractors, and evaluating work performed by contractors, to ensure that the associated risks are acceptable; ensuring that contractors are not exposed to unrecognized hazards or undertake activities that present unknown or intolerable risk.
- Providing training and conducting related activities to ensure reliable human performance at all levels of the organization.
- Recognizing and managing changes.
- Ensuring that units, and the people who operate them, are properly prepared for startups.
- Maintaining a very high level of human performance, particularly among operators, maintenance personnel, and others whose actions directly affect process safety.
- Preparing for and managing emergencies.

This pillar is supported by nine RBPS elements. The element names, along with the short names used throughout the *RBPS Guidelines*, are:

- Operating Procedures (*procedures*), Chapter 10
- Safe Work Practices (*safe work*), Chapter 11
- Asset Integrity and Reliability (*asset integrity*), Chapter 12
- Contractor Management (*contractors*), Chapter 13
- Training and Performance Assurance (*training*), Chapter 14
- Management of Change (*MOC*), Chapter 15
- Operational Readiness (*readiness*), Chapter 16
- Conduct of Operations (*operations*), Chapter 17
- Emergency Management (*emergency*), Chapter 18

The management systems for each of these elements should be based on the company's current understanding of the risk associated with the processes with which the workers will interact. In addition, the rate at which personnel, processes, facilities, or products change (placing demands on resources), along with the process safety culture at the facility and within the company, can also influence the scope and flexibility of the management system required to appropriately implement each RBPS element.

Chapter 2 discussed the general application of risk understanding, tempered with knowledge of resource demands and process safety culture at a facility, to the creation, correction, and improvement of process safety management systems. Chapters 10 through 18 describe a range of considerations specific to the development of management systems that support efforts to develop, implement, and maintain effective systems for managing risk. Each chapter also includes ideas to (1) improve performance and efficiency, (2) track key metrics, and (3) periodically review results and identify any necessary improvements to the management systems that support commitment to process safety.

10

OPERATING PROCEDURES

On December 13, 1994, an explosion occurred shortly before restart of an ammonium nitrate plant in Port Neal, Iowa. Four employees were killed, and an additional 18 employees were admitted to the hospital. Public evacuations were ordered up to 15 miles from the facility, and property damage was estimated at $120 million. The plant had been temporarily shut down 18 hours earlier and operators were preparing to restart the plant at the time of the explosion. The Environmental Protection Agency's (EPA's) Chemical Accident Investigation Team concluded that the explosion resulted from the lack of written procedures for conducting a temporary shutdown on the ammonium nitrate plant (Ref. 10.1).

10.1 ELEMENT OVERVIEW

The RBPS element that ensures proper development, timely maintenance, and consistent use of operating procedures (*procedures*) is one of nine elements in the RBPS pillar of *managing risk*. Section 10.2 describes the key principles and essential features of a management system for this element. Section 10.3 lists work activities that support these essential features, and presents a range of approaches that might be appropriate for each work activity, depending on perceived risk, resources, and organizational culture. Sections 10.4 through 10.6 include (1) ideas for improving the effectiveness of management systems and specific programs that support this element, (2) metrics that could be used to monitor this element, and (3) issues that may be appropriate for management review.

10.1.1 What Is It?

Operating procedures are written instructions (including procedures that are stored electronically and printed on demand) that (1) list the steps for a given task and (2) describe the manner in which the steps are to be performed. Good procedures also describe the process, hazards, tools, protective equipment, and controls in sufficient detail that operators understand the hazards, can verify that controls are in place, and can confirm that the process responds in an expected manner. Procedures also provide instructions for troubleshooting when the system does not respond as expected. Procedures should specify when an emergency shutdown should be executed and should also address special situations, such as temporary operation with a specific equipment item out of service. Operating procedures are normally used to control activities such as transitions between products, periodic cleaning of process equipment, preparing equipment for certain maintenance activities, and other activities routinely performed by operators. The scope of this element is limited to those operating procedures that describe the tasks required to safely start up, operate, and shut down processes, including emergency shutdown. Operating procedures complement *safe work* and *asset integrity* procedures, which are addressed in Chapters 11 and 12, respectively.

10.1.2 Why Is It Important?

A consistent high level of human performance is a critical aspect of any process safety program. Indeed, a less than adequate level of human performance will adversely impact all aspects of operations. Without written procedures, a facility can have no assurance that the intended procedures and methods are used by each operator, or even that an individual operator will consistently execute a particular task in the intended manner.

10.1.3 Where/When Is It Done?

Procedures are normally developed at a facility before an operation is performed. Even if the objective of the operation is to develop or optimize a production method, for example, a pilot plant, a written procedure should be developed that establishes a safe operating envelope and specifies any limiting conditions for operation. Procedures should be updated when a change that affects operating methods or other information contained in the procedures occurs, and procedures should be reviewed periodically to ensure that they remain valid.

10.1.4 Who Does It?

Procedures are often jointly developed by operators and process engineers who have a high degree of involvement and knowledge of process operations. (In this chapter, the term operator is used to describe the person who directly controls the process either via a control system or manipulation of field equipment; note that many facilities use other terms, such as technician, to describe this function.) Operators, supervisors, engineers, and managers are often involved in the review and approval of new procedures or changes to existing procedures. Other work groups, such as maintenance, should also be involved if the operating procedures could potentially affect them. At some facilities, technical writers are used to translate input from subject matter experts, such as experienced operators or engineers, into operating procedures.

10.1.5 What Is the Anticipated Work Product?

The output of this activity is current, accurate, and useful written instructions that apply to normal operations, nonroutine or infrequent tasks, and special high hazard tasks. In many cases, written procedures are developed for otherwise low risk or straightforward tasks that are critical to achieving production goals or when a task includes a series of steps that involve multiple people or departments. This helps ensure that everyone understands their roles/responsibilities in the task and maximizes the efficiency of work processes.

In addition to being a useful guide to operators, procedures help provide critical information to the *training* element and provide the objective standards of performance required for implementing an effective *operations* element. However, operating procedures are generally not training manuals. Procedures should be written in sufficient detail that a qualified worker can consistently and successfully perform the task.

10.1.6 How Is It Done?

To develop an effective set of operating procedures, start by identifying tasks that should be addressed. Once the tasks are identified, determine the expected competence level for personnel who will be assigned to perform the task, and structure each procedure appropriately. In addition to developing a list of tasks (corresponding to a list of procedures), decide on the appropriate procedure format. Using a consistent format with a high degree of structure for each procedure will help operators quickly locate different types of information.

10.2   KEY PRINCIPLES AND ESSENTIAL FEATURES

Safe operation and maintenance of facilities that manufacture, store, or otherwise use hazardous chemicals requires reliable human performance at all levels, from managers and engineers to operators and craftsmen. The *procedures*, *training*, *culture*, and *operations* elements help form the foundation for achieving high levels of human reliability.

The *procedures* element will typically establish requirements for all operating procedures; assign responsibilities for development, use, and periodic review of procedures; and address document control. The *procedures* element also provides guidance on when written procedures are required and defines the interface between various types of written procedures that exist at the facility, for example, operating procedures, safe work procedures, maintenance procedures, quality control procedures, and so forth.

The following key principles should be addressed when developing, evaluating, or improving any management system for the *procedures* element:

- Maintain a dependable practice.
- Identify what operating procedures are needed.
- Develop procedures.
- Use procedures to improve human performance.
- Ensure that procedures are maintained.

The essential features for each principle are further described in Sections 10.2.1 through 10.2.5. Section 10.3 describes work activities that typically underpin the essential features related to these principles. Facility management should evaluate the risks and potential benefits that may be achieved as a result of improvements in this element. Based on that evaluation, management should develop a management system, or upgrade an existing management system, to address some or all of the essential features, and execute some or all of the work activities, depending on perceived risk and/or process hazards that it identifies. However, none of this will be effective if the accompanying *culture* element does not embrace the use of reliable management systems. Even the best management system, without the right culture, will not ensure that procedures are properly developed, maintained, and used.

10.2.1 Maintain a Dependable Practice

Documented, current, and accurate operating procedures help ensure that each shift team operates the process in a consistent, safe manner. Written procedures also help ensure that the same person operates the process consistently from one day (or batch) to the next. Consistent operation, even if it happens to be less than optimal, will lead to consistent results. This enables process engineers and others charged with improving operations to identify opportunities for changes that will improve safety, product quality, yield, throughput, and so forth. Consistent operation will also help facilities determine the impact that changes have on these parameters.

Establish management controls. Just as written operating procedures help ensure consistent operator performance, a written description of the management system for the *procedures* element helps maintain and continuously improve the quality of operating procedures, and therefore the reliability of tasks performed in accordance with those procedures. The written description of the management for the *procedures* element system should clearly specify (1) who has authority to develop, change, and approve operating procedures, (2) how procedures should be developed or changed, (3) methods for communicating changes or new procedures to operators, and (4) each operator's responsibility to maintain proficiency.

Control procedure format and content. A number of different formats have been successfully used for operating procedures (see Section 10.2.3 for additional details on procedure formats). The number of different formats used at a facility is a tradeoff; multiple formats provide procedure writers with the flexibility to use the most appropriate one, but too many different formats can be confusing to operators. In addition to specifying the format, the management controls for operating procedures should also (1) specify the minimum content that must be included in procedures and (2) ensure that all regulatory requirements concerning procedure content are specifically addressed.

Procedures are typically written by many different people. Failure to establish specific standards may make the procedures harder to use because of high variability in format, structure, and content, and could potentially leave an operator without proper written instructions for responding to a process upset or other unsafe condition. Also, because operating procedures are normally developed at the unit level, and personnel often transfer between units, developing a facility-wide standard for operating procedures will help reduce the learning curve for an operator, engineer, or any other worker who transfers from one operating unit to another. If the structure, format, and content of operating procedures are consistent across all units, a newly assigned operator should be able to locate information in operating procedures more quickly and efficiently.

Control documents. Several years ago, an operator charged with lighting a fired heater at a refinery was killed when the heater exploded at startup. The operator had followed a purge procedure that was literally painted on the wall near the heater. The purge procedure was out of date, and the instructions on the wall did not include some critical steps that were contained in the procedure in the manual in the control room.

Controls should be in place to ensure that only current procedures are used. Historically, updating procedure manuals was a very labor-intensive effort. Modern practice is to store the current procedures in a read-only area of the computer network, minimizing the effort required to control documents. This facilitates access by all personnel and protects the documents from loss or unauthorized/inadvertent changes. Printed copies should include an expiration date, which is often the same day it is printed. One current printed copy of each procedure is normally maintained in the control room in the event the computer network is unavailable.

10.2.2 Identify What Operating Procedures Are Needed

The tasks that should be described in written procedures depend on the knowledge, skills, and abilities of qualified workers. Failure to provide the necessary procedures in adequate detail will lead to low human reliability. However, too many procedures, or procedures that contain extraneous information or too much detail, are difficult to use. Nonroutine operating modes warrant particular attention because they often involve much greater risk than routine operations.

Conduct a task analysis. Since procedures exist solely to help improve human performance, a comprehensive list of the tasks people do (or are supposed to do) at the facility is a good starting point for identifying what procedures are needed. In addition, certain regulatory requirements, such as the Occupational Safety and Health Administration's process safety management standard or the Food and Drug Administration's good manufacturing practice regulations, and voluntary standards, such as the International Organization for Standardization's ISO 9000 family of standards, may help specify a minimum scope for the range of tasks and content that should be addressed in operating procedures.

Determine what procedures are needed and their appropriate level of detail. Many routine tasks are properly covered solely by training. For example, procedures often include steps such as "open (a particular) valve". If the proper method to open and close valves is covered in operator training, procedures generally do not specify how to open the valve or how to determine if the valve is fully open. This is a skill that one must demonstrate before becoming a qualified operator. However, if a particular valve has unusual characteristics, such as a mechanical key interlock, then the procedure should specify the special actions required to open the valve.

Procedures may not be needed for situations where a low consequence is associated with failure to properly perform the task, or a very low likelihood of performing a task improperly exists (e.g., little opportunity to make an error, and good feedback to allow the operator to recover if a mistake is made). When determining if a written procedure is needed, assume that, in the absence of a procedure, the task will either not be done or not be done consistently. If the consequence of not performing a task or performing a task in an arbitrary manner is acceptable, a written procedure is probably not required. For example, emptying the trash can in the control room is a task that must be done periodically, but the consequences of not emptying the trash or putting the trash in the wrong dumpster are probably acceptable. Conversely, disposing of adsorbent used to clean up a small spill of a hazardous material is likely governed by a written procedure because failing to dispose of this material in the proper way may (1) expose other workers to hazards and (2) violate a regulatory requirement governing waste disposal.

The level of detail needed in the procedures is a function of (1) the risk associated with the task and (2) the knowledge, skills, and abilities of the person assigned to perform the task. Procedures should include sufficient detail so that the newest or least qualified person can successfully perform the task. Attempts to make a procedure so detailed that absolutely anyone can properly perform the task often result in lengthy, overly complex procedures that will likely not be used. In general:

- Processes or activities that are perceived to be high risk (or in some cases, high hazard) usually dictate a need for greater (1) formality, (2) thoroughness in scope, and (3) level of detail in the governing procedures.
- Facilities with an evolving or undetermined process safety culture may require more prescriptive operating procedures and greater command and control management system features to properly manage risk and ensure good operating discipline.
- Flexible processes/facilities, such as pilot plants, are often accustomed to making a wide range of products with limited process-specific knowledge, although they are typically under the direct supervision of highly trained and experienced engineers and scientists. Such facilities may choose to provide greater flexibility in the sequence of steps or the exact operating conditions, focusing more on safe operating limits and limiting conditions for operation than on the exact process conditions or steps, as these may not be well understood for many products.
- Facilities that have a very formal training program and low employee turnover often find that less detail is required for written procedures. (However, be aware that if historical staff stability is lost, little time or experience may be available to improve procedures for inexperienced personnel.) Conversely, if a facility has a limited training program, much more detailed written procedures may be required to help ensure reliable human performance.

Address all operating modes. One common error in conducting a task analysis is to focus only on activities that occur at the time that the analysis is done. In doing this, the analyst is likely to miss certain nonroutine modes of operation. For example, an analyst assigned to identify the tasks performed by an operator at a single-product continuous plant during a time of routine operation is not likely to observe tasks such as preparation for startup, startup, shutdown, deinventorying and cleaning, and preparation for maintenance. Although most continuous plants operate in the steady-state normal operation mode for greater than 98 percent of the time, the risk associated with other modes normally accounts for well over 2 percent of the total risk. In fact, the risk associated with startup, shutdown, and other nonroutine operations can exceed that of routine operations, even though the risk exposure (in hours per year) for routine operations dwarfs all other operating modes.

10.2.3 Develop Procedures

Once the list of tasks is developed and reviewed, procedures are developed. Procedure development can be a complex activity, particularly if there is disagreement on how a task should be performed. However, identification of differences in how tasks are performed, and the process to resolve the differences and arrive at a single intended method, should improve overall operation of the facility and help reduce risk.

Information must be presented in a consistent and useful format. Fully trained and experienced operators tend to use procedures primarily as reference documents on an infrequent basis, in other words, a procedure is not used as a step-by-step reference each time the task is performed. Consistency allows operators to find information more quickly, which increases the likelihood that procedures will be followed. Clearly, exceptions to this practice exist; some detailed procedures for critical or high-risk operations are used daily, and detailed procedures are routinely used in highly regulated sectors, such as parts of the nuclear and pharmaceutical industries.

Use an appropriate format. The best format for a procedure depends on the task. Table 10.1 lists several procedure formats and includes examples of where each format might typically be used. As mentioned in Section 10.2.1, the number of different formats used in a facility is a tradeoff; more formats provide for greater flexibility, but can lead to confusion.

Different formats can be used within a facility, and are sometimes even found within a single procedure. For example, the best format for railcar unloading may be an outline format, supported by a checklist in the appendix that operators take to the field to verify that each critical step is completed. Safe operating limits, descriptions of safety systems, and troubleshooting guides are often presented in a multi-column format, which is generally a poor choice for operations, such as railcar unloading, that involve sequential steps.

Ensure that the procedures describe the expected system response, how to determine if a step or task has been done properly, and possible consequences associated with errors or omissions. To help ensure consistent performance, operating procedures should specify each step, what system response to expect (if any), and what to do next, which is sometimes based on the system response. For example, if the intent of a particular step is to charge material to a reactor, the operator may be asked to verify that a particular valve opens or to check the flow rate/total flow displayed on the operator console, and to take certain actions if too much, or not enough, material is charged to the reactor.

TABLE 10.1. Procedure Formats

Type	Description
Narrative	Long paragraphs that provide a detailed account of how a task is performed; paragraphs may not be numbered. Widely used, but the narrative format can be very confusing and difficult to follow; this format should be avoided.
Paragraph	Short, numbered paragraphs, typically with a mixture of commands and passive descriptions. Widely used, and better than the narrative format, but more wordy and generally less useful than the outline format.
Outline	Phrases, sentences, and short paragraphs organized using indentation, varied numbering, and logical grouping of information. Often used for sequential or batch operations.
Playscript	Steps grouped according to who performs them or by logical subtasks. Often used for coordinating activities between two operators or operating units, particularly if the order of steps is important.
T-Bar	Two-column format with basic actions in left column and details, notes, and so forth in right column. Can make the procedures longer, but reduces the "noise" typically contained in action steps and helps highlight special details. Particularly useful for combining a step-by-step procedure with a job safety analysis in a manner that minimizes clutter and the potential for confusion.
Multi-Column	A tabular format with multiple compartments of information. Often used for troubleshooting guides or maps that tie other documents/procedures together.
Flowchart	A graphical format that is structured with boxes, diamonds, and arrowheads and contains brief action and conditional statements. Often used for troubleshooting guides or transactional procedures, particularly to display decision steps (this format has been demonstrated to be superior for nuclear plant emergency response actions).
Checklist	Brief step descriptions providing basic actions only, typically with spaces for check marks or initials/signatures. Most often used for simple, repetitive operations, such as hazardous material unloading. Particularly useful if the steps are critical to safe operation, as in critical nonroutine operating tasks such as shutdown prior to a turnaround and restart after a turnaround, or if a record of successful operation is desired (e.g., a completed checklist).

(Adapted from Ref. 10.2.)

Address safe operating limits and consequences of deviation from safe operating limits. Safe operating limits are normally set for critical process parameters such as temperature, pressure, level, flow, or concentration based on a combination of equipment design limits and the dynamics of the process. Safe operating limits are most often specified when the system response may be so severe that the risk of continued operation is unacceptable. In this case, the procedures should include clear, simple instructions for responding to the situation. For example, a reactor may normally operate at 90° C with a safe operating limit of 120° C, based on the potential for a runaway reaction. In this case, the procedures should provide clear, simple instructions to discontinue troubleshooting and take effective measures to stop the reaction if for any reason the reactor temperature reaches 120° C. The actions might include steps to stop reactant flow, set the system to manual cooling at 100% output, add a special material to kill the reaction, and so forth.

Whenever operators investigate and take corrective action, the operator's training, the written procedures, and an understanding of the potential consequences of various courses of action are inextricably linked. Operators equipped with a clear understanding of hazards and supported by a culture that promotes a healthy sense of vulnerability are more likely to recognize dangerous situations and take the appropriate actions. Including the hazards of the process in the written procedures, and highlighting those hazards during training; reemphasizing the hazards at key steps; and clearly stating the consequences of deviating outside of safe operating limits helps operators understand risk. Performance can be further enhanced if the procedures include descriptions of the process control actions, interlocks, and safety systems related to the specific process step.

Address limiting conditions for operation. Limiting conditions for operation are normally facility- or process-specific conditions. An example of a limiting condition for operation may be an instruction to not operate a process unless the flare is operational, and to take certain steps to manage risk – including shutting the unit down – when the flare is out of service. Another common limiting condition for operation is discontinuing the unloading of flammable materials during electrical storms.

Provide clear, concise instructions. Detailed explanations of how the process works should not be embedded in step-by-step operating procedures. Extra verbiage tends to camouflage action steps, making the procedures less useful. However, explanations can be very helpful for training and troubleshooting, and should be included in some sort of written document, such as the training manual, the process technology manual, or as a special section of the procedures manual.

In some cases, a step in a procedure potentially exposes the operator to abnormally severe hazards. For example, unloading railcars containing highly toxic materials can present a significant toxic chemical exposure hazard, particularly if the unloading hose fails. In such cases, highly visible warning, caution, or notice statements should be used to highlight special hazards and direct the operator to take special precautions, for example, pressure testing the hose with nitrogen prior to opening the valve on the railcar or wearing certain personal protective equipment (PPE) during unloading operations. When used judiciously, these statements are a very effective means of communicating an increased level of hazard and helping to manage risk.

Supplement procedures with checklists. A checklist combined with a procedure is a powerful tool that has been proven effective in a variety of settings. For example, pilots review written checklists prior to takeoff even though they check the same equipment prior to each flight (and presumably could recite the checklist verbatim). However, checklists often do not provide specific instructions for what to do if a problem is discovered, rather the checklist normally refers to a written procedure or instructs the user to consult with a particular individual or organization to help resolve the problem. While checklists often augment operating procedures and help reduce the potential for human error, they rarely replace operating procedures.

Make effective use of pictures and diagrams. Pictures and diagrams are a proven means of efficiently communicating information, particularly if the task requires the operator to locate a specific gauge or indicator, switch, valve, and so forth. Identifying such devices via a picture or diagram (with an arrow or other means to clearly identify the particular device) is often more effective than via a written description of the location.

Develop written procedures to control temporary or nonroutine operations. Instructions to deviate from established procedures should be properly documented. At least three broad categories exist for temporary procedures: (1) plant trials or tests, (2) conditional operations, and (3) infrequent operations.

Tests are normally authorized as a temporary change, and appropriate procedures/instructions are included with the change authorization.

Conditional operations are instructions such as: "When the ambient temperature exceeds 90 °F, start the standby cooling water pump." This is a permanent instruction that only applies part of the time. To prevent the loss of institutional memory, these conditional instructions should be incorporated into normal operating procedures or troubleshooting guides.

Infrequent operations include operation with part of the process out of operation (e.g., because of overhaul, inspection, or recurring equipment failure), preparation for certain maintenance or nonroutine activities, and so forth. In general, the initial issuance of temporary procedures should be authorized through the *management of change* element, but the actual activities should be governed by procedures rather than a list of steps in a temporary change authorization. One important advantage of issuing a temporary operating procedure over approving a temporary change request is that the lessons learned from executing the operation in the past are captured and included in the written procedure. This helps manage the risk and improve the efficiency of the operation the next time the same operation is performed.

Group the tasks in a logical manner. Procedures can be divided in a variety of ways. They can be divided by equipment, mode of operation, type of operation, and so forth. No single best way to group tasks exists. For example, it may be best to divide the procedures governing normal operation by each major unit operation, divide preparation for startup by type of equipment (valve alignment, instruments and controls, filter and strainer elements, etc.), and divide the unloading checklists by type of material. Using a logical and consistent approach will help minimize training time and improve the utility of the procedures.

Interlink related procedures. To the degree possible, provide clear linkage between procedures, particularly if the procedures are closely related or if procedures could logically be included in more than one area or procedure manual. For example, procedures for addressing small spills and releases could logically be included in the spill control and countermeasures plan, in a procedure that provides safety information that is applicable to the entire the unit, or in specific operating procedures. Regardless of which option is selected, ensure that pointers are included at other logical locations. This will facilitate use of this type of information without having to include the same information in multiple locations (which can become a procedure maintenance issue). However, excessive cross-referencing makes a procedure extremely difficult to follow and may cause workers to try to work from memory. While it would be appropriate to interlink the overall unit startup procedure to a separate, specific procedure for starting the refrigerant system, it would not be appropriate to send a worker from one procedure to another for a single step, such as "Verify that the refrigerant level is between 70% and 80%."

Validate procedures and verify that actual practice conforms to intended practice. When a procedure is being written or updated, an important task is to ensure that the draft procedure accurately reflects intended practice. To do this, one or more operators should validate that the procedure reflects current practice, and that the steps in the procedure can be safely and reliably performed. This step should also consider who might use the procedure in rare or unusual cases, and evaluate if sufficient detail is included in the procedures. (Recall that procedures should be written so that the newest or least qualified operator can safely and effectively perform the task.) This step often includes field verification by both a subject matter expert and an operator who is qualified to do the job but who is not a subject matter expert (e.g., a relief operator or one of the newest operators in the unit). Procedure validation should be documented.

However, procedures should reflect the proper way to complete a task, not merely the way it has always been done. If an important step is being skipped in practice, for example, verification of a critical process parameter, that step may be left out of the draft procedure or deleted when the procedure is next updated. Therefore, competent technical personnel should independently review new procedures and proposed changes to existing procedures to help prevent the formal adoption of unsafe shortcuts. Validation requires that procedures be compared to (1) existing practice and (2) intended methods.

Once a procedure is validated, it is equally important that everyone follow it. Verification of work practices can be part of a well-structured refresher training program. All of the operators may be asked to review procedures that are particularly critical to safe operation, and field observations may help ensure compliance with procedural requirements. If a strong process safety culture exists at a facility, short cuts or any other discrepancies between the written procedures and normal practice will surface during refresher training, and in some cases operators will describe additional actions that they have found useful. These variances should be reviewed and are often either (1) added to the procedure or (2) specifically prohibited in the procedure (if the shortcut is determined to be unsafe). In either case, changes to procedures should be managed via the *management of change* element and supported with appropriate training.

10.2.4 Use the Procedures to Improve Human Performance

Procedures that are not followed are of little value. Tolerance or endorsement of working solely from memory or using alternatives to approved procedures can lead to highly unpredictable, and sometimes unsafe, operation. To promote their use, procedures should be available to the user at the time and location they are needed.

Use the procedures when training. Using procedures to guide hands-on training helps to ensure that trainees become familiar with the procedure as well as how to correctly perform the task, making it more likely that they will refer to the procedure in the future. In general, people will tend to not spend time looking for a written procedure or other reference unless they believe it exists. Using procedures as a training aid will (1) help familiarize trainees with the content and format of procedures, (2) reinforce the value of the procedures, and (3) allow trainers or other subject matter experts to identify errors or omissions in procedures. However, note that procedures often contain a list of steps that do not explain how the equipment/system works. These sorts of procedures supplement, but do not replace, training manuals. (See Section 14.2.3 for additional information on providing training.)

Hold the organization accountable for consistently following procedures. Too often, operators are held accountable for results based on outputs that they cannot always control, such as output, yield, or product quality, rather than on inputs that they directly control, such as conformance to procedures. In some cases, doing whatever it takes to get the product out the door is even rewarded by management. If adding a little extra catalyst or skipping the final wash step to meet a shipment date is ignored (or viewed as a good thing), operators may be tempted to add a little more catalyst or skip other steps the next time production is running behind schedule. This can quickly lead to unsafe operation. Management should be accountable for establishing a culture in which workers follow procedures. In the absence of a culture of conformance to procedures and standards, it will be difficult, and even hypocritical, for management to hold operators accountable for following procedures.

Ensure that procedures are available. Procedures should be available to operators at all times. One common practice is to store procedures on the facility's computer network and print procedures on demand. While this greatly simplifies document control, operators may not be able to access procedures in the event of a power failure, network failure, and so forth. In almost all cases, a current paper copy of the operating procedure manual should be kept in the control room. This is particularly important because utility failures often cause, or are coincident with, other facility mishaps. Similarly, controlled paper copies of other critical documents should be kept in key locations so that they are always available and accessible. For example, copies of the emergency response plan and material safety data sheets may be kept in the emergency response team's vehicle or assembly area.

10.2.5 Ensure that Procedures Are Maintained

Historically, some organizations treated written procedures as a reference manual. That is, the procedures were not expected to be 100% current or accurate, but they were useful in describing how the process operated and were sometimes useful in troubleshooting. This model has proven inadequate, both for ensuring process safety and for other aspects of effective operation. The accuracy, and hence effectiveness, of static procedures in a dynamic operating environment will decay rapidly with time. Thus, ensuring that procedures are maintained and enforced is vital. If operating procedures are not being following consistently, deviations must be investigated and corrected.

Manage changes. Practices for managing changes to procedures are fully described in Chapter 15. The need to update the content of the training program to coincide with changes to procedures, as well as the need to keep all technical/support personnel, including trainers, informed of changes to procedures, cannot be overemphasized.

Most facilities maintain change logs at the start or the end of each procedure. This helps highlight the overall intent of changes when the revised procedures are first issued and helps remind operators of recent changes, particularly for procedures that are used infrequently. A change log can also help demonstrate that actions generated by other RBPS elements, such as the *risk*, *training*, *incidents*, and *audits* elements, have been addressed. In addition, some facilities draw attention to changes made in the most recent revision by using a different font and/or vertical bars in the margin.

Many management systems that are commonly in place in the process industries require document control for operating procedures and other important documents. For example, many facilities have voluntarily adopted the ISO 9000/ISO 14000 series of standards and/or the American Chemistry Council's Responsible Care Management System® and RC 14001 series of standards. To maximize efficiency and help prevent confusion, a single management system should address all document control requirements that apply to a facility's operating procedures.

Correct errors and omissions in a timely manner. Procedures that are well written in a user-friendly format and at an appropriate level of detail are likely to be used. As operators spot errors or omissions, it is important that these be corrected in a timely manner for two reasons. First, failure to correct an error or omission could lead to a human error. Second, long delays in correcting errors sends a message that "close enough" is acceptable. After all, if the procedure is widely known to be inaccurate and not quickly fixed, why should an operator believe that compliance with the procedure is critical?

Periodically review all operating procedures. Even the most effective *management of change* program, coupled with a culture that embraces consistent use and timely correction of procedures, will not always ensure that procedures are maintained current and accurate. For example, if a number of operating procedures are affected by a change, the team charged with implementing the change may fail to update some of the affected procedures. Hence, even with an effective *management of change* program, procedures can become inconsistent. Procedures should be periodically reviewed to ensure that they are current and accurate. Since operators should be very familiar with the operating procedures, they can often provide helpful insight on how, and how often, procedures should be reviewed (see the *workforce involvement* element). Decisions regarding procedure review methods or interval may also be influenced by regulatory requirements.

The appropriate time interval between reviews varies; but it typically ranges from 1 to 3 years, assuming that in the mean time, the procedures are maintained under *management of change* requirements. The risk associated with executing operating procedures varies widely throughout most facilities. All too often, the review interval, and level of effort applied to the review, is not based on risk. The RBPS approach encourages facilities to consider risk when establishing review cycles and specifying the level of effort that should be applied to the review (e.g., should the review team be limited to one or two operators or should the review team also include a senior engineer or representatives from all shifts?).

Facility management should consider using event-based review periods rather than calendar-based ones. For example, reviewing the unit shutdown procedures immediately prior to a planned turnaround makes more sense than reviewing them on a fixed annual or biannual schedule. Procedure reviews and updates may also follow significant events. For example, lessons learned from a shutdown and subsequent startup should be recorded, and appropriate adjustments should be made to procedures promptly rather than at the next review cycle. Otherwise, procedure errors or issues that prompted concerns may be forgotten and not be addressed. Finally, avoid making all procedure reviews due in the same quarter or month – the resulting workload tends to dilute the effectiveness of review efforts.

10.3   POSSIBLE WORK ACTIVITIES

The RBPS approach suggests that the degree of rigor designed into each work activity should be tailored to risk, tempered by resource considerations, and tuned to the facility's culture. Thus, the degree of rigor that should be applied to a particular work activity will vary for each facility, and likely will vary between units or process areas at a facility. Therefore, to develop a risk-based process safety management system, readers should perform the following steps:

1. Assess the risks at the facility, investigate the balance between the resource load for RBPS activities and available resources, and examine the facility's culture. This process is described in more detail in Section 2.2.
2. Estimate the potential benefits that may be achieved by addressing each of the key principles for this RBPS element. These principles are listed in Section 10.2.
3. Based on the results from steps 1 and 2, decide which essential features described in Sections 10.2.1 through 10.2.5 are necessary to properly manage risk.
4. For each essential feature that will be implemented, determine how it will be implemented and select the corresponding work activities described in this section. Note that this list of work activities cannot be comprehensive for all industries; readers will likely need to add work activities or modify some of the work activities listed in this section.
5. For each work activity that will be implemented, determine the level of rigor that will be required. Each work activity in this section is followed by two to five implementation options that describe an

increasing degree of rigor. Note that work activities listed in this section are labeled with a number; implementation options are labeled with a letter.

> *Note: Regulatory requirements may specify that process safety management systems include certain features or work activities, or that a minimum level of detail be designed into specific work activities. Thus, the design and implementation of process safety management systems should be based on regulatory requirements as well as the guidance provided in this book.*

10.3.1 Maintain a Dependable Practice

Establish Management Controls

1. Develop a written policy describing the management system for the *procedures* element that describes the process for creating, updating, and maintaining operating procedures.
   a. A general written policy exists; written operating procedures are developed by the production organization based on perceived need (and in some cases, in response to an incident).
   b. A written policy specifies requirements for operating procedures, but it does not specifically address all aspects of the management system, such as roles and responsibilities, minimum technical content, under what circumstances written procedures are required.
   c. A written policy addresses all aspects of the management system for the *procedures* element.
2. Address specific roles and responsibilities in the written policy describing the management system for the *procedures* element.
   a. A general written policy states that the production organization is responsible for all aspects of the *procedures* element.
   b. A general written policy states that responsibility for creating and updating procedures belongs to the production organization, and this responsibility is assigned to a specific individual, for example, the department training coordinator or day shift supervisor.
   c. A formal written program document addresses roles and responsibilities for the entire range of activities that comprise the *procedures* element and all of the organizations that should be involved (e.g., operations, training, engineering, safety).

Control Procedure Format and Content

3. Include in the written policy or description of the management system procedure governing the *procedures* element a list or description of acceptable formats/structure for all operating procedures.

   a. A general written policy exists; however, procedures in different areas follow different formats.

   b. A written policy includes sample formats for operating procedures.

   c. A written policy specifies acceptable formats for various types of operating procedures.

4. Provide guidance on content, including what should not be included in operating procedures. Also include guidance on what information should be included in related documents, such as training or process technology manuals.

   a. A written policy provides general guidance on content.

   b. A written policy specifies what content should be included in operating procedures.

   c. In addition to item (b), the written policy includes sample content for different types of operating procedures with detailed examples.

Control Documents

5. Provide a means to ensure that previous versions of procedures are not available or used. Conduct a tool box audit to ensure that personal copies of outdated, marked-up procedures are not used by workers in the field.

   a. Operators keep a current version of the procedures in a binder in the control room and personnel are trained to use these procedures (rather than copies they may have previously copied/printed).

   b. A formal document control program exists. Updated procedures are provided with a cover page that requires that the receiver confirm that they have updated their manual and discarded procedures that have been superseded.

   c. A formal document control program exists. Controlled copies are clearly marked and maintained in the control room for reference; procedures are normally printed on demand and these copies are clearly labeled as uncontrolled with a relatively short expiration date or similar means to ensure that they are not inadvertently used at some future time.

10.3.2 Identify What Operating Procedures Are Needed

Conduct Task Analysis

6. Identify tasks that are performed by each operator or logical group of operators.

     a.   Tasks, and needs for written procedures, are identified based on the production organization's perceived need and, in some cases, in response to an incident.

     b.   Tasks, and needs for written procedures, are based primarily on regulatory requirements.

     c.   Tasks, and needs for written procedures, are based on a task analysis that identified routine tasks performed by operators.

     d.   A thorough task analysis has been performed that addresses all modes of operation. Tasks are identified and procedures are developed as appropriate.

7.   Validate the task list.

     a.   Tasks are validated by management based on perceived needs, including process safety incidents and other incidents that led to loss events, such as equipment failure resulting in a business interruption.

     b.   Tasks lists are developed with minimal input from operators but are validated by the area supervisor.

     c.   Tasks lists are validated jointly by operators and area supervisors.

     d.   A representative number of operators, subject matter experts, and supervisors help validate task lists and decide if procedures are required for each task.

Determine What Procedures Are Needed and their Appropriate Level of Detail

8.   Review the task list to determine which tasks require a written procedure.

     a.   Procedures are developed based on perceived needs, including process safety incidents and other incidents that led to loss events.

     b.   The task list is reviewed and procedures are developed for complex tasks.

     c.   Procedures are developed for all identified tasks, regardless of complexity of the task or risk associated with performing the task incorrectly.

     d.   The task list is reviewed in terms of risk. For example, reviewers ask: "What might go wrong if this task is performed incorrectly?" and "How likely is it to occur and will a written procedure affect the likelihood?" Procedure needs are identified and prioritized based on in part on answers to these risk questions.

9.   Ensure that the *procedures* and *training* elements are coordinated. Simple tasks that are not addressed in procedures should either be (1) covered in training or (2) incorporated into the job qualification

process (and therefore addressed prior to worker selection for a particular position).

    a.   Procedure writers consider the level of training, and necessary enhancements to the training program are suggested to the training department.

    b.   Supervisors and training coordinators consider the level of training for all tasks on the list, regardless of whether a written procedure will be developed, and feedback is provided to the training department.

    c.   The task list is reviewed in terms of risk by a team consisting of operators, supervisors, the training coordinator, process safety personnel, and other key personnel. This team decides whether including a task in the training curriculum, developing a written procedure for it, or both, is the best risk management option.

10. Review the operator training program to determine the necessary level of detail for operating procedures.

    a.   The need to coordinate procedures and training is left up to persons responsible for developing the respective materials.

    b.   Training modules are developed for all tasks, regardless of the complexity of the associated procedures.

    c.   The group assigned to review the task list and determine what procedures will be developed includes representatives from the training department.

    d.   The task list is reviewed in terms of risk, and the review team considers how training can best play a role, along with procedures, in managing risk.

Address All Operating Modes

11. Include all operating modes in the task list, for example, temporary shutdown, shutdown for annual maintenance, emergency shutdown, startup after each type of shutdown, initial startup, temporary operation, and normal operation.

    a.   Procedures address normal operation, startup, and shutdown. For all other modes of operation, the operator is directed to consult with supervision.

    b.   Written procedures generally address all anticipated modes of operation, but the level of detail is based on frequency of performance rather than risk.

    c.   Procedures address all modes of operation, including tasks such as preparing equipment for maintenance. If a new task is required, a written procedure is developed and reviewed by the appropriate personnel prior to the start of the operation.

12. Instill a practice to develop contingency procedures for anticipated emergency operations.

 a. A single shutdown procedure exists; it is intended to be used for normal and emergency shutdowns. Responses to other situations are left to the discretion of supervision.

 b. Emergency shutdown procedures exist; the procedures clearly state when an emergency shutdown should be initiated and how to execute an emergency shutdown.

 c. Procedures exist for the full range of anticipated emergency conditions (e.g., utility interruptions, natural disasters, intentional attacks); the procedures clearly state when they should be implemented; and a clear link exists between troubleshooting procedures, limiting conditions for operation, safe operating limits, and emergency shutdown procedures.

13. Instill a practice to develop procedures for hazardous one-time operations, such as decommissioning.

 a. One-use procedures (e.g., unit decommissioning) are developed only if the project manager determines that specific phases of the operation pose unusual hazards.

 b. One-use procedures (e.g., unit decommissioning) are normally required if the activity requires capital funding.

 c. The scope of the facility's *management of change* element includes one-time events, such as decommissioning and demolition, and written procedures are required as part of the change authorization.

10.3.3 Develop Procedures

Use an Appropriate Format

14. Format procedures in a consistent manner and select the best type of procedure for each task.

 a. All procedures (e.g., operating, maintenance, laboratory, safe work, and other procedures) follow a single template using the paragraph format.

 b. Operating procedures are structured in a standard way throughout the facility.

 c. Operating procedures are structured in a standard way, but authors are allowed to incorporate other formats, such as tables, checklists, or flow charts, if they add value to the procedure.

Ensure that the Procedures Describe the Expected System Response, How to Determine if a Step or Task Has Been Done Properly, and Possible Consequences Associated with Errors or Omissions

15. Ensure that procedures clearly state what to do, and for critical steps or tasks, how to determine if the step or task was completed correctly.

    a.   Procedures include general caution and warning statements, but do not normally include steps to verify critical/intended conditions.

    b.   Procedures typically include a step to verify that each intended condition exists, regardless of the importance.

    c.   Consideration is given to the criticality of properly completing various steps within a procedure, and verification steps are specified as appropriate. Depending on the criticality, some tasks require operator signoff, second-person verification, or supervisor signoff.

16. Describe the consequences associated with errors or omissions in sufficient detail to (1) emphasize the importance of critical steps and (2) make operators aware of what might occur so they can quickly recognize and react to errors or omissions.

    a.   Procedures include caution or warning statements.

    b.   Procedures include consequence statements from hazard analysis documents; however, only in rare cases are the error and the unwanted outcome clearly linked.

    c.   The procedures, troubleshooting guides, and related documents clearly link errors/omissions to specific unwanted outcomes.

Address Safe Operating Limits and Consequences of Deviation from Safe Operating Limits

17. Establish safe operating limits for each process parameter where deviation from the limit is credible and could lead to an unsafe condition. Also, for each safe operating limit, state the potential consequence of exceeding the limit and the steps to avoid deviation or return the process to a safe condition if an excursion outside of the safe operating limits does occur.

    a.   A fairly narrow range of limits has been established; operation within these narrow limits ensures safe operation and helps ensure that yield and product quality targets are met. However, the operating procedures do not directly address safe operating limits, consequences of deviation, and steps to take to prevent or mitigate the consequences of exceeding safe operating limits.

    b.   Safe operating limits can be found in (or sometimes inferred from) the operating procedures, but they are not always addressed in a clear manner.

    c.   Safe operating limits, along with (1) the potential consequences associated with exceeding the limits and (2) the steps operators should take to prevent or mitigate the consequences of exceeding the limits, are clearly described in the operating procedures. These limits are maintained in a special section of the procedures or are otherwise very easy to locate because they are consistently presented in all procedures.

Address Limiting Conditions for Operation

18. Clearly state limiting conditions for each mode of operation, for example, whether or not to stop processing if certain safety systems are not in service.
    a. The limiting conditions for operation are only listed as conditions for startup.
    b. Limiting conditions for operation (both for startup and continued operation) are directly addressed in operating procedures.

Provide Clear, Concise Instructions

19. Refrain from embedding operational steps within long descriptive narratives.
    a. The procedures and training manual are closely integrated such that each operating step tells the operator not only what to do, but specifically how to perform each step. For example, instead of stating "Open valve HV-113", the procedure will specifically describe how to open a ¼-turn ball valve and why it is important to open valve HV-113 at this point in the sequence.
    b. The procedures provide sufficient background and safety-related information; however, action steps are clearly identified.
    c. In addition to item (b), a procedure format is used that allows the operator to easily follow each action step in the procedure while quickly picking out critical safety information, which is presented using warning, caution, and notice statements.

Supplement Procedures with Checklists

20. Include a checklist in the procedure whenever the sequence of operations is important, or when certain steps must be complete prior to moving to the next phase of operation.
    a. Many checklists are developed, but they are normally not supplemented by procedures.
    b. Procedures and checklists are developed in parallel systems, and they are cross checked as part of the periodic procedure review.
    c. Procedures and checklists are co-developed in a manner that the user can access the checklist, the procedure, or both, as needed.

Make Effective Use of Pictures and Diagrams

21. Include pictures and diagrams where appropriate.

a.   Procedures are generally in narrative format, and seldom contain drawings, figures, or pictures.

b.   Procedures are mainly composed of videos and digital pictures that show an operator performing a particular task; very little written guidance is provided.

c.   Figures are used to help improve understanding, but source documents are seldom modified for clarity. For example, scans of P&IDs are used to depict control logic, rather than simplified control diagrams.

d.   Figures and pictures are included when they help improve understanding; figures show only the level of detail needed to clarify the written instructions.

Develop Written Procedures to Control Temporary or Nonroutine Operations

22. Instill a practice to develop temporary procedures, particularly for low frequency recurring activities, in other words, procedures that are not being executed often but are certain to be reused in the future.

a.   The production organization determines if procedures are needed for temporary operations; however, a bias exists against spending time to develop procedures for one-time or infrequent operations.

b.   Nonroutine modes of operation, such as special actions to prepare a vessel for internal inspection, are authorized via the *management of change* element. The change authorization requires at least a minimal written description of the intended activity for each operating mode.

c.   Procedures are generally developed for nonroutine operations, and any changes or improvements to the procedure are noted. These procedures are updated subsequent to the activity and held on file for the next time the activity is conducted.

Group the Tasks in a Logical Manner

23. Develop a procedure numbering or index system that is logical to the end user.

a.   A numbering system designates the content, but little else (e.g., different prefixes are assigned to operating, maintenance, and laboratory procedures, but the numbers that follow the prefix are assigned sequentially).

b.   A master index exists and the procedure number identifies the type of procedure, process area, and possibly other information on a consistent basis.

  c. A comprehensive numbering system has been established that helps operators quickly locate pertinent procedures and related documents, such as checklists; the index includes electronic links to the documents.

Interlink Related Procedures

 24. When procedures are interrelated, provide clear, distinct, but not excessive references.

  a. The front matter of each procedure contains a long list of referenced or related procedures. For example, all procedures reference the facility's master lockout/tagout procedure in case a need arises to perform a lockout/tagout when working on the equipment.

  b. Cross references are provided at appropriate points in the procedures.

  c. Electronic links are provided at appropriate points in the procedures, and the links are maintained.

Validate Procedures and Ensure that Actual Practice Conforms to Intended Practice

 25. Verify that new procedures describing existing operations conform to existing and intended practice.

  a. The responsibility for ensuring that actual practice conforms to written procedures is included in the job description of the supervisor, and the supervisor monitors performance.

  b. All procedures must be verified and approved by an operator or supervisor prior to issue.

  c. All procedures are verified with multiple operators and supervisors from different shifts or work groups prior to issue. This group is intentionally selected to span the range of experience and expertise in the organization in order to ensure that the procedure is consistent with appropriate current practice.

 26. Validate new procedures to ensure that they reflect intended practice.

  a. Procedures are reviewed by operators or supervisors prior to issue, but no technical review of new procedures is performed.

  b. The responsibility for validating new procedures is included in the job description of the process engineer, but no formal tracking system exists.

  c. A formal system is in place to ensure that procedures are validated prior to issue and revalidated periodically.

10.3.4 Use Procedures to Improve Human Performance

Use the Procedures when Training

27. Use procedures as a training aid.
    a. Procedures are the primary written resource for new operator training. The procedures describe the steps needed to perform the task, but do not explain how the overall system operates.
    b. Procedures are provided to trainees, but the trainer primarily uses other training materials and leaves reading/reviewing the procedures to trainees (presumably at some point after the training session).
    c. In addition to review of (or familiarization with) procedures, the training process includes use of quality training materials that supplement the information in the procedures.

Hold the Organization Accountable for Consistently Following Procedures

28. Monitor incident reports for evidence of deviation from established procedures.
    a. Operators are primarily accountable for results, such as yield, quality parameters, and so forth. Operators are held accountable for following procedures if failure to follow procedures leads to a loss event (e.g., safety incident, yield loss, equipment damage).
    b. Operators must log process conditions several times each shift, and these logs are periodically reviewed by supervisors. Failure to maintain intended process conditions is addressed, particularly if the deviation resulted in a loss event, such as off-spec product.
    c. Operators are primarily held accountable for following procedures, whether they are related to maintaining certain process conditions or conforming to other written instructions.
    d. In addition to item (c), an automated system continuously scans the process for unintended conditions and provides a summary report of any discrepancies noted during the shift. Operators are expected to briefly comment on this report, which is reviewed each shift.

29. Reward suggestions for improvement, but do not reward actions that fall outside of procedural limits, regardless of the outcome.
    a. Productivity improvements that are discovered via operator error, such as reduction in the batch cycle time by skipping a step, are acknowledged but not encouraged.
    b. Deviations from procedures are addressed in a consistent manner, regardless of whether the deviations lead to a loss event or an improvement in the production process.

Ensure that Procedures Are Available

30. Make procedures available to operators at all times.
    a. All procedures are maintained in a file drawer in the supervisor's office (or similar location).
    b. Procedures are available in the control room, unloading shack, or other locations where they are needed.
    c. Procedures are available on the network server and can be accessed or printed as long as the computer network is operational.
    d. At least one paper copy of each operating procedure is available to operators in the control room or other location; procedures are also available throughout the facility via the computer network.

10.3.5 Ensure that Procedures Are Maintained

Manage Changes

See the list of work activities related to updating procedures for the *management of change* element (Section 15.3.5).

31. Periodically audit the link between the operating procedures and the training program to determine if the content of the training program (including trainers' knowledge) is up to date.
    a. The training program is based mainly on procedures and on the job training by other operators; in other words, no training program "content" exists outside of written procedures and existing practices.
    b. Training materials are periodically reviewed, and corrections are made as appropriate.
    c. The *management of change* element applies to changes to procedures and helps ensure that training manuals and other written training materials are updated and trainers are informed of changes.
    d. The training materials/program and operating procedures are linked such that changes to procedures would not likely occur without updating the associated training materials.

Use the Procedures; Correct Errors and Omissions in a Timely Manner

32. Instill a practice of identifying errors in procedures and correcting errors in a timely manner.
    a. Errors or omissions in procedures are only noted and corrected in periodic procedure revision efforts.
    b. A system has been established for operators to note errors in procedures on an ongoing basis; however, few errors are ever noted. When found, corrections are issued on errata sheets or

temporary orders. Corrections are periodically incorporated into the procedures.

c. Operators routinely use procedures and are expected to highlight errors and omissions; a high fraction of changes result from routine use rather than periodic reviews or changes. Proposed fixes are reviewed and incorporated into existing procedures on an ongoing basis.

33. Provide a method to quickly make clarifications, correct typographical or grammatical errors, or make other adjustments that improve the procedures.

a. All changes to procedures must be reviewed via the *management of change* element. Therefore, minor corrections are normally made when the procedure is due for annual review/update.

b. The training coordinator is authorized to make corrections to procedures without *management of change* review or further authorization as long as it does not add or remove an action step.

c. In addition to item (b), the production manager must concur with the change, and the production manager decides if operators need to be notified.

Periodically Review All Operating Procedures

34. Periodically revalidate procedures to ensure that they reflect intended practice.

a. Operators are instructed to use the procedures and report any errors or omissions.

b. Procedures are reviewed periodically by operators or supervisors, but no routine technical review of procedures occurs.

c. The responsibility for periodically reviewing procedures to ensure that they are accurate is included in the job description of the process engineer or supervisor, but no follow-up system exists to confirm that this review occurs.

d. A formal system is in place to ensure that procedures are periodically reviewed and revalidated for technical accuracy.

35. Periodically verify that actual practice conforms to the steps listed in the operating procedures.

a. The responsibility for ensuring that actual practice conforms to written procedures is included in the job description of the supervisor. The supervisor monitors performance, but no system exists to periodically verify performance compared to procedures.

b. A formal system exists to ensure that practices are verified (versus simply reviewing procedures); the verifications are staggered throughout the year.

36. Establish the interval for procedure review based on risk or other objective criteria, including input from operators and other affected personnel.

   a. All procedures are reviewed at the same frequency.

   b. All procedures are reviewed at the same frequency, but the intensity of the review is adjusted based on perceived risk.

   c. The review frequency is based on perceived risk and other objective criteria, such as review timing, and the intensity of the review is adjusted based on perceived risk.

37. If discrepancies between actual practice and the procedures are discovered during procedure validation or verification, determine the intended operating practice and either (1) update the procedure and train/inform affected employees (see the *management of change* element for additional information on authorizing changes to procedures) or (2) reinforce the requirement to follow procedures.

   a. When a discrepancy between a procedure and practice is discovered, the production manager decides whether to update the procedure or reinforce the intended practice; changes to procedures are not within the scope of the *management of change* element.

   b. Discrepancies discovered during validation or verification are addressed, but no real effort is made to communicate changes to operators; they are expected to note the changes based on required reading, e-mail notification, or their next usage of the procedure.

   c. Discrepancies discovered during validation or verification are addressed via the *management of change* system, and changes to procedures are reviewed during start of shift meetings or via some other structured system that involves direct contact with each potentially affected operator.

10.4   EXAMPLES OF WAYS TO IMPROVE EFFECTIVENESS

This section provides specific examples of industry tested methods for improving the effectiveness of work activities related to the *procedures* element. The examples are sorted by the key principles that were first introduced in Section 10.2. The examples fall into two categories:

1. Methods for improving the performance of activities that support this element.
2. Methods for improving the efficiency of activities that support this element by maintaining the necessary level of performance using fewer resources.

These examples were obtained from the results of industry practice surveys, workshops, and CCPS member-company input. Readers desiring to improve their management systems and work activities related to this element should examine these ideas, evaluate current management system and work activity performance and efficiency, and then select and implement enhancements using the risk-based principles described in Section 2.1.

10.4.1 Maintain a Dependable Practice

Remember the past. Relearning lessons can be quite expensive. If the facility or company maintains a detailed process technology manual (as described in Chapter 5), specific cross references between the operating procedures and the technology manual can help workers in the operations group answer the question, "Why are we doing it this way?" These cross references may help a facility recognize hidden hazards associated with a change because the individual who is assigned to update those procedure(s) affected by the change is likely to review the referenced sections in the technology manual.

10.4.2 Identify What Operating Procedures Are Needed

Closely examine training and initial operator qualification requirements when identifying the need for operating procedures. Many facilities in the process industries pay a premium wage to attract highly trained and qualified operators or choose to employ only very experienced operators in a specific unit. One benefit this provides is that, while procedures must still tell a highly qualified new operator *what* to do, the facility may decide that less detail is required concerning *how* to perform the task. Too often, procedure writers fail to consider initial qualification requirements for the process unit and devote too many words to how to perform a task rather than simply stating each step with special emphasis on hazards, expected system response, and troubleshooting.

10.4.3 Develop Procedures

Issue generic operating procedures that facilities can customize. Some companies operate identical (or nearly identical) processes at multiple facilities. This provides an opportunity to develop a common set of operating procedures or templates that can be customized as necessary by each facility. In addition to increasing efficiency, this can be a very effective way to share best practices between facilities.

Divide procedures in a logical manner. Attempting to address all aspects of the operation in a single procedure is likely to lead to a very long, hard-to-use document. Dividing procedures into logical, manageable chunks can improve efficiency. For example, batch chemical plants tend to have procedures that describe how to operate a piece of equipment (e.g., standard operating procedures for a reactor) and a second set of formula procedures that govern manufacture of each product (e.g., product-specific batch tickets). Just as one would never include the owner's manual for an oven in each recipe, the operating procedure for a reactor should not be included in the procedure for each material that is produced in the reactor. However, the operator should be very familiar with both types of procedures, and the procedure writer should carefully determine where critical instructions and caution/warning statements should be located. For example, the operator uses a checklist in the batch ticket when producing each lot of material, and the equipment-specific procedure is used only as a reference document when needed, critical instructions and caution/warning statements should probably be included in the batch ticket.

Link instructions that are repetitive. Most companies include certain statements in procedures that are common to a large number of procedures. For example, the PPE requirements for certain tasks involving a particularly toxic material may be different from the standard PPE worn at the facility, but the same for all operations that involve this specific chemical. Including these instructions in each procedure is the right thing to do, but finding and fixing all of these instructions if the standards ever change can be quite difficult. By electronically linking this sort of information in the procedures, the change can be made in one location and it will be replicated in all of the procedures the next time the procedure is used. (Note that this approach requires that all affected hard copy procedure manuals be reprinted after the change is made.)

Make liberal use of diagrams and figures. If a trainer needs to sketch a picture to explain to someone how to perform a procedure, the procedure should probably include pictures, diagrams, or other graphics to aid understanding.

Use tables to separate different types of information. Developing procedures in a tabular format can help structure information to make it easier to find. For example, a procedure could be presented in a six-column format. The first entry in each row is simply the step number. It is followed by the action step, expected system response, what to do if the system does not respond in the expected manner (troubleshooting), warnings and cautions, and a location to record data and/or sign off that the step was complete. The key to this approach is to use a common set of column headers in all operating procedures, or in broad classes of procedures, so operators can quickly locate key information.

Be consistent. Procedure format, symbols, and terminology should be as consistent as possible. For example, a standard set of acronyms will help reduce confusion. The decision to use a notice, caution, or warning statement should be based on objective criteria, not based on the author's general level of concern. Even minor formatting/structural differences can at best annoy the reader and at worst cause confusion (e.g., the reader might wonder if a font change was intended to add emphasis to a section, and if so, why).

Validate procedures for nonroutine operations against actual practice. During turnarounds or other times of very high activity that involve multiple (and intense) nonroutine operations, operators should be encouraged to note instances in which workers had to execute steps in a specific manner that is not fully documented in the procedure. After the activity is over, review these potential changes and revise the procedures to permanently capture what was learned. In some cases, subtle changes do not affect normal operation but can affect startup, shutdown, preparation for maintenance, or other nonroutine activities. For example, a facility may have procedures to prepare two different equipment items for maintenance. Historically, these tasks have been unrelated. However, because of a recent change, task 4 for the first procedure must precede task 7 for the second procedure. This subtle nuance may not be captured during the *management of change* review, but will likely be caught when the equipment is actually prepared for maintenance. If this discovery is not documented at that time it is noted, it is unlikely that it will be identified in a future procedure review, particularly if the two procedures are reviewed independently. Thus, the lesson will have to be relearned.

10.4.4 Use Procedures to Improve Human Performance

Periodically audit conformance to procedures. In addition to (or in lieu of) reading over procedures to ensure that they are current and accurate, take them to the field and validate them via direct observation of one or more operators performing the tasks covered by the procedure.

Eliminate shift-to-shift differences. At some facilities, each shift has slightly different operating techniques. Periodic procedure reviews by operators assigned to each shift can be used to identify these differences, stop unsafe practices, determine the best techniques, and drive continuous improvement.

Use the multi-column format to integrate procedures with the training manual. In many cases, the operating procedures manual and training manuals are separate, self-contained documents. Obviously, keeping the two manuals synchronized as changes are made to the process and to operating methods takes effort. The benefit of maintaining separate manuals is that steps for the operating procedures are not buried in the descriptive text of the training manual. One creative approach to solving this problem is to separate the various types of information related to each step, and store this information in a database application. Types of information that are kept in this sort of database may include:

1. Step number or index.
2. Action step.
3. Physical location of the equipment, including, for example, pictures with arrows showing specifically which valve to open.
4. Expected system response.
5. Steps to take if the process does not respond as expected (troubleshooting).
6. Safe operating limits and limiting conditions for operation related to this step.
7. Warnings, cautions, or notes.
8. Other pertinent remarks.
9. Theory of operation (Why is this step important and what is the system doing?).
10. Equipment description (How does the equipment operate?).

Once this database is created, various documents can be created by selecting different data elements. For example, a simple checklist that can be taken to the field may be created by printing columns 1, 2, and 7. More detailed procedures can be created by printing columns 1, 2, 4, 6, and 7. A troubleshooting guide can be created by printing columns 1, 2, 4, and 5. The training manual may consist of all ten columns. Integrating operating procedures with training manuals in this manner should help reduce the effort required to maintain two manuals as well as the effort required to update procedures, checklists, troubleshooting guides, and related documents for each change.

10.4.5 Ensure that Procedures Are Maintained

Consider risk when establishing the interval for procedure review. Some facilities review all procedures at a set interval, often annually. As discussed in Section 10.2.5, initiating reviews based on events, such as an upcoming turnaround, rather than on calendar time might be more appropriate. In addition, the hazards of the process, potential severity of consequence of failing to follow the procedure, and the inherent risk of the operation should be considered when establishing review intervals. Operators will have unique insight on this issue, so their opinions should be solicited. The review frequency should focus more effort on ensuring that procedures governing higher risk units or processes are kept up to date.

Use an electronic document management system to store/control procedures. Many facilities use an electronic document management system (EDMS) to control drawings and other engineering-related data, but use standard computer operating system features to control procedures, for example, storing the procedures in certain read-only areas on the computer network. Advantages of an EDMS include (1) much better features for tracking the history of changes to the procedure and archiving old procedures, (2) the ability to determine if a procedure is checked out for review or update, and hence prevent two different persons from working on changes in an uncoordinated manner, and (3) the ability to automatically track and report on several of the metrics identified in Section 10.5.

Use an electronic calendaring system to alert persons to upcoming review dates. The need to perform predictable tasks, such as procedure review, should be delegated to the organization with sufficient lead time to get the job done. Some facilities establish preventive maintenance work orders to alert the operations group when procedures are due for recertification.

Simultaneously review procedures for multiple facilities operating the same process. Many companies operate the identical (or nearly identical) process at multiple facilities. In theory, the operating procedures should be largely identical, with obvious exceptions based on equipment tag numbers, facility-specific data, and so forth. Reviewing corresponding procedures for multiple facilities in a single effort can help reduce the overall staff time required for this activity, and more important, it can help identify differences in operating methods, which should help identify and eliminate undesirable or unsafe practices and promote sharing of best operating practices across the entire company.

Assign the primary responsibility for procedure review to operators. At some facilities, engineers or other technical personnel are assigned sole responsibility for periodically reviewing the operating procedures to ensure that they are current and accurate. This requires that the reviewer spend time (1) learning more about exactly how the task should be performed, (2) watching operators (or others) assigned to perform the task to determine if anything is incorrect or missing in the procedures, and (3) resolving discrepancies. In addition, the reviewer needs to evaluate whether the intended audience is likely to be able to understand and follow the procedure. Having operators or supervisors directly participate in periodic procedure reviews helps address these questions in an efficient manner. Directly involving operators and/or supervisors in procedure reviews also helps drive accountability and ownership for the procedures to the user level.

Integrate procedure review with training. At many facilities, operators are assigned to review one or more procedures to determine if they are current and accurate. If the refresher training program is focused mainly on refreshing operators on the procedures that apply to their job, these activities can be combined. Operator(s) assigned to perform a thorough review of a procedure can help prepare and deliver the refresher training on that procedure (or group of procedures) to other operators.

10.5   ELEMENT METRICS

Chapter 20 describes how metrics can be used to improve performance and when they may be appropriate. This section includes several examples of metrics that could be used to monitor the health of the *procedures* element, sorted by the key principles that were first introduced in Section 10.2.

In addition to identifying high value metrics, readers will need to determine how to best measure each metric they choose to track. In some cases, an ordinal number provides the needed information, for example, total number of workers. Other cases, such as average years of experience, require that two or more attributes be indexed to provide meaningful information. Still other metrics may need to be tracked as a rate, as in employee turnover. Sometimes, the rate of change may be of greatest interest. Since every situation is different, the reader will need to determine how to track and present data to most efficiently monitor the health of RBPS management systems at their facility.

10.5.1 Maintain a Dependable Practice

- *Number of standard operating procedures updated per year or staff hours spent updating procedures (per year, quarter, or month, depending on the review cycle).* A downward trend may indicate that more attention or resources needs to be applied to maintaining procedures.
- *Staff hours spent reviewing and approving procedures.* A downward trend may indicate that reviews are becoming perfunctory.

10.5.2 Identify What Operating Procedures Are Needed

- *Number of units that have completed a task analysis to identify procedure needs.* A low number indicates that procedure needs have not been thoroughly analyzed.
- *If task analyses are periodically updated or revalidated, compliance with update/ revalidation schedule.* Schedule compliance (for discrete tasks, such as a review) is one indicator of performance.

10.5.3 Develop Procedures

- *Number of audit or assessment findings related to procedures missing some element of required content.* Although this is a lagging indicator (deficiencies may be discovered years later, depending on the audit cycle and thoroughness of each audit), any finding should prompt a facility to closely examine a wide sampling or all of its procedures to determine if the situation is endemic. In addition to helping a facility determine what procedures need to be updated, this sampling provides metrics as a snapshot in time that can be compared to previous or future reviews.

- *Number of management of change authorizations issued for each unit (per year) to permit temporary operations that recur on a periodic basis; in other words, issuing a temporary management of change authorization rather than a temporary procedure.* By reviewing the management of change log, a knowledgeable individual can quickly determine the number of changes that fall into this category. If the facility intends to develop written procedures for infrequent but recurring tasks, it should see this metric decline over time. Note that the rate of change may be slow if the facility elects to develop procedures as the needs arise.

10.5.4 Use Procedures to Improve Human Performance

- *Number of incident/deficiency reports related to procedures that were unclear, not available, or not widely understood.* An increase in this metric would clearly indicate issues with respect to the procedures element. At most facilities, incidents occur at a very low frequency compared to the use of procedures and the updating of metrics. Therefore, this metric is unlikely to quickly respond to gradual decay in the performance of the management system governing the procedures element.
- *Number of incident investigations that recommend changes to procedures.* Again, an increase in this metric would indicate issues with respect to the *procedures* element, but it is unlikely to respond quickly to changes in management systems or practices.
- *Percent of changes to procedures not covered by MOC authorization.* As described in Chapter 15, facilities must define the scope of their *management of change* element. If the scope includes changes to procedures, one simple way to measure compliance with this requirement, and help ensure that changes to procedures are being properly reviewed, is to compare the number of procedure changes to the number of MOC authorizations related to procedure changes.
- *Percentage of procedures that are annotated in the field.* If (1) procedures are routinely used, (2) key information is missing, and (3) barriers to changing procedures exist, some operators will make handwritten clarifications or notes in the procedures. These notes are not inherently bad; in fact they can be very helpful to operators using the procedure and are normally very helpful to persons charged with periodically reviewing and updating procedures. However, such notations do indicate that a management system issue may exist concerning the method for maintaining current and accurate procedures.

- *Fraction of operators who believe that procedures are current and accurate.* Results of opinion surveys of operators may help provide early indications of changes in the accuracy of procedures. Note that survey participants will likely compare their perception of the current condition of the procedures to their expectations for procedure accuracy, clarity, and content. Thus, positive changes in the *culture* and *operations* elements, with a corresponding increase in expectations for all other RBPS elements, may actually drive this and similar metrics downward with no change or even slight improvements in operating procedures.

10.5.5 Ensure that Procedures Are Maintained

- *The number of procedures that are past due for review.* An increasing number indicates that procedures are in jeopardy of becoming outdated.
- *Mean time to correct/update procedures.* An increase in the time interval between when it is determined that a procedure should be updated and when the revised procedure is issued may indicate that more attention or resources are required for procedure maintenance. In the extreme case, a long lag time for updating procedures may ultimately cause operators and others responsible for maintaining procedures to abandon their efforts to keep procedures current and accurate.
- *Average age of procedures by unit or operation, as well as the most recent and oldest active revision.* Once an effective set of operating procedures has been established, the overall rate of change to the procedures should reflect the rate of change in either (1) design or operation of the unit or (2) management systems that affect the content of operating procedures. If the rate of change in procedures is low compared to either the rate of change for equipment, operating methods, or management systems, the procedures may be out of date. Note that the best benchmark for this metric is likely to be similar units at the facility.
- *Fraction of procedures that are clear, concise, and include all of the required content.* Develop a checklist for procedure content (based on the facility's policy governing this element) and ask an operator from each unit to review several procedures to determine if they are clear, concise, and include:
  o action steps that are clear and, when appropriate, include the expected system response and what to do if the expected response is not observed.
  o safe operating limits, consequences of deviation from limits, and steps to take to maintain the process within the safe operating limits.
  o limiting conditions for operation.
  o checklists (where appropriate).

10.6   MANAGEMENT REVIEW

The overall design and conduct of management reviews is described in Chapter 22. However, many specific questions/discussion topics exist that management may want to check periodically to ensure that the management system for the *procedures* element is working properly. In particular, management must first seek to understand whether the system being reviewed is producing the desired results. If the organization's level of *procedures* is less than satisfactory, or it is not improving as a result of management system changes, then management should identify possible corrective actions and pursue them. Possibly, the organization is not working on the correct activities, or the organization is not doing the necessary activities well. Even if the results are satisfactory, management reviews can help determine if resources are being used wisely – are there tasks that could be done more efficiently or tasks that should not be done at all? Management can combine metrics listed in the previous section with personal observations, direct questioning, audit results, and feedback on various topics to answer these questions. Activities and topics for discussion include the following:

- Review a representative sample of procedures to determine if they follow standards that have been established for content and format.
- When reviewing procedures, check to see if they have been reviewed per the policy, and verify that out-of-date procedures are not present in work areas.
- Observe work practices to determine if workers routinely reference operating procedures. If not, identify the cause. For example, are the procedures
  - o   not user friendly (i.e., a single task spans multiple procedures, or necessary information is available only in referenced documents)?
  - o   too detailed?
  - o   out of date?
  Or, is there a mindset that only inexperienced operators refer to procedures?
- When checklists are used, determine if anyone periodically audits the checklists to verify accuracy/completeness and to ensure that the data recorded on the checklist complies with (1) the associated procedure and (2) management's intent for how the activity should be performed. Review the results of these audits.
- Determine if (1) reviews are being completed in accordance with the schedule and (2) the results of the reviews generally indicate that procedures are being maintained current and accurate
- Review production and quality metrics to determine if significant differences exist in the performance between different shifts, teams, areas, or departments. If so, determine if the outlier(s) are deviating

from approved procedures. (Note that failure to follow procedures could result in exceptionally good performance as well as exceptionally bad performance.)

- Discuss whether the entire organization feels accountable for following procedures. Are good results valued over good performance? How does line management demonstrate to operators that they are accountable for performance (intelligently following procedures and standards) rather than results (doing what it takes to get the product out the door)? Have any operators been rewarded for increased productivity achieved through the use of shortcuts?

Operating procedures need to be current, accurate, complete, concise, understood, and enforced. In addition to helping ensure process safety, procedures help ensure reliable human performance, which should help ensure predictable output, yield, and product quality. Clearly, reliable human performance based on sound procedures will benefit the entire business. If the *procedures* element fails to measure up to the standards that have been established, key managers should carefully examine their commitment to the *procedures* element and take any corrective actions that are needed to achieve the intended level of performance.

10.7   REFERENCES

10.1   *Chemical Accident Investigation Report – Terra Industries, Inc. Nitrogen Fertilizer Facility, Port Neal, Iowa,* U.S. Environmental Protection Agency, Region 7, Emergency Response and Removal Branch, Kansas City, Kansas, issued January, 1996.

10.2   Bridges, William G. and Williams, Thomas R., Create Effective Safety Procedures and Operating Manuals, *Chemical Engineering Progress*, December 1997.

Additional reading

Center for Chemical Process Safety, *Guidelines for Writing Effective Operating and Maintenance Procedures*, American Institute of Chemical Engineers, New York, 1996.

11

SAFE WORK PRACTICES

On the evening of July 6, 1988, an explosion and fire occurred on the Piper Alpha offshore platform, resulting in 167 fatalities and destruction of the platform. The incident was caused by the night crew putting a condensate injection pump into service that, earlier in the day, had been taken out of service for maintenance. The night crew was aware of the maintenance activity and, in fact, had to authorize the electricians to close the switch at the motor control center so that the pump could be returned to service. The evening shift operators were likely told that all of the maintenance work scheduled to be performed by the maintenance group was either complete or had been deferred; however, they were not aware that other work being performed by a contractor was incomplete. In fact, the contract crew had removed a pressure safety valve on the pump discharge line for recertification and was unable to return the valve to service prior to 6 p.m. when they quit for the day. Once the pump was started, a large release of hydrocarbons ultimately led to the disaster (Ref. 11.1).

11.1   ELEMENT OVERVIEW

The RBPS element that helps control hazards associated with maintenance and other nonroutine work is one of nine elements in the RBPS pillar of *managing risk*. Section 11.2 describes the key principles and essential features of a management system for this element. Section 11.3 lists work activities that support these essential features and presents a range of approaches that might be appropriate for each work activity, depending on perceived risk, resources, and organizational culture. Sections 11.4 through 11.6 include (1) ideas to improve the effectiveness of management systems and specific programs that support this element, (2) metrics that could be used to monitor this element, and (3) issues that may be appropriate for management review.

11.1.1 What Is It?

Procedures are generally divided into three categories. Operating procedures, as described in Chapter 10, govern activities that generally involve producing a product. Maintenance procedures, as described in Chapter 12, generally involve testing, inspecting, calibrating, maintaining, or repairing equipment. Safe work procedures, which are often supplemented with permits (i.e., a checklist that includes an authorization step), fill the gap between the other two sets of procedures. Safe work practices help control hazards and manage risk associated with nonroutine work.

In this context, a nonroutine activity is any activity that is not fully described in an operating procedure. Nonroutine does not refer to the frequency at which the activity occurs; rather, it refers to whether the activity is part of the normal sequence of converting raw materials to finished products. Making and breaking connections to unload a railcar would likely be covered by an operating procedure, whereas breaking a connection to remove and calibrate a pressure transmitter would be considered a nonroutine work activity and included in the scope of the safe work practices (*safe work*) element.

Consider a typical work order to calibrate a pressure transmitter with the process in operation. An integrated set of operating, maintenance, and safe work procedures helps prevent accidents and process upsets. For example:

- An operating procedure instructs the operator to place the pressure control loop associated with the transmitter in manual control and return the loop to automatic control when the work is complete. Depending on the hazards present, the operating procedure may also provide instructions on how to clear, drain, and decontaminate the space between the manual valve closest to the transmitter and the transmitter.
- Safe work procedures and permits help ensure communication between the instrument technician assigned to perform the job and production personnel. The procedure or permit should also help ensure that the instrument technician removes the intended transmitter, and that appropriate valves are locked closed (or in the case of a vent line, locked open) prior to removing the transmitter. Safe work procedures may also include steps to help protect against a fire if an enclosure must be opened in an electrically classified area.
- Maintenance procedures help ensure that the calibration work is performed properly and also help protect the instrument technician from a variety of hazards.

Safe work procedures typically control hot work, stored energy (lockout/tagout), opening process vessels or lines, confined space entry, and similar operations. A more comprehensive list of safe work practices is provided in Section 11.2.1. Some facilities also include procedures or practices that protect against standard industrial hazards, such as falling, in the scope of this element. Safe work practices are often required by regulation, regardless of the magnitude of chemical or other hazards present at a facility.

Safe work procedures may be applied to construction work, and should be if the work might affect other operations at a facility. Safe work procedures can also help protect equipment from damage resulting from maintenance, construction, or other nonroutine activities (e.g., excavation near underground lines, lifting over process equipment).

11.1.2 Why Is It Important?

Nonroutine work, such as the simple removal of a pressure safety valve on the Piper Alpha platform for recertification, increases risk and can directly lead to conditions that make a catastrophic accident much more likely. Safe work practices are a critical element in the management of industrial safety. For example, from 1992 to 2002, nitrogen asphyxiation during confined space entry resulted in 80 fatalities in the U.S. (Ref. 11.2). This rate of about 8 fatalities per year caused by exposure to a single inert gas is not significantly different from the rate of 6.5 fatalities per year (or 33 fatalities over a 5-year period) caused by releases of one of 130 highly toxic and flammable materials regulated under the U.S. EPA's risk management program rule (Ref. 11.3). Although this is not a valid method to directly compare risk nor is it proof that inert gases are more dangerous than other hazardous materials, it does demonstrate that safe work procedures are an important safeguard for worker safety as well as for catastrophic accidents such as Piper Alpha.

11.1.3 Where/When Is It Done?

In general, policies related to safe work practices are developed at the corporate level. Procedures and permits that specify how work is to be performed are typically developed by facilities. These procedures and permits are used by operating facilities on a day-to-day basis to control nonroutine work.

11.1.4 Who Does It?

Safety policies are typically issued by a corporate or division safety group. The development of specific procedures and practices is normally assigned to the safety group at each facility, although some companies develop a common set of procedures that is modified/adapted for use at the facility level. Permits are normally issued/authorized by trained operators, supervisors, or safety specialists. Use of safe work procedures and proper execution of nonroutine work is often a shared responsibility between operators, maintenance personnel, and contractors who perform the work.

11.1.5 What Is the Anticipated Work Product?

The main work product of the *safe work* element is an integrated system of procedures and permits that help protect workers from hazards and prevent the sudden release of process materials or energy during nonroutine work activities. This system is often described in a facility-wide policy that addresses management system issues such as (1) the scope of the *safe work* element, (2) roles and responsibilities, and (3) the relationship between safe work procedures/permits and procedures developed for other RBPS elements.

Policies typically state which activities are permitted with no special controls, which activities require special permits, and which activities are prohibited. Procedures provide details on how the work is to be executed. Permits describe job-specific hazards and specify safeguards. Procedures and permits help ensure that workers understand the hazards and take appropriate actions to manage risk when performing nonroutine work activities. Safe work procedures complement operating and maintenance procedures, and also provide objective standards of performance required to effectively implement the *safe work* element.

11.1.6 How Is It Done?

Nonroutine work is controlled through a system of procedures and permits. In many cases, permits are required for each nonroutine job, and permits are updated or reauthorized at the start of each shift.

11.2   KEY PRINCIPLES AND ESSENTIAL FEATURES

Safe operation and maintenance of facilities that manufacture, store, or otherwise use hazardous chemicals requires reliable human performance at all levels, from managers and engineers to operators and craftsmen. Safe work procedures, along with operating procedures, maintenance procedures, and training, form the foundation for achieving high levels of human reliability.

The policy or program-level procedure governing the *safe work* element typically describes the range of activities that are included in the scope of this element and designates the subordinate procedures that govern specific activities, such as confined space entry, lockout/tagout, or line breaking/opening process equipment. The safe work policy also provides guidance on when special permits are required. Finally, it helps define who is responsible for performing certain tasks and who is in control of equipment as it transitions from normal operation to maintenance (or some other mode) and back to normal operation.

The following key principles should be addressed when developing, evaluating, or improving any management system for the *safe work* element:

- Maintain a dependable practice.
- Effectively control nonroutine work activities.

The essential features for each principle are further described in sections 11.2.1 and 11.2.2. Section 11.3 describes work activities that typically underpin the essential features related to these principles. Facility management should evaluate the risks and potential benefits that may be achieved as a result of improvements in this element. Based on this evaluation, the facility should develop a management system, or upgrade an existing management system, to (1) address some or all of the essential features, and (2) execute some or all of the work activities, depending on perceived risk and/or identified process hazards. However, these steps will be ineffective if the accompanying *culture* element does not embrace the use of reliable management systems. Even the best management system, without the right culture, will not sustain safe work practices.

Many regulatory authorities, including the U.S. Occupational Safety and Health Administration, have established highly prescriptive regulations governing many of the activities described in this chapter. Companies should consider regulatory requirements in addition to risk when deciding to adopt (or more specifically, declining to adopt) some or all of the key principles and essential features listed in this section.

This chapter does not provide a comprehensive list of safe work practices or regulatory requirements applicable to facilities located in the U.S. or any other country.

11.2.1 Maintain a Dependable Practice

A written policy should (1) describe the scope of the *safe work* element, (2) identify which safe work procedures and permits govern specific nonroutine work activities, and (3) define the roles and responsibilities for implementing activities associated with the *safe work* element, including who may authorize nonroutine work or how this authority is delegated and controlled. The policy should also specify the physical scope of the *safe work* element (i.e., where certain requirements apply), and how the requirements change over the life cycle of a process. This policy should include a list of references to subordinate safe work procedures that control specific nonroutine work activities.

Define the scope. The scope of the *safe work* element has several aspects. First, a facility needs to decide what types of nonroutine work to control. Work activities that are typically governed by safe work procedures and/or permits are presented in Table 11.1.

Second, a facility needs to decide where the safe work procedures should apply. The hazards may be very different in the facility's finished product warehouse than in the chemical processing area. Typically, some safe work procedures, such as hot work and lockout/tagout, apply throughout the facility, while others, such as vehicle permits and control of access, apply only to high hazard process areas.

Some facilities implement special authorizations to control all maintenance and related work in high hazard process areas. In addition, based on the type of hazard or perceived risk, certain work activities can be authorized only by trained specialists. For example, facilities may decide to allow operations involving the use of ionizing radiation (e.g., taking x-rays of process piping and equipment) to be performed only under the direct supervision of a trained radiation safety specialist.

TABLE 11.1. Activities Typically Included in the Scope of the Safe Work Element

Safe work procedures to control general hazards or protect personnel from a hazard/hazardous environment include:

- Lockout/tagout and/or control of energy hazards.
- Line breaking/opening of process equipment.
- Confined space entry.
- Hot work authorization, including use of equipment that does not conform to electrical classification requirements.
- Access to process areas by unauthorized personnel.
- Access to special areas during normal/routine operation (e.g., rooms in which special hazards may be present, working envelopes for robotic systems).
- Roof access permits.
- Elevated work/fall protection.

Safe work procedures to protect against mishaps that could have catastrophic secondary effects include:

- Excavation in or around process areas.
- Operation of vehicles in process areas.
- Lifting over process equipment.
- Use of other heavy construction equipment in or around process areas.
- Hot tapping lines and equipment.

Safe work procedures to control special hazards include:

- Use of explosives/blasting operations.
- Use of ionizing radiation (e.g., to produce x-ray images of process equipment).

Safe work procedures to prevent unauthorized impairment of safety systems include:

- Fire system impairment.
- Temporary isolation of relief devices.
- Temporary bypassing or jumpering of interlocks.

The scope of the *safe work* element should specifically address construction activities. Construction hazards can extend beyond the work zone and may affect other units at the facility, for example, by accidentally cutting into buried utility or process lines during excavation, or by providing a nearby ignition source). Conversely, construction workers may be exposed to hazards from other units at the facility. Based on risk, some permit requirements for construction activities may be streamlined or set aside entirely for the construction phase because they cover routine construction activities, such as welding. Repeated permitting of the same construction activity each day when clearly no special hazards exist can foster an attitude of complacency toward all safe work procedures.

Specify when in the facility's life cycle the safe work procedures apply. One common misconception is that the risk associated with operating a chemical plant begins when large quantities of chemicals are first introduced to the unit and ends when the unit is shut down. The severity of consequences for many accident scenarios may be proportional to inventory, and many of the hazards are nonexistent or significantly mitigated at early and late stages of the unit's life cycle. However, some hazards, such as falls during construction or exposure to residual materials during decommissioning, exist even when little or no process material is present.

Certain hazards are present primarily in the early and late stages of the life cycle. During the early laboratory- and pilot-scale development stages, hazards may not be well understood, particularly for a new product, new synthesis route, or a new process. Decommissioning also presents a unique range of hazards. Some processes stand idle for years prior to decommissioning. Lighter fractions of mixtures can evaporate, leaving highly viscous liquid or solid materials, making it difficult to completely drain process lines and equipment. In addition, residual process materials that were present when the unit was shut down may have reacted to form other compounds, with the potential for very different hazards.

All of these factors point to the need for effective safe work procedures throughout the life of the process. They also highlight the importance of a flexible system that allows facilities to apply some safe work procedures differently at various points in the life cycle of a unit.

Ensure consistent implementation. Developing a set of safe work procedures helps facilities manage risk associated with nonroutine work. Safe work procedures typically require some form of hazard assessment and a listing of safeguards that must be in place for the work to proceed. This normally requires an integrated system of standing safe work procedures, checklists, temporary operating procedures, maintenance procedures, job-specific work orders, permits, and other procedures that collectively specify:

- The work to be performed.
- The equipment on which the work is to be done.
- The hazards associated with the work.
- Steps to drain, purge, or clean the equipment.
- Safeguards to protect against the hazards, for example, isolation of energy sources and process materials.
- Required monitoring of worksite conditions and worker performance during the course of the nonroutine work.
- Personal protective equipment (PPE) to be used by persons performing the work.
- A method to formally turn control of the equipment over to the group responsible for the work.
- Inspections, tests, or other steps to ensure that the equipment is properly prepared for work.
- Procedures for performing the nonroutine work.
- Steps to return the equipment to service, including prestartup checks, inspections, and tests such as leak checks, shaft alignment, and checking rotation.
- A method to formally return the equipment to the control of the production department.

Involve competent personnel. Although all of the work practices listed earlier in this section are considered nonroutine work, some are much less common and potentially higher risk than others. Authorization and execution of some work clearly requires special expertise. For example, facilities sometimes use explosives to dislodge solidified process material from inside equipment or to break up solids that have fused together in storage. This practice, as well as other special nonroutine work practices can, in the absence of a full understanding of hazards, introduce unacceptable risk. In 1921, a routine practice of using explosives to break up fertilizer materials led to an explosion that killed 500 people.

> *On September 21, 1921, approximately 450 tons of an ammonium nitrate/ammonium sulfate mixture exploded in Oppau, Germany, resulting in approximately 500 fatalities. Until 1921, conventional explosives (e.g., dynamite) were routinely used to break up large clumps of material that would otherwise not come apart. In fact, this technique had been used to break apart ammonium nitrate fertilizers more than 20,000 times prior to September 21, 1921, without causing a detonation of the material (Ref. 11.4).*

Although the example cited in the text box is quite dated, it reinforces the point that some of the most dangerous hazards are unknown (at least to the persons performing the nonroutine work).

While involving competent personnel in reviewing hazards associated with nonroutine work is always important, it is equally important to involve competent personnel in writing procedures, issuing permits, and executing nonroutine work involving inherently hazardous materials and methods. However, do not defer completely to "experts." Combining expert advice with sound reason and unit-specific experience is a powerful means to identify hazards and manage risk associated with nonroutine work. Based on perceived risk, facilities often apply their *management of change* element to certain nonroutine work to help ensure a higher level of scrutiny than might be provided by simply issuing a work permit.

11.2.2 Effectively Control Nonroutine Work Activities

Effective management of risk associated with nonroutine work requires robust systems, thorough training and awareness, a sound culture, and diligence.

Develop safe work procedures, permits, checklists, and other written standards. Written procedures and permits are a practical necessity for effectively and consistently controlling nonroutine work. Safe work procedures typically include checklists that (1) highlight specific hazards that may be encountered while performing the work and (2) specify appropriate safeguards.

A safe work procedure may apply to an entire facility or to a specific piece of equipment. If the hazards are of a general nature, such as hot work, which generally involves a fire hazard, a facility-wide procedure supplemented with a permit/checklist is normally sufficient. If the hazards are specific to a class of equipment or a single equipment item (e.g., fire system impairment), equipment-specific procedures are normally required. In some cases, safe work procedures, permits, and equipment-specific instructions are supplemented with task-specific temporary operating or maintenance procedures as described in Section 10.2.3. Regardless, standardized procedures and permits are beneficial to maintenance and contractor personnel, and others who perform nonroutine work throughout the facility, because they will not have to learn a different set of standards for each unit at the facility.

Safe work procedures should require two-way communication between the owner/user of the equipment (and other potentially affected groups) and the crew assigned to do the work both (1) before equipment is released for maintenance and (2) when maintenance is complete. Safe work procedures normally include a checklist or permit to authorize the work.

Issuing and authorizing work permits ensures two-way communication and establishes written conditions that the crew executing the work must follow. Permits should be authorized by someone who is independent of the work crew and has the training, knowledge, and experience needed to evaluate the hazards of the work. For example, a properly trained first-line supervisor and certain operators may be designated as approvers for hot work permits issued within their unit. Supervisors and operators are typically:

- Responsible for the area in which the work will be performed.
- Very familiar with hazards in the area, and in particular, fire hazards.
- Familiar with the hazards in adjacent areas that might be affected by the hot work, assuming that the hazards extend beyond the unit area, and they are in routine communication with personnel responsible for the adjacent areas.
- Aware of other activities going on at the time that might affect risk associated with the planned work.
- Capable of providing an independent verification that appropriate safeguards are in place to address the hazards of the planned hot work.

Conversely, an operator or first-line supervisor at a multi-unit facility is unlikely to be the best choice for the approval authority for a facility-wide fire system impairment because the potentially impacted area will extend well beyond her area of responsibility. In addition, some permitting systems may require that the authorizer have special training or expertise. For example, a facility may require that excavation permits be issued only by the facility engineer, based on the need to closely check the area for buried underground piping.

Jobsite inspection prior to permit authorization also provides an important safeguard as the permit authorizer may notice an unsafe condition that is not specifically included in the checklist associated with the work permit.

Requiring a representative of the operations group and a representative of the crew assigned to do the work to jointly inspect the work area when the nonroutine work is complete helps prevent incidents when the equipment is returned to service. In addition to inspecting the area, this team is normally responsible for ensuring that valves are properly aligned, blinds are removed from process lines, blinds are reinstalled on vent and drain lines, rotating equipment is turning in the proper direction, sewers and drains are uncovered, tags and locks are removed, barricades have been removed, and so forth. This practice is often incorporated into the *readiness* element. The transition between the *safe work* and *readiness* elements should be closely monitored to ensure that no gaps exist and that the intended degree of overlap or redundancy (if any) is in place.

Open permits should be posted in a prominent location so that operators and supervisors are aware of all open or ongoing work activities. This helps improve decision making regarding how changes that occur over the time that the work permit is valid may affect risk. For example, if two electrical pumps are in a critical service (one being an installed spare) and a backup system is powered by a steam turbine, operations may prudently choose to delay the weekly switch between the electrically driven pumps for a few hours if the turbine-driven pump is out of service.

Facilities should also provide a means to alert personnel to jobs that are not currently being worked on and are not yet complete. In addition to administrative controls, such as notices in designated locations, some facilities establish "department locks" that complement (but do not replace) the lockout/tagout program. For example, production would place a department lock and tag on equipment that is out of service or is not to be operated for some reason. If repair is required, maintenance would also place their department lock once they start to repair the equipment. *Note: the maintenance department lock is in addition to the personal locks placed on the equipment by individual maintenance employees.* As maintenance employees leave for the day or are reassigned to other jobs, they remove their personal lock. However, the maintenance department lock remains on the equipment until all maintenance work is complete and the maintenance crew believes the equipment is ready to return to service. The same method can be used to alert personnel that contractor work is not yet complete. This sort of alert may have helped the crew of the Piper Alpha platform realize that it was not safe to remove locks and return the condensate injection pump to service. (See the beginning of this chapter for a brief description of the Piper Alpha accident.)

Many facilities develop a general permit, sometimes called a work authorization, start work, cold work, control of work, or safe work permit to control all nonroutine work, even if none of the special nonroutine work permits apply. Requiring a general nonroutine work permit helps facilitate good communication between the owner/user of the equipment and the crew assigned to the job.

Some facilities exempt certain low-risk activities from the requirement to obtain a work permit. The list of exempt activities typically includes:

- Maintenance work performed in shop areas after equipment has been properly decontaminated.
- Hot work performed in designated welding areas.
- A limited number of electrical and electronic maintenance activities, provided the activities are not performed in process areas and no workers are directly exposed to high voltage conductors, for example, replacing plug-in modules and controller/computer cards.
- Routine maintenance performed on roads and grounds, as long as traffic is not permanently blocked and the work is well away from process areas.
- Lubrication of in-service equipment.

This list is sometimes expanded to include minor adjustments/repairs to process equipment (e.g., tightening flanges or valve packings) and routine maintenance activities that are very unlikely to affect the process (e.g., vibration monitoring), as long as the operator responsible for the area has verbally authorized entry and the work.

Exempting certain high frequency, low risk activities may help reduce an issuer's perception of the work permit as a "paperwork" exercise. Treating all nonroutine work in the same way, regardless of risk, can cause permit issuers to become complacent. If such complacency becomes entrenched, facilities may begin issuing permits from the office or control room without a full understanding of the work to be performed and/or without inspecting the jobsite.

Train employees and contractors. As with any management system, good performance depends heavily on the competence of personnel. Employees and contractors must know how to effectively authorize and conduct nonroutine work. The *safe work* element contains three aspects of training. These aspects are:

- Understanding the overall system of safe work procedures and associated permits.
- Recognizing hazards typically associated with nonroutine activities.
- Applying procedures and properly filling out permits and checklists.

The labyrinth of safe work guidelines; permit protocols; and equipment specific, job specific, and unit-specific procedures and practices can be very confusing. A flow chart or logic diagram showing the relationship between various procedures and permits is a useful tool for training personnel who issue permits or perform nonroutine work activities covered by permits. If confusion persists, consider simplifying the system of safe work procedures.

Training on how to recognize hazards, and how to recognize when unknown hazards may be present, is much more complex. A worker's ability to recognize hazards depends on his experience, training, and knowledge, and on his having a sense of vulnerability. This ability also ties in with several other RBPS elements, including the *culture, competency, risk, training,* and *process knowledge* elements. Proper execution of these elements, along with a sufficient degree of diligence in evaluating these factors when appointing and training permit authorizers, will help ensure that permit issuers do their part to manage risk associated with nonroutine work activities.

Control access to particularly hazardous areas. Companies have long controlled access of visitors and contractors, regardless of the nature of the work. In addition to protecting business confidential and trade secret information, access control helps (1) prevent a safety or operational incident that could result from inadvertent actions by a visitor or contractor and (2) minimize the number of people in harm's way should an incident occur.

Many companies have extended their access control policies to include employees not directly involved in the operation of the process (sometimes call nonoperational personnel), for example, maintenance workers, technical and laboratory staff, and managers. This change is partly in response to regulatory requirements. Operators should be fully aware, and in control, of all activities within their assigned process areas. Limiting access by nonoperational personnel allows operators to (1) authorize/control entry into the process area and (2) know when workers exit the area. When authorizing entry, operators can inform workers of any special hazards that exist in the area at that time. In addition, the entrant will tell the operator the purpose of the visit, the specific work location, and any intended actions, which will aid the operator in the event of an emergency or unexpected operational issue. During some particularly critical operations, operators (or supervisors) may even deny entry and request that the entrant return after a hazardous or sensitive operation has been completed.

In some cases, expanding the procedure to control access to certain parts of the facility that may not be extremely hazardous under normal conditions may be appropriate. For example, a company may determine that relieving the contents of a runaway batch reaction to the roof of the building rather than a catch tank is acceptable because:

- The roof is normally not occupied.
- The runaway reaction is an infrequent and unexpected event.
- The height of the roof makes it unlikely that process material discharged during a runaway reaction will pose an unacceptable risk to persons on the ground outside the building (or inside the building).

In this case, the company may implement a procedure that prohibits roof access without a permit from the operating team responsible for these reactors. Limiting roof access to times when these reactors are idle should help protect personnel without causing undue burden on maintenance and other activities that require roof access. Likewise, certain facilities do not allow anyone to enter the process area during certain phases of operation without management authorization and, of course, suitable PPE. This practice is normally limited to processes containing materials that are immediately dangerous to human health, even in small quantities.

Enforce the use of safe work procedures, permits, and other standards. Consistent implementation of safe work procedures and permits is achieved by (1) having well designed and widely understood written procedures and permits, (2) providing awareness training to the entire organization, (3) providing more intense training for operators, maintenance personnel, permit authorizers, contractors who perform work under permits, and supervisors, and (4) routinely inspecting worksites.

Because the responsibility for enforcing safe work procedures and issuing/authorizing safe work permits is often shared among dozens or even hundreds of employees at a facility, inconsistent execution of work activities associated with the *safe work* element is possible. Insufficient training, confusing systems, poor operating discipline, or a lack of diligence are all potential causes of inconsistent job execution. The first step in addressing uneven or poor performance is to identify where such conditions exist. Persons conducting routine safety/housekeeping audits should make a point of checking ongoing maintenance work activities to ensure that the proper permits are in place, that the permits specify appropriate conditions, and that workers follow procedures and conform to permit requirements. The frequency of inspections should be based on perceived risk. For example, during normal operations a 30-minute inspection each week may be appropriate, whereas during a shutdown of a large continuous process unit with hundreds of contract employees working at any given time, inspections might be conducted several times each day.

Review completed permits. Permits often contain errors, such as required fields that are not filled in, conditions that do not appear to be appropriate for the job, or persons authorizing the permits or doing the work who have not completed required training. Facilities that simply file completed permits without review (1) forego an opportunity to identify individual training needs or systemic issues and (2) create a file full of permits that do not conform to procedural requirements that will eventually be discovered by an auditor. Permits issued by specific personnel are sometimes consistently illegible. Although the person who issued the permit may be able to decipher the permit and show that it meets the requirements of the governing safe work procedure, illegible permits certainly miss the intent to communicate hazards, safeguards, and conditions for safe operation to persons assigned to perform the work. All these situations can be quickly identified and corrected if permits are reviewed when the activity is complete.

Permit review is often performed by a supervisor, maintenance planner, or other person with general knowledge of the overall policy regarding the *safe work* element, the unit, and the hazards. A review by a knowledgeable person should take only a minute or two per permit if the permits are complete and accurate. Ongoing permit review also provides the metrics needed to track changes in performance over time, and can also help identify refresher training needs.

11.3   POSSIBLE WORK ACTIVITIES

The RBPS approach suggests that the degree of rigor designed into each work activity should be tailored to risk, tempered by resource considerations, and tuned to the facility's culture. Thus, the degree of rigor that should be applied

to a particular work activity will vary for each facility, and likely will vary between units or process areas at a facility. Therefore, to develop a risk-based process safety management system, readers should perform the following steps:

1. Assess the risks at the facility, investigate the balance between the resource load for RBPS activities and available resources, and examine the facility's culture. This process is described in more detail in Section 2.2.
2. Estimate the potential benefits that may be achieved by addressing each of the key principles for this RBPS element. These principles are listed in Section 11.2.
3. Based on the results from steps 1 and 2, decide which essential features described in Sections 11.2.1 and 11.2.2 are necessary to properly manage risk.
4. For each essential feature that will be implemented, determine how it will be implemented and select the corresponding work activities described in this section. Note that this list of work activities cannot be comprehensive for all industries; readers will likely need to add work activities or modify some of the work activities listed in this section.
5. For each work activity that will be implemented, determine the level of rigor that will be required. Each work activity in this section is followed by two to five implementation options that describe an increasing degree of rigor. Note that work activities listed in this section are labeled with a number; implementation options are labeled with a letter.

Many regulatory authorities, including the U.S. Occupational Safety and Health Administration, have established highly prescriptive regulations governing many of the work activities described in this section. Companies should consider regulatory requirements in addition to risk when deciding to adopt (or more specifically, declining to adopt) some or all of the work activities listed in this section.

This chapter does not provide a comprehensive list of safe work practices or regulatory requirements applicable to facilities located in the U.S. or any other country.

11.3.1 Maintain a Dependable Practice

Ensure Consistent Implementation

1. Develop both a policy or program-level governing procedure that describes how nonroutine work is authorized and controlled, and a

supporting system of procedures and permits to authorize and control nonroutine work.

    a.   Safe work procedures are designed to meet regulatory requirements.

    b.   A corporate policy statement is in place that identifies which procedures and permits must be developed, along with examples of each.

    c.   The facility has a program-level governing procedure that describes the *safe work* element, and specifically addresses the scope, roles, and responsibilities for this element.

2.   Include in the policy/governing procedure a list of minimum information required for safe work procedures and permits (see Section 11.2.1).

    a.   Information meets the minimum regulatory requirements.

    b.   No overall program-level governing procedure exists, but safe work procedures are developed based on a consistent format and generally address a common set of issues/requirements.

    c.   The facility has a program-level governing procedure that specifies a minimum set of issues/requirements that must be addressed in each safe work procedure, checklist, and permit.

Define the Scope

3.   State which nonroutine activities are covered by procedures or permits (or which are not covered).

    a.   The scope of the *safe work* element is based on regulatory requirements.

    b.   The scope of the *safe work* element includes practices that have been implemented based on previous incidents as well as regulatory requirements.

    c.   A corporate policy statement is in place that identifies which procedures and permits must be developed.

    d.   The facility's program-level governing procedure lists all of the safe work procedures and explains how they apply to nonroutine activities, with examples.

4.   State in the written safe work procedures if any parts of the facility are exempt from any procedural or permit requirements.

    a.   Generally, no written statement exists regarding where safe work procedures apply, but practices are well established and reinforced with training.

    b.   The program-level governing procedure includes a table or other means to clearly describe where in the facility each safe work procedure does (or does not) apply.

   c. Each safe work procedure includes a scope section that clearly describes where in the facility it does (or does not) apply.

5. Develop special procedures to address special hazards and/or to control work in units that process extremely toxic or otherwise hazardous chemicals.

   a. Corporate policy addresses this issue, but no such special hazards exist at the facility and the requirement is not reflected in the facility's policy or program-level procedure governing safe work.

   b. No written guidance exists, but key personnel at the facility are very mindful of special hazards presented by some nonroutine operations, such as the use of explosives or ionizing radiation. In addition, a supervisor is involved when authorizing nonroutine work in a particularly hazardous process area.

   c. The facility's policy/program-level governing procedure addresses a comprehensive suite of nonroutine work and specifies special requirements, such as temporary change authorizations, for certain high hazard activities.

Specify When in the Facility's Life Cycle the Safe Work Procedures Apply

6. Use safe work procedures to control activities that typically occur before startup or after a unit is permanently shut down. Such activities include construction, decommissioning, and demolition.

   a. Application of the safe work procedures during construction or decommissioning and decontamination is left to the discretion of the project manager.

   b. The program-level governing procedure or policy specifies when safe work procedures apply over the entire life cycle of a unit/facility.

   c. In addition to item (b), the program-level governing procedure or policy identifies specifically which safe work procedures apply throughout the life of a process unit.

Involve Competent Personnel

7. Ensure that persons authorized to approve permits have (1) the training and experience to understand a wide range of hazards and (2) knowledge of well established methods to manage risk associated with the hazards.

   a. Permits can be issued by any operator, maintenance worker, supervisor, or the maintenance planner for the unit.

   b. Permits must be authorized by the area safety representative or the shift representative on the central safety committee.

    c.   Permits can be authorized only by designated personnel who have completed certain training (and sometimes refresher training) requirements.

    d.   A list of designated persons who can authorize each type of permit is provided. A formal process is in place to determine who should be included on each list; determination is based on knowledge, experience, and completion of permit-specific training.

8.   Ensure that persons authorized to approve permits have an appreciation of hazards.

    a.   Permits can be issued by any operator, maintenance worker, supervisor, or the maintenance planner for the unit. All of these persons are normally aware of hazards present within the unit and appropriate safeguards.

    b.   Permits must be authorized by the area safety representative, and these individuals periodically attend conferences and short courses to hone their hazard recognition and safety management skills.

    c.   Permits can be authorized only by experienced personnel who are designated by management. Designated personnel must have completed training that (1) focuses on hazard recognition and (2) emphasizes that permit issuers should "know what they don't know" and, armed with that knowledge, proceed accordingly.

9.   Establish and promote an environment that welcomes questions regarding the safety of all aspects of the operation, including nonroutine activities, even if the nonroutine activities are planned and executed by "experts". (Note: This work activity is closely related to the essential feature for the *culture* element regarding maintaining a sense of vulnerability that is described in Section 3.2.2.)

    a.   A questioning attitude is recognized as a positive attribute, but only when a valid question is posed.

    b.   The facility's culture promotes a "safety first" approach, and all employees are encouraged to question the basis for safety of all work activities.

    c.   The facility's culture promotes a "safety first" approach, and workers are empowered to shut down any work activity that they believe to be unsafe until their concerns are addressed.

11.3.2 Effectively Control Nonroutine Work Activities

Develop Safe Work Procedures, Permits, Checklists, and Other Written Standards

10.   Develop procedures and/or permits to control applicable nonroutine work activities.

    a. Work is authorized verbally, typically in a discussion between the production and maintenance supervisors in a morning meeting. Control is assumed by the individuals performing the work (e.g., locking disconnect switches open).

    b. Most work is authorized verbally in a discussion involving the equipment owner/user (e.g., operator or production supervisor) and the person assigned to perform the work.

    c. The work authorization and control procedures include written permits for certain activities, such as hot work.

    d. An integrated suite of procedures, permits, and checklists, along with physical controls such as locks and barricades, are used to authorize and control nonroutine work. For example, individual locks are supplemented with department locks so that equipment that is shut down for maintenance cannot be returned to service without a maintenance employee removing the maintenance department lock.

11. Require direct communication between the group that operates the equipment or is responsible for the area in which the work will be performed and the group that will execute the nonroutine work.

    a. Control is assumed by the individuals performing the work (e.g., locking disconnect switches open). Persons executing the work often inform others only if they believe other areas may be affected by the work.

    b. Work is authorized by the area supervisor, typically in a discussion that takes place in the control room or in a daily work planning meeting.

    c. Safe work procedures require (1) that permits are issued only by a qualified person representing the organization responsible for operating the equipment and (2) that the process of issuing a permit includes a worksite inspection.

12. Ensure that ongoing work is well communicated to unit operators and other potentially affected employees.

    a. Maintenance employees or other crews assigned to perform the work verbally inform the operator prior to starting the work.

    b. The operator or supervisor issues a written permit that is posted at the worksite.

    c. Permits are issued for all nonroutine work. All open permits are posted at the worksite and in a designated location, such as a specific bulletin board in the control room.

    d. In addition to item (c), a highly visible method has been established to alert operators to incomplete work, including safeguards to prevent operators from returning equipment to service before work is complete.

13. Develop a nonroutine work permit or similar system to control all nonroutine work activities, even if no other safe work procedure or permit applies. This system would be subject to specific exemptions for certain plant areas and/or low hazard tasks.

   a. Work authorization systems are primarily verbal and based on standing practice at the facility.

   b. The system of formal work authorization and control procedures/permits is based on regulatory requirements.

   c. All nonroutine work must be authorized; permits may be authorized only by the area safety representative or the shift safety committee representative.

   d. All nonroutine work must be authorized; permits are authorized by any of a number of trained employees who own, operate, or are otherwise responsible for the area in which the work will take place. Closing the permit when the work is complete requires formal communication between the work crew and the equipment owner/operator.

Train Employees and Contractors

14. Provide awareness training to all employees.

   a. Employees learn about the permit system through on-the-job training.

   b. A written description of the safe work policies, procedures, and permits is provided to new employees during orientation.

   c. New employee training and periodic refresher training addresses the safe work policies, procedures, and permits and includes practical exercises or other means to determine if each trainee has an overall appreciation for the system.

15. Provide additional training to employees who routinely authorize or perform nonroutine work that thoroughly covers the integrated system of safe work procedures and permits.

   a. Only managers, engineers, supervisors, senior operators, and safety technicians are qualified to issue work permits; management assumes that they understand how to issue work permits based on their education, experience, and the general safety training they receive.

   b. Employees receive a one to two hour block of instruction when they are added to the list of authorized permit issuers; instruction focuses mainly on how to fill out permits.

   c. Authorized permit issuers receive initial and periodic refresher training that emphasizes (1) changes in procedures, (2) recent incidents within the facility, company, or industry involving nonroutine work, and (3) common points of confusion or errors that tend to be made consistently across the facility.

16. Ensure that permit authorizers are well aware of, and have a sense of vulnerability toward, special hazards typically encountered while performing nonroutine work.

   a. A practice is in place to not allow particularly unusual nonroutine work (e.g., use of explosives or ionizing radiation) without approval of the facility or unit manager.

   b. Permit authorizers generally participate in hazard reviews and risk analyses, and have a sound appreciation of hazards that are present at the facility during most routine and nonroutine operations.

   c. In addition to participating in hazard reviews, permit issuers routinely review "what went wrong" summaries from other parts of the facility, similar facilities, or other companies in the industry, particularly if the hazard is directly pertinent to the facility.

Control Access to Particularly Hazardous Areas

17. Establish a system to control access to process areas by employees not involved the operation of the process. This system should include a means to ensure proper verbal communication between the operators and other workers who need to enter the controlled area.

   a. Signs are posted on doors or walkways leading to process areas that state, "Authorized Personnel Only".

   b. Contractors must check in with the control room prior to entering a process area; visitors are not allowed in process areas unless they are escorted by an employee.

   c. Managers; maintenance, technical, and laboratory personnel; and other nonoperations personnel are required to notify operators prior to entering a process area, but direct communication is not required.

   d. All nonoperational employees must go to the control room and sign in prior to entering process areas; these employees must also sign out upon leaving.

   e. In addition to item (d), entrants must swipe an electronic badge when entering and exiting the process area. This system maintains a running electronic log that can be accessed from the control room and other critical points at the facility to show who is present in each process area.

18. Control access to areas of the facility in which special hazards are present.

   a. At the discretion of a supervisor or project engineer, particularly hazardous areas may be blocked off with caution tape.

b. Particularly hazardous areas are barricaded and the hazard is identified. Workers rarely cross these boundaries.

c. In addition to item (b), anyone entering the barricaded area must notify operations before entry.

d. In addition to item (c), hard barricades, such as temporary metal walls or fences, are used to seal off particularly hazardous or sensitive areas.

Enforce the Use of Safe Work Procedures, Permits, and Other Standards

19. Establish a system to routinely inspect work areas to determine if (1) safe work procedures are being followed, (2) permit conditions appear to be appropriate, and (3) permit conditions are being followed.

   a. Inspections are routinely carried out by the supervisor for the area in which the work is taking place. The supervisor ensures that corrective actions are taken in a timely manner, but reports only significant violations of safe work procedures or permit conditions.

   b. All managers, supervisors, engineers, and other technical staff members are required to periodically conduct a safety audit, and the audit protocol includes checking on nonroutine work to ensure that safe work conditions are in place.

   c. All managers, supervisors, engineers, and other technical staff members are required to conduct a safety audit each week, and the results of these audits are summarized and trended by the safety department.

   d. Key personnel throughout the facility are trained to perform safety audits, including audits of nonroutine activities, using a well defined method. The results of these audits, known as safe and unsafe acts audits, are reported and tracked to help identify any decrease in the rigor of the *safe work* element.

Review Completed Permits

20. Review completed permits prior to filing or discarding them and, based on the results of the review, take steps to improve accuracy and completeness of permits.

   a. Permits are discarded when they expire or the work is complete.

   b. Permits are collected and filed for a designated period of time once the work is complete.

   c. One copy of each permit is returned to a designated individual (e.g., area safety representative, maintenance planner, lead operator) who reviews the permit to determine if (1) the permit conditions appear to be appropriate, (2) all required fields are completed, and (3) the permit is legible. If discrepancies are noted, feedback is provided to the specific authorizer on how to properly complete a work permit.

d.   In addition to item (c), systemic problems, once identified, are addressed via special training sessions, tool box talks, or procedure reviews.

11.4   EXAMPLES OF WAYS TO IMPROVE EFFECTIVENESS

This section provides specific examples of industry tested methods for improving the effectiveness of work activities related to the *safe work* element. The examples are sorted by the key principles that were first introduced in Section 11.2. The examples fall into two categories:

1.   Methods for improving the performance of activities that support this element.
2.   Methods for improving the efficiency of activities that support this element by maintaining the necessary level of performance using fewer resources.

These examples were obtained from the results of industry practice surveys, workshops, and CCPS member-company input. Readers desiring to improve their management systems and work activities related to this element should examine these ideas, evaluate current management system and work activity performance and efficiency, and then select and implement enhancements using the risk-based principles described in Section 2.1.

11.4.1 Maintain a Dependable Practice

Train personnel to recognize hazards on an ongoing basis. By all accounts, human error is the primary cause of accidents, and many of those errors result from the failure to recognize hazards. For example, almost all workers are well acquainted with asphyxiation hazards, yet workers often willingly enter a confined space with no protective equipment to help a collapsed coworker, resulting in two fatalities rather than one. Ensure that all persons who authorize or perform nonroutine work are acutely aware of hazards.

Look for opportunities to drive out human error, particularly when the system for managing nonroutine work will be stressed. Many safe work procedures work well in periods of low demand. During normal operation, one or two permits per day may be issued for a unit, and nearly all of the field work is performed by the facility's maintenance department. The activity level is low, training and familiarity is high, and workers can focus on a limited range of tasks. However, a system that works well under these conditions may break down when the system is stressed. During a planned shutdown, for instance, dozens of jobs are being performed simultaneously in the unit, more than a hundred permits are issued each day, and hundreds of contract workers are onsite. Conditions are ripe for human errors to occur. Consider implementing additional visual cues during high demand conditions to help reduce mistakes. For example, the facility may assign a unique color to each crew. When issuing a line-breaking permit, the issuer could then use plastic tape corresponding to that crew's color to mark the flanges that are to be broken. Likewise, crews can be instructed to spray paint the handles on any blinds that they install with their designated color. A crewmember who is later asked to pull a blind that is marked with a different crew's color will know that something may be amiss. Or, if blinds remain when the work is complete, it will be easier to determine which job might still be incomplete.

Encourage senior management to take a hands-on approach. Management can play a very visible role in driving efforts to renew or revitalize safe work procedures and practices. For example, a facility that is introducing a new work permit to control all nonroutine work at the facility will likely spend a significant amount of time and effort supporting the development of the procedure and permit, emphasizing its use when it is first rolled out, and working through any glitches that arise. When presenting the new practice at a safety meeting or training session, the unit manager has the opportunity to choose his level of involvement. She can opt to (1) delegate the training to the safety coordinator and attend to other matters, (2) introduce the training, explain why it is important, and turn the meeting over to the safety coordinator, or (3) personally lead the training. These three increasing levels of visible support are likely to result in increasing levels of compliance with the new policy/procedure and corresponding reductions in risk. By involving herself directly in the training, the manager demonstrates her belief in the importance of safe work practices.

11.4.2 Effectively Control Nonroutine Work Activities

Issue and authorize permits at the shop floor level. Some facilities allow only supervisors or safety specialists to issue permits, which can be burdensome in a couple of ways. First, depending on the size of the facility, part or all of a person's time might be spent going from worksite to worksite to issue permits. Second, an additional loss results from the cost of having maintenance or contract personnel waiting on permits. Not only does this affect efficiency, waiting for permit approval may also delay completion of the work and subsequent restart. These delays can directly impact overall down time and, hence, output.

Thus, significant economic drivers exist to allow operators to issue permits. This approach can be very successful; operators are normally keenly aware of the hazards in their area, and they are readily available to issue permits. Having operators issue permits also facilitates communication between the operators and the crew that is assigned to perform the nonroutine work. However, this system only works well if (1) the operators assigned to issue/authorize permits are well trained and (2) the facility's culture empowers the operator to make decisions rather than simply fill out paperwork. In addition, as the number of permit issuers and authorizers increases, control decreases. Facility policies and procedures must clearly define how the permits will be applied and specify minimum safeguards, such as specific PPE, for certain situations. Reviewing permits and ongoing nonroutine work as part of a facility's routine/frequent safety inspection program becomes an important work activity when using this highly distributed approach.

Expand safety observations to peers. Safety inspection programs, such as safe and unsafe acts audits or a program of formal routine worksite inspections, can overload supervisors and technical personnel. One alternative is to enlist the help of operators and maintenance employees in performing these inspections. These employees are often in harm's way should a process incident occur, so they have a vested interest in the safety of their unit. This approach also helps support work activities and programs that are part of the *workforce involvement* and *culture* elements. Most workers at a facility with an advanced process safety culture will support this effort. They recognize the safety program in general, and the *safe work* element in particular, for what it is – a systematic set of policies and procedures designed to make sure that people do not get hurt at work. Finally, just as teaching others forces one to acquire a thorough command of the subject, participating in a formal safety inspection program will sharpen each participant's safety skills, particularly in the areas of hazard recognition and application of appropriate safeguards.

Use multi-copy forms. Another means to promote communication between the operating team and the work crew is to keep a copy of the permit in the control room. Many facilities use multi-copy forms for permits so that one copy can be taken back to the control room and posted in an area designated for open work permits. This allows operators to quickly determine what jobs are ongoing and what equipment is involved. When the work is complete, the copies are removed from this board, signed off, and either discarded or left for the unit supervisor to review. From an efficiency standpoint, however, multi-copy forms can become counterproductive. Some facilities use four-page forms, with copies of completed permits being sent to the safety, maintenance, operations, and engineering departments. Facilities that retain multiple copies of completed work permits in separate files should determine if (1) a valid reason exists to keep any copies of completed work permits beyond the time that they are reviewed and (2) any reason exists to keep multiple copies on file.

Use visual aids. Wherever change is frequent, such as at batch plants, visual reminders of the current status of each piece of equipment can help alert personnel to the hazards associated with nonroutine work. For example, maintaining a status board in the control room with a fixed location for each major equipment item helps keep personnel informed, particularly when multiple nonroutine activities may be in progress at any given time.

Load checklists onto personal digital assistants (PDAs). One advantage of paper-based permit systems is that the permit can become a form of contract between the equipment owner and the crew assigned to perform the work. However, nobody likes to read long contracts with lots of fine print, much of which does not apply to any particular job. One alternative is to develop software applications and install them on PDAs that can be taken into the field. Thus, decision trees could be developed to help permit issuers identify hazards (such as flame or spark producing operations) and launch the issuer into the proper permit/checklist (in this case, hot work). The permits must be printable so that the crew in the field will be able to easily refer to permit conditions and operators or other equipment owners will know which permits are currently active. Advantages of a PDA-based system are: (1) permits can be uploaded to central database files and more readily stored for later review, (2) permits/checklists can be easily added or modified because the permits are printed on demand, and (3) issuers will not be tempted to work from memory if they failed to bring all of the necessary permits/checklists to the field. PDA-based systems can also ensure that no required fields are left blank and that the completed permit will be legible. One drawback to this type of system is the extra time required to print out and distribute the permit after the issuer has gone through the checklists on the PDA.

Develop computer-based training (CBT) modules for permit preparation. Some training, particularly training on procedures that do not normally require decision making, can be efficiently delivered via CBT, and workers can review the CBT modules when needed.

Automate parts of the system used to issue permits. Safe work permits specify conditions that are rarely verified by permit authorizers. For example, a facility may require that persons authorized to issue hot work permits complete certain refresher training requirements every 12 months. Likewise, persons assigned to perform fire watch duty may be required to attend periodic refresher training. However, a permit issuer will rarely verify the training status of the fire watch, beyond asking if the training requirement has been met when issuing the permit. Computers are very good at tracking and managing this type of data, and the database used to issue work permits can be linked to the training/qualification database. Automated systems can also link interrelated permits based on a set of logic rules. For example, the system may be programmed to not allow a hot work permit to be issued if the fire system in the area is impaired, and similarly not issue a fire system impairment permit if any hot work permits for the affected area are still open. Automating the rote portions of the permit system frees the issuer to focus on job-specific conditions and unique hazards.

Take a daily snapshot of permitted work. Most nonroutine work is planned one or more days ahead of time. Nonroutine work can place a strain on the operations group. For example, operators are often tasked to closely monitor contractors or to operate with control loops in manual mode while instruments are being calibrated. Some level of nonroutine work is unavoidable; however, if enough nonroutine work is occurring simultaneously, the potential increase in operational risk could be significant. This is particularly true if other factors are simultaneously placing strain on the organization. For example, the normal operator is on vacation and the field operator is covering the shift on a step up basis, the loader is covering for the field operator, and the senior operator is sick. Under these conditions, the risk associated with otherwise typical nonroutine work activities may become unacceptable. By reviewing all of the planned and unplanned work on a daily basis, along with the staffing and training status for both the production and maintenance departments, managers can evaluate the intensity of higher risk nonroutine work and make better, risk-informed decisions regarding what activities are acceptable given the current level of staffing, training, and experience.

Create procedures for infrequent but repetitive activities. When planning for a major shutdown, determine which of the planned nonroutine activities are not supported with a job-specific procedure or checklist. Consider developing a job-specific procedure to (1) force a pre-job review by highly experienced personnel who may not be readily available once the shutdown is under way, (2) help ensure that the activity is performed in a safe manner, and (3) document how the activity should be performed (which should improve the planning and efficiency of future shutdowns).

This last point requires some further explanation. Even if a permit includes a description of each step that should be performed, permits are seldom retained for very long after a job is complete. Lessons learned during the course of planning and executing an infrequent activity are forgotten years later, particularly if they are unique to one job. Although some activities occur infrequently, such as a 10-year internal visual inspection of a pressure vessel, facilities are certain to undertake them again in the future, and methods that worked well in the past are a good starting point for planning the activity the next time. If these types of procedures are developed, a reliable management system should be established to (1) update the procedures after the activity is performed to capture nuances that were not anticipated by the procedure writer and (2) carefully review the procedures prior to use in the future.

Use real life examples for hazard recognition training, and make it as interactive as possible. A good way to help workers expand their understanding of hazards is to review summaries of previous incidents. Such incidents are recounted in many sources, including: (1) *Process Safety Progress*, (2) several books by Trevor Kletz, (3) *Safety Digests of Lessons*

Learned published by the American Petroleum Institute, and (4) *Process Safety Beacon* published by the CCPS (Ref. 11.5, Ref. 11.6, Ref. 11.7, and Ref. 11.8, respectively). Also, government agencies such as OSHA and the Chemical Safety Board (CSB) produce incident safety alerts or investigation reports (for example, the CSB report noted as Ref. 11.2). These reports contain excellent material for daily safety talks or departmental safety meetings. If time allows, provide a summary of the events leading to the incident and ask the group to identify the hazards and suggest methods to control the hazards. Also, consider having operators and mechanics review any recent events at the facility. Employees who were recently involved in a near miss incident can be particularly effective trainers because of their personal understanding of the importance of the safeguard(s) that helped prevent or mitigate the severity of the incident.

11.5   ELEMENT METRICS

Chapter 20 describes how metrics can be used to improve performance and when they may be appropriate. This section includes several examples of metrics that could be used to monitor the health of the *safe work* element, sorted by the key principles that were first introduced in Section 11.2.

In addition to identifying high value metrics, readers will need to determine how to best measure each metric they choose to track. In some cases, an ordinal number provides the needed information, for example, total number of workers. Other cases, such as average years of service, require that two attributes be indexed to provide meaningful information. Still other metrics may need to be tracked as a rate, as in the case of employee turnover. Sometimes, the rate of change may be of greatest interest. Since every situation is different, the reader will need to determine how to track and present data to most efficiently monitor the health of RBPS management systems at their facility.

11.5.1 Maintain a Dependable Practice

- *Compliance with the training plan for activities related to the safe work element.* A decrease in compliance could indicate that the organization is placing less emphasis on this element. Compliance with the training plan could be measured in terms of the fraction of employees who have completed a specific training module or set of training modules, the fraction of scheduled classes that were actually offered, or some other metric that helps determine if the intended training activities are being conducted in accordance with the plan.

- *Progress toward implementing a new safe work practice or making significant improvements to the existing system.* A lack of progress indicates that additional resources or management emphasis is required. A simple bar chart can be used to track progress on this metric. Gantt charts can be used to graphically illustrate conformance to the schedule when multiple interrelated implementation activities are being measured.
- *Percent of permits completed correctly.* An increase in the fraction of improperly completed permits may indicate a need for additional training. In addition, safe work permits are often kept on file pending periodic review by insurance inspectors or corporate auditors. If permits are not reviewed and corrective actions taken shortly after they are turned in, the facility has (1) failed to take advantage of an opportunity to identify and address training or other issues and (2) unknowingly established the basis for a future audit finding.

11.5.2 Effectively Control Nonroutine Work Activities

- *Number of injuries related to nonroutine work.* An increase in injuries, particularly first-aid events, which occur relatively frequently, may be an early indicator of issues with respect to the *safe work* element.
- *Number of near miss incidents related to nonroutine work.* An increase in this metric might also indicate issues with respect to the *safe work* element. However, near miss incidents normally occur at a relatively low frequency, and this metric may not respond quickly to underlying changes in practices at a facility.
- *Unsafe conditions or permit violations observed during routine audits (on a consistent basis).* Even if proper systems are in place and routine inspections are being performed, an upward trend in permit violations indicates that the system for managing nonroutine work needs attention.
- *Number (or percent) of safe work procedures that are past due for periodic review.* An upward trend would indicate that more attention or resources needs to be applied to reviewing safe work procedures.
- *Number of causal factors identified by incident investigation teams related to failures to properly apply or follow a safe work permit.* An increase in this metric would clearly indicate issues with respect to the safe work element. However, incidents that are formally investigated normally occur at a very low frequency compared to work activities associated with the safe work element; therefore, this metric is unlikely to quickly respond to a decline in adherence to safe work procedures.
- *Percent of scheduled job observations/audits performed.* If a facility elects to widely distribute the authority to issue and authorize safe work permits, the worksites must be routinely inspected to make sure that the

facility's policies and procedures are being properly implemented. A decline in this metric indicates that the management system supporting the *safe work* element may be failing.

- *Frequency of improper shift-to-shift handoffs.* If shift-to-shift communication is facilitated by written status reports, periodically comparing the information in the turnover reports to records of open work permits will help identify lapses in shift-to-shift communication. Increases in this metric indicate that refresher training in shift turnover procedures may be warranted.
- *Number (or percent) of safe work procedures revised each year.* A downward trend in the rate that procedures are revised may indicate that more attention or resources needs to be applied to maintaining safe work procedures.
- *Average time between a request for a permit and when it is issued.* Facilities should take steps to reduce wait times for maintenance or contractor crews. Ultimately, excessive wait times may tempt crews to start without a permit or send a message to the organization that efficient use of time is not important.
- *Average time spent issuing a work permit, for example, total time spent issuing permits divided by the number of permits issued.* A decrease in this metric may indicate that personnel authorized to issue permits are not using proper diligence. An increase in this metric may indicate a loss of efficiency in the facility's system to issue safe work permits. This metric may be difficult to obtain if permits are issued and authorized by a large number of personnel.
- *Staff hours spent writing, reviewing, and approving safe work procedures.* This metric will indicate whether the level of effort is changing with time, but any interpretation should consider the workload or perceived need for changes to safe work procedures.

11.6   MANAGEMENT REVIEW

The overall design and conduct of management reviews is described in Chapter 22. However, many specific questions/discussion topics exist that management may want to check periodically to ensure that the management system for the *safe work* element is working properly. In particular, management must first seek to understand whether the system being reviewed is producing the desired results. If the organization's level of performance for the *safe work* element is less than satisfactory, or it is not improving as a result of management system changes, then management should identify possible corrective actions and pursue them. Possibly, the organization is not working on the correct activities, or the organization is not doing the necessary activities well. Even if the results are satisfactory, management reviews can help determine if resources are being used wisely: are there tasks that could be done more efficiently or tasks that should not be done at all? Management can combine metrics listed in the previous section with personal observations, direct questioning, audit results, and feedback on various topics to help answer such questions. Activities and topics for discussion include the following:

- Review the scope of the safe work procedures and associated permits to determine if nonroutine work activities are being performed that are not adequately controlled.
- Review the scope of the safe work procedures and associated permits to identify activities that are inherently low risk. If permits are routinely being authorized for very low hazard work, determine if this desensitizes permit issuers to the importance of this element.
- Review practices in the field to ensure that permits require two-way communication between the group that owns the equipment or is responsible for the area (typically production) and the group that is authorized to work on the equipment. Ensure that this communication consistently occurs at critical transition times, such as the start of the job, shift change, and the end of the job.
- Review practices that equipment owners use to keep all potentially affected personnel informed of nonroutine work in their area of responsibility. Did the permit authorizer adequately identify hazards and specify reasonable controls? Is the accuracy and completeness improving?
- Examine the balance between work that is governed by temporary operating procedures and that which requires safe work permits. For example, is a unit issuing a safe work permit to authorize monthly hydroblasting of a vessel under a maintenance work order rather than establishing a permanent, approved procedure for this task?

Since nonroutine work occurs at a very high frequency at most facilities, managers can most effectively monitor performance by setting aside time to be out in the process areas observing work as it is performed. Unlike many other RBPS elements, poor performance should be readily apparent to a careful observer.

One potential shortcoming of this direct observation approach is that, although it will likely measure conformance to facility procedures, it will not provide a view of how the facility compares to similar plants. In this respect, second- or third-party safety audits are a good supplement to management review and direct observation. Audits can also be used to identify opportunities to improve the effectiveness of work activities associated with this element.

Finally, managers and supervisors, more than anyone else, need to demonstrate every day that safety is everyone's job. Although this can be said for each RBPS element, observing compliance with many of the RBPS elements cannot be accomplished by simply walking through the facility. Periodic management review sessions are important; however, managers and supervisors can most directly influence the effectiveness of the *safe work* element by their own actions. Managers and supervisors who demonstrate conformance to procedures will likely inspire workers to follow suit. A manager who stops to assess a situation and correct a worker who is not following a prescribed safe work practice is saying, in effect, that the safe work procedures and permits *are* important. Thus, all levels of supervision and management should be mindful that they must take the time to stop, investigate, and correct unsafe acts or unsafe conditions whenever they are observed or risk losing personal credibility and compromising the credibility of the facility's entire *safe work* program.

11.7   REFERENCES

11.1   Lees, Frank P., *Loss Prevention in the Process Industries, Hazard Identification, Assessment, and Control*, 2nd edition, Butterworth-Heinemann, Oxford, England, 1996.

11.2   U.S. Chemical Safety and Hazard Investigation Board, *Safety Bulletin, Hazards of Nitrogen Asphyxiation*, No. 2003-10-B, June 2003.

11.3   Kleindorfer, Paul R., *Accident Epidemiology and the U.S. Chemical Industry: Preliminary Results from RMP*Info*, Center of Risk Management and Decision Processes, The Wharton School, University of Pennsylvania, March 6, 2000.

11.4   Guiochon, Charles, *What Might Have Happened in Toulouse (France) on September 21, 2001*, Société Française de Chimie, 2002.

11.5   Kletz, Trevor, *Case Histories of Process Plant Disasters*, 4th edition, Elsevier Science, Burlington, Mass., 1999.

11.6   Kletz, Trevor, *Lessons from Disaster: How Organizations Have No Memory and Accidents Recur*, Gulf Publishing Company, Houston, Texas, 1993.

11.7   Safety Digest of Lessons Learned – Section 4, Safety in Maintenance, American Petroleum Institute Publication 758, 1981.

11.8   *Process Safety Beacon*, published electronically each month by the American Institute of Chemical Engineers' Center for Chemical Process Safety.

12

ASSET INTEGRITY AND RELIABILITY

Equipment failure has resulted in a large number of accidents within the process industries. Failure to detect conditions that indicated a high likelihood for loss of containment can result in disaster. For example, on the morning of November 19, 1984, a pipe used to transport light hydrocarbons from a refinery to a storage terminal in Mexico City, Mexico, ruptured. Corrosion had gradually weakened the pipe. The light hydrocarbons quickly found an ignition source, triggering a series of fires and explosions, resulting in approximately 500 fatalities (Ref. 12.1).

However, catastrophic accidents due to equipment failure are not limited to undetected (or unaddressed) gradual deterioration, nor are they limited to the process industries. On May 25, 1979, a DC-10 crashed on takeoff at Chicago's O'Hare Airport when the pylon holding the left engine to the wing failed. The National Transportation Safety Board (NTSB) concluded that the damage to the pylon resulted from incorrect maintenance procedures during the replacement of some internal bearings eight weeks before the crash. The replacement procedure called for removal of the engine prior to the removal of the engine pylon. To save time and money, the airline instructed its mechanics to remove the engine together with the pylon all at one time, using a forklift to hold the engine in place. A failure of the forklift's hydraulic system left the engine unsupported during a shift change, damaging the pylon. The damage went unnoticed for several flights, getting worse with each flight. During the final takeoff, the pylon failed, tearing the left engine away from the wing. The resulting crash, which remains the deadliest aviation accident in U.S. history, killed 273 people, including 2 on the ground. Subsequent to the accident, the Federal Aviation Administration grounded the entire DC-10 fleet pending inspection of all of the pylons. Six other DC-10 aircraft operated by two airlines were found with similarly damaged pylons. Both airlines had used the expedient procedure to remove the engine and pylon in a single step (Ref. 12.2).

12.1   ELEMENT OVERVIEW

Asset integrity, the RBPS element that helps ensure that equipment is properly designed, installed in accordance with specifications, and remains fit for use until it is retired, is one of nine elements in the RBPS pillar of *managing risk*. This chapter describes the attributes of a risk-based management system for ensuring the integrity and reliability of critical equipment and safety systems. Section 12.2 describes the key principles and essential features of a management system for this element. Section 12.3 lists work activities that support these essential features, and presents a range of approaches that might be appropriate for each work activity, depending on perceived risk, resources, and organizational culture. Sections 12.4 through 12.6 include (1) ideas to improve the effectiveness of management systems and specific programs that support this element, (2) metrics that could be used to monitor this element, and (3) issues that may be appropriate for management review.

12.1.1 What Is It?

The *asset integrity* element is the systematic implementation of activities, such as inspections and tests necessary to ensure that important equipment will be suitable for its intended application throughout its life. Specifically, work activities related to this element focus on (1) preventing a catastrophic release of a hazardous material or a sudden release of energy and (2) ensuring high availability (or dependability) of critical safety or utility systems that prevent or mitigate the effects of these types of events.

12.1.2 Why Is It Important?

Designing and maintaining equipment that it is fit for its purpose and functions when needed is of paramount importance to process industries. Maintaining containment of hazardous materials and ensuring that safety systems work when needed are two of the primary responsibilities of any facility.

12.1.3 Where/When Is It Done?

Asset integrity activities range from technical meetings involving experts seeking to advance the state-of-the-art in equipment design, inspection, testing, or reliability, to a plant operator on routine rounds spotting leaks, unusual noises or odors, or detecting other abnormal conditions. However, this element primarily involves (1) inspections, tests, preventive maintenance, predictive maintenance, and repair activities that are performed by maintenance and contractor personnel at operating facilities and (2) quality assurance processes, including procedures and training, that underpin these activities. *Asset integrity* element activities occur at many places and extend throughout the life of the facility.

12.1.4 Who Does It?

Asset integrity activities occur at several organizational levels. Industry sponsored technical committees and organizations are continuously working to advance the state of knowledge regarding proper design and inspection, test, and preventive maintenance (ITPM) practices to help ensure that equipment is fit for service at commissioning, and remains fit for service throughout its life. Companies (or business units) often establish centers of excellence in the *asset integrity* field, establish corporate standards, and promote efforts to continuously improve the safety and reliability of process equipment. At an operating facility, the *asset integrity* element activities are an integral part of day-to-day operation involving operators, maintenance employees, inspectors, contractors, engineers, and others involved in designing, specifying, installing, operating, or maintaining equipment.

12.1.5 What Are the Anticipated Work Products?

Asset integrity element work products include:

- Reports and data from initial inspections, tests, and other activities to verify that equipment is fabricated and installed in accordance with design specifications and is fit for service at startup.
- Results from ongoing ITPM tasks, performed by trained or certified personnel and based on written procedures that conform to generally accepted standards, that help ensure that equipment remains fit for service.
- Controlled repairs and adjustments to equipment by trained personnel using appropriate written procedures and instructions.
- A system to control maintenance work, repair parts, and maintenance materials needed for the work to help ensure that equipment remains fit for service.
- A quality assurance program that helps prevent equipment failures that could result from (1) use of faulty parts/materials or (2) improper fabrication, installation, or repair methods (e.g., improperly supporting the engine as discussed at the start of this chapter).

The primary objective of the *asset integrity* element is to help ensure reliable performance of equipment designed to contain, prevent, or mitigate the consequences of a release of hazardous materials or energy. Although proper execution of work activities associated with this element requires a high level of human performance, the ultimate work product for this element is reliable and predictable equipment operation.

12.1.6 How Is It Done?

An effective a*sset integrity* program depends on management ensuring that:

- Equipment and systems are properly designed, fabricated, and installed.
- The unit is operated within the design limits of the equipment.
- ITPM tasks are conducted by trained and qualified individuals using approved procedures and completed as scheduled.
- Repair work conforms to design codes, engineering standards, and manufacturer's recommendations.
- Appropriate actions are taken to address deficiencies, regardless of how they are discovered.

12.2 KEY PRINCIPLES AND ESSENTIAL FEATURES

Safe operation and maintenance of facilities that manufacture, store, or otherwise use hazardous materials requires reliable equipment operation as well as reliable human performance. In particular, safety depends on preventing loss of containment events and maintaining high dependability/availability of safety and utility systems that help prevent or mitigate releases of hazardous materials or energy. Experience shows that safe operations are directly and closely correlated with reliable operations, and conditions that lead to safe and reliable operation generally lead to high overall plant availability and other related benefits.

The policy governing the *asset integrity* element should establish the scope of the element and define roles and assign responsibilities for conducting specific tasks. In addition, this policy normally (1) establishes standards for training and procedures that support this element, (2) defines the standards or technical bases that are used to develop the ITPM program for various types of equipment, and (3) describes the management systems that help ensure that (a) equipment deficiencies are promptly reported and properly managed and (b) equipment is fabricated, installed, and maintained in accordance with specifications and manufacturers' recommendations.

The following key principles should be addressed when developing, evaluating, or improving any management system for the *asset integrity* element:

- Maintain a dependable practice.
- Identify equipment and systems that are within the scope of the *asset integrity* program and assign ITPM tasks.
- Develop and maintain knowledge, skills, procedures, and tools.
- Ensure continued fitness for purpose.

- Address equipment failures and deficiencies.
- Analyze data.

The essential features for each principle are further described in Sections 12.2.1 through 12.2.6. Section 12.3 describes work activities that typically underpin the essential features related to these principles. A facility should evaluate the risks and potential benefits that may be achieved as a result of improvements in this element. Based on this evaluation, management should develop a management system, or upgrade an existing management system, to address some or all of the essential features, and execute some or all of the work activities, depending on perceived risk and/or process hazards that it identifies. However, none of this will be effective if the accompanying *culture* element does not embrace the use of reliable management systems. Even the best management system, without the right culture, will not ensure a high degree of asset integrity and equipment/system reliability.

12.2.1 Maintain a Dependable Practice

Organizations that decide to adopt a formal *asset integrity* program should develop written policies or procedures to guide the implementation and execution of the program. These documents should address the scope, roles and responsibilities, standards, and other aspects of the management system for this element.

Develop a written program description/policy. Facilities should develop a written policy that:

- Governs implementation of the *asset integrity* element at the facility/unit level.
- Identifies corporate procedures, industry standards, and regulations that establish the minimum acceptable standard of care for inspection, test, and maintenance activities.
- References related documents and subordinate procedures that guide day-to-day execution of *asset integrity* work activities.
- Specifies requirements for *asset integrity* work activities, such as element-specific records retention requirements.
- Defines roles and responsibilities for the *asset integrity* element.

Although a large part of the *asset integrity* work activities is directed at ITPM tasks, these activities are only one aspect of the element. Even the best ITPM program will fail unless all of the following conditions exist:

- Equipment is properly designed, fabricated, and installed.
- Repairs are made by trained personnel using proper procedures/methods.
- Repair parts and maintenance materials conform to specifications.
- A process is in place to correct deficiencies and apply the lessons learned from deficiencies or near miss incidents to other equipment/systems.

These issues should be clearly addressed in the policy or procedure that governs the *asset integrity* element.

Determine the scope of the asset integrity element. Equipment that contains hazardous materials, or safety/utility systems that help prevent or mitigate the effects of a catastrophic release of a hazardous material or a sudden release of energy, should be included in the scope of the *asset integrity* element. Some companies limit the list of equipment covered by the *asset integrity* element as much as possible based on a narrow interpretation of regulatory requirements. However, a risk-based approach to process safety requires a more holistic view, basing decisions regarding the scope of this element on the risk associated with equipment failure and/or failure modes. Regardless, the policy governing this element should clearly state management's intent for including or excluding equipment from the *asset integrity* program.

Base design and ITPM tasks on standards. The premise of using standards is that we can all benefit from each others' experiences, particularly when the learning often comes from low-frequency, high-consequence events. (In this chapter, the term standards will be used to include codes, engineering standards, recommended practices, company- or industry-specific standards, and regulations as defined in Chapter 4.) Widely accepted standards usually reflect the experience of an entire industry or business sector. When designing systems, purchasing equipment, and planning ITPM tasks, it makes sense to start with recognized standards. Widely accepted standards are tried and true; they generally define the standard of care within the industry. In addition, they are periodically updated to reflect new information from all stakeholders (equipment designers, manufacturers, users, etc.). In some jurisdictions, regulators have directly adopted industry standards as legally enforceable requirements.

Involve competent personnel. Developing and maintaining competence in the wide variety of standards, test and inspection methods, and generally acceptable engineering and maintenance practices requires significant time and expense. Maintaining this level of expertise essentially requires that companies appoint technical experts within their company or leverage the expertise of contractors and consultants. (This is an extension of the concept of the "technology steward" first introduced in the *competency* element,

Chapter 5.) Assigning someone at each facility to thoroughly research the range of standards that may apply and keep abreast of changes to existing standards as well as publication of new standards would generally be an inefficient use of resources. Assigning this responsibility to a central or shared resource, or to a network of technical personnel (one person has responsibility for API, ASME, and national board publications, another has responsibility for NFPA publications, and so forth), makes good business sense. This practice also facilitates networking throughout the company or business unit, sharing best practices, procedures, experience, and information on particularly good (or particularly bad) experiences. Smaller companies often rely on external experts, including qualified contractors hired to perform specific inspections. Regardless of who performs the work, efforts to maintain competence should be closely integrated with the *competency* and *standards* elements.

Update practices based on new knowledge. Simply assimilating information published in standards may not be an adequate basis for establishing the best mix of *asset integrity* element work activities for several reasons:

- The lag time between discovery of new information and incorporation into standards is typically several years, and waiting may not be prudent. The incident involving the damaged pylons on the DC-10 fleet is a good example of how rapid reaction can prevent similar accidents. If the regulators had not promptly grounded the entire DC-10 fleet and mandated 100% inspections of the pylons, six additional aircraft could have lost an engine before any maintenance or inspection standard could have been revised.
- Some failure modes are very service- or process-specific. Some committees charged with updating standards that address a class of process equipment, such as pressure relief devices, may not address a narrowly focused, service-specific issue.
- In most cases, the development of new or improved ITPM methods precedes their incorporation into standards by several years. Newer technologies can sometimes provide better information and better value.
- Failures or unexpected conditions discovered during ITPM tasks should cause a company (or industry sector) to ask, "Can that happen here?" Based on the risk, each company should decide how to apply the lessons learned from one facility or type of process throughout the company.

Companies can keep abreast of recent occurrences that will eventually affect standards in several ways, ranging from regularly reading industry journals to attending industry conferences and working meetings for industry associations. Many changes to ITPM tasks result from (1) examination of equipment failures, (2) incident investigation or risk analysis team recommendations, and (3) manufacturer notifications. Still others are a consequence of corporate objectives for its process safety program. For example, a change in a company's risk tolerance, described in Section 9.2.1, may require a review of ITPM tasks independent of any other changes.

Thus, companies and facilities should actively seek new knowledge related to its asset integrity program. Based on the new information, facilities should either (1) implement additional ITPM tasks, (2) change the frequency or scope of existing ITPM tasks, (3) apply existing activities to different equipment/systems, or (4) reduce the level of effort by altering work activities, frequency, or scope.

Establish a means to disseminate information and facilitate continuous improvement in practice at the unit level. Work activities that add to the corporate knowledge base generally filter through a single person or a small committee. Management needs to study this information to determine what it means for all corporate facilities, or more important, what it means for each specific unit at their facility. This information should also be put into terms that the maintenance supervisor or lead inspector at each facility can understand. This communication should also provide specific guidance on what, if any, action that facilities are expected to take with respect to the new information.

Integrate the asset integrity element with other goals. Efforts to implement an ITPM program independent of efforts to maintain equipment will be at best inefficient and, at worst, could lead to increased risk. For example, assume a facility implements separate initiatives to improve process safety, product quality, and yield; also assume that a particular instrument is important to all three initiatives. These three teams, working independently, could potentially establish three different preventive maintenance work orders to periodically calibrate this particular instrument. Each time an instrument technician tests or calibrates a sensor, some chance occurs that a procedural error will leave the instrument in a less reliable condition than it was prior to calibration. Failure to integrate the recommendations from these three initiatives could actually increase risk associated with instrument failure. Moreover, redundant maintenance is inefficient. Even if the initiatives require different ITPM tasks, an integrated approach to scheduling and performing maintenance tasks normally minimizes cost and downtime.

12.2.2 Identify Equipment and Systems that Are Within the Scope of the Asset Integrity Program and Assign ITPM Tasks

The scope of the *asset integrity* element includes the physical equipment that provides containment and safety/utility systems that are designed to prevent or mitigate the effects of loss of containment or a sudden release of energy.

Identify equipment/systems for inclusion in the asset integrity element. Define, in clear unequivocal terms, what equipment should be included in the reliability program. This is best done on the basis of function rather than equipment description. For example, one primary objective of facilities that process highly hazardous materials is to maintain containment – keep the hazardous materials inside the pipes and vessels and keep everything else out (oxygen, contaminants that could lead to a runaway reaction, etc.). Another objective is to ensure that safety systems and critical utility systems that help prevent or mitigate incidents (1) are capable of performing their intended function and (2) will operate when needed. Thus, maintaining the performance of the pumps and fans for a scrubber that should mitigate a release of toxic gas, and maintaining the required concentration of caustic in the scrubber bottoms, are activities that fall under the *asset integrity* element, even though the related work activities may be performed by operators and described in operating rather than maintenance procedures.

Develop an ITPM plan. Once the equipment and systems are identified, the next step is to identify the tasks that should be performed. These tasks are largely driven by standards, manufacturers' recommendations, and facility/corporate equipment history. The ITPM plan normally specifies a frequency for each task, along with other supporting information, such as procedures and inspector qualifications. Note that risk plays a key role in developing the ITPM plan. Based on the consequence of failure of a pump, the appropriate ITPM task may range from (1) continuous condition monitoring, for example, monitoring a critical fluid temperature or vibration analysis for rotating equipment to (2) periodically lubricating the pump and doing a cursory visual inspection at that time to (3) simply running the pump to failure (no ITPM). In addition, the likelihood of a particular failure, and the likelihood that a particular ITPM task could help predict the failure, shapes the ITPM plan. We intuitively base ITPM decisions on risk. For example, most automobile owners run their headlights to failure but routinely inspect brake systems. This is due to (1) the inherent redundancy with headlights, (2) the relative simplicity of predicting hydraulic line or brake pad failure compared to predicting light bulb failure, and (3) the severity of consequences that can result from catastrophic brake system failure.

Many facilities use a standards-based risk management system to identify safety instrumented systems (SISs), which are systems designed to achieve or maintain safe operation of a process in response to an unsafe process condition. Examples of these management systems include the Instrumentation, Systems, and Automation (ISA) Society's ISA 84.01 series of standards and the International Electrotechnical Commission's (IEC's) standards IEC 61508 and IEC 61511. These management systems typically rely on (1) results from the *hazard identification and risk analysis* element and (2) the company's risk tolerance to identify SISs and determine the required risk reduction that the SIS must provide. The required risk reduction is expressed as the safety integrity level (SIL) for the safety instrumented function (SIF) associated with the SIS. In practice, the SIL specifies an upper limit for the probability of failure on demand of the SIF. The probability of failure on demand (and thus SIL) depends on failure modes, component reliability, system architecture, and testing/maintenance activities. Generally, for a given design, longer test intervals lead to higher probability of failure on demand. Thus, the ITPM tasks associated with a SIS should not be modified without special analysis.

Update the ITPM plan when equipment conditions change. Facilities frequently add or remove equipment, or change equipment services. This activity is managed via the *management of change* element, placing a demand on the *asset integrity* element to update the list of equipment included in the *asset integrity* program, assign ITPM tasks, and possibly develop new procedures or train personnel on new ITPM tasks.

12.2.3 Develop and Maintain Knowledge, Skills, Procedures, and Tools

Successful execution of work activities for the *asset integrity* element depends on trained workers using the right tools and executing activities in accordance with written procedures. In this context, tools include both (1) devices used to conduct inspections, tests, and repairs and (2) systems used to schedule ITPM tasks and store/analyze the large volume of data generated by these activities.

Develop procedures for inspection, test, repair, and other critical maintenance activities. As described in Chapter 10, written procedures help ensure reliable human performance. Some procedures will be used every time that the job is performed, while others are used as a reference for troubleshooting or when the job does not go as planned. Regardless, failure to document how a job should be performed can lead to erratic performance as well as the potential for injury to persons performing the work. Written procedures should govern activities such as:

- Inspection and tests
- Critical repairs
- Preparation for maintenance
- Control of equipment/safe work

While operating procedures tend to be self-contained and unit-specific, procedures governing ITPM tasks and other *asset integrity* element activities are often equipment-specific and are composed of a suite of distinct procedures, such as:

- *Safe work* practices to control nonroutine work, including lockout/tagout and line breaking, which normally apply throughout the facility.
- Equipment removal and decontamination procedures, which are normally chemical- or service-specific.
- Repair or overhaul procedures (typically the original equipment manufacturer's [OEM's] maintenance manual, references to the OEM's manual, or site-developed repair procedures based on the OEM's manual).
- Procedures for reinstallation, which are normally specific to equipment and service.
- General procedures/practices governing equipment inspection and testing, such as checking rotation and alignment, checking for leaks, and conducting performance checks. These tasks tend to be specific to equipment and service.
- Safe work procedures to return control of the equipment to the operations group.

This group of written procedures is often documented in a job plan that is specific to a maintenance task. For example, consider a relatively common task – removing a 2-in. by 3-in. pressure safety valve in fuming sulfuric acid service for test and refurbishment, and returning the valve to service. The valve is equipped with block valves on the inlet and discharge piping. The activity is planned to coincide with a shutdown (when the vessel will be open and offline), so the vessel can be opened to provide an alternate relief path. In this hypothetical case, the following nine activities and training requirements apply, none of which are specific to this particular pressure safety valve or vessel.

1. *Safe work* procedures governing removal of a pressure safety valve from a vessel that is off line dictate that an alternate relief path be provided. Thus, operations confirms that the vessel has been deinventoried and cleaned, and uses the safe work procedure governing line breaking/opening process equipment to authorize removal of a 2-in. blind flange on the top of the vessel.
2. Line breaking procedures specify (1) how the pressure safety valve is to be isolated from the vessel and the vent header prior to removal, (2) that temporary support be provided for the inlet and discharge lines, and (3) how the flanges on the pressure safety valve are to be broken.
3. Rigging associated with lowering the relief valve to the ground is addressed in mechanic training and is not specifically covered by procedure.

4. Since the valve is in fuming sulfuric acid service, a special safe work procedure that applies to all work on equipment in fuming sulfuric acid service dictates (1) what PPE must be worn during line breaking and (2) how to clean equipment prior to (a) removing it from the unit and (b) sending it offsite. Once the pressure safety valve is cleaned in accordance with this procedure, it is sent to an outside shop for testing and refurbishment.

5. The valve repair shop maintains a "VR" stamp, which requires it to maintain procedures governing initial pop testing, refurbishment, and final testing in accordance with standards and manufacturers' recommendations.

6. Once the valve has been removed, other maintenance procedures specify visual inspection of the inlet and discharge piping by facility maintenance personnel. These procedures also specify the conditions under which the isolation valves should be opened, or how the piping should be inspected if they are not opened.

7. Once the valve is returned from the valve repair shop, maintenance mechanics reinstall the valve based on training that is generic to installing all flanged equipment.

8. A procedure specific to relief valve maintenance governs application of car seals to the inlet and outlet block valves.

9. Finally, safe work procedures govern the return of control of the pressure safety valve and the associated vessel to the operations group, opening the block valves on the inlet and outlet lines, and closing the alternate relief path.

None of these activities is specific to the particular pressure safety valve. By structuring the procedures in this manner, a group of procedures linked into a single job plan can adequately control all activities associated with this task while maintaining the flexibility to "mix and match" these and similar procedures to a wide variety of maintenance activities.

In addition, safe work procedures typically govern ownership of equipment through the maintenance cycle. For example, the production group may own the equipment until it is locked out, at which point the maintenance group assumes responsibility for the equipment. Other facilities dictate that the production groups maintain control over equipment unless it is physically removed from the process area. Although no one best means to control the transition of responsibility exists, failure to establish clear accountability can be a contributing factor to an incident when equipment is either removed from service or returned to service before all of the necessary steps have been taken. (This issue is similarly addressed under the *safe work* and *operations* elements; see Sections 11.2.2 and 17.2.3, respectively.)

Train employees and contractors. Many equipment fabrication, installation, test, inspection, preventive maintenance, and repair activities require specialized knowledge. Developing and implementing training programs to help ensure a high level of performance are addressed in Chapter 14. Contract employers should have similar training programs in place. Because workers who are involved in these activities typically work in multiple units at a facility (or in the case of contractors, multiple facilities), it is important that they also be trained on an overview of processes in units where they are assigned to work and the hazards that exist in each unit. Providing this information to workers helps them (1) recognize abnormal or hazardous conditions and (2) protect themselves and fellow workers from these hazards.

Ensure that inspectors hold appropriate certifications. Some of the standards that apply to ITPM tasks require that workers hold specific certifications. Examples include pressure vessel and piping inspectors, welders, workers who perform nondestructive testing, and in some locations, electricians and other craftspeople. Many certification programs require periodic retesting or ongoing practice in a particular field or activity. These types of certifications are an effective means to ensure that workers maintain their proficiency in a particular area, and that they are periodically exposed to the newest practices and technologies.

Provide the right tools. The importance of having the right tools, knowing how to use the tools properly, and maintaining tools in serviceable condition cannot be over-stressed. This is particularly true for nondestructive testing and other advanced test/inspection methods. This extends beyond just having a particular testing tool in the tool kit and knowing how to use it; it depends on knowing which test or group of tests will provide the best results in a particular situation.

For example, consider a situation in which tube failure in a heat exchanger can lead to release of a toxic chemical into a utility system, resulting in toxic exposure hazards at remote units. Although the facility can readily conduct a hydrostatic pressure test on the tube bundle, a simple pressure test may be inadequate to detect cracked or thin tubes. Management should determine whether knowing the exact condition of the tubes is critical to safety and, if so, select the best tool for determining this, such as eddy current testing. If the facility does not have the tools and training to perform eddy current testing and interpret the data, appropriate contractors should be hired to do the work. If the answer is important to obtain, it is important to use the right tool.

Another critical tool is a computerized maintenance management system (CMMS). Although a manual system of maintenance plans, schedules, and equipment history is possible to operate, the large number of ITPM actions and the vast amount of corresponding data virtually mandate use of a CMMS. A CMMS helps (1) plan and schedule ITPM tasks, (2) store information such as equipment history, repair data, and maintenance procedures, and (3) identify chronic failures. For the CMMS to be an effective tool, facility employees should be able to enter data efficiently, and easily create reports of equipment history. These reports, along with other information, can help managers make risk-informed decisions on issues such as (1) replacing versus repairing equipment or (2) adding, deleting, or changing ITPM tasks.

Software is also needed to analyze data. Over time, tests and inspections generate a significant amount of data, and the rate of change is as important, or more important, than the absolute value of a measurement. If all data are collected and recorded on paper, analyzing the data, predicting future conditions, and planning future testing based on the data become very time-consuming tasks.

12.2.4 Ensure Continued Fitness for Purpose

Regardless of the procedures, tools, and other conditions, the ultimate measure of success for the *asset integrity* element is ensuring that equipment remains fit for its intended purpose, at least until its next scheduled inspection. In the case of safety systems, equipment needs to be available when needed, and capable of operating at a specified level of performance for a specified mission time. This performance is normally achieved via an integrated combination of ITPM tasks and quality assurance measures.

Conduct initial inspections and tests as part of plant commissioning. Equipment, systems, and facilities should be inspected/tested during fabrication and installation. In many cases, multiple inspections are required throughout the course of fabrication and installation to ensure that the equipment conforms to specifications. Just as inspecting the foundation prior to framing a house is a good idea, proper tests and inspections at each step of the fabrication and installation process can help minimize project cost overruns and schedule slips that result from poor quality of work (or inadequate design/specifications).

Many companies inspect large, special-order items at the fabricator's shop. This helps ensure that equipment is fabricated as designed, avoiding the need for field modifications or difficult decisions regarding whether equipment that does not conform to specifications is "good enough."

Initial testing extends to tests of controls, alarms, and interlocks. Written procedures should guide these tests and inspection activities. In most cases, these procedures become the basis for ITPM procedures used throughout the life of the facility. In addition to calibrating and testing control loops, critical software should also be tested or validated.

Initial test and inspection data should be retained as it provides valuable baseline information used to determine the rate of change in critical characteristics of process equipment, for example, wall thickness.

Conduct tests and inspections during operations. Continue to monitor the condition of equipment during operation using appropriate tests and inspection methods. Although tests and inspections should be based on standards, the activities and/or the interval are often adjusted based on equipment history, results of ITPM tasks, and consequence of equipment failure. Using a widely applicable method, such as visual inspection, to identify anomalies and target specific areas for more rigorous testing or inspection techniques maximizes the value of the ITPM program. For example, a vessel inspector may order radiographic testing on a specific area based on results from visual inspection.

Execute calibration, adjustment, preventive maintenance, and repair activities. Some activities should be performed periodically in lieu of testing or regardless of test results. This is normally a function of the cost of the activity versus the cost of condition monitoring, and the likelihood of a failure if an activity is not done or is delayed. For example, we change oil based on a mileage interval, but replace tires based on condition monitoring, partly because the cost of oil analysis exceeds the cost of oil replacement (an inexpensive activity), and the cost of tire inspection is far less than the cost of tire replacement (a moderately expensive activity).

Plan, control, and execute maintenance activities. All maintenance activities should be controlled, and to the maximum degree possible, planned. Planned maintenance activities are more efficient than unplanned activities, and it is much more likely that planned maintenance will be performed correctly, using the proper parts and procedures (planners typically ensure that these are in place prior to assigning the job to the maintenance crew). Planning the work facilitates approval of nonroutine work requests and coordination between the equipment owner/user (production) and the crew performing the work (maintenance).

Ensure the quality of repair parts and maintenance materials. Continued safe operation depends on maintaining integrity. The life of most process equipment is quite long. Equipment is repaired, overhauled, or rebuilt rather than replaced. Therefore, repair parts and materials must conform to specifications.

Information that indicates what repair parts should be used for each equipment item should be readily available. Creating a bill of material or list of spare parts for each equipment item helps prevent (or uncover) inadvertent changes. For example, designating in the CMMS the proper seal for each pump will help prevent an inadvertent change, and it will also help prevent perpetuation of a previous unintended change, for example, if a pump seal had been changed in error previously, simply using a "like for like" strategy when replacing the seal will perpetuate the unauthorized change.

Ensure that overhauls, repairs, and tests do not undermine safety. Just as specific tests, inspections, and other work activities help ensure that new equipment is properly fabricated and installed at commissioning, certain work activities are needed to help ensure that new or overhauled equipment conforms to specifications. These work activities should be integrated with the readiness reviews described in Chapter 16 that are part of the *readiness* element.

12.2.5 Address Equipment Failures and Deficiencies

Although the notion of taking action based on inspection and test data is intuitive, in practice this can become an issue. At some facilities, inspection files are full of unheeded recommendations. Causes of this vary, including recommendations that lack specificity, unavailability of the downtime necessary to complete the repair work, insufficient funding in the maintenance budget, and lack of awareness of the need to carefully review inspection/test reports.

Promptly address conditions that can lead to failure. Deficient equipment that is included in the scope of the *asset integrity* element should be promptly repaired or shut down, or steps should be taken to ensure safety pending repair.

In some cases, the equipment is expected to degrade progressively with time and prompt repair or replacement is not a viable option, for example, corrosion of a pressure vessel or process piping. In these cases, facilities assess the risk of continued operation and may choose to conduct very frequent inspections to help establish the short-term rate of degradation. In other cases, equipment is rerated or operated at less than full capacity pending replacement. Even if equipment is temporarily rerated pending repair or replacement, frequent inspections may be appropriate to closely monitor the equipment condition.

In other cases, the rate of degradation is unpredictable. Thus, frequent inspections and rerating do little to help understand the likelihood of catastrophic failure. In this situation, only two of the three risk questions posed in Chapter 9 – What can go wrong? and How bad could it be? – need to be answered to determine if continued operation is acceptable; the likelihood of a failure must be assumed to be very high.

Review test and inspection reports. Review test and inspection reports, and either (1) repair any deficiencies that are noted by the inspector or (2) document why repairs are not needed. This can be problematic for tests and inspections performed by contractors as there is often a substantial delay between when the field work was completed and the reports arrive. In addition, some reports do not include a summary of recommendations. Address this issue when developing the scope of work or specification for inspection services and ensure that inspection reports are not only complete, but timely and easy to use. Also, if the facility has established any standards for report forms or electronic data structures, ensure that this information is provided to the inspection firm. In all cases, a mechanism should be in place for immediately reporting any situations that an inspector discovers that might pose a risk of imminent danger.

Examine results to identify broader issues. Unexpected test or inspection results should cause one to question if other equipment may be deficient. For example, if a significant loss of vessel thickness has occurred, the piping to or from the vessel may also be thin, particularly if it is made from the same material. Likewise, if a valve fails as a result of poor metallurgy, other valves procured from the same vendor or in the same lot/shipment are immediately suspect and should be checked.

Investigate chronic failures using a structured methodology. Some failures have a systemic, but hidden, cause. Based on the potential consequence of failure and the ongoing costs of repair/replacement, facilities sometimes apply their incident investigation techniques to chronic equipment failures. However, these chronic root cause investigations should be chosen carefully. Too many investigations, or investigations of random failures, can be a poor use of resources. Investigating all equipment failures can also lead to perfunctory investigations, even when underlying cause is truly systemic.

Plan maintenance and repair activities. Avoid the temptation to schedule maintenance and repairs for the "next time the equipment breaks down." Making a risk-informed decision to continue to operate equipment with some sort of temporary repair to a critical component pending receipt of repair parts or until the process can be safely shut down is not uncommon. However, repairs should be planned and carried out as soon as possible. One contributor to risk is length of time that a temporary (presumably less than optimal) repair is in place. Extending the service time to the next breakdown increases the risk. A risk that might be marginally tolerable if the permanent repair is made within a week might be considered intolerable if the repair is delayed for two months.

In addition, planning to perform work in conjunction with unplanned breakdowns often fails. First, the chance that the breakdown will occur at night or on a weekend is greater than 75%. Thus, the maintenance crew may be without critical technical or logistical support when the next breakdown occurs. Also, production personnel and the shift maintenance crew may become so focused on getting the process running again that they may not take the time to perform the repair work to acceptable standards. Finally, planned maintenance work takes less time and is more likely to permanently fix the problem.

12.2.6 Analyze Data

The value of knowing the condition of a piece of equipment at a specific point in time is somewhat limited. The intent of the *asset integrity* element is to determine with a reasonable degree of certainty that the equipment (1) is currently fit for service and (2) will continue to be fit for service, at least until the next scheduled inspection or test. Thus, data collection **and** analysis are generally needed.

Collect and analyze data. In many cases, the rate of change in inspection or test data is at least as important as the data itself. As mentioned in Section 12.2.3, it is important to be able to:

- Efficiently collect and store data.
- Determine the rate of change (and in some cases, if the rate is accelerating, decelerating, or remaining constant).
- Project the condition into the future and predict when the equipment will no longer meet performance specifications or other critical limits.
- Identify other anomalies.

Although data can be collected and analyzed manually, doing so is much more labor intensive and efforts expensive to sustain over the long term.

Adjust inspection frequencies and methods. In some cases, inspection and test data indicate the need for additional activities, changes to the interval for an activity, or discontinuation of an activity.

Conduct additional inspections or tests as needed. Screening tests or inspections can often be performed with the intent of detecting abnormal conditions. Once detected, other, more specialized methods may be needed to analyze the extent of the condition failure, and in some cases to help devise a plan to restore the equipment to acceptable condition. These situations are normally apparent to the person doing the ITPM task (see the related discussion in Section 12.2.4). However, in some situations, the need for additional or more rigorous testing may become apparent only through analysis of compiled data.

Plan replacements or other corrective actions. In many cases, test and inspection data can help guide the capital expenditure program. Improved understanding of the rate of change in the condition of equipment helps facilities plan replacement or other corrective actions, which helps the facility manage risk and the business manage its capital expenditure program.

Archive data. Determine appropriate record retention requirements for test and inspection reports. The retention time should depend on the value that might be derived from the data. For example, records of a routine function test of an interlock will likely be of little value beyond proving to an audit team that the test was performed (note that this is a valid requirement, but the company may decide to only keep records for the two or three most recent tests). Retention requirements for data that are derived from condition monitoring efforts, such as vibration or thermography data, should be based on technical requirements associated with the test method, and should be sufficient to demonstrate to auditors that the tests are being performed as scheduled.

Some data should be kept for the life of the process. For example, thickness data, radiograph test results (i.e., X-ray films), photographs of conditions that are of special concern to an inspector, and so forth are often needed to evaluate the rate of change in equipment condition. In addition, these data can sometimes prove invaluable to other efforts. For example, the condition of an internal component might be identified as a possible cause of a sudden change in product quality, yield, or output. If radiographs of that part of the process were taken during a previous inspection (prior to the onset of the quality, yield, or output issue), the investigation team might request that new films be shot, providing direct evidence of the credibility of the postulated mechanism.

12.3 POSSIBLE WORK ACTIVITIES

The RBPS approach suggests that the degree of rigor designed into each work activity should be tailored to risk, tempered by resource considerations, and tuned to the facility's culture. Thus, the degree of rigor that should be applied to a particular work activity will vary for each facility, and likely will vary between units or process areas at a facility. Therefore, to develop a risk-based process safety management system, readers should perform the following steps:

1. Assess the risks at the facility, investigate the balance between the resource load for RBPS activities and available resources, and examine the facility's culture. This process is described in more detail in Section 2.2.
2. Estimate the potential benefits that may be achieved by addressing each of the key principles for this RBPS element. These principles are listed in Section 12.2.
3. Based on the results from steps 1 and 2, decide which essential features described in Sections 12.2.1 through 12.2.6 are necessary to properly manage risk.
4. For each essential feature that will be implemented, determine how it will be implemented and select the corresponding work activities described in this section. Note that this list of work activities cannot be comprehensive for all industries; readers will likely need to add

work activities or modify some of the work activities listed in this section.

5. For each work activity that will be implemented, determine the level of rigor that will be required. Each work activity in this section is followed by two to five implementation options that describe an increasing degree of rigor. Note that work activities listed in this section are labeled with a number; implementation options are labeled with a letter.

> *Note: Regulatory requirements may specify that process safety management systems include certain features or work activities, or that a minimum level of detail be designed into specific work activities. Thus, the design and implementation of process safety management systems should be based on regulatory requirements as well as the guidance provided in this book.*

12.3.1 Maintain a Dependable Practice

Develop a Written Program Description/Policy

1. Develop a written policy that describes the work activities related to the facility's *asset integrity* element.

   a. No overall policy or program-level governing procedure exists, but it is widely understood that the facility maintenance manager is responsible for all activities associated with the *asset integrity* element.

   b. The written policy procedure governing the *asset integrity* element states regulatory and corporate requirements. The policy specifically assigns responsibility for all activities associated with the *asset integrity* element to the facility maintenance manager.

   c. A written policy or program-level procedure describes how the *asset integrity* element is implemented. It addresses the scope of the element, lists key work activities, and assigns roles and responsibilities for each work activity.

Determine the Scope of the Asset Integrity Element

2. Define the scope of the *asset integrity* element in terms of (1) how equipment is identified as being covered and (2) those units or areas in which the policy does and does not apply.

   a. Equipment that helps prevent a catastrophic release of a hazardous material (principally by maintaining containment) is included in the *asset integrity* element. Safety and utility systems that mitigate the effects of catastrophic accidents are generally not addressed under the *asset integrity* element, but may be addressed elsewhere based on corporate, regulatory, or insurance company requirements.

b. Equipment that contains hazardous materials and safety/utility systems that help prevent or mitigate the effects of a catastrophic release of a hazardous material or a sudden release of energy are included in the *asset integrity* element.

c. In addition to item (b), the *asset integrity* element uses a risk-based approach to equipment identification that addresses a variety of different types of losses, for example, business interruption, long-term harm to the environment, loss of good will within the community.

Base Design and ITPM Tasks on Standards

3. Ensure that appropriate personnel are aware of, and apply, requirements and recommended practices contained in standards.

a. Standards are referenced in design documents at a high level, for example, specifications require that piping systems be installed in accordance with ASME B31.3, but no additional detail is provided to help facility personnel determine if fabricators and contractors are following standards.

b. Requirements are applied to the design and installation of new equipment, and some standards are used to specify ITPM tasks, such as, pressure vessel inspections are to conform to API 510 (Ref. 12.3).

c. Facility personnel have a high degree of awareness of standards, and standards are routinely used to determine appropriate design, installation, and ITPM tasks. In many cases, requirements are translated into facility procedures so that access to, and detailed knowledge of, the original source standard is not necessary.

4. Develop a company- or facility-wide standard that summarizes applicable design, test, and inspection requirements for each type of equipment.

a. Employees at each facility must identify and apply appropriate standards when developing ITPM plans.

b. Corporate engineering standards reference external standards that govern design, installation, and ITPM tasks for each type of equipment.

c. The company has a comprehensive guideline that facilities can use to better understand requirements included in standards. The guideline points users to specific sections of referenced standards where they can find more detailed information.

Involve Competent Personnel

5. Actively seek information on new developments in design, test, and inspection requirements.

    a. Engineers attempt to stay abreast of changes to standards through technical journals. Once identified, changes normally result in more detailed investigation and, if applicable, a change in practice.

    b. Personnel who play a key role with respect to the *asset integrity* element periodically attend short courses; sometimes these courses address new thinking in the area of design, test, and inspection requirements for various types of equipment.

    c. In addition to item (b), the company has appointed one or more stewards to maintain its knowledge of standards, stay abreast of changes, and help facilities properly implement the design, test, and inspection requirements for each type of equipment.

    d. In addition to item (c), the company actively promotes and supports participation on technical committees that develop and issue standards that are closely related to its operations. Personnel who participate on these committees monitor the state-of-the-art in design, test, and inspection practices, and disseminate this information via formal networks or working groups within the company.

6. Assign specific individuals within the company the responsibility of monitoring changes to standards or new standards that apply to the facilities that the company operates.

    a. An up-to-date set of applicable standards is available in the corporate library, and all facility engineering and maintenance managers are notified via e-mail when new editions are purchased.

    b. An engineer within the corporate engineering group is assigned to review new editions of standards, notify potentially affected persons of any changes, and help facilities revise their practices or procedures, when appropriate.

    c. In addition to item (b), the company actively promotes participation on technical committees that develop and issue standards that are closely related to its operations.

Update Practices Based on New Knowledge

7. Based on information collected from all sources, take appropriate action to improve design, test, and inspection practices.

    a. ITPM practices are generally changed only as a result of external stimuli, such as supplier notifications or external audits.

b. In addition to item (a), outputs from other RBPS elements (e.g., *management of change, incidents*) often result in changes to ITPM practices.

c. In addition to item (b), facility management actively promotes efforts to remain abreast of the latest technical developments involving ITPM practices, to pilot test newly developed practices, and to apply innovative but proven practices.

Establish a Means to Disseminate Information and Facilitate Continuous Improvement in Practices at the Unit Level

8. Establish networks within the company to facilitate dissemination of information related to the *asset integrity* element, including:
   - changes to (or new) standards
   - newly discovered failure modes or test/inspection methods
   - equipment failures and equipment history
   - effective practices

   a. Employees pass along this type of information to their counterparts at other facilities via informal networks. For example, maintenance managers at sister plants call each other frequently and proactively share information and best practices.

   b. The central engineering group monitors this type of information. New information is broadcast via e-mail, the company's intranet, or a bulletin board-type application.

   c. The corporate technology steward assigned to monitor the latest advances in design, test, and inspection methods (possibly for a limited range of equipment) regularly visits each facility to review existing and planned activities, and to ensure that regulations, standards, and corporate requirements are properly addressed.

Integrate the Asset Integrity Element with Other Goals

9. Ensure that work activities associated with the *asset integrity* element are integrated with initiatives to reduce downtime, increase output and yield, and improve quality via improved equipment reliability.

   a. Management focuses the effort applied to the *asset integrity* element on specific equipment/systems in an attempt to ensure that the most important ITPM tasks are performed on the most critical equipment. "Scope creep" is discouraged.

   b. Some scope creep is permitted for the *asset integrity* element, but the equipment reliability, asset integrity, quality improvement, and similar programs are generally segmented. A consolidated effort happens only when the individuals responsible for the separate programs decide to work together.

c. Work activities supporting the *asset integrity* element are executed by the same group that supports similar initiatives for systems that are outside of the scope of the *asset integrity* element. Although efforts are made to distinguish between equipment that is within and outside of the scope of the *asset integrity* element, this group tends to consolidate activities when ITPM tasks are assigned.

d. Work activities supporting the *asset integrity* element are closely integrated into all of the maintenance and operational improvement efforts at the facility.

10. Periodically review the ITPM plan for each type of equipment (or equipment item) to determine if redundancy exists or if the activities could be accomplished more efficiently if they are performed in a specific sequence.

a. No requirement exists to periodically review maintenance plans; however, plans are often reviewed on an episodic basis, for example, based on a cost-cutting initiative, after an incident, or after an equipment failure.

b. From time to time teams are chartered to improve equipment reliability. These teams often review maintenance plans, but normally focus on a limited set of failure modes, such as, failure modes that lead to poor product quality, failure modes that reduce plant output, and so forth.

c. Maintenance plans are periodically reviewed, mainly to "level" the activities between various equipment items (i.e., assign the same activity at standard intervals for a given type of equipment).

d. Maintenance plans are periodically reviewed; the focus is to ensure that the rigor in ITPM tasks and intervals is appropriate given the postulated failure modes and risk associated with the failure modes. Risk-based changes that reduce ITPM tasks are as important as identification of new activities. This allows the facility to focus on the "important few."

12.3.2 Identify Equipment and Systems that Are Within the Scope of the Asset Integrity Program and Assign ITPM Tasks

Identify Equipment/Systems for Inclusion in the Asset integrity Element

11. List each equipment item that is included in the scope of the *asset integrity* element.

a. All major equipment items (vessels, tanks, pumps shutdown systems, etc.) included in the scope of the *asset integrity* element are listed in a document such as a spreadsheet or database application, or hand marked on a set of piping and instrumentation diagrams.

b.   Equipment included in the scope of the *asset integrity* element is clearly designated in the CMMS.

c.   In addition to item (b), a checklist embedded in the *management of change* program includes specific questions to help the user determine if a proposed change might affect data stored in the CMMS and includes a step to confirm that CMMS records are updated, if applicable.

Develop an ITPM Plan

12. For each item that is identified, develop an ITPM plan that is based on standards, manufacturers' recommendations, equipment history, internal requirements, and the expected consequence(s) of failure of the specific equipment item.

a.   Although no ITPM plan has been established, preventive maintenance tasks are generally assigned to equipment in the CMMS based on the facility's understanding of applicable standards as well as ITPM work activities that have proven sufficient for similar equipment that is in service at the facility.

b.   An ITPM plan is developed for each type of equipment and used as a guideline when assigning ITPM tasks.

c.   In addition to item (b), all exceptions to the facility's ITPM plan are documented, for example, substituting external thickness testing for internal visual inspection of a storage tank, including the rationale for the exception.

13. Establish preventive maintenance work orders in the CMMS based on the ITPM plan.

a.   Work orders are entered in the CMMS for tasks performed by the maintenance department.

b.   Work orders are entered in the CMMS for tasks performed by the maintenance department and inspection contractors.

c.   Work orders are entered in the CMMS for all ITPM tasks (except daily activities performed by operators), including inspections of emergency response equipment normally performed by ERT personnel and infrequent ITPM tasks performed by operators.

14. For SISs, develop an ITPM plan that helps ensure that each system is capable of meeting the specified risk reduction requirement for each SIF.

a.   No SISs are present at the facility.

b.   Changes to the ITPM interval or rigor of testing for SISs are reviewed and approved through the *management of change* element. However, the facility has no special means to indicate which equipment or ITPM tasks are associated with SIFs.

c. In addition to item (b), ITPM tasks associated with SIFs are flagged in the CMMS to alert personnel to special requirements that apply when testing is delayed, ITPM tasks are changed, or SISs are modified.

Update the ITPM Plan when Equipment Conditions Change

15. Include action steps in the change authorization process to:
  * Update the ITPM plan.
  * Establish ITPM tasks in the CMMS.
  * Establish bills of material for various repairs, and link the repair parts (even if they are a non-stock item) to the equipment item.

  a. The *management of change* procedure includes an instruction to update applicable maintenance records.

  b. The *management of change* procedure includes a checklist with questions for each data element in the CMMS that is normally affected by a change.

  c. The *management of change* procedure includes a detailed checklist, requiring that the change request indicate specifically which records need to be updated. The requestor must affirmatively state that the records have been updated before the change can be closed.

  d. In addition to item (c), the *management of change* element includes periodic audits to independently confirm that maintenance records have been updated as intended.

12.3.3 Develop and Maintain Knowledge, Skills, Procedures, and Tools

Develop Procedures for Inspection, Test, Repair, and Other Critical Maintenance Activities

16. Develop procedures and checklists to guide ITPM tasks.

  a. The procedures are fairly general checklists that list the tasks but include specific steps only if they are deemed unusual or complex, or if an incident occurred as a result of failure to properly execute the procedure.

  b. The procedures list specific steps or actions to be performed, and include a place for the inspector to initial or record data for critical steps. The procedures also include caution, warning, and notice statements as needed.

    c.   In addition to item (b), the procedures list acceptable ranges for each data item that is recorded.

    d.   In addition to item (c), the procedures require second-person verification for items that are very critical to process safety.

17. Identify critical repair activities.

    a.   The maintenance group has a general understanding regarding which repairs are critical, and this work is only assigned to highly experienced and skilled maintenance employees.

    b.   A list of critical maintenance activities is kept. For each activity, a corresponding procedure and specific training modules, as needed, are included.

    c.   Critical maintenance procedures are included in the training curriculum for maintenance employees, and only trained employees are assigned to perform this work.

18. Once the critical repairs are identified, determine if the manufacturer's maintenance manual, along with safe work procedures, adequately control and govern repair work. If not, develop written procedures to fill gaps that are identified.

    a.   OEM manuals are maintained in a central file. Maintenance personnel are instructed to consult with supervision (or the reliability engineer) if they believe that manufacturer's manuals and safe work procedures do not adequately describe the tasks to be performed.

    b.   OEM manuals are supplemented with facility-specific procedures and checklists. The facility-specific procedures include steps to make the equipment safe for maintenance, checklists, places to sign off or record critical data, and steps to properly return the equipment to service.

19. For all ITPM and critical repair activities, develop job plans that list:

- The procedures to be applied (typically in the order they are to be used).
- Repair parts and maintenance materials that are needed.
- Special tools that will be required.
- Special calibration requirements, such as whether an instrument used for calibration must be traceable to a national or international standard.
- Certification requirements for personnel involved in doing the work.

    a.   No job plans are in place for inspection and test activities; however, OEM manuals are generally available that specify repair parts and special tools.

  b. In addition to OEM procedures, job plans are developed for ITPM and critical maintenance repair activities.

  c. In addition to (b), work orders for ITPM and critical repair tasks include a place where the crew performing the work can note if anything was left out of the job plan. These notes are reviewed by maintenance planners or supervisors, and appropriate changes are made to the job plans.

Train Employees and Contractors

Work activities related to training programs are addressed in Chapter 14.

Ensure that Inspectors Hold Appropriate Certifications

20. Become familiar with requirements related to special certifications for inspectors and ensure that inspectors hold the certifications listed in applicable standards.

  a. The facility depends on statements by inspection firms that they employ qualified inspectors.

  b. The facility evaluates each inspector's qualifications, but recognizes equivalent training and experience in lieu of some of the certification requirements listed in the respective standards.

  c. The facility requires that all contract inspectors hold any certifications listed in the respective standards but recognizes equivalent training and experience for its own maintenance employees.

  d. The facility requires that all inspectors hold appropriate certifications listed in the respective standards.

Provide the Right Tools

21. Provide tools and training required to conduct tests and inspections.

  a. Specialized tools are available for routine jobs. Training is normally provided by vendors' service engineers.

  b. The tasks listed in the ITPM plan drive a comprehensive review of the need for tools and training and they are provided when needed.

22. Provide tools and training required to store test data and equipment history in a manner that it can be easily analyzed.

  a. Test/inspection data and equipment history are archived, but they cannot be readily searched or manipulated electronically.

  b. Test data and equipment history are archived electronically, but customized reports or searches can be done only by designated power users who are generally not available at the facility level.

    c.   Test data and equipment history are archived electronically, and customized reports or searches can generally be done by facility personnel.

12.3.4 Ensure Continued Fitness for Purpose

Conduct Initial Inspections and Tests as Part of Plant Commissioning

23. Conduct initial inspections to:
- Confirm that equipment has been properly fabricated and installed.
- Collect baseline data that can be used to determine the rate of change during subsequent tests or inspections.

    a.   Initial inspections are normally conducted only when a specific concern arises or as a result of a previous incident.

    b.   Initial inspections are conducted if they are specified by the project engineer.

    c.   Company standards require that fabricators provide initial inspection reports for major equipment items, such as pressure vessels and storage tanks.

    d.   The inspection group is involved in determining what initial inspections are needed and gathers baseline data regardless of any concern about whether the equipment is fit for service. For example, baseline thickness readings are collected for all critical piping system components.

Conduct Tests and Inspections

24. Conduct tests and inspections in accordance with the ITPM plan.

    a.   ITPM tasks are entered into the CMMS, but some activities are delayed or skipped because equipment is not available.

    b.   ITPM tasks are entered in to the CMMS and generally conducted in accordance with the schedule.

    c.   In addition to item (b), the facility has established a procedure to authorize late or skipped ITPM tasks; these situations must be justified and presented to management for review and approval.

Execute Calibration, Adjustment, Preventive Maintenance, and Repair Activities

25. Schedule and perform routine preventive and predictive maintenance activities in accordance with the ITPM plan.

    a.   Calibration, adjustment, preventive maintenance, and repair activities are assigned based on previous incidents or historical rates of failure at the facility.

    b. Calibration, adjustment, preventive maintenance, and repair activities are generally based on OEM recommendations, but are performed more frequently if indicated by the historical failure rate at the facility.

    c. In addition to item (b), the facility has established a procedure to authorize late or skipped calibration, adjustment, preventive maintenance, and repair activities; these situations must be formally approved by management.

26. Schedule and perform equipment overhauls based on the results of condition monitoring activities or based on other criteria, such as time in service or operating cycles.

    a. Overhauls are scheduled based on production demands.

    b. The production scheduling process takes the need for equipment overhaul into account and plans for maintenance shutdowns.

Plan, Control, and Execute Maintenance Activities

27. Plan maintenance work, including repairs and ITPM work.

    a. At the start of each shift, maintenance supervisors determine what work orders can be executed (i.e., parts and materials are in stock, the equipment is available). Jobs are assigned based on priority and whether they can be performed at that time.

    b. A weekly planning meeting between production and maintenance is held to finalize the list of jobs for the following week. Based on this agreed-to list, ITPM and repair jobs are scheduled.

    c. In addition to item (b), each significant maintenance job is planned. At the start of the job, maintenance personnel are issued a package that includes parts or a bill of materials, special procedures or instructions, work permits that will need to be filled out, safety equipment, and other items that will be needed to complete the work.

28. Authorize all unplanned repair work.

    a. Most day shift work is planned and scheduled by the maintenance supervisor; off-shift maintenance crews generally make repairs or adjustments based on verbal requests by production operators.

    b. No formal system has been established to authorize unplanned maintenance work; however, general practice is for maintenance personnel to notify their supervisor or the shift supervisor before proceeding with anything beyond minor adjustments. A work request is entered for any significant unplanned work to capture equipment history.

    c. Unplanned repair work can be authorized only by the shift supervisor (or at larger facilities, the maintenance supervisor for the shift). A work request is entered for any significant unplanned work to capture equipment history.

Ensure Quality of Repair Parts and Maintenance Materials

29. Develop specifications for critical repair parts and maintenance materials.
    a. Specifications are available, but purchasing has the authority to override them or at least require that maintenance or engineering justify use of particularly expensive parts or materials.
    b. Specifications are available. There is no intent to substitute items that do not conform to specifications, but substitution is allowed without the requestor's authorization.
    c. Specifications are available, and substitution is allowed only if the requestor specifically permits this when placing the purchase request. This does not preclude the purchasing agent from asking the requestor if a substitute part/material would meet the user's requirement.

30. Ensure that vendors supply parts and materials that conform to specifications.
    a. In most cases, receiving personnel check the packing slip against the purchase order.
    b. Stockroom personnel confirm that the part is correct when placing it in its designated storage location.
    c. In addition to item (b), maintenance personnel are well aware of *management of change* requirements and are trained to check part numbers against the part being replaced.
    d. In addition to item (c), critical parts or raw materials are flagged in the purchasing system and placed in a special "QC hold" area near the receiving dock. An inspector or other trained individual releases them for transfer to the stock room based on tests to confirm material composition (positive material identification), material hardness, or other critical characteristics.

31. Ensure that the storeroom for maintenance parts and materials is well organized and controlled.
    a. Repair parts are mainly kept in a stock room that is not manned or controlled, but parts are sorted in a logical manner, such as by type of equipment, material of construction.
    b. Repair parts are maintained in a central stock room that is staffed or otherwise controlled. Parts are labeled with a stocking number that ties back to a bill of material or spare parts list maintained in the CMMS.

   c.   In addition to item (b), critical parts are labeled with a bar code that
        is scanned when the part is issued. The person requesting the part
        enters the equipment item or work order number and the inventory
        control system double checks that the part is authorized for use on
        the equipment or for issue against the work order.
32. Associate repair parts with equipment items.
   a.   Maintenance personnel are instructed to check the replacement part
        to ensure that it has the same part number as the one being
        replaced.
   b.   A bill of material or parts list for each equipment item is stored in
        the CMMS; the CMMS issues parts against the work order based
        on the bill of materials.
   c.   Parts are issued to work orders that are tied to equipment items,
        and the data tying parts to an equipment item is stored in the
        CMMS. The CMMS alerts the person issuing the part any time
        that parts requested for a work order or equipment item do not
        match the bill of material in the CMMS.

Ensure that Overhauls, Repairs, and Tests Do Not Undermine Safety
33. Include steps in preventive maintenance, repair, and overhaul
    procedures to ensure that equipment is fit for service when it is turned
    over to the production team.
   a.   The decision to perform any tasks associated with final testing is
        left up to the maintenance crew assigned to the job. However, final
        testing for most equipment is covered in training and a practice is
        in place to perform appropriate tasks.
   b.   The *readiness* element includes a checklist with options for
        checking rotation, leak testing, confirming calibration and
        operability of instruments or interlocks, and so forth. Based on the
        work that was performed, the maintenance crew selects the
        appropriate tasks and initials when they are complete.
   c.   Final testing tasks are specified on the job plan. The maintenance
        crew assigned to the job initials each task when it is completed.

12.3.5 Address Equipment Failures and Deficiencies

Promptly Address Conditions that Can Lead to Failure
34. If equipment that is included in the scope of the *asset integrity*
    element is found to be deficient, promptly remove the equipment from
    service or implement appropriate safeguards to ensure safe operation
    pending repair or replacement.

a. Prudent steps are generally taken to help ensure safety if deficient equipment is not promptly shut down. However, no formal process is in place to approve continued operation of deficient equipment.

b. A formal process exists to approve continued operation with deficient equipment in service. However, this does not extend to deficiencies that result from ITPM or repair activities, such as impairment of a safety system or disabling of an interlock.

c. All deficiencies are reported, and a formal process exists to approve continued operation with deficient equipment in service, regardless of the cause of the deficiency.

d. In addition to item (c), deficiencies are reviewed frequently (at least weekly) by management and the decision to continue to run equipment that is deficient is reviewed in light of how it, coupled with all of the other deficient or out-of-service equipment, impacts risk associated with facility operation.

Review Test and Inspection Reports

35. Review test and inspection reports and either (1) repair deficiencies noted by the inspector or (2) document why repairs are not needed.

a. The responsibility to review test and inspection reports is specifically assigned to one or more knowledgeable persons. However, no formal system is in place that specifies how suggested repairs should be documented or what must be done if the reviewer believes that an inspector's comment, concern, or recommendation does not need to be addressed.

b. The responsibility to review test and inspection reports is specifically assigned to one or more knowledgeable persons and a standard method exists for documenting completion of repair work. However, no formal system is in place that specifies what must be done if the reviewer believes that an inspector's comment, concern, or recommendation does not need to be addressed.

c. The policy or management system procedure governing the *asset integrity* element clearly states (1) who is responsible for reviewing test and inspection reports, (2) how repair work should be documented, (3) how cases should be documented if recommended repairs will not be made, and (4) who can approve a decision to decline to address a recommendation contained in a test or inspection report.

Examine Results to Identify Broader Issues

36. If the rate of change in equipment condition is faster than anticipated, (1) determine if other equipment exists that might also be affected by

the same conditions that caused the unexpected condition or deficiency and (2) conduct appropriate inspection or test activities to determine if the equipment is still fit for service.

a. ITPM tasks are conducted as planned, and deficiencies are addressed in a timely manner. Further actions are left to the discretion of the maintenance or engineering departments. However, no formal process has been established that requires that the test or inspection methods or intervals for equipment in similar service be reviewed.

b. ITPM tasks are conducted as planned, and deficiencies are addressed in a timely manner. In addition, similar equipment or equipment in similar service is examined to help understand the extent of the problem. For example, if a vessel and the piping leading to/from the vessel are fabricated from similar materials, and an inspector finds that the vessel wall has corroded more than expected, the response includes thorough testing and inspection of inlet and outlet piping.

Investigate Chronic Failures Using a Structured Methodology

See Section 19.3.5 for work activities associated with trending, analyzing, and investigating chronic failures.

Plan Maintenance and Repair Activities

37. Plan repair and maintenance activities so that they will be executed in a timely manner with adequate technical and logistical support.

a. Temporary repairs are authorized via a formal process, such as the *management of change* procedure. Permanent repairs are planned, but execution of noncritical maintenance work is often delayed until the next breakdown.

b. Temporary repairs are authorized via a formal process, such as the *management of change* procedure. An expiration date is assigned to temporary repairs, and every effort is made to execute a permanent repair prior to the expiration date for the temporary repair.

c. Temporary repairs are authorized via a formal process, such as the *management of change* procedure. Permanent repairs are generally made expeditiously. Appropriate consideration is given to risk when deciding whether to take an unplanned shutdown versus delaying plans to make a permanent repair.

12.3.6 Analyze Data

Collect and Analyze Data

38. Establish a means to efficiently collect and analyze data, and highlight anomalies.

   a. Data are mainly recorded on hand-written field notes or in word processing applications.

   b. Data are manually transferred to a spreadsheet or similar application, but not in a manner that facilitates analysis of any change in equipment condition over time.

   c. Data are manually transferred to a software application that can be used to (1) project equipment life, (2) identify chronic failures, and (3) evaluate test or inspection intervals.

   d. Data are downloaded directly from inspection devices to software applications designed to analyze the information, project equipment life, and produce other reports.

   e. In addition to recording data as specified in items (a) through (d), personnel are vigilant for indications of abnormal component parts. Parts showing abnormal wear patterns are turned in for further inspection and, in some cases, formal investigation.

Adjust Inspection Frequencies and Methods

39. Based on the results of tests and inspection activities, make appropriate adjustments to the inspection or test interval.

   a. Evaluation of ITPM results is generally limited to determining if the current condition of the equipment is acceptable.

   b. Results of ITPM tasks are sometimes used to adjust intervals.

   c. ITPM results are periodically reviewed to evaluate the rate of change in conditions. That data, along with an understanding of failure modes, is used to adjust ITPM task intervals or methods.

Conduct Additional Inspections or Tests as Needed

40. Implement a tiered approach to conducting inspections.

   a. A uniform approach is used. For example, all pumps are included in the vibration monitoring program.

   b. ITPM tasks are tailored and staged. Although visual inspection and other simple methods are the primary methods, other methods are used when indicated.

   c. In addition to item (b), consideration is given to the consequence of failure. Based on risk, additional ITPM tasks may be scheduled regardless of any other ITPM results.

Plan Replacements or Other Corrective Actions

41. Use the results of tests and inspections to plan overhaul, replacement, or other corrective actions.

    a. ITPM tasks are conducted as planned. Although the results are not used to plan future capital budgets, overhauls, and so forth, equipment that is discovered to be deficient or worn out is repaired or replaced in a reactive but timely manner.

    b. The results of ITPM tasks are considered when developing the capital plan or planning for unusual maintenance expenses in the upcoming year, but no formal process exists to ensure that this occurs.

    c. The facility maintains a register or database that lists the projected retirement date (or date for the next major overhaul) for each major equipment item. The dates are based on the results of ITPM tasks, updated regularly, and used to plan capital and maintenance budgets.

Archive Data

42. Save inspection data so that they are readily accessible, for example, file data and reports by equipment item rather than by the year the inspection was performed or the contractor who performed the inspection.

    a. Reports listing the results of ITPM tasks are kept on file, but are filed by when the activity was performed, the contractor that performed the work, or some other criteria that is not directly related to the equipment.

    b. Reports listing the results of ITPM tasks are filed by an equipment identifier, such as a tag number.

    c. Most of the results of ITPM tasks are kept in electronic databases and an entry in the notes section of each record in the CMMS states where any other records can be found (e.g., maintenance files, piping databases, equipment files in the engineering department).

12.4   EXAMPLES OF WAYS TO IMPROVE EFFECTIVENESS

This section provides specific examples of industry-tested methods for improving the effectiveness of work activities related to the *asset integrity* element. The examples are sorted by the key principles that were first introduced in Section 12.2. The examples fall into two categories:

1. Methods for improving the performance of activities that support this element.
2. Methods for improving the efficiency of activities that support this element by maintaining the necessary level of performance using fewer resources.

These examples were obtained from the results of industry practice surveys, workshops, and CCPS member-company input. Readers desiring to improve their management systems and work activities related to this element should examine these ideas, evaluate current management system and work activity performance and efficiency, and then select and implement enhancements using the risk-based principles described in Section 2.1.

12.4.1 Maintain a Dependable Practice

Develop a corporate standard for ITPM tasks. The ITPM plan described in Section 12.2.2 should be based on standards, manufacturers' recommendations, and equipment history of the particular equipment in similar service at the facility (or sister facilities within the company/industry). However, reviewing all of this data and defining a rational set of requirements takes a significant amount of research. More consistent and efficient ITPM plans can be achieved by centralizing this function within a company or business unit and developing a set of ITPM tasks that (1) are based on all of this information and (2) are specific to the range of service conditions at operating facilities. This approach also facilitates sharing of best practices between facilities. Corporate standards typically address the following topics:

- Equipment identification/prioritization.
- ITPM guidelines for each type of equipment
  - pressure vessels
  - storage tanks
  - heat exchangers
  - piping systems
  - relief and vent devices
  - rotating hardware (likely further subdivided into pumps, compressors, agitators, etc.)
  - instruments
  - control systems
  - interlocks and emergency shutdown systems
  - utility systems (likely further divided by type of utility)
  - mitigation systems
  - emergency response equipment
- Inspection and test requirements for each type of equipment.

- Templates for procedures and checklists for each type of test or inspection.
- Training and certification requirements for each category of ITPM tasks.
- Quality assurance standards/inspection checklists for each type of equipment.
- Sample deficiency management procedures.
- An audit protocol for specific ITPM tasks associated with the *asset integrity* element.

12.4.2 Identify Equipment and Systems that Are Within the Scope of the Asset Integrity Program and Assign ITPM Tasks

Provide a clear link from risk analysis efforts to the ITPM plan. Companies spend considerable effort identifying hazards and analyzing risks. One byproduct of these efforts is a list of safeguards that help prevent or mitigate the effects of accident scenarios that are analyzed. Depending on the technique, the consequences (or sometimes even the safeguards) have a risk ranking associated with them. This information can provide clear direction to a group charged with developing an ITPM plan. Some facilities also perform layer of protection analyses to better understand the risk associated with certain accident scenarios, and this technique directly provides a list of independent protection layers (IPLs) along with the maximum acceptable probability of failure for each IPL. These data, along with knowledge of equipment history and vendor-supplied failure rates, if available, can be used to design an ITPM strategy to help achieve the target reliability.

Formally integrate hazard identification with developing ITPM plans. Some companies use formal hazard identification/screening tools to assign units to various "levels of rigor" within their process safety program. These tools often take the form of a matrix that lists inherent hazards of various materials on one axis (extremely toxic material which, if released, presents a substantial danger to the public, extremely flammable material stored above its normal boiling point, flammable liquid, etc.) and the quantity of material (in mass units) on the other axis. The intersecting cells provide additional guidance on what ITPM tasks should be assigned. Thus, for a pressure vessel above a certain capacity that contains a highly toxic gas that is liquefied under pressure, a facility may choose to (1) implement ITPM tasks that far exceed those listed in applicable pressure vessel inspection standards and (2) apply the minimum standards-based inspection requirements for storage tanks in noncorrosive flammable liquid service.

Depending on risks, consider running certain equipment to failure. At some companies, the general belief is that all equipment failures can and should be prevented. Although detecting and preventing equipment failures by spending more time and effort on activities associated with the *asset integrity* element is normally feasible, many failures (or failure modes) do not lead to an unacceptable outcome. When developing ITPM plans, consider:

- The consequences associated with the failures/failure modes that the ITPM task will help prevent.
- The expected failure rate.
- The likelihood that the ITPM task will help detect or prevent the failure.
- The risk introduced by the ITPM task, for example, potential to damage the equipment, making an error when placing the equipment back in service, and so forth.
- The total cost of the activity, including lost production time, costs to prepare the equipment for the ITPM task, impact on product quality/yield, and so forth. This consideration is particularly pertinent for failure modes that result in equipment damage or downtime but are unlikely to lead to an unacceptable safety consequence.

On that basis, develop an ITPM plan that makes sense from a technical and business perspective.

Base test intervals for instruments on mean time to failure (MTTF) data. Some companies test or calibrate all instruments of a particular type at a fixed interval. Because of the ISA 84.01 series of standards, manufacturers of instrumentation and control equipment are publishing MTTF data. These data can be used to establish test or calibration intervals based on engineering analysis and acceptable failure rates rather than solely on judgment and experience. (Note that setting the test interval at the MTTF can lead to relatively high failure rates; however, MTTF data can be used to adjust test intervals to optimize the value delivered by *asset integrity* work activities.)

12.4.3 Develop and Maintain Knowledge, Skills, Procedures, and Tools

Take advantage of training offered by local colleges. Many local colleges, particularly community colleges, are keen to develop job-specific curricula, particularly if good employment opportunities exist in the field. For example, many facilities now hire instrument technicians from these types of programs rather than maintaining an internal apprenticeship program.

Ask equipment suppliers to provide training. Equipment suppliers often provide short, hands-on training modules (e.g., 1 hour to 1 day) on specific equipment. This training can be a particularly effective means to update the skills and knowledge of trained maintenance employees. In many cases, suppliers will provide this training at little or no additional cost if the request is bundled into the initial purchase of a major equipment item.

Use inspectors from one facility to audit programs and practices at another facility. Auditing the *asset integrity* element requires deep knowledge of standards. Audit teams lacking such knowledge often hire a third-party expert to review that part of the program. An alternative to this approach is to use certified and experienced inspectors from a sister facility. This also promotes cross-learning; both the audited facility and the facility that loaned the inspector to the audit team are likely to benefit from this approach. However, the inspector assigned to do the audit should (1) have a deep knowledge of standards and be aware of how they are applied by other companies in the industry and (2) be trained in auditing methods.

12.4.4 Ensure Continued Fitness for Purpose

Implement a risk-based inspection program. In recent years, facilities have applied risk-based analytical techniques in an effort to develop more performance-based *asset integrity* programs. In fact, some inspection standards, such as American Petroleum Institute (API) 510 (Ref. 12.3) and API 570 (Ref. 12.4) now include provisions for determining inspection requirements based on risk. Also, the API, the American National Standards Institute (ANSI), and other organizations that publish standards have developed standards and recommended practices that encourage the use of risk-based techniques to define inspection and testing requirements, including:

- API-580, Risk-Based Inspection, and API-581, Base Resource Document – Risk-based Inspection (Ref. 12.5).
- ANSI/ISA-84.00.01-2004 Part 1 (IEC 61511-1 Mod), Functional Safety: Safety Instrumented Systems for the Process Industry Sector – Part 1: Framework, Definitions, System, Hardware and Software Requirements (Ref. 12.6).

Many organizations find that using risk analysis techniques to determine ITPM tasks provides several benefits, including:

- Assurance that a structured, systematic, and technically defensible approach is being used to make decisions.
- Improved knowledge of the system operation and the cause/effect relationships that result from specific equipment failures.
- Confidence that resources are being focused on the most important failures by explicitly assessing the risk and then assigning ITPM resources to those areas in which they will be the most effective.

Engage operators in the asset integrity element. Some facilities seem to erect a wall between the maintenance and production departments. Production is responsible for getting product out the door; maintenance is responsible for keeping the equipment running properly. In reality, operators can often sense when equipment is not operating properly. Instituting a sense of equipment ownership among production team members, and having operators frequently patrol areas looking for unusual conditions, will help detect abnormal conditions before they become equipment failures. (Similar ideas are included as work activities or improvement ideas in the *culture, involvement,* and *operations* elements in Chapters 3, 6, and 17, respectively.)

Assign some testing and preventive maintenance tasks to operators. Certain tasks should be assigned to operators, particularly activities that do not require special tools or highly specialized training. The *operations* element includes several essential features that address this point, including "monitor equipment status" and "maintain instruments and tools" (see Section 17.3.3). In fact, some ITPM tasks will provide more useful results if they are performed by operators. For example, facilities often have backup generators to provide power to critical equipment, and operators are expected to start and monitor this type of equipment in the event of an offsite power failure. One typical ITPM task for an emergency generator is to periodically start it, run it for a specified time, and monitor the gauges to ensure that all systems are operating properly. Assigning this task to an operator who may have to start and operate the generator during a power failure (1) provides a real hands-on refresher training opportunity, particularly if the task is rotated among all of the operators in a structured manner, (2) frees up maintenance personnel for tasks that require more specialized training and equipment, and (3) provides the facility with a better overall view of the dependability of the generator.

Implement systems for continuous condition monitoring. Advances in technology have made it technically feasible and cost effective to implement continuous condition monitoring for some types of equipment. For example, vibration sensors can be permanently mounted on critical rotating equipment and excessive vibration characteristics (e.g., amplitude, frequency, and rate of change of key parameters) can be alarmed on the operator's console, providing more timely warning of potential failures. Collecting real-time condition monitoring information and alarming an unacceptable condition or a high rate of change on the operator's console can help identify and mitigate the effects of failures that were previously considered random events.

12.4.5 Address Equipment Failures and Deficiencies

Consider ITPM tasks at the design stage. Designers and project engineers often overlook maintenance and ITPM needs at the design stage. Conducting a design review with experienced maintenance personnel to address issues such as (1) accessibility, (2) isolation for maintenance work, and (3) reliability, and spending the money to implement their recommendations, can provide significant benefits over the life of a unit.

12.4.6 Analyze Data

Trend results. Review test data periodically to detect changes in equipment condition. If equipment life is decreasing over time, critically examine the maintenance program to determine if seemingly unrelated failures have a common cause, such as a change in maintenance strategy, parts supplier, or process operations. Use this information to make better decisions regarding common cause factors, as well as the frequency and rigor of ITPM tasks.

Fully utilize equipment reliability analysis tools in CMMS software. Many companies purchase and deploy very capable software systems to manage maintenance activities. The initial rollout of these systems is normally very chaotic. The focus during commissioning is on maintaining continuity of operations. Tools that can help users analyze equipment reliability, repair history, and so forth are planned for "Phase II," which is scheduled for sometime after the work order processing, preventive maintenance scheduling, inventory, time accounting, and other basic CMMS functions are operating as intended. Too often, the Phase II initiatives never occur. Thus, users charged with understanding causes of downtime or other reliability-related issues often develop simple spreadsheets to help analyze data. Although this effort might be an expedient means to achieve short-term results, it is expensive to maintain. Implementing modules that are inherent to the CMMS supports efforts to measure and drive improvements in equipment reliability and helps ensure that these efforts are sustained.

Invest in software that can help monitor equipment condition by analyzing process data. Many facilities have implemented distributed control systems with data historian modules that provide the capability to capture large amounts of data at a very low cost. However, assigning engineers or others to periodically review the data simply to look for indications that the performance of process equipment has degraded can be very cost prohibitive. Software packages can monitor equipment on a virtually continuous basis and can be programmed to alert reliability or process engineers of a significant change in performance. For example, a trend indicating that a pump is requiring increasingly more power to drive the input shaft (at a given output requirement) may indicate that some component is wearing out or that adjustments/preventive maintenance activities are needed.

Use proven tools to analyze reliability and failure data. Many tools that were originally developed to support product quality initiatives can be used to analyze other types of data as well. Facilities that have already learned to use tools such as the six sigma process, statistical quality control (more specifically, control charts), and similar methods often find that adapting these techniques to analyzing data generated by the *asset integrity* element is much more effective than developing a new method.

12.5   ELEMENT METRICS

Chapter 20 describes how metrics can be used to improve performance and when they may be appropriate. This section includes several examples of metrics that could be used to monitor the health of the *asset integrity* element, sorted by the key principles that were first introduced in Section 12.2.

In addition to identifying high value metrics, readers will need to determine how to best measure each metric they choose to track. In some cases, an ordinal number provides the needed information, for example, total number of workers. Other cases, such as average years of experience, require that two or more attributes be indexed to provide meaningful information. Still other metrics may need to be tracked as a rate, as in the case of employee turnover. Sometimes, the rate of change may be of greatest interest. Since every situation is different, the reader will need to determine how to track and present data to most efficiently monitor the health of RBPS management systems at their facility.

12.5.1 Maintain a Dependable Practice

- *Display a simple chart showing which facilities or units have fully implemented specific programs or practices.* Sometimes simply displaying this information, and reviewing it in management team meetings, will spur management interest/support.

12.5.2 Identify Equipment and Systems that Are Within the Scope of the Asset Integrity Program and Assign ITPM Tasks

- *Number of equipment items included in the asset integrity program.* Although this number has very little meaning in isolation, it could be used as a basis to compare *asset integrity* programs, particularly if the company operates similar processes at multiple facilities. The metric is likely to be more meaningful if the data are further subdivided by type of equipment, such as pressure vessels, storage tanks, safety interlocks, pumps, agitators, instrumented control loops, and so forth.
- *Number of ITPM work orders (per month or quarter) that apply to equipment that is no longer present at the facility.* A higher than expected number may indicate a weak link between the asset integrity and management of change elements. If the ITPM plan and preventive maintenance work orders in the CMMS are not updated when equipment is removed from service, it is quite likely that they are not updated when new equipment is installed.

12.5.3 Develop and Maintain Knowledge, Skills, Procedures, and Tools

- *Number of inspectors/maintenance employees holding each type of required certification.* A decline in this metric may be a leading indicator of skill gaps or a higher than acceptable backlog for ITPM tasks.

12.5.4 Ensure Continued Fitness for Purpose

- *Number (or percent) of overdue ITPM tasks.* A high number (or rate) of overdue ITPM tasks may indicate resource constraints or that equipment is not being made available for scheduled maintenance.
- *Number of emergency/unplanned repair work orders per month.* One of the primary objectives of the *asset integrity* element is to reduce unplanned/breakdown maintenance work. Although many unplanned failures will not involve equipment included in the scope of the *asset integrity* element, an increase in this metric may be a leading indicator of an overall slip in the effectiveness of the maintenance program at the facility.
- *Work order backlog for the inspection group, in other words, planned activities that are not yet past due.* Similar to the number of past-due ITPM tasks, a backlog may indicate resource constraints. However, this metric may be a better leading indicator than the number of past-due ITPM tasks.
- *Total time charged to ITPM tasks each month/quarter.* A decline in the amount of time that is charged to these activities may indicate a change in focus for the maintenance department. Note that changes could be cyclical by design or could be an intended result (e.g., an effort to rationalize redundant or unnecessary calibration activities should result in a decline in time spent on ITPM tasks).

12.5.5 Address Equipment Failures and Deficiencies

- *Number of temporary repairs currently in service (deferred maintenance items).* This metric is another leading indicator of risk. It may be a particularly useful measure of efforts to plan maintenance if the metric is limited to repairs that are scheduled to be completed outside of major turnarounds (e.g., when parts arrive) versus repairs that must be made during an extended outage or turnaround.
- *Total number of deferred repairs, such as known deficiencies that will be addressed at the next turnaround.* Note that this metric will often increase linearly over time until the next maintenance shutdown, when it drops off sharply. However, the rate of increase could be a leading indicator of risk.

- *Average time to address/correct deficiencies.* This can be another leading indicator of risk and may help indicate if a step change has occurred in the ability to quickly repair equipment. However, at a continuous plant, this metric may be heavily influenced by a few deficiencies that are scheduled to be repaired at the next turnaround. Facilities may need to exclude "turnaround jobs" from this metric to provide a meaningful trend line.

12.5.6 Analyze Data

- *Number (or percent) of ITPM tasks that uncover a failure.* Clearly, one objective of the *asset integrity* element is to discover and correct hidden failures before they lead to catastrophic accidents. However, an increase in this metric may indicate that risk associated with equipment failure is gradually increasing.
- *Equipment reliability (or availability).* Similar to the previous metric, a decrease in reliability (or availability) may indicate that risk associated with equipment failure is gradually increasing.

12.6 MANAGEMENT REVIEW

The overall design and conduct of management reviews is described in Chapter 22. However, many specific questions/discussion topics exist that management may want to check periodically to ensure that the management system for the *asset integrity* element is working properly. In particular, management must first seek to understand whether the system being reviewed is producing the desired results. If the organization's level of *asset integrity* is less than satisfactory, or it is not improving as a result of management system changes, then management should identify possible corrective actions and pursue them. Possibly, the organization is not working on the correct activities, or the organization is not doing the necessary activities well. Even if the results are satisfactory, Management review can help determine if resources are being used wisely – are there tasks that could be done more efficiently or tasks that should not be done at all? Management can combine metrics listed in the previous section with personal observations, direct questioning, audit results, and feedback on various topics to help answer these questions. Activities and topics for discussion include the following:

- Compare the plan that is used to assign ITPM tasks at the facility to corporate standards. Does the plan contain any blanket exceptions? If so, review the justification(s) for these exceptions to ensure that they are still valid.

- Review individual exceptions to the ITPM plan that have been authorized since the last management review. (Note, this item should only address tasks not being performed, not individual ITPM tasks that are currently past due or were performed late.) As a group, are these exceptions justified or does the facility have an underlying tendency to not perform certain ITPM tasks (or not perform a variety of ITPM tasks in a certain unit)?
- Review inspection or test results that have been obtained using new/trial methods. Did these methods deliver the expected benefits and should their use be extended to other areas/equipment at the facility?
- Review the special certifications that maintenance department personnel hold. Has this changed since the previous management review? If any personnel holding critical certifications are planning to retire soon, review plans to train their replacements.
- Determine if failure to comply with the ITPM work schedule is more often caused by random events or systemic issues. For each systemic issue that is identified, determine if it was reviewed in advance to assess the risk.
- Determine the dominant causes of past-due ITPM tasks that appear to be random.
  o Is equipment not available?
  o Are intervals for ITPM tasks intentionally set low because a widely held belief exists that tasks are not normally completed on time (i.e., the "snooze button" approach to scheduling)?
  o Was the maintenance or inspection group unprepared to conduct the activity when equipment was available?
- Review which inspection activities are routinely contracted out and which are performed by maintenance department employees. Does this make sense from a risk management and business perspective?
- Review metrics related to the fraction of maintenance work that is planned. If the work is planned, the OEM parts and shop manuals are much more likely to be available, and the work can be scheduled when a fully staffed and trained work crew is available. Furthermore, an increase in unplanned maintenance work may reflect an underlying increase in the rate of unplanned breakdowns.
- Review new maintenance programs and work methods that are designed to reduce maintenance costs. What, if any, impact has the cost cutting had on equipment reliability, the fraction of ITPM tasks completed as scheduled, or other key metrics for this element?
- Examine Pareto charts showing chronic equipment failures and determine if the facility has a practice to report and investigate these failures as chronic (possibly near miss) incidents.

Reliable equipment, coupled with reliable human performance, is critical to managing risk. In addition, both are necessary conditions for reliable operation. The management review for this element often starts with questions related to metrics such as equipment availability or performance. However, the management review must delve deeper, examining the quality of work activities that underpin the *asset integrity* element. Just as the lack of a catastrophic process safety incident in the past 10 years is not necessarily a reliable predictor of the likelihood of a catastrophic incident in the next year, the lack of catastrophic equipment failure does not indicate that the *asset integrity* element is fully functional. Piping may be about to fail as a result of corrosion, nozzles on vessels may be cracked, safety systems that mitigate the consequences of an incident may have already failed. Thus, the management review for this element must examine the details, asking hard questions such as, "How many ITPM tasks have we missed?" and "What temporary repairs or deficient equipment are currently in service?" Also, maintenance and contract personnel can make errors that compromise the integrity of equipment. Are the training programs and procedures adequate? Answering these and other tough, but fundamental, questions is critical to understanding the true health of the *asset integrity* element.

12.7   REFERENCES

12.1   Lees, Frank P., *Loss Prevention in the Process Industries, Hazard Identification, Assessment, and Control*, 2nd edition, Butterworth-Heinemann, Oxford, England, 1996.

12.2   NTSB Report Number: AAR-79-17 (NTIS Report Number: NTISUB/E/104-017), American Airlines, Inc., DC-10, N110AA, Chicago International Airport, Chicago, IL, May 25, 1979, adopted on 12/21/1979.

12.3   API 510, *Pressure Vessel Inspection Code: In-Service Inspection, Rating, Repair, and Alteration*, ANSI/API 510-2006, American Petroleum Institute, 9th edition, June 2006.

12.4   API 570, *Piping Inspection Code: Inspection, Repair, Alteration, and Rerating of Inservice Piping Systems*, ANSI/API 570-2000, American Petroleum Institute, 2nd edition, October 1998.

12.5   *Risk-Based Inspection*, API-580. American Petroleum Institute, Washington, DC, 2000.

12.6   *Functional Safety: Safety Instrumented Systems for the Process Industry Sector – Part 1: Framework, Definitions, System, Hardware and Software Requirements ANSI/ISA-84.00.01-2004 Part 1 (IEC 61511-1 Mod)*, The International Society for Measurement and Control, Research Triangle Park, North Carolina, 2004.

Additional reading

Base Resource Document – Risk-based Inspection, API-581. American Petroleum Institute, Washington, DC, 2000.

Center for Chemical Process Safety, *Guidelines for Mechanical Integrity Systems*, American Institute of Chemical Engineers, New York, 2006.

13

CONTRACTOR MANAGEMENT

> The Piper Alpha explosion and fire was described in Chapter 11, *Safe Work Practices*, where it was noted that the failure to properly follow a safe work procedure allowed a massive release of hydrocarbon that led to the disaster. The subsequent investigation indicated that the contract supervisor responsible for the related maintenance job had not been properly trained in the safe work procedure. Furthermore, the investigation inferred that inadequacies in the emergency response training given (or in some cases, not given) to contractors on the oil platform likely contributed to the high loss of life in the accident.

13.1 ELEMENT OVERVIEW

Implementing practices to ensure that contract workers can perform their jobs safely, and that contracted services do not add to or increase facility operational risks, is one of nine elements in the RBPS pillar of *managing risk*. This chapter addresses the responsibilities of the contracting company and the contract employer in implementing a *contractor management* program. Section 13.2 describes the key principles and essential features of a management system for this element. Section 13.3 lists work activities that support these essential features, and presents a range of approaches that might be appropriate for each work activity, depending on perceived risk, resources, and organizational culture. Sections 13.4 through 13.6 include (1) ideas for improving the effectiveness of management systems and specific programs that support this element, (2) metrics that could be used to monitor this element, and (3) issues that may be appropriate for management review.

13.1.1 What Is It?

Industry often relies upon contractors for very specialized skills and, sometimes, to accomplish particularly hazardous tasks – often during periods of intense activity, such as maintenance turnarounds. Such considerations, coupled with the potential lack of familiarity that contractor personnel may have with facility hazards and operations, pose unique challenges for the safe utilization of contract services. *Contractor management* is a system of controls to ensure that contracted services support both safe facility operations and the company's process safety and personal safety performance goals. This element addresses the selection, acquisition, use, and monitoring of such contracted services.

Contractor management does not address the procurement of goods and supplies or offsite equipment fabrication functions that are covered by the *asset integrity* quality assurance function. While the most significant contractor safety challenges typically involve workers located closest to process hazards or involved in high-risk occupations, such as construction work, the safety needs of contractors providing simpler and more routine tasks, such as janitorial or groundskeeping services, must also be addressed in the *contractor management* program.

13.1.2 Why Is It Important?

Companies are increasingly leveraging internal resources by contracting for a diverse range of services, including design and construction, maintenance, inspection and testing, and staff augmentation. In doing so, a company can achieve goals such as (1) accessing specialized expertise that is not continuously or routinely required, (2) supplementing limited company resources during periods of unusual demand, and (3) providing staffing increases without the overhead costs of direct-hire employees.

However, using contractors involves an outside organization that is within the company's risk control activities. The use of contractors can place personnel who are unfamiliar with the facility's hazards and protective systems into locations where they could be affected by process hazards. Conversely, as a result of their work activities, the contractors may expose facility personnel to new hazards, such as unique chemicals hazards or x-ray sources. Also, their activities onsite may unintentionally defeat or bypass facility safety controls. Thus, companies must recognize and address new challenges associated with using contractors. For example, training and oversight requirements will be different from those for direct-hire employees. Thus, companies need to carefully select contractors and apply prudent controls to manage their services (Ref. 13.1).

Only by working together can companies and contractors provide a safe workplace that protects the workforce, the community, and the environment, as well as the welfare and interests of the company (Ref. 13.1).

13.1.3 Where/When Is It Done?

Contractor management begins well before the issuance of any service contract. Systems must be established for qualifying candidate firms based upon not only their technical capabilities, but also their safety programs and safety records.

Orientation and training of contractor personnel must be accomplished before they begin work. Responsibilities for this training must be defined, with some training often provided by the contract employer and some by the contracting company. The boundaries of authority and responsibilities must be clearly set for any contractor that works at the facility.

Periodic monitoring of contractor safety performance and auditing of contractor management systems is required. At the end of each contract period, retrospective evaluation of a contractor's safety performance should help determine whether the particular contractor is retained or considered for future work.

13.1.4 Who Does It?

While some responsibilities for implementing the *contractor management* element are assigned to contractor personnel, many tasks are assigned to company staff at either the facility or corporate level. Specific delegation of responsibilities between contractor, corporate, and facility personnel needs to be resolved prior to the start of a contractor-company relationship. Company groups having delegated roles and responsibilities could include operations, maintenance craft groups, facility or corporate safety, and perhaps the process safety group. The *contractor management* program is often managed by the procurement function, so purchasing personnel should be involved when appropriate.

13.1.5 What Is the Anticipated Work Product?

The anticipated work products for the *contractor management* element include:

- The creation of a list of pre-qualified candidate contract firms.
- The selection of specific contractors with strong safety programs and good safety records.

- The preparation of contract employers and their employees to safely provide their services based upon an understanding of relevant risks, facility safety controls and procedures, and their personal safety responsibilities.
- The safe delivery of the contracted services, with improved quality and productivity.
- Appropriate documentation of the contractor screening and selection process, the contractor's safety performance during the performance of its services, and any other issues relevant to the evaluation of the contractor for potential selection for future services.

Outputs of the *contractor management* element can also be used to facilitate the performance of other elements. For example, contract workers must be trained to properly implement *safe work practices*, and contract workers should be effectively integrated into the *workforce involvement* program.

13.1.6 How Is It Done?

The scope of contracted services can encompass a broad spectrum, ranging from contracting with an individual to provide a very specialized service to contracting with a large firm (perhaps with many subcontractors) who will provide hundreds of workers with diverse skills for a major construction project or maintenance turnaround. A contract firm could be onsite for only a few hours, never to return, or could have a continuing presence at a facility for decades. Some contractors will be directly exposed to the process and its hazards, while in other situations, such as new project construction adjacent to an operating unit, effective controls will be required to isolate the contractors from process hazards. Finally, some contract service companies have a very stable workforce, while others have a high rate of turnover.

The *contractor management* element must be well defined and flexible enough to handle this gamut of potential circumstances. The responsibilities and procedures for implementing these activities should be documented in a written program description with specific criteria for screening and selecting contract service providers.

13.2 KEY PRINCIPLES AND ESSENTIAL FEATURES

Safe operation and maintenance of facilities that manufacture, store, or otherwise use hazardous chemicals requires controls over the use of contracted services. That is, facilities should implement management systems to (1) consider contractor safety-related qualifications in the contractor selection process, (2) ensure that contractors are made aware of facility and operational hazards and that facility personnel are aware of any potential hazards introduced by contractor work activities, and (3) maintain high standards of safety performance during the conduct of the contracted services.

The following key principles should be addressed when developing, evaluating, or improving any management system for the *contractor management* element

- Maintain a dependable practice.
- Conduct element work activities.
- Monitor the *contractor management* system for effectiveness.

The essential features for each principle are further described in Sections 13.2.1 through 13.2.3. Section 13.3 describes work activities that typically underpin the essential features related to these principles. Facility management should evaluate the risks and potential benefits that may be achieved as a result of improvements in this element. Based on this evaluation, the facility should develop a management system, or upgrade an existing management system, to address some or all of the essential features, and execute some or all of the work activities, depending on perceived risk and/or process hazards that it identifies. However, these steps will be ineffective if the accompanying *process safety culture* element does not embrace the use of reliable management systems. Even the best management system, without the right culture, will not ensure the safe delivery of contracted services.

13.2.1 Maintain a Dependable Practice

When a company identifies or defines an activity to be undertaken, that company likely wants the activity to be performed correctly and consistently over the life of the facility. In order for the *contractor management* practice to be executed dependably across a company or facility involving a variety of people and situations, the following essential features should be considered.

Ensure consistent implementation. For consistent implementation, the *contractor management* element must be documented to an appropriate level of detail in a procedure or a written program addressing the general management system aspects discussed in Section 1.4. This program should address:

- Requirements for pre-screening candidate firms and for subsequent selection of specific contractors, including explicit criteria for evaluating past safety performance and safety program adequacy as part of the selection process.

- Orientation and training requirements for contractor personnel, defining who is responsible for performing the training and identifying the records that must be maintained.
- Coordination with the facility security procedures and emergency response procedures to address issues related to contractor access and evacuation. These issues may be particularly challenging in circumstances such as major construction projects or turnarounds requiring a large contractor presence at the facility.
- Compliance with specific company, facility, or regulatory requirements. Responsibility for each associated work activity should be identified and designated, as appropriate, to the company or the contractor. Recordkeeping requirements (e.g., injury and illness logs, environmental release reports) should be similarly identified and designated.
- Appropriate protection for proprietary information, for example, requirements for the execution of confidentiality agreements.
- Audit and oversight requirements for monitoring contractor safety performance, including the conduct of an end-of-contract evaluation of the contractor's overall safety performance. Appropriate records should be retained for reference when making selections for future contracts.

Many companies find it appropriate to assign one individual to coordinate the *contractor management* program at the facility. For small facilities, this individual may perform most or all of the program oversight roles. For larger facilities, or more complex *contractor management* systems, additional staffing may be required to fulfill responsibilities in the field. The owner of the *contractor management* element may be responsible for qualifying and selecting contractors, or this function may be conducted at the regional or corporate level. The *contractor management* element owner could be located in any of several functional groups, but close ties will be needed with the procurement function, with the facility safety function, and likely with operations and maintenance crafts groups. Considering the diversity of tasks involved, management must clearly define the roles and responsibilities for the facility and corporate staff members who are implementing the *contractor management* program.

Identify when contractor management is needed. The *contractor management* program documentation should explicitly define, or provide guidance for identifying, the contract services to which the *contractor management* program applies. Typical protocols address the potential safety significance of the contract services to be provided, and the nature and severity of hazards to which the contract employees will be exposed. For example, contract maintenance within a process unit would fall under the *contractor management* program, contract janitorial services in the administration building remote from the process probably would not, and contract janitorial services in a process unit control room may or may not.

The *contractor management* program should ensure that its requirements are extended down to any subcontractor firms that the contract company engages to provide onsite services under the primary contract. The company's role in approving subcontractors should be established in the primary contract language. The company's responsibilities for oversight (discussed below) apply to both contractors and subcontractors.

Involve competent personnel. Company personnel implementing the *contractor management* program will require a diverse mix of expertise. Staff members would require knowledge of (or, in some cases, at least an awareness of) the following topics:

- Process safety management principles.
- Conventional worker safety principles.
- Construction safety principles.
- Any local, regional, or national regulations relevant to the three items above.
- Facility processes, hazards, safety systems, and safety procedures.
- Procurement practices and regulations.

While responsibilities may be distributed to a number of individuals, each must be properly trained to appropriately fulfill their respective functions and responsibilities.

Ensure that practices remain effective. Records should be maintained to substantiate the qualifications of each selected contractor. These records might include information regarding (1) the contractor's safety performance, such as prior injury and illness statistics, (2) other performance data, such as experience modification rates calculated for a workers' compensation insurance rating, and (3) the contractor's safety program. Records documenting that certain firms were not selected because of their safety performance or safety programs might also be important (e.g., should the company ever need to substantiate or defend its selection process).

Companies should seek the advice of corporate legal counsel to determine appropriate retention periods for records generated under the *contractor management* element. Companies usually retain records documenting the contractor's performance throughout the life of the contract; for example, injury and illness logs, incident investigations, safety inspection results, and other performance audit records would be kept. In addition, regulations may require that certain types of records be retained, such as OSHA PSM requirements for incident investigation reports.

13.2.2 Conduct Element Work Activities

Appropriately select contractors. Many companies maintain a list of pre-screened contract firms as potential bidders for future contracts. Such a list (often maintained and managed by the purchasing function) can be based upon preliminary information supplied by potential contractors, and supplemented with a more detailed evaluation during the final qualification process. Contractors previously used by the company, whose past performance was satisfactory, could automatically be included on this list.

Company requirements and expectations regarding contractor safety performance and safety program contents should be included in bid requests and made part of the final contract language. Expectations should be specific and explicit, going beyond general requirements such as compliance "with all applicable regulations" (Ref. 13.1). To avoid duplication of efforts, the company should clearly identify which safety responsibilities it will assume during the contract. Table 13.1 lists information that the American Petroleum Institute (API) suggests requesting from bidders (a more detailed list is provided in Appendix B of the reference).

Procedures for final selection of contractors must address both the functional capabilities and the safety qualifications of the candidate firms. Candidates should be evaluated against the criteria established in the facility *contractor management* program description. When a list of pre-screened candidates is used, updates to relevant information regarding safety performance and safety programs should be required as part of the bid package.

TABLE 13.1. Safety Program and Performance Information Useful in Evaluating
          Potential Contractors
_____
- Injury and illness incidence rates.
- Experience modification rates, if applicable.
- The contractor's safety staffing plan, describing the onsite person(s) appointed by the contractor who will be responsible for safety, and their expertise and authority.
- A description of the safety orientation program that will be provided to all contractor employees on site.
- The contractor's enforcement and disciplinary action program for safety violations.
- The contractor's policy and programs regarding alcohol, controlled substances, and weapons.
- A list of safety equipment that will be provided by the contractor.
- A narrative describing the contractor's perspective on the hazards of the job, and the steps that will be taken to eliminate or minimize the potential for accidents.
- A description of the contractor's programs to comply with applicable regulatory requirements.
- A description of the contractor's employee training program.
_____
(Ref. 13.1.)

Deciding whether to extend an existing contract should involve addressing the same considerations as in the original selection, based upon relevant experience with the contractor during the current contract. Records should be sufficiently detailed, and retained long enough, to support such decisions.

Situations may arise in which the only qualified, or available, contractor needed for essential or urgent work does not meet the facility *contractor management* program criteria for safety performance or safety program content. Companies typically have a waiver provision in their *contractor management* program for such situations. Approvals of such waivers are typically made above the level of the *contractor management* element owner, and are contingent on establishing a program of special attention for that particular contractor while they are working at the facility. For example, more frequent inspections and audits may be warranted or, in the extreme, continuous surveillance may be required. A risk assessment may be warranted when making the decision to grant a waiver or when defining the nature of the special oversight program for the contractor.

Establish expectations, roles, and responsibilities for safety program implementation and performance. Most companies require that contractor safety standards be comparable to those of the facility. Even if company requirements and expectations regarding safety performance were communicated in the bid solicitation, these issues should be reemphasized and reviewed in greater detail with the selected contractor. Such discussions may be handled in a pre-job meeting held with the management of the contract firm. This meeting should (1) provide both sides with a clear, mutual understanding of safety performance expectations, (2) formalize the respective roles and responsibilities of the company and the contractor, and (3) resolve any interface issues, such as issues related to work permit approvals, first aid responsibilities, incident reporting and investigation, response to emergencies, and provision of basic and specialized safety equipment.

Opportunities for contractor personnel involvement in the RBPS program should be identified and such participation should be promoted (see Chapter 6, Workforce Involvement).

Ensure that contractor personnel are properly trained. Employees of onsite contractors will require considerable training on facility-specific topics such as:

- Facility hazards.
- Facility safety and security rules.
- Facility safety systems and procedures.
- Emergency response procedures.
- Facility security procedures, including any contractor access restrictions (e.g., areas of the facility that are off limits).
- Relevant safe work practices (see Chapter 11, Safe Work Practices).

- RBPS program fundamentals, including the expected participation of contractor personnel in the *workforce involvement* program
- Other individual safety responsibilities

Of course, contractors are also responsible for ensuring that their personnel safely and effectively perform the job-specific tasks they are assigned. If certifications or special skills are required, the contractor is also responsible for ensuring that their personnel possess such certifications and maintain their special skills.

Training requirements might not be identical for all contractor personnel. When valid bases for distinctions exist, a protocol should be developed for identifying training requirements based upon issues such as the type of work that the contractor personnel will be performing and the hazards in the work area. For example, all contractor personnel will require at least an awareness level of training on relevant subjects, while some will likely require detailed training. The contract workforce should also be integrated into any relevant emergency response training and drills (see Chapter 18). Requirements for appropriate refresher training should also be established; long-term contractors may require refresher training on certain topics on an annual basis, or an otherwise appropriate interval.

In determining training requirements, the company should keep in mind the concept of similarly situated employees. Consistent with this concept, contractor personnel performing the same functions as a facility employee should receive training comparable to that given the facility employee.

Contract employers are increasingly employing workers who do not speak the language predominantly used at the facility. Training requirements and materials, as well as subsequent procedures for administering such training, must address such bilingual contexts.

Responsibilities must be defined for administering the training. While contractors may provide certain training, site-specific training is often provided by the facility, which may also elect to provide some general training, such as fall protection, that would otherwise be conducted by the contractor.

Records documenting what training was given, when it was given, who gave the training, and who received the training should be maintained. Certain regulatory requirements may specify that documentation include the means used to verify that the training was understood, for example, through testing or a documented supervisory evaluation. Any such requirements should be reflected in the *training* program.

Fulfill company responsibilities with respect to safety performance. The company must supply appropriate information to the contractor to ensure that the contractor can safely provide the contracted services. As a minimum, the facility's safety policies, procedures to control work, and general work rules should all be provided to the contractor with sufficient lead time so that the contractor can train their personnel. Information needed for any training that the company delegates to the contractor should also be provided to the contractor well in advance of the start of work. Such information could include an overview of the process and its hazards, and actions related to reporting and responding to emergencies. Depending on the tasks that the contractor will perform, additional technical information (e.g., P&IDs) may have to be supplied.

Although contractors normally have the primary responsibility for promptly informing the company of any safety issues encountered in the conduct of their work, the company has a reciprocal responsibility to make this responsibility clear to each contractor and to promptly respond to and resolve such issues and concerns. Similarly, contractors must inform the company of special hazards associated with their work so the company can inform its employees and other potentially affected contractors. The company has the responsibility to carefully evaluate the hazards and ensure that the risk is controlled to a tolerable level.

Contractor activities may have an impact on other RBPS elements, and the company should plan and respond accordingly. One common example of such an impact, pertaining to the *knowledge* element, involves material safety data sheets (MSDSs) for chemicals that the contractor may bring into the facility in support of their services. The contractor should be required to inform the facility of any hazards introduced to the facility, and the facility should in turn provide this information to potentially affected employees.

Appropriate communication should be maintained between facility and contractor representatives with regard to work scheduling and associated safety matters. Facility representatives should participate in job planning meetings, as appropriate.

The company should maintain relevant records of contractor work activities, in accordance with any corporate records retention guidelines. Such records could include completed work permits and job plans. Regulatory requirements may dictate that the company maintains records of contract employee injuries and work-related illnesses, in addition to any such records maintained by the contract employer. (Caution: general corporate records policies do not supersede specific regulatory requirements such as required retention periods; any such inconsistencies must be resolved.)

Certain contractor activities, for example, bringing temporary office trailers, tool trailers, or lunch facilities into the facility during activities such as turnarounds, pose the potential to compromise existing safety systems and procedures. Issues often arise related to siting, compliance with area electrical classification, access of emergency vehicles to newly congested areas, and potential impacts of contractor activities on surrounding process units. Facility representatives must apply *management of change* to control the siting and configuration of such temporary additions.

13.2.3 Monitor the Contractor Management System for Effectiveness

Audit the contractor selection process. Documentation of the contractor screening and selection process should be periodically audited by an authority independent of this process. The audit should seek to confirm whether the contractor safety program and performance information required by the *contractor management* program was reliably obtained and thoroughly reviewed during the selection process, and whether the selections made were supported by the content of this information. Where waivers of selections were determined to be necessary, the audit should determine whether such waivers were properly authorized, and whether appropriate controls had been implemented to address the deficiencies prompting the waivers.

Monitor and evaluate contractor safety performance. While contractors have a responsibility to monitor the action of their employees and to enforce the safety performance requirements, the ultimate responsibility for ensuring the safety of its facility rests with the company. Facility management should establish systems for monitoring the conduct of contractor activities. Such systems should be proportional to risk; in other words, less frequent monitoring may be warranted (1) for contractors with strong safety programs and good safety records, especially if a longstanding relationship exists between the contracting company and the contractor, or (2) in situations where the risks associated with the contracted activity, or to which contractors are exposed, are low. In contrast, contractors with which the company has little prior experience, or a firm with a marginal safety record, would warrant more frequent monitoring.

The company should conduct random, unannounced inspections of contractor activities. Such inspections should include all shifts for round-the-clock activities. Adherence to safety procedures, compliance with work permit requirements, and maintenance of safe working conditions should be evaluated. Formal audits should look at contractor records, such as training and craft skills qualification records.

Company representatives should be alert to, and monitor contractor response to, incidents and near misses involving contractor activities or employees. The company should confirm the adequacy of investigations and appropriate followup to findings for any investigations handled by the contractor. The company should assume the lead responsibility for investigating any near miss or incident having actual or potential process safety significance.

If contractor safety performance is determined to be substandard, the company must take appropriate remedial actions to improve performance. Problems and their resolutions should be documented. If substantive safety problems cannot be resolved, the company should dismiss the contractor from the facility and terminate the contract.

At the conclusion of the contract, the facility should conduct a comprehensive review of the contractor's performance, including the manner and degree to which it achieved its safety responsibilities. Records of such reviews will serve as important input to any consideration of whether to use this contractor again in the future.

By providing close oversight of contractor activities and safety performance, the company reinforces its interest and commitment to safety. Frequent communications on safety issues, and reinforcement of the facility's expectations, allows the contractors a better understanding of, and an opportunity to subscribe to, the company's safety culture.

13.3 POSSIBLE WORK ACTIVITIES

The RBPS approach suggests that the degree of rigor designed into each work activity should be tailored to risk, tempered by resource considerations, and tuned to the facility's culture. Thus, the degree of rigor that should be applied to a particular work activity will vary for each facility, and likely will vary between units or process areas at a facility. Therefore, to develop a risk-based process safety management system, readers should perform the following steps:

1. Assess the risks at the facility, investigate the balance between the resource load for RBPS activities and available resources, and examine the facility's culture. This process is described in more detail in Section 2.2.
2. Estimate the potential benefits that may be achieved by addressing each of the key principles for this RBPS element. These principles are listed in Section 13.2.
3. Based on the results from steps 1 and 2, decide which essential features described in Sections 13.2.1 through 13.2.3 are necessary to properly manage risk.
4. For each essential feature that will be implemented, determine how it will be implemented and select the corresponding work activities described in this section. Note that this list of work activities cannot be comprehensive for all industries; readers will likely need to add work activities or modify some of the work activities listed in this section.
5. For each work activity that will be implemented, determine the level of rigor that will be required. Each work activity in this section is followed by two to five implementation options that describe an

increasing degree of rigor. Note that work activities listed in this section are labeled with a number; implementation options are labeled with a letter.

Note: Regulatory requirements may specify that process safety management systems include certain features or work activities, or that a minimum level of detail be designed into specific work activities. Thus, the design and implementation of process safety management systems should be based on regulatory requirements as well as the guidance provided in this book.

13.3.1 Maintain a Dependable Practice

Ensure Consistent Implementation

1. Develop a program for implementing the *contractor management* element.

    a. The *contractor management* program has been implemented in an ad hoc manner.

    b. A *contractor management* program has been developed and documented, addressing basic regulatory requirements.

    c. The *contractor management* program has been expanded to address issues such as integration of the contractor personnel into the RBPS *workforce involvement* program, *culture* initiatives, and so forth.

2. Establish an element owner at the facility level for the *contractor management* program.

    a. Oversight of the *contractor management* program is provided on an ad hoc basis.

    b. Oversight of the *contractor management* program is formalized and assigned to a responsible party.

3. Define roles and responsibilities for corporate or facility staff who oversee the *contractor management* program.

    a. Roles and responsibilities are informally understood.

    b. Roles and responsibilities are formally defined.

    c. Roles and responsibilities are reinforced through an active accountability system.

Identify When Contractor Management Is Needed

4. Determine the scope of the *contractor management* program application.

    a. The scope of application of the *contractor management* program is informally understood.

     b. The physical scope of application and services covered under the *contractor management* program are documented.

     c. Item (b), and the scope of application is periodically reevaluated and updated as appropriate.

5. Ensure that *contractor management* program requirements are communicated to subcontractors.

     a. The responsibility for extending *contractor management* program requirements to subcontractors is left with the prime contract company.

     b. Facility staff members ensure that *contractor management* program requirements are communicated to subcontractors.

Involve Competent Personnel

6. Train facility staff on their role in administering the *contractor management* program.

     a. Informal training is provided.

     b. Formal training is provided.

     c. Item (b), and refresher training is provided as warranted.

Ensure that Practices Remain Effective

7. Maintain records substantiating contract award decisions for successful and unsuccessful bidders.

     a. Records are maintained for the duration of the contract, but are disposed of shortly thereafter.

     b. Auditable records are maintained as required to support legal and programmatic needs, for example, for subsequent review and possible identification of program improvement opportunities.

8. Maintain records of contractor safety performance during the contract, including inspection and audit results, injury statistics, and incident investigation findings.

     a. Records are maintained for the duration of the contract, but are disposed of shortly thereafter.

     b. Auditable records are maintained as required to support legal and programmatic needs, for example, for subsequent review and possible identification of program improvement opportunities.

9. Maintain records of contractor workforce involvement in RBPS management system implementation.

     a. Records are maintained for the duration of the contract, but are disposed of shortly thereafter.

b.   Auditable records are maintained as required to support legal and programmatic needs, for example, for subsequent review and possible identification of program improvement opportunities.

13.3.2 Conduct Element Work Activities

Appropriately Select Contractors

10. Develop and maintain a list of pre-screened candidates.

a.   A list of pre-screened candidates is informally maintained.

b.   A list is developed and periodically updated, and it reflects the results of post-contract evaluations.

11. Include company safety expectations in the request-for-bid package sent to candidates.

a.   Safety goals are addressed in a general context in the bid package, but specific expectations are not detailed.

b.   Safety expectations are explicitly outlined in the bid package.

c.   Item (b), and bidders are required to explicitly include descriptions of their plans to meet safety expectations in their proposals.

12. Select contractors based upon their functional capabilities, past safety performance, and soundness of their safety programs.

a.   Contractor evaluations focus predominantly on technical capabilities.

b.   Contractor evaluations involve a balanced assessment of technical capabilities, past safety performance, and soundness of their safety programs.

c.   Item (b), and contractor evaluations include interviews with contractor staff who will be responsible for overseeing functional and safety performance.

13. Consider the company's past experience with the contractor when making decisions for future contracts or contract extensions.

a.   An informal system exists for considering past performance in the selection process.

b.   A formal system exists for considering past performance in the selection process.

c.   Item (b), and significant emphasis is placed on past performance in the selection process.

14. Provide a carefully controlled waiver policy to address situations in which the only available contractor for a particular service does not meet minimum *contractor management* program requirements for safety program and performance.

    a.   An informal waiver policy exists.

    b.   A formal waiver policy, which includes provisions for enhanced oversight of targeted contractors, has been implemented and is closely monitored.

    c.   Item (b), and waivers are permitted only in extreme circumstances.

Establish Expectations, Roles, and Responsibilities for Safety Program Implementation and Performance

15. Conduct a pre-job meeting with the selected contractor to address safety issues, clearly establish a mutual understanding of company expectations for contractor safety performance, delineate safety program roles and responsibilities for the company and the contractor, and discuss contract workforce involvement in RBPS management system implementation.

    a.   An informal pre-job meeting is conducted.

    b.   A detailed pre-job meeting is conducted.

    c.   Item (b), and understandings reached during the pre-job meeting are formally documented and become part of the contract package.

Ensure that Contractor Personnel Are Properly Trained

16. Identify appropriate training requirements for contractor personnel.

    a.   Training requirements are addressed in an ad hoc fashion.

    b.   General training requirements are identified, but without differentiating the types of work to be performed or the areas in which the work will be performed.

    c.   Training requirements are explicitly defined as a function of the risks associated with the type of work to be performed and the areas in which the work will be performed.

    d.   Item (c), and training requirements are periodically reviewed and updated, as warranted.

17. Define who is to provide the various types of training.

    a.   Training responsibilities are addressed on an ad hoc basis.

    b.   Training responsibilities are formally defined and documented.

    c.   Item (b), and training effectiveness is monitored and training responsibilities are modified accordingly, when required.

18. Provide and confirm awareness training for contractor personnel.

    a.   Initial training is provided.

    b.   Item (a), and refresher training is provided, as warranted.

    c. Item (b), and contractor performance and incidents data are reviewed and serve as input into any modifications to the training program.

19. Provide and confirm detailed training for contractor personnel.

    a. Initial training is provided.

    b. Item (a), and refresher training is provided, as warranted.

    c. Item (b), and contractor performance and incidents data are reviewed and serve as input into any modifications to the training program.

20. Retain records of training given, including the means used to confirm understanding, as appropriate.

    a. Training records are maintained in an informal manner.

    b. Records are maintained for the duration of the contract, but are disposed of shortly thereafter.

    c. Auditable records are maintained as required to support programmatic and legal needs.

21. Identify required certifications for special qualifications, such as for welders, heavy equipment operators, non-destructive testing technicians, and so forth, and ensure that required documentation of such certifications is maintained.

    a. Certification records are maintained in an informal manner.

    b. Records are maintained for the duration of the contract, but are disposed of shortly thereafter.

    c. Auditable records are maintained as required to support programmatic and legal needs.

22. Ensure that similarly situated contractor personnel receive training comparable to that given to their facility employee counterparts.

    a. Contract employees receive comparable training in some, but not all, circumstances.

    b. Contract employees receive the same training as similarly situated facility employees.

23. Include the contract workforce in any relevant emergency response drills.

    a. Contractor personnel participate in emergency response drills.

    b. Contractor personnel are involved in planning and conducting emergency response drills.

Fulfill Company Responsibilities with Respect to Safety Performance

24. Provide appropriate and sufficient information to contractors on safety policies, relevant procedures used to control work, and general facility safety rules.

    a. Information is provided to contractors, who are expected to inform their employees.

    b. Facility personnel train contractor personnel in facility policies and procedures.

    c. Item (b), and formal systems have been established for periodically updating information provided to contractor personnel.

25. Promptly respond to and resolve safety issues identified by contractors.

    a. Facility staff responds to contractor personnel safety concerns on an ad hoc basis.

    b. Formal systems have been established for responding to contractor personnel safety concerns.

    c. Item (b), and responses to contractor personnel safety concerns are monitored for timeliness and appropriateness.

26. Participate in preparing and evaluating job plans.

    a. Facility staff members participate in preparing and evaluating job plans for those jobs entailing the most significant risks.

    b. Facility staff members generally participate in preparing and evaluating job plans.

    c. Facility staff members always participate in preparing and evaluating job plans.

27. Maintain records of contractor injuries and illnesses.

    a. Facility staff is notified of contractor injuries and illnesses.

    b. Facility staff maintains only those records of contractor injuries and illnesses required by regulations or corporate policies.

    c. Facility staff maintains comprehensive records of contractor injuries and illnesses, exceeding regulatory requirements.

28. Maintain records of safety-related work, such as completed work permits.

    a. Records of safety-related work are maintained only in the event of an incident or in response to potential legal actions.

    b. Records of safety-related work are maintained for the duration of the contract, but are disposed of shortly thereafter.

    c. Records of safety-related work are maintained to support auditing necessary to improve system performance.

29. Maintain control over contractor temporary facilities.
    a. Ad hoc controls are maintained over contractor temporary facilities.
    b. Formalized and effective controls are maintained over contractor temporary facilities.
    c. Item (b), and such controls are integrated into the facility *management of change* process.
30. Integrate contractor-supplied information into the *process safety knowledge* program.
    a. Contractor-supplied information is integrated into the *knowledge* program on an ad hoc basis.
    b. Formal systems are established for integrating contractor-supplied information into the *knowledge* program.

13.3.3 Monitor the Contractor Management System for Effectiveness

Audit Contractor Selection Process
31. Implement a program for auditing the contractor selection process.
    a. Informal reviews of the contractor selection process are conducted.
    b. Formal audits of the contractor selection process are conducted, but are coordinated at the facility level.
    c. Formal audits of the contractor selection process are coordinated at the corporate level.

Monitor and Evaluate Contractor Safety Performance
32. Conduct unannounced field inspections of contractor work activities.
    a. Unannounced field inspections of contractor work activities are occasionally conducted.
    b. Unannounced field inspections of contractor work activities are frequently conducted on all shifts.
    c. Rigorous field inspections of contractor work activities are conducted, and include higher level facility managers on the team.
33. Attend contractor safety meetings.
    a. Facility staff members occasionally attend contractor safety meetings.
    b. Facility staff members frequently attend contractor safety meetings.

    c.   Higher level facility managers often attend contractor safety meetings.

34. Periodically audit contractor records.

    a.   Facility staff audits contractor records on an ad hoc basis.

    b.   A formal program exists for auditing contractor records.

    c.   Item (b), and results of contractor records audits are reviewed with upper management.

35. Review contractor incident and near miss investigation practices and reports.

    a.   Facility staff reviews contractor incident and near miss investigation practices and reports on an ad hoc basis.

    b.   A formal program is in place for reviewing contractor incident and near miss investigation practices and reports.

    c.   Item (b), and results of these reviews are discussed with upper management.

    d.   Contractor incidents are integrated into the facility incident program and investigated using teams with both facility and contractor representation.

36. Implement corrective actions to address contractor safety performance problems, as required.

    a.   Corrective actions are limited to the most severe problems or infractions.

    b.   Corrective actions are generally implemented to address problems or infractions.

    c.   Contractor personnel are held to the same standards of performance as facility staff.

37. Conduct and document an end-of-contract evaluation of contractor performance.

    a.   End-of-contract evaluations are conducted for larger contracts.

    b.   End-of-contract evaluations are conducted for larger contracts and before contract renewals or extensions.

    c.   End-of-contract evaluations of contractor performance are always conducted.

13.4   EXAMPLES OF WAYS TO IMPROVE EFFECTIVENESS

This section provides specific examples of industry tested methods for improving the effectiveness of work activities related to the *contractor management* element. The examples are sorted by the key principles that were first introduced in Section 13.2. The examples fall into two categories:

1. Methods for improving the performance of activities that support this element.
2. Methods for improving the efficiency of activities that support this element by maintaining the necessary level of performance using fewer resources.

These examples were obtained from the results of industry practice surveys, workshops, and CCPS member-company input. Readers desiring to improve their management systems and work activities related to this element should examine these ideas, evaluate current management system and work activity performance and efficiency, and then select and implement enhancements using the risk-based principles described in Section 2.1.

13.4.1 Maintain a Dependable Practice

Include checklists and standardized prequalification and evaluation forms in the contractor management program documentation. Standardized forms will help ensure consistent, thorough evaluations. Examples of such forms are provided in References 13-1, 13-2, and 13-3.

Use a risk matrix when evaluating and defining the administrative controls appropriate for a particular contracted service. A company may establish graduated screening, qualifications, and capabilities requirements proportional to the safety significance of the contracted services. For example, higher risk activities may necessitate greater detail in training or more specialized control of access to the worksite. The API suggests that the following factors be considered when evaluating the risk associated with contractor activities, and provides a risk matrix to use in the evaluation (Ref. 13.2).

- Nature and location of work.
- Potential for exposure to work site hazards.
- Potential for the contractor to create hazards.
- Duration of work.
- Contractor's experience and expertise in performing similar work.

13.4.2 Conduct Element Work Activities

Participate in joint programs for contractor evaluation and contractor personnel training. In some more heavily industrialized areas, companies have found it advantageous to collectively sponsor a consolidated contractor evaluation and training program, which may include sharing of end-of-job performance evaluations. Sometimes this function is provided by independent third party organizations that provide successful students with credentials that they can take from facility to facility. Some of these institutions also provide controlled-substance screening and preliminary background check services. Such organizations can often be located by searching the internet for phrases such as "Contractors' Safety Council" or through referrals from other area companies.

Use third-party organizations to pre-screen contractor firms. Subscription services are available to pre-screen contractor firms against commonly accepted standards, and to act as a clearinghouse for referrals to companies seeking a contractor with particular expertise.

Provide a system for pre-screening candidate contractor firms at the corporate level. Some companies find that assigning the responsibility for pre-screening to a centralized corporate group is most efficient. The facility staff can then simply select a qualified contractor from the corporate list.

Develop a set of corporate standard procedures, requirements, and so forth, for contractor screening and selection to be used at all company facilities, across all regions. These standards will help ensure consistency across the company (particularly if screening and selection is a facility responsibility).

Interview contractor management to determine their value for, and commitment to, high standards of safety performance. The contractor safety program included with a bid package is a description on paper. Interviews with contractor management can provide greater insight into the emphasis placed on implementing the program.

Share results of contractor evaluations between company **facilities.** Sharing these results can help ensure that one facility does not accept a contractor that was rejected for valid reasons at another facility.

Establish a roles and responsibilities matrix to formalize agreements on safety matters between the company and the contractor. Such a matrix will help ensure that critical safety responsibilities are not overlooked.

Establish long-term alliances between the company and contractors. Where other business considerations permit, the advantages of long-term associations with particular contract firms include: (1) reducing the screening burden, (2) providing an opportunity for the contractor to better learn the safety expectations of the company, and (3) helping the contractor more effectively integrate into the company safety culture.

Conduct meetings with potential bidders prior to bid submission. Use such meetings to review specific safety issues and requirements with potential bidders to ensure that the bidders know what is expected of them, and to ensure that necessary information on safety programs and performance are included in subsequent bid packages.

Attend, or even help conduct, contractor safety meetings. Having a facility representative attend contractor safety meetings to reinforce company safety expectations, to answer any questions from workers, and to gather issues for resolution could help improve communications and enhance the effectiveness of the safety program.

Establish close linkages between the facility and contractor safety program. This might include establishing the same performance targets and same monitoring programs. Some companies integrate contractor safety performance into the facility's safety performance program.

Ensure, where appropriate, that inviolable safety requirements and expectations have been established. Contractor personnel must be made aware of those safety requirements that, if not adhered to, will result in the immediate removal of an individual from the facility (or, possibly all workers associated with that contract employer).

Negotiate contract terms that provide incentives for superior safety performance or, alternatively, place a portion of the contract company's compensation at risk, dependent upon safety performance. Such contract terms will reinforce safety expectations.

Require the contractor to establish a program to ensure that contractor personnel are fit for duty and are not compromised by external influences. External influences include substance abuse, excessive fatigue, and emotional issues. Programs that monitor for, or help prevent, these conditions are particularly important for contractors working in high risk areas.

Use computer-based training and testing. Companies are increasingly using computer-based media to supplement videotaped and live presentations for contractor orientation and other training. Computerized tests can draw upon a random selection of questions to reduce cheating and ensure that retests evaluate more than the student's ability to remember the correct answer to the question missed previously. Computerization also assists in the maintenance of training and testing records.

Encourage facility employers to take responsibility for providing orientation and training for contract employees. Some companies find that the thoroughness and consistency of facility-specific training is better if the facility provides the training, rather than providing the information to contract employers and expecting them to correctly instruct their employees.

Work with the primary facility contractor (general contractor) to develop their training skills or to develop a contract training center. Some companies find it effective to use the long-term, primary facility contractor to conduct the training for all contract companies on the facility.

Combine proof of orientation and access control. Access cards or badges should be issued only after successful completion of orientation or training. Training should be renewed periodically as a condition of continued facility access. Companies are increasingly concluding that hardhat stickers denoting successful completion of training indicate only that the hardhat has been trained, and not necessarily the person who is currently wearing the hardhat.

Encourage contractors to use external training resources and programs. These resources include programs provided by the National Center for Construction Education and Research, and other similar organizations.

Integrate the training of contractor personnel with facility staff. For similarly situated employees (i.e., contractors performing the same functions as facility employees), the training needs may be identical or very similar. Integrating the training of contractor personnel with facility staff is often the most cost-effective approach.

Involve contractor workers in RBPS management system implementation, such as safety observation programs. See Chapter 6, Workforce Involvement.

Ensure that contractor personnel have a means to report safety concerns. The reporting systems could be the same, or parallel to, systems that facility employees use to report safety concerns or suggestions. Such a system will facilitate tracking and responding to concerns.

Involve contractor personnel in facility safety promotions and in the celebration of facility safety achievements. Such integration may enhance the degree to which contractor personnel engage in the facility's safety efforts. This is particularly relevant if the facility and contractor safety programs and metrics are closely linked.

13.4.3 Monitor the Contractor Management System for Effectiveness

Emphasize to facility employees the importance of their role in monitoring contractor safety performance. Facility employees must be aware of their responsibility to intercede or report to management when they observe an unsafe act or condition related to contractor activities.

Use third-party safety inspectors to monitor contractor activities. An independent inspector might be hired to monitor performance of a contractor that required a waiver of established selection criteria as a result of past safety performance or safety program deficiencies. Some companies also use third-party personnel to supplement company resources and monitor all contractors with respect to safety performance, housekeeping practices, and so forth.

Ensure that contractor safety statistics are reviewed in regular facility management safety performance reviews. Facility management needs to be aware of contractor safety statistics so they can prompt any needed improvements in contractor safety performance.

13.5 ELEMENT METRICS

Chapter 20 describes how metrics can be used to improve performance and when they may be appropriate. This section includes several examples of metrics that could be used to monitor the health of the *contractor management* element, sorted by the key principles that were first introduced in Section 13.2.

In addition to identifying high value metrics, readers will need to determine how to best measure each metric they choose to track. In some cases, an ordinal number provides the needed information, for example, total number of workers. Other cases, such as average years of experience, require that two or more attributes be indexed to provide meaningful information. Still other metrics may need to be tracked as a rate, as in the case of employee turnover. Sometimes, the rate of change may be of greatest interest. Since every situation is different, the reader will need to determine how to track and present data to most efficiently monitor the health of RBPS management systems at their facility.

13.5.1 Maintain a Dependable Practice

- *Frequency of required waivers of qualification requirements for past safety performance or current safety program.* This indicates whether waivers are being issued too frequently. Is it too easy for a company with a poor safety record or a weak safety program to gain access to the facility?
- *Percentage of contracted firms that, based upon post-job evaluation, would be considered for future contracts.* A low value could indicate that the contractor screening process is not sufficiently discerning. A high value usually indicates that the screening process is effective, yet an unusually high value could indicate that the post-job evaluation is not sufficiently discerning.

13.5.2 Conduct Element Work Activities

- *Percentage of required contractor training conducted on schedule.* A low value indicates a gap that needs to be filled, and it should be investigated to identify the underlying reason the training is not occurring on a timely basis.
- *Percentage of contractor-related incidents or near misses that were subjected to a root-cause analysis.* A low value indicates that the contractor is not fully committed to learning from incidents and near misses, which is a clear warning that its management is not fully committed to safety.
- *Frequency of, and percentage attendance for, contractor safety meetings.* A low value may indicate that the contractor is not fully committed to its safety program.

- *Number of safety program improvement suggestions contributed by contractor personnel.* A high value usually indicates significant engagement by the contractor workforce in the facility safety program.
- *Number of open contractor safety suggestions, in other words, those not yet resolved by a company representative, or average age for unresolved suggestions.* High values would indicate a lax response, which will discourage the continued involvement of contractor personnel in the program.

13.5.3 Monitor the Contractor Management System for Effectiveness

- *Safety performance metrics for contractor companies.* Such metrics include injury or illness rates, and near miss and incident rates. High values may indicate inadequacies in (1) the contractor's safety program, (2) the contractor's safety culture, or (3) the company's oversight of contractor activities. However, a high ratio of near miss to incident reports typically indicates aggressive efforts to improve safety.
- *Percentage of incidents and near misses investigated by the facility that had root causes related to contractor activities.* A high value may indicate inadequacies in (1) the contractor's safety program, (2) the contractor's safety culture, or (3) the company's oversight of contractor activities.
- *Relevant statistics monitoring compliance with safe work practice procedures for contractor involved jobs.* See Chapter 11, Safe Work Practices.
- *Number or frequency of negative findings in job safety evaluations, field inspections, audits of safe work practice implementation, and other safety-related audits.* A high value may indicate inadequacies in (1) the contractor's safety program or (2) the contractor's safety culture. While a high value might also indicate that the company's oversight of contractor activities is sufficient to find the problems after they occur, it may not be adequate to detect precursors and prevent the problems before they occur.

13.6 MANAGEMENT REVIEW

The overall design and conduct of management reviews is described in Chapter 22. However, many specific questions/discussion topics exist that management may want to check periodically to ensure that the management system for the *contractor management* element is working properly. In particular, management must first seek to understand whether the system being reviewed is producing the desired results. If the organization's level of *contractor management* is less than satisfactory, or it is not improving as a result of management system changes, then management should identify possible corrective actions and pursue them. Possibly, the organization is not working on the correct activities, or the organization is not doing the necessary activities well. Even if the results are satisfactory, management reviews can help determine if resources are being used wisely: are there tasks that could be done more efficiently or tasks that should not be done at all? Management can combine metrics listed in the previous section with personal observations, direct questioning, audit results, and feedback on various topics to help answer these questions. Activities and topics for discussion include the following:

- Determine whether all applicable requirements are being met/enforced with regard to the selection of contractors and subcontractors.
- Based upon intra-company benchmarking and a review of facility practices, determine whether facilities make consistent decisions about acceptability/qualifications of contractors. Have any inconsistent decisions been made with respect to a particular contractor? (This management review item would be most appropriate at the division or corporate level.)
- Review contractor safety performance metrics to identify any negative trends. If any exist, have they been investigated, and have remedial actions been implemented?
- Determine whether facility management periodically meets with all contractor representatives to identify, discuss, and resolve safety issues. Review the typical nature and content of these discussions as well as the status of any open action items that result from these discussions.
- Based upon attending or reviewing the minutes from contractor safety meetings, determine whether the content is of consistently high quality, and whether it addresses relevant topics.
- Conduct periodic inspections of worksite conditions (including housekeeping), and monitor contractor work practices. What do the observations indicate about contractor safety attitudes?
- Participate in, or review reports from, contractor incident investigations to determine if they are of sufficient rigor and quality. Are root causes determined, and do recommendations adequately address the causes? Are recommendations being addressed in a timely fashion?
- Discuss contractor safety performance with facility staff members most familiar with contractor activities. Do they perceive performance to be improving or degrading? What gaps, if any, do they believe should be addressed?
- Evaluate the activities of facility staff responsible for directly monitoring contractor activities. Are they spending enough time in the field monitoring, and at an appropriate frequency? Is the level of

oversight by facility staff appropriate to the perceived level of risk of the contractor activities?

- Determine whether company required audits of contractor activities are being completed. Are the audits completed on schedule? Are results being followed up (recommendations closed)?
- Based upon discussions with contract workers, evaluate whether they have an adequate awareness of (1) process hazards to which they might be exposed, (2) emergency response procedures, and (3) their personal responsibility for safe behaviors.

The use of contracted services will likely become increasingly common within the chemical process industries. A well-designed *contractor management* program, implemented in a risk-appropriate fashion, can help ensure that contract workers are effectively integrated into the facility's risk control activities.

13.7 REFERENCES

13.1   API Recommended Practice 2220, *Improving Owner and Contractor Safety Performance,* September 1991.

13.2   API Recommended Practice 2221, *Manager's Guide to Implementing a Contractor Safety Program*, June 1996.

13.3   CCPS, *Contractor and Client Relations to Assure Process Safety*, 1996.

Additional reading

International Association of Oil & Gas Producers, HSE Management – *Guidelines for Working Together In a Contract Environment*, Report No. 6.64/291, 1999.

HSE, *Use of Contractors – A Joint Responsibility*, INDG368, ISBN 0 7176 2566 4, 2002.

14

TRAINING AND PERFORMANCE ASSURANCE

A flash fire followed by a boiling liquid expanding vapor explosion (BLEVE) occurred within the liquefied petroleum gas (LPG) tank farm area of a refinery near Lyon, France. Eighteen people were killed and 81 were injured (Ref. 14.1). The gas storage facility was destroyed and the fire spread to nearby liquid hydrocarbon storage tanks. The incident was started by a worker improperly draining water from the bottom of one of the LPG storage spheres. LPG flashing through the upstream drain valve (in a double block valve arrangement) that was being used to regulate the flow in the drain line caused both the upstream and downstream valves to freeze open, releasing LPG to the atmosphere through the open drain. If the worker had regulated water flow with the downstream valve, the upstream valve would not have frozen open when LPG began to flash. This upstream valve, then, might have been used to stop the flow. This incident illustrates why the management system must ensure that workers are trained in the correct procedures and why their satisfactory performance must be periodically verified.

14.1 ELEMENT OVERVIEW

Training workers and assuring their reliable performance of critical tasks is one of nine elements in the RBPS pillar of *managing risk*. In the context of process safety management systems, this chapter describes the meaning of training, the attributes of a good training system, and the steps an organization might take to implement a robust training program. Section 14.2 describes the key principles and essential features of a management system for this element.

Section 14.3 lists work activities that support these essential features, and presents a range of approaches that might be appropriate for each work activity, depending on perceived risk, resources, and organizational culture. Sections 14.4 through 14.6 include (1) ideas for improving the effectiveness of management systems and specific programs that support this element, (2) metrics that could be used to monitor this element, and (3) issues that may be appropriate for management review.

14.1.1 What Is It?

Training is practical instruction in job and task requirements and methods. It may be provided in a classroom or workplace, and its objective is to enable workers to meet some minimum initial performance standards, to maintain their proficiency, or to qualify them for promotion to a more demanding position. Performance assurance is the means by which workers demonstrate that they have understood the training and can apply it in practical situations. Performance assurance is an ongoing process to ensure that workers meet performance standards and to identify where additional training is required.

14.1.2 Why Is It Important?

A consistently high level of human performance is a critical aspect of any process safety program; indeed, a less than adequate level of human performance will adversely impact all aspects of operations. Without an adequate training and performance assurance program, a facility can have no confidence that work tasks will be consistently completed to minimum acceptable standards, in accordance with accepted procedures and practices.

14.1.3 Where/When Is It Done?

Training takes place both in the workplace and the classroom, and it should be completed before a worker is allowed to work independently in a specific job position. Refresher training is provided on an ongoing basis thereafter as needed. Ideally, training is based on needs analyses that define the minimum acceptable knowledge, skills, and abilities (KSAs) required for a worker in a specific position. These analyses also include any requirements imposed by regulations, codes, industry standards, or company policies. The training program should then be developed to bridge the gap between what is demanded of a qualified job applicant (e.g., basic reading and writing skills) and what is required to succeed in a specific job. The performance assurance system then tests the trained workers initially, and periodically thereafter, to demonstrate that they possess the required KSAs and are qualified to work independently.

14.1.4 Who Does It?

Each work group should define the KSAs necessary to work successfully in their department. A human resources group often manages the overall process because it affects hiring, job placement, and retention decisions. A department or designated individual(s) usually plans and coordinates training activities. The training itself is typically conducted by subject matter experts (SMEs) or outside specialists who have been trained and qualified as trainers. Qualified peers, trainers, managers, human resource personnel, or third parties, depending on the exact nature of the testing, may conduct performance assurance testing.

14.1.5 What Is the Anticipated Work Product?

The output of this activity is a set of job performance standards and a list of initial and ongoing training needs for each job position. A set of training materials should be developed for each training need that will be met by internal resources. For other training needs, a list of approved suppliers should be developed. A training record should be provided for each worker showing that person's training needs, the dates on which initial training and any refresher training was satisfactorily completed, and a schedule of future training classes. In addition, the work group should document an appropriate approach for verifying performance and provide examiners with the resources (model tests, observation checklists, etc.) necessary to test workers. Outputs of the *training* element can also be used to facilitate the performance of other elements. For example, teaching employees about company history will reinforce the *culture* element, and teaching employees to recognize hazards will improve the *risk*, *management of change*, and *operations* elements.

14.1.6 How Is It Done?

To develop an effective set of training materials, start by identifying the jobs and tasks that must be performed. Once the tasks are identified, determine the KSAs for personnel who will be assigned to perform the task, and the KSAs of individuals qualified to apply for the job. Then identify the gaps and develop or procure training materials and programs that will enable workers to achieve the required KSAs. Also, use the list of required KSAs to develop testing methods that will reasonably ensure that workers have the required performance competencies. Finally, validate the training approach and develop acceptable norms for new workers by testing workers who are currently considered qualified.

14.2   KEY PRINCIPLES AND ESSENTIAL FEATURES

Safe operation and maintenance of facilities that manufacture, store, or otherwise use hazardous chemicals requires qualified workers at all levels, from managers and engineers to operators and craftsmen. Training and other performance assurance activities are the basis for achieving high levels of human reliability. In this context, training broadly includes education in specific procedures governing operations, maintenance, safe work, and emergency planning and response, as well as in the overall process and its risks.

Work activities are commonly categorized as skill-based, rule-based, or knowledge-based, and each category requires a different type of training and performance assurance. Consider a common task – driving a vehicle. When one first attempts to drive, the activity is predominantly knowledge-based. The driver must consciously follow traffic regulations and actively work to maintain speed and distance, remember to signal before turning, and perform the other myriad actions associated with driving. With practice, one develops rules, such as automatically accelerating to maintain speed driving up a hill or increasing following distance at higher speeds. Eventually, with continued practice, the actions associated with driving become second nature and the driver expends little active thought on the activity of getting from here to there. Once mastered, these skills can be applied to most any vehicle one might choose to drive.

Likewise, in a much more complex and hazardous process plant, (1) some activities should be second nature to a qualified worker, such as starting a centrifugal pump, (2) some activities are governed by procedural rules, such as making a batch, and (3) some activities require extensive knowledge, such as troubleshooting a temperature excursion. The challenge is to determine the training needs for workers in various job positions and to ensure their ongoing competence to perform necessary tasks.

Current, job-relevant training helps ensure reliable worker performance, which both prevents process safety incidents and improves the response to those incidents that do occur. Periodic testing and refresher training as needed provides assurance that workers are maintaining the required KSAs. Figure 14.1 illustrates the sequence of tasks in the overall *training* system.

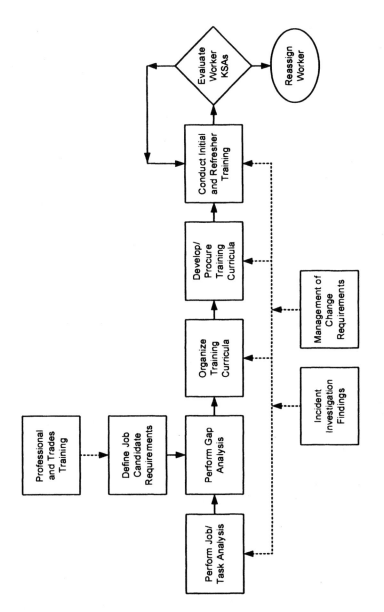

FIGURE 14.1. Training System Tasks

The following key principles should be addressed when developing, evaluating, or improving any management system for the *training* element:

- Maintain a dependable practice
- Identify what training is needed
- Provide effective training
- Monitor worker performance

The essential features for each principle are further described in Sections 14.2.1 through 14.2.4. Section 14.3 describes work activities that typically underpin the essential features related to these principles. Facility management should evaluate the risks and potential benefits that may be achieved as a result of improvements in this element. Based on this evaluation, the facility should develop a management system, or upgrade an existing management system, to address some or all of the essential features, and execute some or all of the work activities, depending on perceived risk and/or process hazards that it identifies. However, these steps will be ineffective if the accompanying *culture* element does not embrace the pursuit of excellence in performance. Even the best management system, without the right culture, will not produce reliable workers.

14.2.1 Maintain a Dependable Practice

A documented training program is fundamental to achieving reliable worker performance. The jobs and tasks that workers are expected to perform must be identified, workers must be selected and trained, and their performance must be monitored on an ongoing basis. The management system must be designed to accomplish those objectives consistently over the life of the process.

Define roles and responsibilities. The written description of the training and performance assurance system will identify the essential tasks that must be performed, and management must clearly define who is responsible for those tasks. Each department is typically responsible for performing the job/task analyses that define the KSAs for workers in that department. Managers and the human resource department typically define the minimum qualifications for a job applicant. The responsible department (or, if delegated, the training department) must then develop training programs that will enhance (1) the KSAs of job applicants so they can successfully perform the work or (2) the KSAs of job incumbents to meet additional job requirements. Each department is then typically responsible for identifying appropriate testing methods and minimum acceptable standards for new and incumbent workers. Depending on the nature of the testing, various groups will be responsible for administering and evaluating the tests within human resource department guidelines. For those that fail to meet minimum standards, the responsible department typically provides remedial training. The organization must have a policy addressing those who cannot meet or maintain minimum acceptable performance standards.

Validate program effectiveness. Performance assurance testing, periodic auditing, and root cause incident investigation all provide insight into the effectiveness of the training program. For example, if an unacceptable percentage of incumbent workers fail a performance test, the management system should require more frequent refresher training or implement a more effective alternative training method, for example, field training instead of classroom training. Auditing may discover that workers are not actually following the methods prescribed in training, which could be due to a lack of knowledge (training deficiency) or a lack of operational discipline (management deficiency). Each time an incident investigation identifies "less than adequate training" as a root cause, management should investigate whether the finding reflects (1) an isolated deficiency, (2) a systemic problem that requires corrective action, or (3) a situation unrelated to training, such as the management deficiency mentioned above.

Control documents. Training materials must be controlled and updated when changes occur. Changes should be controlled through the *management of change* element, and the change review process should identify situations that warrant updating the training materials. Training and performance assurance records should also be kept to enable each worker's qualifications for various work assignments, as well as any needs for additional or refresher training, to be readily determined. Performance assurance test records may contain confidential information and must be controlled; however, workers should have access to their own training records.

14.2.2 Identify What Training Is Needed

Training is needed when a gap exists between a worker's or job applicant's current KSAs and those required for successful performance of the job. Performance testing is a common way to discover those gaps and identify training needs.

Conduct a job/task analysis. Since the purpose of training is to maintain or improve human performance, a good starting point for identifying training needs is to determine what workers do, or are supposed to do, at the facility. The core activity is job/task analysis, which defines the jobs and tasks that workers are expected to perform. In addition, certain regulatory requirements (e.g., OSHA's process safety management standard or the FDA's good manufacturing practice regulations) and voluntary standards (e.g., the ISO 9000 family of standards) specify a minimum scope for the tasks and content of training programs. Based on this analysis, the required KSAs for each job can be determined and documented. For example, to add catalyst, a worker may be required to lift a 10 kg bag to the top of the reactor; hence, only a worker who is physically capable of lifting that amount could perform the job.

Determine minimum requirements (or essential elements) for job candidates. Determining minimum requirements (1) sets the baseline from which training needs are assessed and (2) is an important management task that affects the labor pool and the wage scale. For example, a job may require the worker to drive a forklift. Management must decide whether it is willing to pay a higher wage to attract applicants who are already qualified forklift operators, or whether to offer a lower wage and train workers to operate a forklift. Obviously, the latter decision dictates that the company must have a forklift operator training program. Similarly, a company may decide that only workers with three years of field experience would be considered for a position as a control room operator; that decision directly affects the training required to qualify a worker for that position. Management may decide to outsource a particular position rather than train workers to do it; however, whether direct-hire or contractor, workers in a given position must be trained to equivalent levels and demonstrate similar performance.

Determine what training is needed. Given the job requirements and the minimum qualifications of applicants, the type and extent of training can then be determined using a simple gap analysis. Part of that analysis involves deciding whether to provide "skill" training versus "rule" or "knowledge" training, and that directly relates to how the procedures are written. For example, procedures often include steps such as "start pump X," but do not specify how to position the inlet and outlet valves or how to start the driver. If starting a pump is a skill that a worker must demonstrate before becoming a qualified operator, then writing the procedure to simply say "start pump 101" is acceptable. Otherwise, the procedure should state the specific, detailed steps required to start pump 101. At the other extreme, executing a complicated procedure cannot be considered a skill, even if some workers do it every day. "Unload the railcar" is not a valid step in a procedure, regardless of the worker's training. A written procedure should describe the rules for railcar unloading in sufficient detail so that the newest or least qualified operator following those steps can successfully perform the task. The training program should be structured so workers can meet this minimum standard. Training workers for tasks requiring extensive knowledge is a difficult challenge because successful performance typically requires judgment, which evolves from both successful training and practical experience.

In addition to the training required to accomplish specific tasks, the training program should provide workers with an overview of the process and an understanding of its hazards. This helps workers recognize dangerous situations and respond in ways that will protect themselves and others. Visitors and contractors also need this training, or special versions of it, before they enter the facility.

Finally, the type of training depends on whether the worker is new to the position or incumbent. New workers must be completely trained in all the required KSAs, while experienced workers may only need to be refreshed on areas in which they have a low performance score or have not performed the task for an extended period. As part of the *workforce involvement* element, worker input on the content and frequency of training should be sought.

Group training into logical programs. Training programs can be organized in a variety of ways. They can be divided into modules based on type of equipment, mode of operation, type of activity, and so forth. However the programs are organized, a logical path(s) should exist for progression through the training modules for the various roles or jobs in a plant, such as managers, technical personnel, chemical operators, maintenance workers, and environmental technicians. In addition, logical points at which comprehension and/or skills can be demonstrated should be noted. Using a logical and consistent approach will help minimize training time and improve its effectiveness.

Manage changes. Practices for managing process changes, and the additional or revised training that they may precipitate, are fully described in Chapter 15. These practices should also apply to managing changes in the content, materials, or format of training courses.

14.2.3 Provide Effective Training

Training may be delivered by a variety of means in various venues, but the objective is always the same – help workers to be successful in their jobs. Success means that the worker can accomplish the required task to a specified standard of performance.

Develop or procure training materials. Adults usually derive more value from training if they are actively engaged in the training process. Thus, workers being trained in pump alignment should actually align a pump. However, shutting down the boiler feedwater pump just so students can align it may not be practical, so simply walking through the tasks in the field may have to suffice. Traditional classroom training is still the best way to provide some types of training, although the "instructor" may be a computer and the lesson may be broken into short modules that the worker can complete as time permits.

On-the-job training is a popular format because it allows the worker to be productive while learning the job. A common difficulty is that training opportunities more often result from happenstance than from a set plan. For example, a worker who was on shift when the heat exchanger started leaking might learn how to isolate and drain the exchanger, while another worker on a different shift might learn how to return the heat exchanger to service. Therefore, on-the-job training must be accompanied by a training plan that lists key skills that each worker must demonstrate to the satisfaction of a trainer, who could be a supervisor or co-worker.

Consider timing. Training, particularly refresher training, is best scheduled just before the task will be performed. Thus, the best time for refresher training on unit shutdown procedures is just before a planned shutdown. However, many procedures, such as emergency shutdown, may be demanded at any moment, and workers must always be ready to perform the task. In those cases, a refresher training session should be scheduled frequently enough to ensure reliable performance. Ideally, informal training for unusual situations is an everyday activity. "What-if" scenarios should be posed to workers randomly, at opportune times, and the selected workers should walk-through their diagnosis and response actions.

Interweave related topics. To the degree possible, training modules should include interrelated topics. For example, training for handling small spills and releases could logically include refresher training on the hazards of various spilled materials and on the proper way to wear a respirator. When training modules are combined, multiple objectives should be clearly defined.

Ensure that training is available. A frequent byproduct of plant staff reductions is the reduction or elimination of time during normal working hours for developing, delivering, or receiving training. Training becomes an overtime activity that is disliked by both workers and management. Computer technology can mitigate this problem by delivering short modules of individual training to workers on demand. The computer can also ask questions, administer quizzes, and record results. This approach is not a panacea, but for some types of training, it is effective, minimizes the cost, and makes training available when it best fits each worker's schedule.

14.2.4 Monitor Worker Performance

Performance assurance requires that workers do more than regurgitate the correct answers to a few questions on a written quiz. The only meaningful criterion for measuring training effectiveness is whether workers can successfully perform their jobs. Written quizzes, field observations, quality audits, and so forth are simply tools that management can use to gauge current worker performance. Poor performance might be the result of inadequate training, or it might result from other factors such as substance abuse, illness, or physical decline. The performance assurance system should be designed to detect unacceptable performance, regardless of cause, so that appropriate corrective actions can be implemented.

Qualify workers initially. Each job position should have a defined list of KSAs that are essential to successful performance of the job. The performance assurance system should be designed to verify those KSAs before the worker is qualified to work independently. This is often a phased qualification – after some minimum KSAs are successfully demonstrated, the person is allowed work in the facility as an apprentice under the close supervision of a qualified worker. Once apprentices have completed their on-the-job training, they should demonstrate their proficiency in a set of skills or tasks to an independent reviewer to earn full qualification. This approach has the added benefit of gauging the performance of the mentor who, presumably, was satisfied with the apprentice's performance before recommending qualification.

Test workers periodically. Qualified workers should be tested periodically to verify that they possess the required KSAs to successfully perform the job. These tests fall naturally into three categories that mirror the job requirements. Most companies rely on periodic medical testing to verify basic physical ability – strength, hearing, eyesight, respiratory capacity, and so forth. This testing should also look for other factors that may adversely affect performance, such as substance abuse or a hidden medical condition. Skill testing is also relatively straightforward. By definition, a skill is something done competently, so performance problems should quickly become evident in everyday work activities. Thus, many companies forego testing frequently applied skills, such as opening a valve or calibrating a meter. However, unused skills will deteriorate, so requiring workers to perform certain tasks periodically (e.g., start the emergency generator) to remain qualified may be necessary.

Assuring worker performance on knowledge-based tasks is more difficult. Can the worker diagnose the cause of the upset? Does the worker recognize which procedure applies in this situation? Simple written tests alone are seldom satisfactory gauges of this knowledge. In critical jobs, some companies require periodic testing in a simulator to challenge the worker's ability to handle situations that the worker will hopefully never see in real life. Occasionally, full-scale drills are required to test the integrated abilities of the entire work crew.

Review all qualification requirements periodically. Even the most effective *management of change* program, coupled with a culture that embraces training, will not ensure that all qualification requirements are current and complete. In many cases, changes are evolutionary – incumbent workers learn new things and adapt to new situations, but job descriptions may not reflect new KSAs, for example, how to set up a new gauging device or how to secure a hot work permit during the off-shift when no supervisor is available. If the usual worker is absent, the qualified replacement worker from another plant area may not know all the nuances of the job. Thus, in conjunction with performing periodic procedure reviews, evaluating worker qualification requirements and updating them as necessary is prudent.

14.3 POSSIBLE WORK ACTIVITIES

The RBPS approach suggests that the degree of rigor designed into each work activity should be tailored to risk, tempered by resource considerations, and tuned to the facility's culture. Thus, the degree of rigor that should be applied to a particular work activity will vary for each facility, and likely will vary between units or process areas at a facility. Therefore, to develop a risk-based process safety management system, readers should perform the following steps:

1. Assess the risks at the facility, investigate the balance between the resource load for RBPS activities and available resources, and examine the facility's culture. This process is described in more detail in Section 2.2.
2. Estimate the potential benefits that may be achieved by addressing each of the key principles for this RBPS element. These principles are listed in Section 14.2.
3. Based on the results from steps 1 and 2, decide which essential features described in Sections 14.2.1 through 14.2.4 are necessary to properly manage risk.
4. For each essential feature that will be implemented, determine how it will be implemented and select the corresponding work activities described in this section. Note that this list of work activities cannot be comprehensive for all industries; readers will likely need to add work activities or modify some of the work activities listed in this section.
5. For each work activity that will be implemented, determine the level of rigor that will be required. Each work activity in this section is followed by two to five implementation options that describe an increasing degree of rigor. Note that work activities listed in this section are labeled with a number; implementation options are labeled with a letter.

Note: Regulatory requirements may specify that process safety management systems include certain features or work activities, or that a minimum level of detail be designed into specific work activities. Thus, the design and implementation of process safety management systems should be based on regulatory requirements as well as the guidance provided in this book.

14.3.1 Maintain a Dependable Practice

Define Roles and Responsibilities

1. Develop a written procedure for managing the *training* element that describes the process for creating, updating, and maintaining training materials.
    a. General guidance applies to all elements.
    b. Detailed guidance addresses training requirements for each job function.
    c. Detailed guidance addresses training requirements for each job function and describes the process for creating, updating, and maintaining training materials.
    d. Performance is continuously monitored by supervisors and co-workers, and behaviors indicating KSA deficiencies are corrected.
2. Include specific roles and responsibilities in the management system procedure governing the *training* element.
    a. Responsibilities are assigned on an ad hoc basis and are generally controlled by the work group.
    b. A written description of the requirements exists, but responsibilities for creating and updating training materials have not been assigned.
    c. A formal written program document addresses the scope, roles, responsibilities, and so forth for creating, updating, and maintaining training materials.
3. Describe the interfaces between the *training* element and other RBPS elements, especially the *management of change* and *procedures* elements.
    a. Procedures are used as training materials, are updated on an ad hoc basis as part of *management of change*, and are generally verified during *operational readiness* reviews.
    b. A written description of the interfaces exists, but responsibilities for performing required tasks have not been assigned.
    c. A formal written program document defines the interfaces; implementation of the program is routinely monitored.
4. Define the qualifications for a trainer.
    a. Trainers are chosen on an ad hoc basis; no formal training of the trainers exists.
    b. Trainers are chosen on an ad hoc basis from those who have been trained as trainers.
    c. Trainers must be trained as trainers, must be proficient in the subject taught, and must maintain their status based on ongoing training and student feedback.

Validate Program Effectiveness

5. Identify metrics by which training program effectiveness will be judged.
   a. Effectiveness is judged by the trainee.
   b. Effectiveness is judged by an impartial observer.
   c. Effectiveness is judged by student test and performance scores at the end of the training.
   d. Effectiveness is judged by worker results in their jobs, for example, lower error rates, better process stability, fewer incidents, and so forth.
6. Periodically evaluate whether workers are retaining the necessary KSAs to perform their jobs.
   a. Job performance is reviewed after an incident.
   b. KSA retention is adequate if workers can test out of scheduled refresher training.
   c. Performance is continuously monitored by supervisors and co-workers, and behaviors indicating KSA deficiencies are corrected.
7. Periodically evaluate work practices in the field to verify that they are consistent with training.
   a. Evaluations are performed after incidents.
   b. Evaluations are performed before procedures are recertified as accurate.
   c. Evaluations are performed on all shifts before procedures are recertified as accurate and after significant changes.
8. Review incident investigation results and correct any root causes related to training program deficiencies.
   a. Those involved in incidents are retrained. Training is relied upon to reduce human errors instead of eliminating error-likely situations.
   b. Training materials are revised based on incident investigation findings.
   c. The training program is revised to address deficiencies identified as root causes of incidents.

Control Documents

9. Maintain a library of current, approved training materials.
   a. Individual instructors maintain their own files.
   b. Individual departments maintain their own files.
   c. A central library is maintained.
   d. A central library is maintained, materials are reviewed on a regular basis, and updates are controlled.
10. Provide a means to ensure that training materials are updated to reflect changes in the process.

     a. Training materials are updated occasionally.
     b. Training materials are updated periodically.
     c. Training materials are updated as part of *management of change.*

11. Provide a means to track worker training records, such as completed courses, performance records, projected needs, and so forth.
     a. Workers are responsible for keeping their own records.
     b. Various departments maintain records of workers they have trained.
     c. A central system maintains each worker's complete training record.
     d. A central system maintains each worker's complete training record and forecasts training needs for individuals and the facility.

14.3.2 Identify What Training Is Needed

Conduct a Job/Task Analysis

12. Identify jobs and tasks that are performed by each worker (or logical group of workers).
     a. Workers do whatever needs to be done.
     b. General job descriptions exist for an area or department.
     c. Specific job descriptions exist that define the duties for each position.
13. Identify the KSAs required to successfully perform each job and task.
     a. Workers request training as needed.
     b. Written lists of KSAs for an area or department exist.
     c. Written lists of KSAs for each job position exist.
14. Solicit worker input on the training required to successfully perform each job and task.
     a. Worker training suggestions are considered if offered.
     b. Worker training suggestions are solicited for an area or department.
     c. Worker training suggestions are solicited for each job position.

Determine Minimum Requirements (or Essential Elements) for Job Candidates

15. Identify the outside labor sources on which the company can rely to provide workers who have already been trained (e.g., unions, trade schools).
     a. Outside labor sources are identified.
     b. Preferred labor sources for skill training, such as welders and technician, are selected.
     c. Preferred labor sources for skill and process safety training are selected.
16. Determine what skills the company will attempt to buy from the labor pool versus those that will be developed internally.

    a.   Hiring plans are developed, but often change and affect training workload.

    b.   Provisional hires must demonstrate required KSAs or their employment in that position is terminated.

17. Determine what oral and written language skills are required.

    a.   Language skills are specified.

    b.   Language skills are specified and tested.

18. Determine whether job positions will be direct hires or contractors, and if contractors, who will be responsible for their training.

    a.   Hiring plans often change and affect training workload.

    b.   Contractors must demonstrate required KSAs or employment for that position is terminated.

19. Determine whether job positions will be outside hires or internal promotions.

    a.   Hiring and promotion plans often change and affect training workload.

    b.   Minimum experience requirements are specified and enforced.

Determine What Training Is Needed

20. Perform gap analysis between job candidates' skills and required skills.

    a.   General training needs are identified.

    b.   Specific training needs are identified.

21. Perform gap analysis between job incumbents' skills and required skills.

    a.   General training needs are identified.

    b.   Specific training needs are identified.

22. Categorize training requirements for knowledge-based, rule-based, and skill-based tasks, consistent with the written operating and maintenance procedures.

    a.   Training depth generally matches procedure writing detail.

    b.   Corresponding training modules exist for each skill assumed by procedure writers; in other words, a training module exists for "Using a scale," which corresponds to procedural instructions to "Weigh the sample."

23. Identify who must be trained in the hazards of the process and the depth of that training.

    a.   Only those entering a process unit unaccompanied are trained.

    b.   Only those entering a process unit are trained.

    c.   Anyone on site who may affected by an emergency is trained.

    d.   Anyone who may affected by an emergency is trained. Offsite personnel are informed or trained as necessary via the *outreach* and *emergency* elements.

24. Identify who must be trained in process safety.

    a.   Only operators and selected maintenance workers are trained.

    b.   Operations, maintenance, and engineering personnel are trained.

    c.   Operations, maintenance, engineering, and management personnel are trained.

    d.   Anyone who routinely works at the facility is trained.

Group Training into Logical Programs

25. Organize training modules into logical courses of study.
    a. Prerequisites are identified for individual training modules.
    b. A training program is designed specifically for various job functions, such as technical personnel, chemical operators, maintenance workers, supervisors, managers, and so forth.

Manage Changes

(See the list of work activities in Chapter 15, Management of Change)

14.3.3 Provide Effective Training

Develop or Procure Training Materials

26. Identify how each training module can best be presented (e.g., live, videotape/DVD, interactive computer) and where the training should take place (e.g., in a classroom, shop, lab, simulator, operating unit, offsite venue).
    a. Training media and location are based on availability.
    b. Training media and location are based on objectives.
    c. Training media and location are based on objectives and feedback from post-training performance.

27. Prepare or acquire course plans, presentations, and exercise materials.
    a. Materials are prepared ad hoc, from whatever is available.
    b. Materials are prepared to support an objective-based lesson plan.
    c. Materials are regularly updated based on student evaluations, post-training performance, and lessons learned from both in-house and published incident investigations.
    d. Item (c) and remedial training materials are customized to specifically target areas in which performance was deficient.

28. Review quality of course plans, presentations, and exercise materials.
    a. Materials are reviewed by the instructor.
    b. Materials are reviewed by an impartial observer.
    c. Materials are independently reviewed and pilot-tested on a sample of the target audience.

Consider Timing

29. Identify what training must be completed before a worker or visitor can enter the facility.
    a. Overview training on process hazards is required for those entering a process area.
    b. Overview training on process hazards, emergency procedures, and general facility rules is required for all entering the facility.
    c. Training and a demonstrated understanding of process hazards, emergency procedures, and general facility rules is required for all entering the facility.
30. Identify what training must be completed before a worker can begin on-the-job training.
    a. Training on process hazards, emergency procedures, and general facility rules is required for access.
    b. General employee training for unescorted access to the area is required.
    c. General employee training is required plus classroom training in the specific activities in which the worker will be involved.
31. Identify when refresher training must be performed (both to retain skills and to meet any regulatory requirements).
    a. Refresher training is scheduled to meet regulatory requirements.
    b. Refresher training is scheduled to meet regulatory requirements and process activities, such as unit startup.
    c. Refresher training is scheduled to meet regulatory requirements and process needs, and, as necessary, based on lessons learned from incidents and near misses.
32. Instill a practice of challenging workers randomly with "What-if" scenarios and having them walk through their response.
    a. Abnormal situation training is limited to scheduled classes.
    b. Supervisors challenge workers to manage abnormal situations at any time.
    c. Workers challenge each other to manage abnormal situations at any time.

Interweave Related Topics

33. Identify which training modules can accomplish multiple training objectives.
    a. Separate training modules are developed for each training objective.
    b. Training modules are designed to meet the needs of several departments.
    c. Training modules are designed to meet multiple training objectives.

Ensure that Training Is Available

34. Provide training when needed.
    a. Training is primarily an overtime activity.
    b. Worker schedules are adjusted so they can attend training classes.
    c. Training is available on demand when workers need it.

14.3.4 Monitor Worker Performance

Qualify Workers Initially

35. Develop methods for testing job applicant qualifications.
    a. Applicants are tested for general KSAs.
    b. Applicants are tested for specific, job-relevant KSAs.
36. Develop methods for testing trainee progress toward, and achievement of, minimum acceptable performance standards.
    a. Instructors appraise trainees' progress.
    b. Trainees are tested for specific, job-relevant KSAs at defined points in the training program.
37. Identify remedial training requirements for those who fail to qualify or lose their initial qualification.
    a. Employees are retrained indefinitely until they demonstrate the required KSAs.
    b. Employees are given a defined probationary period to demonstrate the required KSAs.
    c. Employees are suspended from hazardous work duties until they demonstrate the required KSAs.

Test Workers Periodically

38. Identify KSAs that require periodic testing to assure performance.
    a. The same KSAs are tested after initial and refresher training.
    b. After refresher training, a unique set of KSAs that reflects the expectations for an experienced worker is tested.
39. Identify methods for testing experienced workers.
    a. Workers are tested for factual knowledge.
    b. Workers must demonstrate ability to do the required work.
    c. Workers must demonstrate ability and understanding of the required work.
40. Establish a metric that will help alert management to any rapid increase in error rates.
    a. Written tests scores are the only metric.
    b. Performance tests under simulated conditions are scored.
    c. Workers must score 100% on critical tasks and are coached until they achieve satisfactory proficiency.

Review All Qualification Requirements Periodically

41. Develop a database or matrix of qualification requirements and periodically update it.
    a. Initial qualification requirements are defined.
    b. Qualification requirements are defined and regularly updated.

14.4 EXAMPLES OF WAYS TO IMPROVE EFFECTIVENESS

This section provides specific examples of industry tested methods for improving the effectiveness of work activities related to the *training* element. The examples are sorted by the key principles that were first introduced in Section 14.2. The examples fall into two categories:

1. Methods for improving the performance of activities that support this element.
2. Methods for improving the efficiency of activities that support this element by maintaining the necessary level of performance using fewer resources.

These examples were obtained from the results of industry practice surveys, workshops, and CCPS member-company input. Readers desiring to improve their management systems and work activities related to this element should examine these ideas, evaluate current management system and work activity performance and efficiency, and then select and implement enhancements using the risk-based principles described in Section 2.1.

14.4.1 Maintain a Dependable Practice

Develop subject matter experts as trainers. Having a "talking head" read prepared material to a class is not an effective training method. SMEs are much more credible and engaging for adult learners because they can explain and enhance the concepts using real-life examples from the workplace. They can also critique course materials and update them as necessary to match current practice. The SMEs should themselves be trained as trainers to help ensure the effectiveness of their presentations.

14.4.2 Identify What Training Is Needed

Train a backup pool of workers. Occasionally, illness, planned vacations, and so forth can result in more ongoing absenteeism than can reasonably be covered with overtime. To prepare for such occurrences, facilities may find it beneficial to maintain a trained, flexible pool of workers who can fill a variety of job roles.

Select training methods based on hazards and potential consequences. Training tools range from simple self-study workbooks to complex, full-scale drills. Applying complex tools when simple ones would suffice is inefficient, but simple tools cannot effectively prepare workers to manage some complex situations. One way to make the training process more efficient is to perform an initial screening based on hazards and potential consequences. If, for example, the screening does not identify any severe hazards (e.g., runaway reactions), or if the potential consequences are inherently limited (e.g., because of small quantities of reactants), then a simple lecture or demonstration may be sufficient to ensure that risks are tolerable. This prevents wasting resources by overtraining for minor risks. Conversely, a screening that identifies serious hazards or major consequences clearly indicates the need for rigorous simulations and exercises. This prevents wasting resources on ineffective training, which will require extensive remedial effort.

Automate training administration. Software tools can simplify and improve the performance of training administration. Once a database is loaded with the workers' names, job positions, and training requirements, the software can then track each worker's current status and upcoming training needs. Queries can quickly show which of the available workers is currently qualified to perform a task or when a training session should be scheduled to maintain worker qualification. Software can also be written to automatically notify workers and managers when training is due.

14.4.3 Provide Effective Training

Use standard templates for training modules. Development of a new training module requires significant resource expenditure, particularly if those involved do not routinely develop training. By having a standard training module template, subject matter experts can more quickly prepare required training materials. This also helps ensure that the materials can be effectively presented with the audiovisual, demonstration, and workshop equipment that will be available to the trainer. Standardized materials can be shared among multiple facilities to further improve efficiency.

Collaborate with others for training. Some types of training are common to many facilities with hazardous chemicals. By collaborating with other facilities or other companies, a facility can ensure that a trained labor pool is available for a fraction of the cost. For example, a company might develop a library of maintenance training modules that are accessible to all facilities. The Contractors Safety Council of Texas City is an example of organizations collaborating to ensure that trained contractors are available in the area, which then allows site-specific training to be much more streamlined.

Conduct virtual and computer-based training. Offering training to every shift, as well as to those who work at remote facilities or who may have been away from the facility because of illness or vacation, can be very expensive.

Having everyone physically at the same facility at the same time is not always necessary for some types of training. In many cases, internet or videoconference technology is more efficient for involving trainers and/or students from several locations without the time and expense of travel. The session can be recorded and played back on demand by those who could not attend the live session. However, what is gained by efficiency can be quickly lost if the remote participants are distracted by other activities or the virtual meeting equipment limits their participation.

Computer-based training (CBT) offers similar efficiencies. Employees can study CBT modules as their schedule permits, and the computer can verify understanding of one topic before proceeding to the next. CBT is ideal for refresher training – it can allow the worker to demonstrate mastery of the material by passing a test without spending time reviewing basic information, and it can be delivered just before a task when the worker needs it. CBT can be updated as needed at any time, and all workers have immediate access to the new information. However, CBT also has its drawbacks. It cannot answer the same range of questions as a live instructor, and it limits educational interactions with co-workers.

Schedule refresher training based on risk. Some organizations schedule refresher training under the assumption that everyone needs the same training. Allowing workers to test out of scheduled training improves overall efficiency, but the testing can still be a time burden to those who know the material. A better system is more individualized and considers both the criticality of the task and the worker's prior performance when scheduling refresher training sessions (within any regulatory constraints). For example, a worker who routinely uses a self-contained breathing apparatus (SCBA) and who scores high in previous training sessions may not be scheduled for refresher training for three years. On the other hand, individuals who do not use SCBAs routinely or did not score highly in previous training sessions may be scheduled for annual refresher training.

Combine refresher training with procedure reviews. Procedures should be updated regularly, and workers should have their training refreshed periodically. By having workers review procedures as they undergo refresher training, both requirements can be met simultaneously.

Educate workers in unit operations. When workers understand unit operations, they tend to be more diligent about following procedural steps because they understand the consequences of deviating from them. They also tend to be better at recognizing deviations and troubleshooting the causes of the deviations. The responsible engineer should prepare the technical content of the training, but experienced operators should generally present the material with support from engineering. Technical training should be offered on a regular basis as part of the refresher training program.

Provide simulator-based training. Experiential learning is one of the most effective adult training methods. Yet many situations exist that workers must be trained to handle that rarely, if ever, occur. Simulators provide a way to challenge workers with both routine and abnormal tasks that allows them to discover the best ways to handle situations without posing any risk to process operations.

Share lessons learned in incident investigations. Lessons learned from thorough incident investigations are most valuable when they are shared with other workers who may face similar situations. These lessons are particularly powerful training tools because they are derived from real plant situations and may have seriously threatened or affected co-workers. Such training not only improves the effectiveness of those participating in risk studies for the *risk* and *management of change* elements, but also helps maintain a sense of vulnerability that encourages adherence to best practices in the *operations* and *safe work* elements.

14.4.4 Monitor Worker Performance

Automate screening for refresher training. Software tools can significantly reduce the refresher training load. When a worker is due for refresher training, an automated system can send the worker an e-mail reminder with an electronic link to a computer-based training module. If the worker completes the quiz satisfactorily, then the system records the successful performance. Otherwise the worker is scheduled for an appropriate alternative means of refresher training.

Schedule training during low vacation periods. Many workers schedule vacations to coincide with other events (school holidays, hunting season, etc.). By scheduling training during low vacation periods, the company can maximize attendance and minimize the need for make-up sessions.

14.5   ELEMENT METRICS

Chapter 20 describes how metrics can be used to improve performance and when they may be appropriate. This section includes several examples of metrics that could be used to monitor the health of the *training* element, sorted by the key principles that were first introduced in Section 14.2.

In addition to identifying high value metrics, readers will need to determine how to best measure each metric they choose to track. In some cases, an ordinal number provides the needed information, for example, total number of workers. Other cases, such as average years of experience, require that two or more attributes be indexed to provide meaningful information. Still other metrics may need to be tracked as a rate, as in the case of employee turnover. Sometimes, the rate of change may be of greatest interest. Since every situation is different, the reader will need to determine how to track and present data to most efficiently monitor the health of RBPS management systems at their facility.

14.5.1 Maintain a Dependable Practice

- *Percentage of incidents with training and performance root causes.* An upward trend may indicate deficiencies in the training element.
- *Number of subject matter experts providing training.* Considering the number of qualified SMEs, this metric shows who is actually doing the training. If only a few individuals are doing most of the work, it may identify the most talented and capable resources, or it may identify units, departments, or managers whose absence indicates a lack of support.
- *Number of exceptions to training requirements.* A high number may identify managers who are willing to accept more risk than the company's risk tolerance guidelines.
- *Percentage change in the training budget.* Sudden decreases in this metric may indicate a shift in management priorities and a declining investment in worker skills.
- *Percentage of course deliveries audited.* Decreases in this metric may indicate a lack of management interest in the quality of training content and delivery.
- *Number of qualified personnel in defined process safety management roles.* Risk analysis leaders, management of change leaders, procedure writers, trainers, incident investigation leaders, and so forth all need specific training to be qualified. Decreases in these numbers below minimum staffing requirements may indicate imminent backlogs or breakdowns in process safety management systems, perhaps due to lack of resources.

14.5.2 Identify What Training Is Needed

- *Number of workers of each type whose training is overdue.* This metric provides an indication of resource constraints or a lack of management support for the program. In regulated situations, it also highlights where obligations are not being met (e.g., 3-year refresher training for operators).
- *Percentage of workers whose training is overdue.* An upward trend in this metric may indicate that required training is unavailable, that management will not authorize time for workers to participate, or that training is offered at inopportune times.
- *Percentage of workers who miss a scheduled training session.* A high number or upward trend may indicate (1) a lack of management

accountability or (2) that training is being given lower priority. A spike in this metric may indicate scheduling conflicts with, and preemption by, other plant situations (e.g., a shutdown), personal priorities (e.g., school vacations), or unforeseeable events (e.g., a flu outbreak).

- *Percentage of training sessions that are offered on schedule.* A downward trend or unexpected drop may indicate that training is being given lower priority.
- *Number of training sessions of each type scheduled.* This metric provides an indication of staff, material, and facility resource requirements. A downward trend or unexpected drop may indicate that required training is not being performed.

14.5.3 Provide Effective Training

- *Time spent in training for individuals, shifts, departments, and job functions.* Unexpected decreases in this metric may indicate a declining investment in worker skills or an increasing effectiveness of training delivery. Upward spikes in this metric may indicate major changes in plant processes, staffing levels, or work assignments.
- *Percentage of workers who believe training is appropriate and effective.* A downward trend in this metric may indicate that required training is repetitive or presented poorly. Management should consider options for reinvigorating the training materials or allowing workers to test out of training.
- *Average test scores for classes, individuals, shifts, departments, and job functions.* Changes in this metric over time indicate changes in training effectiveness, hiring standards, staff experience, and/or staff competence.
- *Time spent on CBT modules.* For an individual module, a downward trend may indicate that the material is so well known that everyone can test out, or that individuals are not really studying the material.

14.5.4 Monitor Worker Performance

- *Percentage of workers who require remedial training.* An upward trend in this metric may indicate that the original training material or delivery was poor, or that the time interval for refresher training is too long.
- *Percentage of workers who miss a particular test question.* A spike in this metric may indicate that the question is poorly worded, that the subject was poorly covered, or that the workers' practical experience conflicts with the "right" answer.
- *Percentage of workers who test out of a training module.* An upward trend in this metric may indicate an increase in experience and

competence, that the test is not challenging, or that the test answers are being shared.

- *Number of errors during simulator training.* An abnormally high number may indicate that the original training was poor or that the time interval for refresher training is too long. Errors may also indicate problems in the human-machine interface, the procedures, or the work practices.

14.6   MANAGEMENT REVIEW

The overall design and conduct of management reviews is described in Chapter 22. However, many specific questions/discussion topics exist that management may want to check periodically to ensure that the management system for the *training* element is working properly. In particular, management must first seek to understand whether the system being reviewed is producing the desired results. If the organization's *training* results are less than satisfactory, or they are not improving as a result of management system changes, then management should identify possible corrective actions and pursue them. Possibly, the organization is not working on the correct activities, or the organization is not doing the necessary activities well. Even if the results are satisfactory, management reviews can help determine if resources are being used wisely – are there tasks that could be done more efficiently or tasks that should not be done at all? Management can combine metrics listed in the previous section with personal observations, direct questioning, audit results, and feedback on various topics to help answer such questions. Activities and topics for discussion include:

- Investigate whether any regulatory required training is imminent or overdue.
- Investigate any unexpected delays in scheduled training; for example, delays due to a lack of material, a lack of leadership or expert resources, or poor planning.
- Verify that training and development needs are specifically discussed during performance reviews.
- Review the number of courses required for a given job and any similarities in their content.
- Determine whether appropriate SMEs are preparing and delivering training material.
- Determine whether the mix of training delivery methods, such as lecture, on-the-job, CBT, and so forth, is appropriate and effective.
- Verify that the latest version of applicable resources (e.g., checklists, course templates) is being used.
- Review the amount of overtime required to conduct necessary training.

- Verify that course materials are being routinely updated to address new knowledge, performance issues, lessons learned from incident investigations, student feedback, and so forth.
- Evaluate whether long delays occur when making simple updates to training materials, such as correcting the wording of a slide.
- Verify that peers are observing and coaching peers on an ongoing basis, outside of formal training classes.
- Review the performance scores of different teams, areas, and departments, and investigate any inconsistencies with the performance results.

Regardless of the questions that are asked, the management review should try to evaluate the organization's depth on several levels. Is the thoroughness of training sufficient? Does performance assurance extend beyond routine operational tasks? When things go wrong, are key employees likely to have the knowledge required to make prudent decisions? Also, is there sufficient depth in key personnel? If the entire organization depends on a single individual to train the workforce, what will happen if that person suddenly resigns or falls ill? Management reviews provide an opportunity for the organization to honestly assess its depth. Management can then take action to address any concerns before (1) experiencing a loss event resulting from insufficient knowledge or poor execution and (2) losing the skilled workforce that is has worked long and hard to develop. Any weaknesses revealed by the management review should be resolved as described in Chapter 22.

14.7   REFERENCES

14.1   Lees, Frank P., *Loss Prevention in the Process Industries, Hazard Identification, Assessment, and Control*, 2nd edition, Butterworth-Heinemann, Oxford, England, 1996.

15

MANAGEMENT OF CHANGE

The *management of change* (*MOC*) element has been called the minute-by-minute process risk assessment and control system. The significance of *MOC* – or the lack of it – was never more visible than in the Flixborough accident. This watershed event involved a temporary modification to piping between cyclohexane oxidation reactors. In an effort to maintain production, a temporary bypass line was installed around the fifth of a series of six reactors at a facility in Flixborough, England, in March of 1974. The bypass failed while the plant was being restarted after unrelated repairs on June 1, 1974, releasing about 60,000 pounds of hot process material, composed mostly of cyclohexane. The resulting vapor cloud exploded, yielding an energy release equivalent to about 15 tons of TNT. The explosion completely destroyed the plant, and damaged nearby homes and businesses, killing 28 employees, and injuring 89 employees and neighbors.

The temporary modification was constructed by people who did not know how to design large pipes equipped with bellows – the design work for the change was treated more like pipefitting than large diameter, high pressure piping system design. As was said in the official report: "…they did not know that they did not know." An effective *MOC* system would have discovered the design flaw before the change was implemented, thus averting the disaster.

15.1 ELEMENT OVERVIEW

Managing changes to processes over the life of a facility is one of nine elements in the RBPS pillar of *managing risk*. This chapter describes the management practices involving (1) the recognition of change situations, (2) the evaluation of hazards, (3) the decision on whether to allow a change to be made, and (4) necessary risk control and follow-up measures. Section 15.2

describes the key principles and essential features of a management system for the *MOC* element. Section 15.3 lists work activities that support these essential features and presents a range of approaches that might be appropriate for each work activity, depending on perceived risk, resources, and organizational culture. Sections 15.4 through 15.6 include (1) ideas for improving the effectiveness of management systems and specific programs that support this element, (2) metrics that could be used to monitor this element, and (3) management review issues that may be appropriate for *MOC*.

15.1.1 What Is It?

The *MOC* element helps ensure that changes to a process do not inadvertently introduce new hazards or unknowingly increase risk of existing hazards (Refs. 15.1 and 15.2). The *MOC* element includes a review and authorization process for evaluating proposed adjustments to facility design, operations, organization, or activities prior to implementation to make certain that no unforeseen new hazards are introduced and that the risk of existing hazards to employees, the public, or the environment is not unknowingly increased. It also includes steps to help ensure that potentially affected personnel are notified of the change and that pertinent documents, such as procedures, process safety knowledge, and so forth, are kept up-to-date.

15.1.2 Why Is It Important?

If a proposed modification is made to a hazardous process without appropriate review, the risk of a process safety accident could increase significantly.

15.1.3 Where/When Is It Done?

MOC reviews are conventionally done in operating plants and increasingly done throughout the process life cycle at company offices that are involved with capital project design and planning. *MOC* reviews should be done for bona fide "changes" – not for replacements-in-kind (RIKs).

15.1.4 Who Does It?

An individual originates a change request. Qualified personnel, normally independent of the *MOC* originator, review the request to determine if any potentially adverse risk impacts could result from the change, and may suggest additional measures to manage risk. Based on the review, the change is either authorized for execution, amended, or rejected. Often, final approval for implementing the change comes from another designated individual, independent of the review team. A wide variety of personnel are normally involved in making the change, notifying or training potentially affected employees, and updating documents affected by the change.

15.1.5 What Is the Anticipated Work Product?

The main product of an *MOC* system is a properly reviewed and authorized change request that identifies and ensures the implementation of risk controls appropriate to the proposed change. Ancillary products include appropriate revisions or updates involving other RBPS activities, such as modifying process safety information, and change communication/training. Outputs of the *MOC* element can also be used to facilitate the performance of other RBPS elements. For example, approved change requests are necessary to determine when some *readiness* activities are performed.

15.1.6 How Is It Done?

Organizations usually have written procedures detailing how *MOC* will be implemented. Such procedures apply to all work that is not determined to be RIK. The results of the review process are typically documented on an MOC Review form. Supplemental information provided by system designers to aid in the review process is often attached to the MOC review form. Once the change is approved, it can be implemented. Potentially affected personnel are either informed of the change or provided more detailed training, as necessary, prior to startup of the change. Follow-on activities, such as updates to affected process safety information and to other RBPS elements, are assessed to identify which are required before startup, and which may be deferred until after startup. All such activities are tracked until completed.

Higher risk situations usually dictate a greater need for formality and thoroughness in the implementation of an *MOC* protocol, for example, a detailed written program that specifies exactly how changes are identified, reviewed, and managed. Companies having lower risk situations may appropriately decide to manage changes in a less rigorous fashion, for example, through a general policy about managing changes that is implemented via informal practices by trained key employees.

Facilities that exhibit a high demand rate for managing changes may need greater specificity in the *MOC* procedure and a larger allocation of personnel resources to fulfill the defined roles and responsibilities. Lower demand situations can allow facilities to operate an *MOC* protocol with greater flexibility. Facilities with a sound process safety culture may choose to have more performance-based *MOC* procedures, allowing trained employees to use good judgment in managing changes in an agile system. Facilities with an evolving or uncertain process safety culture may require more prescriptive *MOC* procedures, more frequent training, and greater command and control management system features to ensure good *MOC* implementation discipline.

15.2   KEY PRINCIPLES AND ESSENTIAL FEATURES

Safe operation and maintenance of facilities that manufacture, store, or otherwise use hazardous chemicals requires robust process safety management systems. The *MOC* element establishes a formal (normally documented) authorization process for all changes that are not RIK. It helps prevent changes to facility design, activities, operations, organization, or policies that (1) introduce unforeseen new hazards or (2) unknowingly increase the risk associated with existing hazards. It also helps to communicate approved changes to potentially affected employees and ensures that process safety knowledge, procedures, training materials, and other affected documentation is updated. The following key principles should be addressed when developing, evaluating, or improving any system for the *MOC* element:

- Maintain a dependable practice.
- Identify potential change situations.
- Evaluate possible impacts.
- Decide whether to allow the change.
- Complete follow-up activities.

The essential features for each principle are further described in Sections 15.2.1 through 15.2.5. Section 15.3 describes work activities that typically underpin the essential features related to these principles. Facility management should evaluate the risks and potential benefits that may be achieved as a result of improvements in this element. Based on this evaluation, management should develop a management system, or upgrade an existing management system, to address some or all of the essential features.

15.2.1 Maintain a Dependable Practice

When a company identifies or defines an activity to be undertaken, that company likely wants the activity to be performed correctly and consistently over the life of the facility. To dependably execute an *MOC* system across a facility involving a variety or people and situations, the following essential features should be incorporated.

Establish consistent implementation. For consistent implementation, the *MOC* policy should be documented to an appropriate level of detail in a procedure or a written program addressing the general management system aspects discussed in Section 1.4 of Chapter 1. The scope of the *MOC* system should be well defined to help ensure that all types of change situations are addressed and all significant sources of potential changes are monitored. Specific procedures for review of changes should be defined to an appropriate level of detail.

Involve competent personneL Even within an explicit *MOC* practice, performance will only be as good as the people that are involved in conducting the reviews. All personnel in the facility should have a basic awareness of the *MOC* system so they will know how to interact properly within it and not cause problems by circumventing the system. Moreover, people that are assigned specific *MOC* duties may need more detailed training, for example, on specific process/equipment hazards, hazard evaluation techniques, and so forth. In addition, the *MOC* element should have someone assigned to monitor *MOC* activity who is able to respond to *MOC* questions, issues, and emergencies.

Keep MOC practices effective. Once an *MOC* system is in place, periodic monitoring, maintenance, and corrective action are needed to keep it operating at optimum performance and efficiency. To ensure effectiveness, companies should implement some of the *MOC* system performance and efficiency indicators discussed in Section 15.5, and set up data collection and analysis systems to evaluate the real-time status of the *MOC* system. Process safety personnel should use *MOC* audits to monitor longer-term management practices, which sometimes reveal gaps in the *MOC* practice that require corrective action.

15.2.2 Identify Potential Change Situations

Modifications cannot be evaluated until they are identified. Facilities should implement effective means to identify the types of modifications that are anticipated for the facility/activity and the sources/initiators of these modifications. In order for an *MOC* system to address all potentially significant change situations, the following essential features should be considered.

Define the scope of the MOC system. An *MOC* system should address all of the types of changes that can reasonably be foreseen. Anticipated change types can be identified by (1) searching historical records, such as maintenance work order files, incident reports, hazard/risk studies, audits and design reviews, (2) observing work, and (3) interviewing workers. The *MOC* authorization form should be designed to remind workers of the *MOC* system's scope, for example, using check boxes for the most frequent types of changes. When unusual change types are possible, the form should include an "Other" change category to accommodate special situations within the scope of the *MOC* system.

In addition, the system should (1) identify the physical areas where the *MOC* element applies and (2) address anticipated levels of urgency for implementation, for example, emergency change requests; the duration of the change, such as temporary changes; and any specific differences in *MOC* practices. Evolved systems may document the technical basis for recognized change types and any exemptions/exclusions from the *MOC* program.

Manage all sources of change. Pathways for getting work accomplished should be noted and monitored to ensure that each work stream is being screened for potential change situations. To help reduce inadvertent circumvention of the *MOC* system, examples of RIK versus change decisions should be generated for each significant type of change addressed in the *MOC* program. These examples can be used in awareness-level and detailed training for employees.

15.2.3 Evaluate Possible Impacts

Once potential change situations are identified, they can be evaluated using an appropriate level of scrutiny to determine whether the change introduces a new hazard or increases the risk associated with an existing hazard. For facilities to adopt and implement appropriate review protocols for relevant change types, the following essential features should be considered.

Provide appropriate input information to manage changes. A thorough review of possible adverse impacts of proposed changes must be based on accurate facility drawings and procedures. Facilities should establish a list of information and data that must be submitted with change requests. This list can be documented in the *MOC* program, the review procedure, or in another manner, such as in the *knowledge* element. Sometimes this list of inputs will vary based upon the types of changes and the perceived risk associated with those change categories.

Apply appropriate technical rigor for the MOC review process. An appropriate review process should be established for each anticipated type of change. In low risk situations, only one simple review process may be necessary. In more complicated situations, multiple review processes may be defined using different review protocols. A review protocol should include the required number and disciplines of reviewers and the order of the review (series, parallel, or team-based reviews). The design of these review protocols should consider the following issues:

- Type of change.
- RIK versus change determination.
- Risk significance of the change (to identify the proper depth of review).
- Process or business need, and the technical basis for the change.
- Schedule for implementing the change.
- Duration of the change (temporary or permanent).
- Type of information needed to review the change.
- Expertise needed to review the change.
- Acceptable methods for evaluating risk.
- Tools/techniques available to change reviewers.
- Risk tolerance guidelines.

At a minimum, an *MOC* review protocol should address the technical basis for the change and evaluate the potential impacts on safety and health. In some cases, such as complex situations or those with a higher perceived risk, specific hazard evaluation techniques or lists of issues may be required.

Ensure that MOC reviewers have appropriate expertise and tools. Normally, the *MOC* reviewers will be independent of the person who requested the change. Beyond that, reviewers should have appropriate technical experience commensurate with the change type and perceived risks to facilitate a thorough risk review. Appropriate review personnel may be specified by name, discipline, or facility department/function. Complex change types may require reviewers with special skills, experience, or formal training.

15.2.4 Decide Whether to Allow the Change

Once a change has been reviewed and its risks evaluated, management can then decide whether to (1) approve the change for implementation as requested, (2) require amendment to the change request or implementation process, or (3) deny the change request. For facilities to adopt and implement appropriate *MOC* approval protocols, the following essential features should be considered.

Authorize changes. Concurrent with or following the risk review step, one or more people approve the change for implementation, place restrictions on the request, or reject it for cause. The number and type of individuals designated to do this should be explicitly defined. Generally, the personnel who are the designated change approval authorities should not be the same person(s) who requested the change.

Ensure that change authorizers address important issues. The decision on the authorization protocol should require consideration of the following issues:

- Type of change
- Significance of the change
- Timing/urgency for the change
- Robustness of the process used to review the specific change

In addition, the personnel designated to approve changes for implementation should have access to the following items:

- Change review package
- Results of the risk evaluation
- Risk control measures identified/required via the review

- Understanding of the follow-up items associated with change implementation, both prior to and following startup

15.2.5 Complete Follow-up Activities

Once a change is authorized, it is released for implementation. Typically, the execution of a change is performed via work practices under other RBPS elements (*asset integrity, procedures, safe work*, etc.) by facility staff or contractors involved in design, engineering, or construction. Prior to the startup of the change (exposure of personnel to the modified situation, which could introduce new hazards or increase risk), the *MOC* procedure or reviewers/authorizers may require that drawings and procedures be updated, affected personnel be trained, required risk control measures be implemented, and so forth.

Action items are occasionally deferred until after startup, for example, the installation of heat tracing on bypass piping commissioned in summertime. Such deferred items should be carefully tracked to completion. For facilities to ensure that approved MOCs are properly concluded, the following essential features should be considered.

Update records.   Accurate, up-to-date process knowledge is the foundation for recognizing and managing future changes. If the appropriate records are not updated within a reasonable time, confusion may arise that can increase the risk of operating and maintenance activities. Some companies require that all records be updated prior to change implementation; some allow records to be modified in the interim between approval of the change and startup of the change. And still others will allow some updates to take place after startup of the change occurs.

Companies that do not control document changes will waste resources and increase risk. Facilities should define (as a function of change type or urgency/duration, if necessary) and enforce appropriate records updating practices (types of records and deadlines). Facilities should also specify whether such updating is performed as a part of the *MOC* process or as a task ascribed to another RBPS element, such as *procedures*. Historically, many facilities have struggled to maintain effective control-of-change documentation. Thus, designing management system overlap and feedback control loops into the *MOC* system and related records management system elements may be appropriate.

Communicate changes to personnel. Changes that are not understood by employees can quickly lead to incidents. Once changes are approved, communication of appropriate change details to potentially affected employees should begin. Facilities should have a process that (1) identifies who needs to be informed or trained on the change, (2) defines the content and extent of the communication, ranging from simple awareness to detailed training,

(3) provides the means for implementing the change in a timely fashion, and (4) ensures that the change was understood to the extent needed by the employees or contractors.

Enact risk control measures. Some change requests are approved with risk control measures proposed as a part of the change request or as a contingent approval requirement. Facilities should consider how to identify, track, implement, and monitor these risk control measures. Special risk control measures associated with temporary or emergency change requests should also be addressed. Otherwise, the risk evaluation basis used by the change reviewers and authorizers would be compromised.

Maintain MOC records. One work product of the *MOC* system is a reviewed and authorized change request. However, if records of the *MOC* review process are not kept, there is no easy way to review, evaluate, or audit the *MOC* system, or to easily update or revalidate hazard identification and risk analysis (HIRA) studies. Facilities should decide what types of records to keep and develop a retention policy. *MOC* records are increasingly being kept electronically with typical word processing, spreadsheet, or database applications, or with standalone *MOC* workflow software. Simple systems operated with few changes may still find paper-based records systems adequate.

15.3   POSSIBLE WORK ACTIVITIES

The RBPS approach suggests that the degree of rigor designed into each work activity should be tailored to risk, tempered by resource considerations, and tuned to the facility's culture. Thus, the degree of rigor that should be applied to a particular work activity will vary for each facility, and likely will vary between units or process areas within a single facility. Therefore, to develop a risk-based process safety management system, readers should perform the following steps:

1. Assess the risks at the facility, investigate the balance between the resource load for RBPS activities and available resources, and examine the facility's culture. This process is described in more detail in Section 2.2.
2. Estimate the potential benefits that may be achieved by addressing each of the key principles for this RBPS element. These principles are listed in Section 15.2.
3. Based on the results from steps 1 and 2, decide which essential features described in Sections 15.2.1 through 15.2.5 are necessary to properly manage risk.

4. For each essential feature that will be implemented, determine how it will be implemented and select the corresponding work activities described in this section. Note that this list of work activities cannot be comprehensive for all industries; readers will likely need to add work activities or modify some of the work activities listed in this section.

5. For each work activity that will be implemented, determine the level of rigor that will be required. Each work activity in this section is followed by two to five implementation options that describe an increasing degree of rigor. Note that work activities listed in this section are labeled with a number; implementation options are labeled with a letter.

Note: Regulatory requirements may specify that process safety management systems include certain features or work activities, or that a minimum level of detail be designed into specific work activities. Thus, the design and implementation of process safety management systems should be based on regulatory requirements as well as the guidance provided in this book.

15.3.1 Maintain a Dependable Practice

Establish Consistent Implementation

1. Establish and implement procedures to manage changes.
   a. A simple written procedure applies to "any" change.
   b. A simple written procedure applies to designated types of changes.
   c. A detailed written program applies to designated types of changes.
   d. In addition to item (c), multiple change review protocols are defined.

2. Assign a job function as the owner of the *MOC* system.
   a. A part-time *MOC* owner, local or offsite, is assigned.
   b. Multiple, part-time *MOC* owners are assigned across the facility.
   c. A single, full-time *MOC* owner is designated.

3. Define the technical scope of the *MOC* system so that the types of changes to be managed are unambiguous and the sources of changes are monitored.
   a. The scope is informally defined and understood.
   b. Multiple change types are generally defined.
   c. Plant areas where *MOC* applies, as well as the different types of changes, are generally defined.
   d. In addition to item (c), changes are distinguished from RIKs using context-sensitive examples for each type of change.

Involve Competent Personnel

4. Define the *MOC* roles and responsibilities for various groups of personnel.
   a. A generic policy makes *MOC* the responsibility of all.
   b. An individual in each unit is responsible for *MOC* in that area.
   c. A single person is responsible for *MOC*.
   d. All *MOC* roles/responsibilities are assigned to job functions/departments.
5. Provide awareness training and refresher training on the *MOC* system.
   a. Ad hoc, informal training is provided.
   b. The *MOC* practice is broadcast once (e.g., via e-mail) to everyone.
   c. *MOC* initial awareness training is provided once to affected personnel.
   d. *MOC* initial awareness and refresher training are provided to affected personnel.
6. Provide detailed training to all affected employees and contractors who are assigned specific roles within the *MOC* system.
   a. Ad hoc *MOC* training is provided.
   b. Structured initial training is provided to key *MOC* personnel.
   c. Structured initial and refresher training are provided to key *MOC* personnel.

Keep MOC Practices Effective

7. Keep a summary log of all *MOC* reviews, including the items that must be included on an *MOC* review form, to aid day-to-day management of the *MOC* process.
   a. An *MOC* log is kept by a few people.
   b. An *MOC* log is kept in a notebook in each unit.
   c. An *MOC* log is kept electronically by the *MOC* coordinator.
   d. In addition to item (c), the *MOC* active/inactive log is accessible to all affected personnel.
8. Establish and collect data on *MOC* performance and efficiency indicators.
   a. A few data items are collected.
   b. Basic *MOC* activity data are collected.
   c. *MOC* performance indicators are collected annually.
   d. *MOC* performance and efficiency indicators are collected regularly.
9. Include the results of the *MOC* performance indicators when performing internal management reviews of *MOC* practices and the *MOC* system.
   a. Ad hoc, informal internal reviews are done.

b. Internal *MOC* reviews are performed, but no analysis of performance information is provided to help reviewers target *MOC* improvement opportunities.

c. *MOC* performance information and analysis are provided to internal reviewers annually.

d. *MOC* performance and efficiency information and analysis are provided to reviewers for regular monthly/quarterly reviews.

15.3.2 Identify Potential Change Situations

Define the Scope of the MOC System

10. Determine the types of changes to be addressed in the program:

- RBPS management system
- Plant layout or equipment location/arrangement
- Facility and equipment
- New chemicals
- Software
- Procedures
- Process technology
- Process knowledge
- Process controls
- Chemical specifications and suppliers
- Job assignments (individual, shift, or staff)
- Personnel and organization
- Policies
- Building locations and occupancy patterns
- Others

a. Only basic change versus RIK is defined.

b. Only "hardware" and "procedure" categories are defined.

c. Several change categories are defined.

d. A complete list of change types and applicable process locations is defined.

11. Document the rationale for not addressing specific types of changes in the *MOC* program.

a. General rationale is understood by the *MOC* owner.

b. General rationale is understood by key *MOC* personnel.

c. Specific rationale is documented.

d. Specific rationale is documented and accessible to all affected personnel.

12. Develop a list of areas, departments, and activities to which the *MOC* system applies.

a. The written MOC policy states where *MOC* applies.

b. In addition to item (a), facility personnel are told once about applicable *MOC* areas.

c. A list/drawing of facility areas where *MOC* applies is maintained and well communicated.

d. The *MOC* system applies to all facility areas, processes, and equipment.

Manage All Sources of Change

13. Monitor change sources for unrecognized change.
    a. Plant personnel are told to watch out for *MOC* abuse.
    b. Plant personnel are told once about change types.
    c. Key unit personnel are informed about change types.
    d. Key *MOC* personnel are assigned to periodically monitor change sources.

14. Develop specific examples of changes and RIK for each category of change, and use these in employee awareness training to minimize the chance that the *MOC* system will be inadvertently bypassed.
    a. Notional examples are developed.
    b. A few generic change types are described.
    c. Specific change/RIK examples are developed for a few categories.
    d. Multiple examples of changes/RIKs are developed for all categories or types of change. The examples are documented in different manufacturing areas and updated based on *MOC* performance.

15.3.3 Evaluate Possible Impacts

Provide Appropriate Input Information to Manage Changes

15. Identify the types of information necessary to properly evaluate changes within the scope of the *MOC* system.
    a. An informal understanding exists among reviewers regarding the information needed to review a change.
    b. A generic list of input information is provided to *MOC* reviewers.
    c. A checklist of key inputs for reviewing changes is provided and enforced.
    d. A series of checklists that specify key inputs for each type of change is provided and enforced.

Apply Appropriate Technical Rigor to the MOC Review Process

16. The written *MOC* procedures should include the use of an *MOC* review form and should ensure that the following items are addressed prior to any change:
    - The technical basis for the proposed change
    - Impact of the proposed change on safety and health
    - Authorization requirements for the proposed change
    a. A simple review form addresses the three key review issues for all types of changes.
    b. The same detailed review form is used for all types of changes.
    c. A multi-part review form is used to address predetermined issues for each type of change.

17. Use appropriate analytical techniques, including qualitative hazard evaluation methods, to review the potential safety and health impacts of the change.
    a. Reviewers use their judgment.
    b. A generic safety issues/hazard checklist is provided for use in the review of all changes.
    c. Change-specific hazard review tools are provided.
    d. Formal hazard evaluation methods are required under specified circumstances.
18. Identify issues that must be addressed in a review, commensurate with the level of complexity and significance of the proposed change, regardless of the technique used. Specify quality parameters for the review results.
    a. Reviewers have an informal understanding of the issues that are common to all changes.
    b. Each reviewer assembles a specific list of issues.
    c. A simple, generic list of issues is provided.
    d. A list of issues is provided for each type of change.
19. If temporary changes are permitted, the *MOC* review procedure should address the allowable length of time that the change can exist. and the procedure should include a process to confirm removal of temporary changes or restoration of the change to the original condition within the time period specified in the approved change request.
    a. The same generic time period is given for all change types.
    b. A time period is specified; follow-up is provided on an ad hoc basis.
    c. A time period is specified; the change originator must confirm restoration.
    d. A time period is specified; an independent party must confirm restoration within the specified time.
20. If emergency changes are permitted, the *MOC* review procedure should define (1) what constitutes an emergency change and (2) the process for evaluating and authorizing the emergency change.
    a. A general definition of emergency change is provided, and it is up to the originator to decide when to use the procedure.
    b. A specific definition of an emergency change request is provided.
    c. Specific circumstances are defined under which the emergency changes can be made.
    d. In addition to item (c), specific requirements for a special emergency change review process are defined.

Ensure that MOC Reviewers Have Appropriate Expertise and Tools

21. *MOC* reviews should be performed by qualified personnel.
    a. Anyone can do it, based upon the *MOC* awareness training provided.
    b. Specific persons are designated as reviewers.
    c. Specific job functions are specified as reviewers.
    d. Reviewer qualifications are specified, and designated personnel are approved.
22. For each type of change, provide a description of the disciplines needed on an *MOC* review.
    a. An informal practice exists for senior personnel to be involved.
    b. General disciplines are specified without regard to change type.
    c. The same specific disciplines are required for any change.
    d. Requirements for specific disciplines are customized for each change type.
23. Each review should involve someone who is qualified in risk analysis.
    a. Reviewers should have some familiarity with risk evaluation methods.
    b. Reviewers should have some experience with risk evaluation methods.
    c. Risk evaluation training is provided to key personnel.
    d. Qualified HIRA leaders must lead team-based reviews for certain types of changes.
24. Reviewers should have access to, and be trained in the use of, company risk tolerance criteria.
    a. Reviewers are asked to use their judgment.
    b. Informal risk tolerance guidelines are provided.
    c. *MOC* reviewers are provided with, and trained on, company risk tolerance guidelines.

15.3.4 Decide Whether to Allow the Change

Authorize Changes

25. Each change should be authorized by a person(s) with designated approval responsibilities. Sometimes this function is satisfied by the *MOC* reviewers; sometimes the approvers are different from the *MOC* reviewers.
    a. An unspecified "authorizer" is required.
    b. A senior manager, such as the facility manager, must approve all changes.
    c. Specific job functions or named personnel are designated to approve changes.
    d. Varying levels of approval are defined for different types of changes.

26. Develop a list of responsibilities for those who are authorized to approve changes.
    a. Authorizers use their judgment and experience, based upon awareness training.
    b. Authorizers are provided a general list of duties.
    c. Authorizers are provided a list of responsibilities specific to each type of change.
27. In the *MOC* procedure, include the need for backup personnel when designated authorizers are not available.
    a. An informal understanding exists regarding *MOC* backup personnel.
    b. No backup personnel are mentioned, but general instructions are provided about backup personnel.
    c. A specific senior person is designated as the "universal backup."
    d. Specific personnel/job functions are designated as backups.

Ensure that Change Authorizers Address Important Issues

28. The *MOC* procedure should guide authorizers in making *MOC* approval, modification, or rejection decisions.
    a. An informal understanding exists about what options the authorizers have.
    b. Authorizers are provided general guidance.
    c. Authorizers are provided a specific list of options.
    d. Authorizers are provided lists of options that are specific to each type of change.
29. Authorizers should have access to the company's risk evaluation guidelines and risk tolerance criteria guidance.
    a. Authorizers are asked to use their experience and judgment.
    b. Informal risk tolerance guidelines are provided.
    c. *MOC* authorizers are trained on company risk tolerance guidelines.
    d. Formal risk tolerance criteria exist, and authorizers are trained to follow them.

15.3.5 Complete Follow-up Activities

Update Records

30. Update all process knowledge prior to startup of the change.
    a. The MOC procedure does not address this requirement, but management emphasizes the need to keep information up to date.
    b. Red-lined information is kept on file pending periodic updates; incomplete items are tracked regularly until they are brought up to date, reviewed, and approved.

    c.   Personnel monitor the update of information; a maximum length of time (e.g., 90 days) is allowed after implementation of the proposed change.

    d.   All information must be updated before the change is placed in service.

Communicate Changes to Personnel

31. Communicate changes to personnel.
    a.   Communication/training occurs informally.
    b.   A formal system exists for informing/training operating personnel.
    c.   A formal system exists for informing/training all potentially affected personnel.
    d.   In addition to item (c), a formal system also exists for informing/training contractors.

32. Document that the training was completed.
    a.   Ad hoc training documentation is kept.
    b.   Sign-in rosters are kept listing the names of who was trained and the date the training occurred.
    c.   A formal system exists for determining who needs to be trained and documenting how and when each person was trained.
    d.   In addition to item (c), a formal system exists for documenting that the training was understood.

Enact Risk Control Measures

33. Create a system to resolve *MOC* review action items and to document their completion.
    a.   An informal system of action item follow-up exists.
    b.   A formal system exists for tracking action items to completion.
    c.   A formal system exists for tracking action item to completion and enforcing requirements for those that must be completed before the change is implemented.

34. Confirm that temporary changes are removed from service and that conditions are properly restored to normal operation.
    a.   Informal measures confirm restoration after a temporary change.
    b.   Formal measures confirm restoration after a temporary change.
    c.   For each temporary change, an individual is designated to confirm restoration to normal operation after a temporary change has been removed from service.
    d.   A single person/job position is assigned to confirm restoration to normal operation for all temporary changes that have been authorized for the unit/facility.

35. If emergency changes are permitted, ensure that the normal *MOC* procedures are completed within a designated time.

    a.   Emergency *MOC* review risk control measures are specified.

    b.   Emergency *MOC* reviews are followed by completing the normal *MOC* procedure within a specified time period.

    c.   Emergency *MOC* reviews are accompanied by temporary risk control measures until completion of the normal *MOC* procedure within a specified time period.

Maintain MOC Records

36. *MOC* review packages, which contain materials and information used by reviewers and authorizers when performing the review, are retained in accordance with the *MOC* policy/procedure.

    a.   *MOC* records are kept until the change is commissioned.

    b.   *MOC* records are kept for a year or two.

    c.   Extensive *MOC* records are kept until the next HIRA revalidation. Review packages are retained for a specified period, perhaps 1 to 5 years, to support other RBPS work activities.

    d.   Extensive *MOC* records are archived electronically for the life of the process.

15.4   EXAMPLES OF WAYS TO IMPROVE EFFECTIVENESS

This section provides specific examples of industry tested methods for improving the effectiveness of work activities related to the *MOC* element. The examples are sorted by the key principles that were first introduced in Section 15.2. The examples fall into two categories:

1. Methods for improving the performance of activities that support this element.

2. Methods for improving the efficiency of activities that support this element by maintaining the necessary level of performance using fewer resources.

These examples were obtained from the results of industry practice surveys, workshops, and CCPS member-company input. Readers desiring to improve their management systems and work activities related to this element should examine these ideas, evaluate current management system and work activity performance and efficiency, and then select and implement enhancements using the risk-based principles described in Section 2.1.

15.4.1 Maintain a Dependable Practice

Periodically evaluate/monitor the demand rate for MOC reviews and backlog status. Increases in the monthly *MOC* origination rate, review time, or backlog may indicate a transient spike in demand that requires additional resources. An increase in backlog alone may indicate a bottleneck in the system. In that case, further investigation into the causes may reveal issues such as inadequate training of new workers, a lack of qualified reviewers, or inadequate preparation for *MOC* reviews.

Provide periodic refresher training on MOC to share lessons learned. A variety of personnel will be involved in *MOC* reviews. Each person learns something from the reviews they do. These collective lessons can provide rich content for refresher training to help improve the performance of all *MOC* reviewers.

Use sampling methods to objectively evaluate the quality of MOC reviews. Facilities collect various data for a number of purposes: generating performance metrics, developing inputs for management reviews, or identifying work product populations for formal audits. Because most facilities perform more *MOC* reviews than they could realistically afford to examine in detail, sampling of these *MOC* data becomes very important. Appropriate sampling methods can help facilities objectively examine the results of *MOC* reviews while they may also aid in the discovery of systemic problems.

Periodically observe the MOC review process to ensure that it conforms to the MOC procedure. Sometimes management practices can experience "scope creep." This creep occurs whenever people decide to stretch the procedure to address things for which it was not originally intended, such as quality impacts, environmental impacts, security impacts, and so forth. In *MOC*, this can also happen when complete designs (i.e., beyond the physical scope of the actual change) are submitted for *MOC* review, when changes are beyond the defined physical scope of the system, or when unnecessarily complex review techniques are applied to simple changes. Any of these situations can result in a waste of resources that could be better used in addressing more significant process safety issues. Periodic, random *MOC* work observations can help ensure that the *MOC* work is being carried out as defined by the procedure.

Implement MOC software-based solutions so that most MOC review packages/results are transmitted electronically. Some *MOC* delays may result partly from the "lag time" in transmitting *MOC* packages along the review and authorization path. Some companies have found that electronic communication of *MOC* information (e.g., through e-mail) to be a very efficient solution. Other companies have implemented specialized *MOC* software that not only helps the communication process, but "automates" the *MOC* workflow, tracks *MOC* data, and trends of *MOC* performance and efficiency indicators. Providing *MOC* review packages via email transmission or through an intranet is a relatively low-cost approach. Customized *MOC* workflow software requires a greater initial investment in development, customization, personnel training, and pilot testing, but often proves more cost-effective in high demand situations.

15.4.2 Identify Potential Change Situations

Periodically review MOCs to ensure that they are not RIKs, and vice versa. If a large number of changes are routed through the *MOC* system that actually qualify as RIK, then the facility is wasting resources. Conversely, if changes are improperly classified as RIKs, then activities involving unknown hazards or intolerable risk could be introduced to the facility. The former issue can be monitored by examining a group of *MOC* requests each quarter to determine what fraction were actually RIK. The latter situation can be monitored by sampling work orders, completed HIRA action items, and so forth to determine if any had been improperly classified as RIKs. If either fraction gets too high, additional training or refresher training may be warranted.

15.4.3 Evaluate Possible Impacts

Specify essential issues that must be addressed in MOC reviews for each type of change. For frequent types of changes, develop a checklist summarizing the minimum safety and health issues that reviewers are expected to address. This list can be gleaned from previous reviews of similar changes. Providing such a checklist should help ensure that risks are not overlooked.

Develop a specification for a minimum technical basis description. Poor *MOC* reviews can sometimes be traced to missing or inadequate information about the proposed change. Just stating in the *MOC* program that the technical basis for the change must be provided is rarely enough to enable a quality review. Listing the types of information that should be included and providing examples of good and inadequate technical basis descriptions greatly enhances *MOC* awareness and detailed training.

Provide a list of necessary information for each change type. Change requests cannot be properly reviewed if the review packages are missing critical information. Inadequate description of the change, insufficient detail or types of drawings, or simple lack of knowledge can hamper an *MOC* review. Developing a simple list of the standard information that should be supplied with each change request, supplemented by the specific lists for each change type, can improve the effectiveness of the *MOC* process.

Provide copies of existing drawings with proposed changes already noted. For certain types of changes, simply describing the proposed change does not fully communicate the change intent. For process equipment, controls, or physical change requests, such descriptions should be accompanied by a revised drawing so that the full ramifications of the proposed change can be evaluated by reviewers.

Provide copies of existing procedures with proposed changes already noted. For certain types of changes, simply describing the proposed change does not fully communicate the change intent. For operating or maintenance procedure changes, supplying a revised procedure to reviewers helps them to evaluate the full ramifications of the proposed change.

Provide hazard evaluation training to MOC reviewers. The ability to identify and evaluate hazards associated with change requests is an acquired skill. *MOC* reviewers given the responsibility to anticipate risks should have a questioning attitude. For complex situations, team-based reviews are often used to improve confidence in the completeness of the hazard analysis. In either case, the quality of a hazard review typically improves when the reviewer is trained in hazard identification, basic risk concepts, and risk analysis techniques.

Ensure that the experts involved in the MOC review process are appropriate for the change type. Some facilities experience a wide range of change types. A review process that specifies a common list of functional disciplines for the review team, such as industrial hygiene, information technology, and human resources, may not ensure that persons with the appropriate technical expertise review each change. By specifying particular skills and experience levels for each change type, a company can help ensure that knowledgeable personnel are making the risk judgments and that others are not spending time reviewing changes that have no connection to their functional discipline.

Ensure that adequate personnel are available to support MOC reviews. A lack of qualified personnel to conduct *MOC* reviews will increase the *MOC* backlog, reduce the thoroughness of reviews, or tempt employees to circumvent the *MOC* system in order to expedite necessary or urgent process changes. This is especially true if a facility regularly has a large number of changes that must be reviewed. Facilities should consider the rate at which they need to process change requests in the *MOC* system and seek to have adequate personnel available to minimize backlog.

Determine whether a series, parallel, or team-based review process is appropriate for each change type. Simple changes are often reviewed consecutively by relatively few people. Parallel review paths are defined (1) when companies are concerned about review time/holdup and (2) when they believe that the separate, non-sequential reviews will not diminish the risk analysis quality. More complex changes often demand multiple minds working in a team-based environment using a formal risk evaluation method. Companies seeking to optimize review quality and applied effort should consider allowing flexibility when selecting the appropriate review path for each change based on a classification of the change or by designating different review paths for appropriate change classes.

Consider reducing the number of people involved in MOC reviews. Changes must be reviewed by independent parties to help ensure that risks are identified. On the other hand, having too many independent parties involved in the review process can be wasteful. Too many reviewers can also lead to lower quality reviews because each person in the review path believes someone else will surely spot any unacceptable risk. *MOC* quality problems caused by lackadaisical reviews may not be evident until a root cause analysis of an incident determines that a poor quality *MOC* review was the root cause of the incident. To prevent these situations, a company should review the number of disciplines/job functions/personnel required to be in the review chain as a function of change type and (1) reassess the optimum/minimum number of people needed or (2) specialize/simplify the *MOC* review process for certain types of changes that occur frequently at the facility that do not need a broad look by a larger number of people.

For series review MOC procedures, consider adopting a parallel review process for certain types of changes. Some *MOC* procedures require reviews or authorizations to be done in a specific sequence. If delays occur at any point in the review, then the entire *MOC* system can bog down. A slow system that does not meet the needs of the production or maintenance organization can lead to people taking shortcuts around the system in order to get work done. Some companies track efficiency metrics, such as elapsed time between change origination and approval or *MOC* backlog, to help identify if the facility is experiencing these types of delays. If chronic bottlenecks are discovered, a company should take steps to fix them, or restructure the *MOC* review procedure to allow parallel reviews. A parallel review process sometimes requires a more robust authorization/approval step because approvers are often asked to confirm that nothing has been missed by the hazard reviewers even though they no longer see the previous reviewers' ideas/concerns. Alternatively, a company may decide to use a series review path for certain change types and a parallel review path for others.

Provide more experienced people as MOC reviewers. Less experienced personnel are likely to take longer and provide less insightful *MOC* reviews, particularly for the more complex change situations. Experienced personnel who have a broad understanding of hazards and experience in evaluating risk normally conduct the MOC reviews more effectively. Companies that have difficulty in completing *MOC* reviews on time or are spending inordinate amounts of time on the reviews should examine the experience levels of the people involved in the less efficient *MOC* reviews. Note that using more experienced reviewers may provide a short-term solution to the efficiency problem, but may create a bottleneck over the longer term unless additional experienced personnel are trained to be *MOC* reviewers.

15.4.4 Decide Whether to Allow the Change

Adopt a single-step, combined review/approval process for low risk significance MOCs. One approach to simplifying the *MOC* process is to combine the risk review and authorization steps. As reviewers complete their work, they document their concerns and/or authorize the change in a single review step. Once all of the required reviews are complete, the change is approved for implementation. Typically, if this option is chosen, the reviews are completed is a specified sequence, not in parallel, especially if only two or three reviewers are involved.

Streamline MOC reviews for frequent change types. Identify the categories of change that consume the most resources. Determine if certain functional disciplines specified in the *MOC* procedure consistently failed to contribute to the review results because the type of change did not impact their technical area. For these situations, the review process can be streamlined by limiting the review to only those functional disciplines that have historically been relevant for that type of change. These streamlined reviews should help reduce the amount of time spent on future *MOC* reviews.

15.4.5 Complete Follow-up Activities

Provide a final MOC quality check by MOC authorizers or MOC coordinator. Having all of the *MOC* records pass before a single person/function can help ensure that the system is consistently working as intended. Having one or two closely coordinated people performing a final check on *MOC* review packets may help detect anomalies that can be quickly corrected, which has a positive, reinforcing feedback effect on the entire *MOC* process.

15.5   ELEMENT METRICS

Chapter 20 describes how metrics can be used to improve performance and when they may be appropriate. This section includes several examples of metrics that could be used to monitor the health of the *MOC* element, sorted by the key principles that were first introduced in Section 15.2.

In addition to identifying high value metrics, readers will need to determine how to best measure each metric they choose to track. In some cases, an ordinal number provides the needed information, for example, total number of workers. Other cases require that two or more attributes be indexed to provide meaningful information, such as averaging employees' length of service to obtain a unit's average years of experience per employee. Still other metrics may need to be tracked as a rate, employee turnover, for example. Sometimes, the rate of change may be of greatest interest. Since every situation is different, the reader will need to determine how to track and present data to most effectively monitor the health of RBPS management systems at their facility.

15.5.1 Maintain a Dependable MOC Practice

- *The number of MOCs performed each month.* An unexplained drop may indicate that the system is being circumvented or that the system is backlogged; unexplained rises may mean that it is being overused.
- *The monthly average in the percent of work requests classified as a change.* A rise or fall may indicate that new people are involved in the decision making process; *MOC* training should be reviewed to ensure all employees are following proper change/RIK classification guidelines.
- *The percentage or variation in the number of changes processed on an emergency basis.* A high or increasing number may indicate that people are abusing a less extensive emergency change request system to avoid the work of the normal change review process.
- *The percentage of personnel involved in the MOC system who believe the system is effective.* A low or decreasing number may indicate problems with the *MOC* process that require closer examination.
- *The difference between the percentages of senior managers and routine users who believe the MOC program is effective.* A large gap may indicate that the system may not be working and that management is being lulled into complacency.
- *The average backlog of MOCs/active MOCs.* An increase may indicate a resource problem or an efficiency problem.
- *The average amount of calendar time taken between MOC origination and authorization.* A high or increasing amount may indicate that efficiency improvements or more resources are needed.
- *The average number of staff-hours per MOC from the time the MOC is originated until the time the MOC is approved for implementation.* A high or increasing number may indicate that an efficiency improvement is needed.

15.5.2 Identify Potential Change Situations

- *The percentage of work orders/requests that were misclassified as RIKs (or were not classified) and were really changes.* A high or increasing number may indicate the need for *MOC* refresher training for those people involved in change/RIK classification decisions.
- *The ratio of identified undocumented changes to the number of changes processed by the MOC program.* A high or increasing number indicates that the MOC is being circumvented.

15.5.3 Evaluate Possible Impacts

- *The percentage of changes that were reviewed within the MOC system but were reviewed incorrectly.* A high or increasing number indicates a need for refresher training in *MOC* or risk analysis techniques.
- *The percentage of recent changes that involved the use of backup MOC personnel.* A high or increasing number indicates the need for a re-evaluation of the capacity of the change review step and possibly an increase in designated resources.

15.5.4 Decide Whether to Allow the Change

- *The percentage of changes that were properly evaluated, but did not have all authorization signatures on the change control document.* A high or increasing number indicates the need for MOC refresher training.

15.5.5 Complete Follow-up Activities

- *The percentage of MOCs reviewed that were not documented properly.* A high or increasing number indicates the need for *MOC* refresher training.
- *The percentage of MOCs for which the drawings or procedures were not updated.* A high or increasing number indicates the need for *MOC* refresher training or a deficiency in the *knowledge* element.
- *The percentage of MOCs for which the workers were not informed or trained.* A high or increasing number indicates the need for MOC refresher training or a deficiency in the *training* element.
- *The percentage of temporary MOCs for which the temporary conditions were not corrected/restored to the original state at the deadline.* A high or increasing number indicates the need for MOC refresher training.

15.6 MANAGEMENT REVIEW

The overall design and conduct of management reviews is described in Chapter 22. However, many specific questions/discussion topics exist that management may want to check periodically to ensure that the management system for the *MOC* element is working properly. In particular, management must first seek to understand whether the system being reviewed is producing the desired results. If the organization's level of quality of *MOC* reviews is less than satisfactory as evidenced by process incidents precipitated by changes, or it is not improving as a result of management system changes, then management should identify possible corrective actions and pursue them. Possibly, the organization is not working on the correct activities, or the organization is not doing the necessary activities well. Even if the results are satisfactory, management reviews can help determine if resources are being used wisely – are there tasks that could be done more efficiently or tasks that should not be done at all? Management can combine metrics and indicators listed in the previous section with personal observations, direct questioning, audit results, and feedback on various topics to help answer these questions. Activities and topics for discussion include the following:

- Review the completeness and quality of a sample of *MOC* reviews for each operating area. Discuss any significant gaps.
- Talk to people in each operating area and find out if the *MOC* process is being used as intended. Is there evidence of circumvention?
- Review recent audit findings that addressed *MOC* and determine the status of all corrective actions. Are any items overdue or backlogged?
- Determine if any incidents occurred or if any trends were found that listed *MOC* failure as a root cause or contributing factor.
- Ask employees their opinion of how effective the new employee and contractor training is in regard to *MOC*. Query employees on their knowledge of *MOC* to monitor refresher training.
- Review the staffing, resources, and time spent to address *MOC* issues. Examine any major changes from previous years.

In addition, the *MOC* element owner should be able to explain any trends or anomalies in the current metrics. If any major projects to address known *MOC* gaps are ongoing, a briefing should be prepared for the management review committee.

The results of a management review of *MOC* activities should demonstrate that leadership at the facility is aware of and values *MOC*, and is intent on ensuring that all changes are properly evaluated prior to implementation. The management review process should also shine a bright light on efforts to perform, document, collect, and maintain *MOC* information and metrics, including improving the efficiency of work activities that support this element. In addition, an effective management review process educates the entire leadership team on the importance of *MOC* and the role it can play in helping to identify hazards, manage risk, and sustain the business.

15.7   REFERENCES

15.1   *Guidelines for Managing Change*, AIChE Center for Chemical Process Safety, Wiley, New York, NY, 2007.

15.2   *Guidelines for Managing Process Change*, American Chemistry Council, Arlington, VA, 1993.

16

OPERATIONAL READINESS

A refinery was in the midst of a major turnaround of one of its units, which involved extensive lock-out-tag-out (LOTO) issues for multiple pieces of equipment. Procedures were in place for LOTO, initial line breaking, and all other relevant safety processes. One unique aspect of this unit turnaround was that several pieces of equipment included in the process boundary were only operated on an occasional basis when feedstocks were changed and additional heating and/or cooling were needed.

All pre-startup checks were made per the established checklists, and the unit restarted smoothly. Unfortunately, six months later, when one of the occasionally used heat exchangers was valved into service, a major leak of flammable gas ensued. The resulting fire caused a significant outage and equipment damage before being brought under control. It was discovered that the startup checks for this equipment had not been performed as a part of the overall turnaround, because this particular piece of equipment was not going to be used immediately. Although individual processes were in place to perform pre-startup checks, no overall operational readiness process had been established to ensure that the entire unit was ready to run.

16.1 ELEMENT OVERVIEW

Ensuring the safe startup of processes over the life of a facility is one of nine elements in the RBPS pillar of *managing risk*. This chapter describes the management practices for performing pre-startup reviews of (1) new processes, (2) processes that have been shut down for modification, and (3) processes that have been administratively shut down for other reasons. These practices verify the operational readiness (*readiness*) of a process and help to ensure that the process is safe to restart. Section 16.2 describes the key principles and essential features of a management system for this element.

Section 16.3 lists work activities that support these essential features and presents a range of approaches that might be appropriate for each work activity, depending on perceived risk, resources, and organizational culture. Sections 16.4 through 16.6 include (1) ideas for improving the effectiveness of management systems and specific programs that support this element, (2) metrics that could be used to monitor this element, and (3) management review issues that may be appropriate for *readiness*.

16.1.1 What Is It?

The *readiness* element ensures that shut down processes are verified to be in a safe condition for re-start. This element addresses startups from all types of shut down conditions and considers the length of time the process was in the shut down condition. Some processes may be shut down only briefly, while others may have undergone a lengthy maintenance/modification outage, or they may even have been mothballed for an extended period. Other processes may have been shut down for administrative reasons, such as a lack of product demand; for reasons unrelated to production at all; or as a precautionary measure, for example, because of an approaching hurricane. In addition to the shutdown duration, this element considers the type of work that may have been conducted on the process (e.g., possibly involving line-breaking) during the shutdown period to help focus the *readiness* review prior to startup.

The *readiness* element in these *Guidelines* is defined more broadly than the OSHA process safety management pre-startup safety review element in that it specifically addresses startup from all shutdown conditions – not only those resulting from new or changed processes (Refs. 16.1 and 16.2).

16.1.2 Why Is It Important?

Experience has shown that the frequency of incidents is higher during process transitions such as startups. These incidents have often resulted from the physical process conditions not being exactly as they were intended for safe operation. Thus, it is important that the process status be verified as safe to start.

16.1.3 Where/When Is It Done?

Readiness reviews are conducted prior to startup on new processes, and on existing processes that were shut down for any reason. A review involves some or all of the following concerns and activities: (1) confirming that the construction and equipment of a process are in accordance with design specifications, (2) ensuring that adequate safety, operating, maintenance, and emergency procedures are in place, and (3) ensuring that training has been completed for all workers who may affect the process. Also, for new processes, it confirms that an appropriate risk analysis has been performed, and that any recommendations have been resolved and implemented. Modified processes should have undergone a *management of change* (*MOC*) review. For all startups (including those after minor, short-term shutdowns not involving any changes), *readiness* reviews ensure that the process is safe to be released to operations by examining issues such as the equipment lineup, leak tightness, proper isolation from other systems not yet ready for startup, and cleanliness.

16.1.4 Who Does It?

Simple *readiness* reviews may involve only one operator, maintenance person, or engineer. More complex startups, such as those for new units, entire plants, or a large unit after an extended shutdown, may involve many people, from all of the disciplines that are typically a part of a large capital project.

16.1.5 What Is the Anticipated Work Product?

The output of the *readiness* activity is either (1) an affirmation that the process is ready to safely start up and authorization to do so or (2) a list of actions that must be taken to make the process ready. Many times the *readiness* process and startup authorization is documented on a form, which creates an audit trail to ensure that all required actions have indeed been completed. For complex startup situations, reviewers may use several different checklists of information and necessary actions to confirm the status of all of the items reviewed. Outputs of the *readiness* element can also be used to facilitate the performance of other elements. For example, a walkdown of a process prior to startup can identify deficient equipment conditions that can provide input to the *asset integrity* element.

16.1.6 How Is It Done?

Readiness reviews of simple startups may involve only one person walking through the process to verify that nothing has changed and the equipment is ready to resume operation. Complex reviews may extend over many weeks or months as engineering, operations, and maintenance personnel verify equipment conformance to design intent, construction quality, procedure completion, training competency, and so forth. Typically, extensive checklists, multi-stage verification, and multiple functional sign-offs are required for startup authorization.

Higher risk situations usually dictate a greater need for formality and for thoroughness in scope and level of detail; for example, greater detail in the checklists used to guide the evaluation. Less rigorous approaches, such as those using a simple checklist or no checklist at all, may be adequate for lower risk processes.

Facilities whose operations are very dynamic may require *readiness* practices that are very flexible. Facilities with solid process safety cultures can often rely on more performance-based pre-startup tools, since a strong culture provides greater confidence that employees will check out the system thoroughly according to the designed startup procedure. However, in practice, those facilities with a sound process safety culture normally embrace the use of comprehensive checklists to guide and document *readiness* reviews. Facilities with an evolving or undetermined process safety culture may require more detailed, prescriptive approaches and tools to provide greater command and control management system features to ensure good performance.

16.2   KEY PRINCIPLES AND ESSENTIAL FEATURES

Safe operation and maintenance of facilities that manufacture, store, or otherwise use hazardous chemicals requires robust process safety management systems. The primary objective of the *readiness* element is to ensure that processes that have been shut down are safe to restart. The readiness element covers the management practices dealing with performing pre-startup reviews of (a) new processes, (b) processes that have been shut down for modification, and (c) processes that have been administratively shut down for other reasons. The following key principles should be addressed when developing, evaluating, or improving any system for the *readiness* element:

- Maintain a dependable practice.
- Conduct appropriate *readiness* reviews as needed.
- Make startup decisions based upon *readiness* results.
- Follow through on decisions, actions, and use of *readiness* results.

The essential features for each principle are further described in Sections 16.2.1 through 16.2.4. Section 16.3 describes work activities that typically underpin the essential features related to these principles. Facility staff should evaluate the risks and potential benefits that may be achieved as a result of improvements in this element and, on that basis, develop a management system, or upgrade an existing management system, to address some or all of the essential features.

16.2.1 Maintain a Dependable Practice

A written program that documents the intentions of the *readiness* element is key to the long-term success of *readiness* activities. Defining roles and responsibilities, where and when *readiness* activities should be carried out, the technical issues that should be addressed, and the necessary technical expertise of personnel is critical to having an effective *readiness* system. Records should be maintained concerning *readiness* activities so that performance and efficiency can be periodically evaluated.

Ensure consistent implementation. Formal procedures for performing readiness reviews will help ensure dependable, high quality reviews. Designating an owner of the readiness system and defining the readiness system roles and responsibilities will help ensure that personnel know what they are supposed to do.

Determine types of and triggers for the readiness practice. Many types of startup situations can present themselves. Thinking through these situations in advance will assist the facility in preparing to conduct a readiness review without taxing its resources or unnecessarily delaying production.

Determine the scope of readiness reviews. The same readiness procedure may not apply to all facility areas. Identifying areas and situations for which the review applies and determining the content/issues to be addressed for each type of startup situation ahead of time will help ensure thorough and efficient readiness reviews.

Involve competent personnel. All personnel should be trained on the *readiness* program. Awareness training on this system should be provided to employees and contractors, and detailed training should be given to personnel who are assigned specific roles within the *readiness* element.

Ensure that readiness practices remain effective. Maintaining an archive of readiness data and metrics will provide valuable information for internal audits and management reviews of readiness practices.

16.2.2 Conduct Appropriate Readiness Reviews as Needed

Quality *readiness* reviews depend upon accurate input information and sufficient personnel expertise and resources. The review process should be thorough, yet flexible enough to be appropriate for simply restart situations as well as more complex startups of new processes. Appropriate tools should be used, and records should be created to document the results of each review.

Provide appropriate inputs. High quality *readiness* reviews cannot be performed unless the necessary process design, construction, inspection, training, and risk information is made available to the review team. Having a list of required information for each type of startup will help ensure thoroughness and efficiency.

Involve appropriate resources and personnel. Defining the readiness roles and responsibilities for various groups of personnel, and providing awareness training and refresher training to all employees and contractors will help ensure that effective *readiness* reviews are performed. Quality *readiness* reviews cannot be completed without competent personnel representing the necessary disciplines and having the requisite experience. Some *readiness* reviews may require specialized training and tools to access and inspect complex process equipment.

Apply an appropriate work process. *Readiness* reviews must provide sufficient confidence that the process is safe to start. Depending upon the type of startup, the specific items that must be addressed may vary considerably. Preparations that are necessary for safe startup might include, but are not limited to, the following items:

- Construction and equipment is verified to be in accordance with design specifications for new or modified facilities.
- Process control, emergency shutdown, and safety systems have been tested.
- Equipment is properly isolated from other systems not yet ready for startup.
- Equipment has been cleaned or flushed, where appropriate, and cleaning materials have been removed.
- Equipment lineup has been verified as secure and has been released to operations for startup.
- Safety, operating, maintenance, and emergency procedures are in place and are adequate.
- Emergency response equipment is in place and training has been completed.
- Training of each employee involved in operating or maintaining a process has been completed.

For new facilities, a process hazard analysis has been performed and recommendations have been resolved or implemented before startup. Modified facilities meet the requirements described in MOC documentation, and the team has verified that processes idled through administrative shut downs have not changed or been degraded. Facilities should design the appropriate review process based on risk considerations. Then, a *readiness* review with the appropriate level of detail can be performed for (1) new facilities, (2) modified facilities implementing MOC, and (3) facilities that are being started after being shut down for a long period.

Perform element work in a diligent manner. *Readiness* review personnel should use appropriate tools (such as checklists) to conduct, document the basis for, and record the results of a *readiness* review.

Create element work products. Preparing the *readiness* documentation, including the review completion form and the readiness basis/rationale, will provide a proper record of the *readiness* review.

16.2.3 Make Startup Decisions Based upon Readiness Results

The results of each *readiness* review should drive action – either deciding that the startup may safely proceed or establishing conditions that must be met prior to startup. The *readiness* results and startup information should be broadly communicated to all potentially affected personnel.

Consider important issues affecting the startup. If issues are discovered that require action, determine whether the actions must be completed prior to startup or if they can be deferred until after startup. If an action can be deferred, document the rationale for this (including any substitute actions, if required) and the deadline for completing the action. Each deferred action should be authorized by explicitly identified individuals, or representatives of departments or functions, as specified in the written program. The level of authorization should be commensurate with the risk incurred by the deferral decision.

Communicate decisions and actions from the readiness review. Communicate the results of the *readiness* review to potentially affected personnel, including contractors, and coordinate with other affected groups outside the reviewed unit, such as maintenance, emergency response, administration, and so forth.

16.2.4 Follow Through on Decisions, Actions, and Use of Readiness Results

Readiness reviews may establish conditions that must be met prior to startup; completion of these conditions should be tracked and documented. Modifications to process safety knowledge and records should be completed.

Enact risk control measures. If a *readiness* review identifies items that must be addressed prior to or during startup, management must assign the responsibility for ensuring that each item is completed to one or more individuals. Some actions may require *MOC* review before implementation. All *readiness* review action items should be documented and tracked to completion.

Update process safety knowledge and records. Following the completion of a *readiness* review and the resolution of all action items, relevant process safety information and records should be updated and the *readiness* review should be closed.

Maintain element work records. Keeping an archive of all *readiness* reviews for new facilities will preserve information that can be used to continuously improve the *readiness* process.

16.3 POSSIBLE WORK ACTIVITIES

The RBPS approach suggests that the degree of rigor designed into each work activity should be tailored to risk, tempered by resource considerations, and tuned to the facility's culture. Thus, the degree of rigor that should be applied to a particular work activity will vary for each facility, and likely will vary between units or process areas within a single facility. Therefore, to develop a risk-based process safety management system, readers should perform the following steps:

1. Assess the risks at the facility, investigate the balance between the resource load for RBPS activities and available resources, and examine the facility's culture. This process is described in more detail in Section 2.2.
2. Estimate the potential benefits that may be achieved by addressing each of the key principles for this RBPS element. These principles are listed in Section 16.2.
3. Based on the results from steps 1 and 2, decide which essential features described in Sections 16.2.1 through 16.2.4 are necessary to properly manage risk.
4. For each essential feature that will be implemented, determine how it will be implemented and select the corresponding work activities described in this section. Note that this list of work activities cannot be comprehensive for all industries; readers will likely need to add work activities or modify some of the work activities listed in this section.
5. For each work activity that will be implemented, determine the level of rigor that will be required. Each work activity in this section is followed by two to five implementation options that describe an increasing degree of rigor. Note that work activities listed in this section are labeled with a number; implementation options are labeled with a letter.

> *Note: Regulatory requirements may specify that process safety management systems include certain features or work activities, or that a minimum level of detail be designed into specific work activities. Thus, the design and implementation of process safety management systems should be based on regulatory requirements as well as the guidance provided in this book.*

16.3.1 Maintain a Dependable Practice

Ensure Consistent Implementation

1. Establish and implement procedures to perform *readiness* reviews.
   a. The unit manager defines the *readiness* reviews procedure for each startup.
   b. A detailed written checklist governs startups of new units.
   c. In addition to item (b), a general written procedure governs all startups.
   d. A detailed written program with formal checklists governs all types of startups.
2. Assign a job function as the owner of the *readiness* system.
   a. The process owner serves as the *readiness* system owner.
   b. The *readiness* system owner, local or offsite, serves part time.
   c. The *readiness* system has multiple owners across the facility.
   d. The *readiness* system has a single, full-time owner.
3. Define the *readiness* roles and responsibilities of various groups of personnel.
   a. Roles/responsibilities are informally accepted.
   b. *Readiness* is the informal duty for several people.
   c. *Readiness* roles/responsibilities are assigned to job functions/ departments.
   d. A single person is responsible for *readiness.*

Determine Types and Triggers for the Readiness Practice

4. Determine the types of *readiness* reviews that are needed and when to conduct them.
   a. Ad hoc *readiness* reviews are defined, and triggers are established.
   b. *Readiness* reviews are performed only for major startups.
   c. *Readiness* reviews are defined, and triggers are established for startup after changes.
   d. *Readiness* reviews are defined for all types of startups, and triggers are established.

Determine the Scope of Readiness Reviews

5. Determine the areas of the facility in which the *readiness* procedure applies. Also identify areas/situations where it does not apply.
   a. A *readiness* review is performed whenever someone believes one is needed.
   b. *Readiness* reviews are applied only in selected facility areas.
   c. *Readiness* reviews are applied in all facility areas with a standard scope.

    d.   *Readiness* reviews are applied in all areas with a scope appropriate to the type of startup.

6.   Determine the content/issues to be addressed for each type of startup situation:
   - New facility.
   - A replacement-in-kind maintenance activity.
   - Modified facility/equipment.
   - Startup of facility/process/system following normal shutdown after short period of time for administrative or precautionary reasons.
   - Startup of facility/process/system following normal shutdown after long period of time, for example, resuming production of inactive products or mothballed process.
   - Startup following a nonroutine shutdown (i.e., shutdowns not covered by operating procedures).

    a.   Scope is developed ad hoc.
    b.   Scope is developed for simple *readiness* review types.
    c.   Scope is established for major startups.
    d.   Scope is established for all *readiness* review types.

Involve Competent Personnel

7.   Provide training on the *readiness* system to employees and contractors.
       a.   Informal awareness training is provided.
       b.   *Readiness* practice is broadcast once (e.g., through e-mail) to everyone.
       c.   *Readiness* awareness initial training is provided once to affected personnel.
       d.   *Readiness* awareness initial and refresher training are provided to affected personnel.

8.   Provide detailed training to personnel who are assigned specific roles within the *readiness* system.
       a.   Informal detailed *readiness* training is provided.
       b.   *Readiness* practice is broadcast once (e.g., through e-mail) to everyone.
       c.   Detailed training is provided once to key personnel implementing *readiness* activities.
       d.   Detailed/refresher training is provided to key personnel implementing *readiness* activities.

Ensure that Readiness Practices Remain Effective

9.   Collect *readiness* data.
       a.   *Readiness* data are kept for a brief period.

    b. *Readiness* data are kept until the next startup for complex *readiness* reviews only.

    c. *Readiness* status log is kept by the *readiness* coordinator and is accessible to all affected personnel.

    d. *Readiness* status log is kept electronically and is accessible via company network.

10. Establish and collect metrics data on the *readiness* element.

    a. Metrics data are collected on an informal basis.

    b. Basic *readiness* activity data are collected.

    c. *Readiness* performance indicators are collected regularly.

    d. *Readiness* performance and efficiency indicators are collected regularly.

11. Provide input to internal audits of *readiness* practices based upon learnings from the metrics data.

    a. Informal internal reviews are performed.

    b. Internal *readiness* audits are performed, but no analysis of performance information is provided to help reviewers target specific areas.

    c. *Readiness* performance information is provided to internal reviewers annually.

    d. *Readiness* performance and efficiency information is provided to reviewers for regular monthly/quarterly management reviews.

16.3.2 Conduct Appropriate Readiness Reviews as Needed

Provide Appropriate Inputs

12. Create a list of the necessary information that should be provided to participants of *readiness* reviews.

    a. Informal lists are created.

    b. A basic restart checklist is created for startup after a change.

    c. A list of *readiness* input information is developed for new construction *readiness* reviews.

    d. Detailed lists of necessary input information are developed for each type of *readiness* review.

Involve Appropriate Resources and Personnel

13. Provide personnel for each *readiness* review.

    a. Limited planning for *readiness* review is done; whoever is available is assigned.

    b. Operations personnel are designated for *readiness* reviews.

    c. Multi-discipline group are assigned to *readiness* reviews.

    d. Appropriate disciplines are assigned for each type of *readiness* review based on need.

14. Provide *readiness* review tools.

a. Limited tools are provided.
b. A basic restart checklist is provided.
c. Detailed checklists appropriate for the *readiness* review type are provided.
d. In addition to item (c), electronic data recording tools are provided.

Apply Appropriate Work Process

15. Have *readiness* reviews confirm that preparations have been completed prior to the introduction of hazardous substances into a new process or before the restart of an existing process.
    a. *Readiness* reviews are performed on an ad hoc basis.
    b. *Readiness* reviews are performed by walking around and looking at the equipment.
    c. *Readiness* reviews confirm the status of items that have caused previous problems.
    d. *Readiness* reviews address a comprehensive list of issues.
16. Perform a *readiness* review for facilities that are being started.
    a. *Readiness* reviews for performed for some startups selected on an ad hoc basis.
    b. *Readiness* reviews are performed for major startups.
    c. *Readiness* reviews are performed for changes and new processes.
    d. *Readiness* reviews are performed for all types of startups.

Perform Element Work in a Diligent Manner

17. Use tools, including checklists, to conduct and document the basis for the *readiness* review.
    a. *Readiness* reviews are performed informally, but not documented.
    b. *Readiness* reviews are performed and documented on ad hoc basis.
    c. *Readiness* reviews are performed and documented using checklists.
    d. *Readiness* reviews are performed and documented using electronic checklists and data gatherings tools.

Create Element Work Products

18. Prepare *readiness* documentation, containing the review completion form and the readiness basis/rationale.
    a. Informal documentation is created.
    b. A *readiness* review form is completed and signed off.
    c. A *readiness* review form and checklists are completed and signed off.
    d. A *readiness* review form and checklist are completed, and the startup decision rationale is documented.

16.3.3 Make Startup Decisions Based upon Readiness Results

Consider Important Issues Affecting the Startup

19. If issues are discovered that require action, ensure that actions are completed.
    a. Pre-startup action items are addressed during the *readiness* reviews.
    b. *Readiness* review action items are documented and tracked to completion.
    c. *Readiness* review action items are completed, or the rationale for post-startup completion is documented, and the actions are tracked to completion.
20. Authorize startups based upon *readiness* review results as specified in the written program.
    a. Informal startup signoff or permission is required.
    b. Restart is authorized by the operator in attendance.
    c. Restart is authorized by the unit supervisor.
    d. Restart authorized is based on multi-department approvals.

Communicate Decisions and Actions from the Readiness Review

21. Communicate *readiness* review results to personnel.
    a. Informal communication/training on changes occurs for some personnel.
    b. A formal system exists for informing/training operating personnel on *readiness* review results.
    c. A formal system exists for informing/training all potentially affected personnel on *readiness* review results.
22. Coordinate with other potentially affected groups outside the subject unit, such as maintenance, emergency response, administration, and so forth.
    a. Informal, ad hoc coordination occurs.
    b. A formal system exists within the affected unit to coordinate with all personnel in the unit.
    c. A formal coordination system exists between the affected unit and other areas.

16.3.4 Follow Through on Decisions, Actions, and Use of Readiness Results

Enact Risk Control Measures

23. Create a system to address *readiness* review action items and to document their completion.
    a. An informal practice exists for documenting the completion of action items, for example, using unit or operator logs.

   b.   *Readiness* review action items are tracked in a spreadsheet.
   c.   *Readiness* review action items are integrated into the overall facility action item tracking system.

Update Process Safety Knowledge and Records
   24. Update *process knowledge.*
      a.   *Process knowledge* is updated on an ad hoc basis.
      b.   *Process knowledge* is updated using a formal procedure.
      c.   All *process knowledge* must be updated, and completion of the updates is tracked, including post-startup updates.

Maintain Element Work Records
   25. Retain *readiness* review records.
      a.   Some records are informally kept.
      b.   New process *readiness* review records are kept by project personnel.
      c.   *Readiness* review records are kept according to the facility retention policy.

16.4   EXAMPLES OF WAYS TO IMPROVE EFFECTIVENESS

This section provides specific examples of industry tested methods for improving the effectiveness of work activities related to the *readiness* element. The examples are sorted by the key principles that were first introduced in Section 16.2. The examples fall into two categories:

   1.   Methods for improving the performance of activities that support this element.
   2.   Methods for improving the efficiency of activities that support this element by maintaining the necessary level of performance using fewer resources.

These examples were obtained from the results of industry practice surveys, workshops, and CCPS member-company input. Readers desiring to improve their management systems and work activities related to this element should examine these ideas, evaluate current management system and work activity performance and efficiency, and then select and implement enhancements using the risk-based principles described in Section 2.1.

16.4.1   Maintain a Dependable Practice

Combine the readiness practice with MOC. The *MOC* and *readiness* elements are closely linked, and many companies have found it efficient to include a *readiness* review as one of the last steps of an MOC. This facilitates those *readiness* reviews that are performed as a result of a change, but this does not account for all other types of startup situations.

16.4.2 Conduct Appropriate Readiness Reviews as Needed

Send company personnel to vendor facilities to conduct in-situ inspections for fabrication or assembly of custom equipment. Some startups involve new processes or a major equipment installation that was fabricated many months before field erection. Having inspectors visit fabrication locations can be a more effective way to discover defects than during a *readiness* review. In-situ inspections also help to ensure that appropriate documentation will be available for the subsequent *readiness* review and provide the baseline data for comparison with as-installed conditions.

Develop startup-type specific checklists for readiness reviews. The most common tool for performing *readiness* reviews is the checklist. However, because of the range in complexity of *readiness* reviews, using the same checklist for all reviews can be inefficient. Tailoring the checklists to specific *readiness* review situations aids the review teams in conducting efficient reviews.

Enter readiness review checklists into personal digital assistants (PDAs). Some facilities have increased their efficiency by having personnel enter their *readiness* review results into a PDA or similar device. If the *readiness* review checklists are loaded onto the PDA, then the review personnel can perform the review without having to use hard-copy checklists. Doing so also enables workers to download the review data and generate reports electronically.

16.4.3 Make Startup Decisions Based upon Readiness Results

Classify readiness review action items based upon completion urgency. Some actions may not be able to be completed, or may not need to be completed, prior to actual startup of the process. Categorizing *readiness* review action items according to urgency can help avoid unnecessary startup delays. Note: deferral of action item completion carries with it tolerance of risk; this risk should be considered when making and documenting an action item deferral decision.

16.4.4 Follow Through on Decisions, Actions, and Use of Readiness Results

Track readiness review action items to completion using the facility action item tracking system. Most facilities will already have an action item tracking system; incorporating *readiness* review items into this system will allow easy monitoring of status.

16.5   ELEMENT METRICS

Chapter 20 describes how metrics can be used to improve performance and when they may be appropriate. This section includes several examples of metrics that could be used to monitor the health of the *readiness* element, sorted by the key principles that were first introduced in Section 16.2.

In addition to identifying high value metrics, readers will need to determine how to best measure each metric they choose to track. In some cases, an ordinal number provides the needed information, for example, total number of workers. Other cases require that two or more attributes be indexed to provide meaningful information, such as averaging employees' length of service to obtain a unit's average years of experience per employee. Still other metrics may need to be tracked as a rate, employee turnover, for example. Sometimes, the rate of change may be of greatest interest. Since every situation is different, the reader will need to determine how to track and present data to most effectively monitor the health of RBPS management systems at their facility.

16.5.1 Maintain a Dependable Practice

- *Number of incidents that occur during startup.* A high number or increasing rate might indicate that *readiness* reviews are not being conducted in a careful manner or that MOCs were not performed well.
- *Number of spurious shutdowns after startup.* A high number or increasing rate might indicate that *readiness* activities were not effective.
- *Number of improperly assembled pieces of equipment found during readiness reviews.* A high number would indicate that *readiness* reviews were effective, but they may also indicate deficiencies in maintenance or construction practices.
- *Number of personnel trained prior to startup.* A high number or percentage usually indicates that *readiness* activities are being performed on schedule.
- *Staff-hours expended on readiness reviews.* An abnormally high number might indicate more difficult startup situations or the need to improve efficiency.
- *Duration of startup.* A high number might indicate that *readiness* reviews were not performed well.
- *Amount of off-spec product or loss of raw material as a result of startup problems.* A high number might indicate that *readiness* activities were not performed well or that *MOC* activities did not address quality hazards.
- *The number of people trained per year on readiness.* A high number or percentage indicates an active *readiness* program.

16.5.2 Conduct Appropriate Readiness Reviews as Needed

- *Number of startups for which readiness reviews were not performed.* A high or increasing number indicates that the *readiness* program is being circumvented and that remedial awareness training might be needed.
- *Number of readiness reviews performed.* A high or increasing number indicates an active program. A low number, combined with knowledge of a high rate of *MOC* reviews, may indicate that the *readiness* system is being bypassed.

16.5.3 Make Startup Decisions Based upon Readiness Results

- *The number of readiness reviews for which authorizations to restart were not found.* A high number or percentage would indicate that readiness reviews are not being finished or documented properly.
- *The number of startups deferred as a result of problems found during readiness reviews.* A high or increasing number may indicate that the *MOC* review process is not thoroughly identifying hazards and managing risk.

16.5.4 Follow Through on Decisions, Actions, and Use of Readiness Results

- *Number of issues during startup that should have been discovered during the readiness review.* A high or increasing number indicates poor *readiness* review performance.
- *Number of action items overdue.* A high or increasing number indicates that the *readiness* program is not diligent about conducting follow-up activities.
- *Time from readiness review to completion of all action items.* A high or increasing length of time might indicate the need to improve efficiency or that the *MOC* program is ineffective.

16.6 MANAGEMENT REVIEW

The overall design and conduct of management reviews is described in Chapter 22. However, many specific questions/discussion topics exist that management may want to check periodically to ensure that the management system for the *readiness* element is working properly. In particular, management must first seek to understand whether the system being reviewed is producing the desired results. If the organization's level of quality of *readiness* reviews is less than satisfactory as evidenced by problem-plagued startups, or it is not improving as a result of management system changes, then management should identify possible corrective actions and pursue them.

Possibly, the organization is not working on the correct activities, or the organization is not doing the necessary activities well. Even if the results are satisfactory, management reviews can help determine if resources are being used wisely – are there tasks that could be done more efficiently or tasks that should not be done at all? Management can combine metrics and indicators listed in the previous section with personal observations, direct questioning, audit results, and feedback on various topics to help answer these questions. Activities and topics for discussion include the following:

- Review a sample of *readiness* reviews for each operating area to evaluate their thoroughness and quality.
- Review any significant gaps that were identified in recent *readiness* reviews.
- Determine whether the *readiness* process is being used.
- Ask employees if any evidence exists of the facility attempting to circumvent the *readiness* process.
- Review any recent audit findings that addressed *readiness*; determine if any corrective actions are overdue.
- Determine if any incidents or trends identified *readiness* failure as a root cause or contributing factor. Have the associated management system failures been identified and addressed?
- Review data on new employee and contractor training for the *readiness* element.
- Ask employees questions about the *readiness* process training.
- Evaluate resource utilization to address *readiness* issues.

In addition, the *readiness* element owner should be able to explain any trends or anomalies in the metrics. Finally, if any major projects to address known *readiness* gaps are ongoing, a briefing should be prepared for the management review committee.

The results of a management review of *readiness* activities should demonstrate that leadership at the facility is aware of and values *readiness* and is intent on ensuring that all processes are reviewed prior to startup. Management reviews of *readiness* activities that uncover chronic, serious process safety deficiencies should result in a thorough evaluation of *MOC* and related RBPS elements. In addition, an effective management review process educates the entire leadership team on the importance of *readiness* and the role it can play in helping to identify hazards, manage risk, and sustain the business.

16.7   REFERENCES

16.1   *Pre-startup Safety Review Element,* Process Safety Management of Highly
        Hazardous Chemicals (29 CFR 1910.119(i)), *U.S. Occupational Safety and
        Health Administration, May 1992. www.osha.gov*
16.2   *Guidelines for Performing Pre-Startup Reviews,* AIChE Center for Chemical
        Process Safety, Wiley, New York, NY, 2007.

17

CONDUCT OF OPERATIONS

On January 21, 1997, an explosion and fire occurred at a refinery hydrocracker unit in Martinez, California, resulting in one death, 46 worker injuries, and an order for the surrounding community to shelter in place (Ref. 17.1). Operators initiated a temperature excursion while attempting to recover from a process upset and ignored the erratic readings of a temperature data logger because it had a history of unreliability. Radio transmissions from a field operator attempting to verify the temperatures locally were garbled. Thus, the control room operators did not depressurize the unit as required by procedure, and the overheated effluent piping burst. This incident illustrates how weaknesses in the conduct of operations can lead to tragedy. Continued operations in the face of unreliable or incomplete process information ultimately led to unreliable performance that exceeded safe operating limits.

17.1 ELEMENT OVERVIEW

Developing and sustaining high standards in the conduct of operations is one of nine elements in the RBPS pillar of *managing risk*. This chapter describes the concept of conduct of operations, the attributes of a reliable system for conducting operations, and the steps an organization might take to formalize the conduct of operations. Section 17.2 describes the key principles and essential features of a management system for this element. Section 17.3 lists work activities that support these essential features, and presents a range of approaches that might be appropriate for each work activity, depending on perceived risk, resources, and organizational culture. Sections 17.4 through 17.6 include (1) ideas for improving the effectiveness of management systems and specific programs that support this element, (2) metrics that could be used to monitor this element, and (3) issues that may be appropriate for management review.

17.1.1 What Is It?

Conduct of operations (*operations*) is the execution of operational and management tasks in a deliberate and structured manner. It is also sometimes called "operational discipline" or "formality of operations", and it is closely tied to an organization's culture. Conduct of operations institutionalizes the pursuit of excellence in the performance of every task and minimizes variations in performance. Workers at every level are expected to perform their duties with alertness, due thought, full knowledge, sound judgment, and a proper sense of pride and accountability (Refs. 17.2 and 17.3).

17.1.2 Why Is It Important?

A consistently high level of human performance is a critical aspect of any process safety program; indeed a less than adequate level of human performance will adversely impact all aspects of operations. As the complexity of operational activities increases, a commensurate increase in the formality of operations must also occur to ensure safe, reliable, and consistent performance of critical tasks.

17.1.3 Where/When Is It Done?

Like culture, conduct of operations applies everywhere workers perform tasks – from the boardroom to the plant floor. It applies every time a worker performs a task throughout the life of a facility or an organization because it is an ongoing commitment to reliable operations.

17.1.4 Who Does It?

Conduct of operations applies to all work activities, not just those of the operations department. Thus all workers, employees, and contractors are included. Each work group should define the framework of controls necessary to ensure that tasks for which it is responsible are performed reliably. The manager of each work group is responsible for the conduct of operations in his group, but overall responsibility rests with the facility manager. The human resources group is often involved with the process because it includes fitness-for-duty, progressive discipline, salary, bonus, and retention decisions.

17.1.5 What Is the Anticipated Work Product?

The output of this activity is a policy describing the organization's overall expectations for worker conduct and specific procedures and goals for implementing those policies. A defined framework of controls should be established that implements a defense-in-depth strategy to ensure that process operations remain within safe operating limits. A clear chain of command, defined authority, and accountability for reliable work performance in accordance with approved procedures and work practices should also be established. Outputs of the *operations* element can also be used to facilitate the performance of other elements. For example, monitoring equipment status will improve *asset integrity*, and near miss reports will enhance the effectiveness of the *incidents* element.

The ultimate product is the execution of each worker's tasks, from the boardroom to the shop floor, in a disciplined, consistent manner that safely delivers the goods and services required to meet the organization's objectives.

17.1.6 How Is It Done?

To develop an effective *operations* program, an organization must start with an honest statement of its objectives and risk tolerance. Considering the outputs of other elements, the organization can then formulate an *operations* policy and document it, along with the implementing procedures. However, the program cannot be merely words on paper. Workers must be trained in the policies and procedures so that they understand the goals and expectations, the lines of authority, and their personal accountability. They must apply good reasoning and judgment (founded upon a sound process safety culture) in all situations, but particularly when action is required in situations not specifically addressed by policy or procedure.

Beyond that, the most critical, ongoing requirement is that management lead by example. If a procedure instructs workers to shut down the process under defined emergency conditions, but management praises operators who "ride it out" and avoid a shutdown, then operational discipline will suffer. *Operations* tolerates no deviation from approved procedures, even if the outcome of a deviation is inconsequential or desirable. Thus, management must hold workers accountable for their actions in all circumstances to avoid the normalization of deviation.

17.2   KEY PRINCIPLES AND ESSENTIAL FEATURES

Safe operation and maintenance of facilities that manufacture, store, or otherwise use hazardous chemicals requires reliable human performance at all levels, from managers and engineers to operators and craftsmen. Experience shows that safe operations are directly and closely correlated with reliable operations, because the same human errors that lead to poor quality and low productivity also lead to safety and environmental incidents.

Conduct of operations helps ensure reliable performance by establishing observable standards of behavior and holding workers at every level of the organization accountable for their performance. In this context, it broadly includes the execution of tasks associated with operations, maintenance, safe work, and emergency planning and response, as well as the overall management of the process and its risks. It includes a defense-in-depth philosophy that acknowledges that even the best worker will occasionally err, so the system must be designed to detect and correct errors before a loss event occurs. When properly implemented, the *operations* program works with design, training, maintenance, and engineering to provide a series of diverse, robust barriers against losses.

The *operations* policy will typically state what activities are prohibited, what activities are permitted with no special controls, and what procedures govern all other activities. Its procedures attempt to document the work practices – those widely understood, but often unwritten, ways that work is accomplished at a facility. Formalizing those work practices in procedures allows them to be taught and enforced consistently throughout the organization.

The following key principles should be addressed when developing, evaluating, or improving any management system for conduct of operations:

- Maintain a dependable practice.
- Control operations activities.
- Control the status of systems and equipment.
- Develop required skills/behaviors.
- Monitor organizational performance.

The essential features for each principle are further described in Sections 17.2.1 through 17.2.5. Section 17.3 describes work activities that typically underpin the essential features related to these principles. Facility management should evaluate the risks and potential benefits that may be achieved as a result of improvements in this element. Based on this evaluation, management should develop a management system, or upgrade an existing management system, to address some or all of the essential features, and execute some or all of the work activities, depending on perceived risk and/or process hazards that it identifies. However, these steps will be ineffective if the accompanying *culture* element does not embrace the pursuit of excellence. Even the best management system, without the right culture, will not produce reliable operations.

17.2.1 Maintain a Dependable Practice

A documented *operations* program is fundamental to maintaining reliable worker performance. The procedures governing worker activities and interactions must be documented, workers must be trained, and their performance must be monitored on an ongoing basis. The management system must be designed to accomplish those objectives and to consistently provide positive feedback for desired behaviors over the life of the process. A good organization and effective administration establishes the framework for the *operations* activities to build upon.

Define roles and responsibilities. The written description of the *operations* program should identify the organization's objectives and the procedures that must be followed to achieve those objectives. Each procedure should identify those responsible and accountable for its execution, and clear lines of authority for making decisions should be established. In general, decisions should be made at an organizational level commensurate with the risk involved – the higher the risk, the higher the level of management authorization. In addition to intra-department lines of authority, the inter-department interfaces and lines of authority must also be defined; sometimes the roles, responsibilities, and interfaces may change as a project moves through its life cycle. For example, a process engineer may be much more heavily involved in the day-to-day operation of a pilot plant than of a production unit. Each department is typically responsible for ensuring that the work of its employees and contractors conforms to the program requirements. For those workers who fail to meet expectations, the owning department is typically responsible for any coaching or remedial training, and the human resources department typically participates in progressive discipline of those who will not meet or maintain minimum acceptable standards.

Establish standards for performance. Management should develop operating goals that support facility and organization goals. These goals are useful for measuring operating effectiveness, but their real value is in improving performance. Thus, the goals should be challenging, but realistically achievable and aligned with management's expectation that workers will follow procedures; complete required logs; maintain a clean, well-organized workplace; and so forth. All levels in the organization should provide input toward establishing these performance goals so they all feel some commitment toward achieving them. Ideally, each goal is directly under the control of the individual or group responsible for achieving it. For example, a goal to reduce the number of reportable spills by 20% is preferable to a goal to reduce environmental emissions by 20%, because the second goal may not be achievable even if operations were perfect. The goals should be incrementally measurable so the workers can see progress toward their achievement, and successes should be recognized and celebrated.

Validate program effectiveness. Daily observations, periodic auditing, and root cause incident investigation all provide insight into the effectiveness of the *operations* program. For example, if a worker engages in casual radio communications, co-workers should correct that behavior immediately. Auditing may discover that an entire work group is not actually following the methods prescribed in training, which could be due to a lack of knowledge (training deficiency) or lack of operational discipline (management deficiency). Each time an incident investigation identifies an issue related to the conduct of operations, such as inadequate management of shift turnover or poor teamwork, management should investigate whether the issue is (1) an isolated deficiency or (2) indicative of a systemic problem. Either case requires appropriate corrective action because a fully effective *operations* program should prevent even an isolated occurrence from escalating to a reportable incident.

17.2.2 Control Operations Activities

The control of operations activities is the heart of the *operations* element. The management system must establish clear expectations for every operations activity – from following procedures to controlling access. In particular, reliable communication between workers, shifts, and work groups helps to ensure that all operations activities are planned and controlled.

Follow written procedures. Written procedures should be established for both normal and abnormal situations. (See the *procedures* element described in Chapter 10 for more details.) A cornerstone of operational discipline is a requirement to follow the written procedures. However, strict adherence must not be blind adherence – the ideal is "thinking" compliance, where the worker executes the procedure as written unless doing so would create an adverse condition. In that case, the worker should stop and get permission to deviate from the standard procedure and initiate the process to change or update the procedure, as appropriate. When checklists are provided, the worker should have the checklist at the actual work location and check off tasks as they are completed. When independent verifications are specified, they must be performed – no matter how experienced or reliable the worker.

Follow safe work practices. Written procedures should be developed for maintenance and nonroutine activities. (See the *safe work* element described in Chapter 11 for more details.) Adherence to safe work practices is a fundamental requirement. Workers must be trained in the procedures, and their use must be enforced. If existing safe work practices cannot be followed, work must be stopped until an appropriate alternative method can be devised and approved.

Use qualified workers. Workers should be trained and qualified for the tasks they are expected to perform. (See the *training* element described in Chapter 14 for more details.) Workers must be closely supervised while undergoing on-the-job training. If an upset occurs, unrelated on-the-job training activities should be suspended so that the qualified operators can devote their full attention to the emergency.

Assign adequate resources. An understaffed plant is anathema to reliable operations. When workers are overloaded, many will begin to cut back on tasks that they perceive to be lower priority. Crosschecks may be skipped, operator rounds may be hurried, and paperwork might be incomplete. Chronic understaffing also results in worker fatigue, especially if excessive overtime is required to fill the gaps. Thus, management should define the maximum acceptable working hours per day and per period of consecutive days (logical blocks of time, usually based on shift rotation schedules and labor laws).

In companies striving for operational excellence, senior management must provide adequate resources throughout the organization. In addition to the operating staff, adequate technical, maintenance, and administrative resources must be provided to support them. The staff also needs adequate material resources, such as appropriate work areas to conduct facility activities, and reliable, capable equipment, such as computers, communication devices, diagnostic instruments, and so forth, with which to efficiently perform their jobs.

Formalize communications between workers. Everyone depends on communication to exchange information, and reliable communication is essential to reliable operation. The information must be transmitted and received accurately, completely, and in a timely manner, and it must be understood. For oral communications, (1) the sender should identify the recipient and himself, (2) the sender should transmit the message clearly and concisely, (3) the recipient should repeat the message back to the sender, (4) the sender should confirm or retransmit the message until it is repeated back correctly, (5) the recipient should report completion of the task or any difficulties encountered, and (6) the sender should acknowledge the report. Obviously, public address or paging systems should not be used for critical communications unless an independent means also exists to confirm that the message was received and understood.

Management should have a structured method for communicating guidance and short-term information to the operations staff. For example, operators may need to know the schedule of special activities during a plant shutdown, or they need to be aware of recent changes to policies, procedures, equipment status, production runs, and so forth. The system for issuing timely orders to operators should (1) define a standard format for orders, (2) identify who has authority to issue such orders, (3) distinguish between daily and ongoing (standing) orders, (4) describe how orders should be posted and acknowledged, and (5) describe how orders should be cancelled or superseded.

Urgent matters should be orally communicated to workers in a pre-job or pre-shift briefing. For less urgent matters, a prioritized list of documents should be provided for workers to review during their next shift, so everyone will get the same message, even if they are away from the plant for several days or weeks. This required reading can satisfy the *management of change* element's requirements to inform workers about some changes, but it should not be used as a substitute for training workers as required by the *training* and *management of change* elements.

Formalize communications between shifts. Many incidents have occurred because of a breakdown in communications between shifts. (See the description of the Piper Alpha accident in Chapter 11, Safe Work Practices.) Having workers maintain a written record of operations in a structured logbook is one of the most common and reliable ways to ensure the oncoming workers understand the status of all equipment and work activities when they take control. The logbook is also useful to orient workers arriving mid-shift, perhaps to share the workload in a crisis.

Logs should be established for each key operating position, and written guidance or a standard form for recording information should be developed. The logs must be recorded in a timely manner, and the entries must be legible. If an error is made, a standard method for corrections should be determined that does not obliterate the erroneous entry. At shift change, the outgoing shift should orally brief the incoming shift, in addition to noting its final entry and transferring the written logs to them. Supervisors should periodically review the logs for accuracy and completeness to emphasize their importance – both for inter-shift communication and for building a useful database of performance information.

Formalize communications between work groups. Written communication between work groups will help to minimize the potential for misunderstanding. This is particularly important during nonroutine activities such as startups, shutdowns, construction, and partial plant outages. Typical documents include purchase orders, batch sheets, work orders, and safe work permits. The facility should adopt (1) standard forms to help ensure that each work group gets all the information it needs and (2) specific contact points to help ensure that documents are properly routed. To handle abnormal situations, designated individuals should be appointed to handle emergency communications and to notify appropriate external work groups, authorities, or stakeholders as specified by the *emergency* element.

Adhere to safe operating limits and limiting conditions for operation. Management should establish an explicit policy requiring that (1) processes be operated within their safe operating limits and (2) limiting conditions for operation be enforced without exception. Safe operating limits are normally set for critical process parameters, such as temperature, pressure, level, flow, or concentration based on a combination of equipment design limits and the dynamics of the process. Limiting conditions for operation are specifications for critical systems that must be operational and critical resources that must be available to start the process or continue normal operation. Critical systems often include fire protection, flares, scrubbers, emergency cooling, and thermal oxidizers; critical resources normally involve staffing levels for operations, the fire brigade, and the emergency response team. In addition, workers must understand that any deviation outside of safe operating limits or otherwise unauthorized operation must be reported. These sorts of situations normally constitute a near miss incident and should be investigated, even if satisfactorily resolved.

Control access and occupancy. Management should establish clear rules governing access to operator work areas and process areas. Anyone desiring entry to a process area should first notify the responsible operator, state his intentions, and get the operator's permission before proceeding. This helps ensure that the entrant has the proper permits and protective equipment and will not interfere with or be endangered by other activities in the area. Visitors should also notify the operator when they leave the area. In the event of an emergency, this will help ensure that everyone is accounted for and that emergency responders are not unnecessarily endangered looking for someone who is not there. To minimize both safety and security risks, operators should maintain surveillance of the area and challenge anyone who is present without permission.

Access to control rooms and other operator work areas should be similarly controlled. The mere presence of unnecessary personnel interferes with the free movement of the operations staff, and the distraction is exacerbated by loud conversations, noises, and irrelevant questions. The visitors may cause operational upsets by using their radios near sensitive electronics or leaning/sitting on control panels if adequate seating is not available.

17.2.3 Control the Status of Systems and Equipment

Maintaining a keen awareness of the status of process systems and equipment at all times enables operators to perform their duties reliably. Thus, the human-machine interface should be designed and maintained in a manner that facilitates the collection of information. In addition, the administrative system should make it clear who is in control of the equipment at any given time and responsible for maintaining safe conditions.

Formalize equipment/asset ownership and access protocols. For each stage of a project's life cycle, some work group should be designated as its owner. For example, during construction, the contractor may be responsible for the equipment; during initial startup, engineering may be in control; and during production, the operations group is normally in control. Workers should accept total responsibility for maintaining their work area in the best possible operating condition.

The *operations* program should require two-way communication between the owner/user of the equipment and anyone else wanting access to the equipment both (1) before equipment is released and (2) when work is complete. Maintenance work is often controlled by *safe work* practices, and a similar permit system can be used to authorize and control all other access. Inspecting the job site and issuing a permit prior to authorizing workers to start their job activities provides an important safeguard against a variety of safety and operational incidents. However, sometimes minor adjustments or repairs to process equipment (e.g., tightening flanges or valve packing) or routine maintenance activities that are unlikely to affect the process (e.g., vibration monitoring) are allowed as long as the person responsible for the area has verbally authorized entry.

Requiring a representative of the owning group and a representative of the executing group to participate in any post-job testing and to jointly inspect the work area when the job is complete helps prevent incidents when the equipment is returned to service. In addition to inspecting the area, this *readiness* team is normally responsible for ensuring that valves are properly aligned, plugs and caps are reinstalled on vent and drain lines, rotating equipment is turning in the proper direction, sewers and drains are uncovered, tags and locks are removed, barricades have been removed, and so forth. For safety-critical equipment, such as fire protection systems or pressure relief systems, the *operations* program typically requires an independent verification that the system is properly aligned for service.

Monitor equipment status. Operators must maintain a current knowledge of the status of their equipment and support systems. In addition to monitoring the information available in the control room, operators should physically inspect their equipment on regular tours or rounds. Local readings confirm the accuracy of remote instrument readings and reveal incipient problems that require attention, such as abnormal configurations, odd noises, or unusual smells. Operators should record information on a standard checklist that clearly indicates the acceptable range for each parameter so that abnormalities can be spotted immediately. Supervisors should periodically review the tour reports for accuracy and completeness to emphasize their importance, to verify that the noted information is consistent with other data, and to ensure that abnormalities are resolved in a timely manner.

Workers should also look for opportunities to monitor the condition of equipment as part of normal operations or planned shutdowns. For example, when switching between two centrifugal pumps, operators should look for reverse rotation of the idle pump, which would indicate that the check valve is not properly seated. Or, rather than start the spare pump manually, operators could trip the primary pump and verify that the autostart controls work properly, assuming that operating procedures include the option (or a specific step) to opportunistically test the autostart function. These opportunistic test results should be documented as part of the *asset integrity* element.

Maintain good housekeeping. "Dirt" (grit, grime, dust, spillage, stains, rust, trash, etc.) is the enemy of reliable operations. Not only does it cause premature equipment failure, it masks incipient problems and constantly contradicts any management exhortations to excellence. To instill a sense of ownership and pride in the workforce, management should clearly state its housekeeping expectations and enforce them (Refs. 17.4 and 17.5). Working areas should be kept clean and organized, trash should be removed, damaged equipment should be repaired, paint and surface finishes should be maintained, abandoned equipment should be demolished, and so forth. Management should (1) audit the work areas periodically to ensure that housekeeping standards are being maintained and (2) provide the resources to correct any situations that are beyond the ability or authority of the operations staff.

Maintain labeling. Clear, accurate equipment labeling is essential to reliable operation. Management should adopt or develop facility-wide standards for labeling and color coding. Operators should be responsible for monitoring the condition of posted operator aids and labels on instruments and equipment in their work area. Management should audit the work areas periodically to ensure that labeling standards are being maintained and provide the resources to promptly repair or replace labels and operating aids as requested by operations.

Maintain lighting. Operators need well-lit work areas so they can identify equipment, read instruments, and see developing problems. Operators should assess lighting conditions in their work areas, and management should correct any deficiencies that are noted, such as dimness, shadows, glare, and so forth. Operators should monitor the condition of lighting elements as part of their normal rounds to ensure that burned out bulbs are replaced promptly.

Maintain instruments and tools. Operators rely on their installed and portable instruments to provide data, their tools to perform tasks, and their communication equipment to coordinate work activities. Facility management should develop a program, coordinated with the *asset integrity* element, to maintain the human-machine interface equipment in good working order. Instruments and analyzers should be calibrated on a regular basis and the last valid date should be evident on the device. Operators must be able to trust instrument indications until they are proven false, so if an operator notes any instrument that is damaged or operating erratically, a standard method should be in place to report it, remove it from service, and get it quickly repaired.

17.2.4 Develop Required Skills/Behaviors

The focus of the *operations* element is on maintaining reliable operations activities. Obviously, developing and maintaining a workforce with the necessary knowledge, skills, and abilities is central to achieving this objective. In addition, the process safety culture must continually reinforce desired behaviors, such as a questioning attitude and attention to detail.

Emphasize observation and attention to detail. Equipment and procedures rarely fail without warning. However, if those warning signs are not noted and corrected, the situation may deteriorate rapidly. Thus, it is vital that management stress the importance of (1) observation as operators conduct their normal activities and (2) reporting abnormalities so they can be investigated and corrected. On-the-job training should emphasize the proficient use of workers' basic senses to observe process and equipment conditions.

For equipment, the warning may be an odd noise, an unusual or conflicting instrument reading, or a drip seeping out from under insulation that could alert operators to an imminent failure. These conditions should be noted in operator logs and followed up with appropriate corrective action. For procedures, the warnings may be more subtle, because errors in their execution do not always produce an immediately observable condition. Thus, in addition to observing carefully, workers must be attentive to details such as inconsistencies in raw material numbers on a batch sheet versus the pallet delivered from the warehouse or inconsistencies in the written procedure versus their on-the-job training.

Keen observation also offers security benefits in addition to operational benefits. Should there be an attempt to sabotage process equipment, observant operators may be able initiate countermeasures and disrupt such plans.

Promote a questioning/learning attitude. The best operators conduct their work activities with technical inquisitiveness. They continually ask themselves questions about the performance and condition of the system and then answer the questions by using what they see, hear, and smell. When their senses indicate something abnormal during surveillance, they follow up and involve others as appropriate to resolve the discrepancy. When their senses indicate something abnormal during execution of a procedure, they stop until the observed condition is explained or corrected.

Management should foster a climate, as described in the *culture* element, in which workers are encouraged to ask questions. Encouraging a questioning attitude in every person involved with the process, including operators, maintenance employees, technicians, supervisors, engineers, managers, and contractors, will help ensure that abnormal conditions are identified and corrected before an incident occurs. Those who raise legitimate questions must be acknowledged and praised, even if their concerns prove unwarranted. The answers to worker questions should be shared with everyone, both to improve their diagnostic skills and to demonstrate management's interest in resolving worker concerns.

Train workers to recognize hazards. Hazards may arise in any aspect of operations. Even well trained, observant operators with well-defined safe operating limits may not recognize when an unusual circumstance or combination of activities may create hazardous conditions. Thus, management should train workers on how to recognize hazards, and how to recognize when unknown hazards may be present. One's ability to recognize hazards is a function of experience, training, knowledge, and having a sense of vulnerability. This is related to several other RBPS elements, including the *culture, competency, risk, training,* and *knowledge* elements. Proper execution of these elements, and an ongoing emphasis on learning from incidents, will help operators understand hazards and manage risk associated with nonroutine conditions or work activities.

Self-reliance is usually an asset, but overconfidence can lead workers to ignore warning signs of hazardous situations. Management must emphasize the value of teamwork in risk analysis and support those who request help from co-workers before authorizing or engaging in unusual activities.

Train workers to self-check and peer-check. In all situations, workers should take deliberate actions with an expectation of specific results. For example, when a process pump is started, the operator should observe a pressure rise at the discharge. If the expected result is not observed, the worker should stop and reassess the situation before proceeding. If the worker can identify the cause of the unexpected result, appropriate corrective action can be taken as defined by procedure; otherwise, the worker should stop until help from co-workers or supervisors can be obtained. When the worker is relying on the actions of others to put equipment in a specific state, the same approach is desirable. If possible, the worker should verify that the equipment is in the expected state or at least confirm that the co-worker completed the expected actions. If there is any doubt or inconsistency, the worker should stop until the anomaly is resolved.

Establish standards of conduct. Management should establish professional standards of conduct for all workers. Workers are expected to arrive at their workstations on schedule and work cooperatively with peers and other work groups. Workers must be honest in their recording of data and reliable in their dealings with others to foster an atmosphere of mutual trust and teamwork. Operators should not be distracted from their duties by non-work-related activities such as reading magazines, watching television, playing games, or surfing the internet. Where workstations are clustered, such as a control room, workers must be particularly mindful that their behavior does not distract others. Behaviors that disrupt the workplace, such as fighting, horseplay, or discrimination, cannot be tolerated, and constructive ways to resolve disputes between individuals or work groups must be developed. Workers should follow the defined chain of command when situations require decisions that exceed their individual authority.

17.2.5 Monitor Organizational Performance

Long-term excellence in performance cannot be achieved with fear and intimidation. The workers must perceive real rewards flowing from their achievements, and the intangible satisfactions, such as peer recognition, supervisor recognition, and team victory, are often more important than pure financial rewards. Thus, the objective of monitoring organizational performance is primarily to provide a gauge of progress and achievement, not to provide an excuse for punishment. Poor performance must be analyzed so that appropriate corrective actions can be implemented; however, poor performance is more often an indication of management system weaknesses than of worker failings.

Maintain accountability. Workers must understand what is expected of them and be held accountable for their actions. On the positive side, accountability means that a worker should be recognized, and perhaps rewarded, for exhibiting the desired behavior, particularly if doing the wrong thing would be so much easier. For example, dumping a batch is a difficult decision for an operator, even when the criteria to do so are clearly met. In such cases, management must overtly recognize and support the worker's decision to stay within safe operating limits despite the immediate costs.

On the negative side, accountability means that the organization's progressive discipline system must be invoked when training, coaching, and peer pressure fail to correct a worker's unacceptable behavior. The expedient approach is often to ignore such behavior because it may provoke the worker or strain labor relations. Unfortunately, this approach simply encourages others to behave similarly until the situation is intolerable. When fairly applied, workers and their representatives will support the progressive discipline system to correct or eliminate behavior that jeopardizes everyone's goals.

With all work activities, workers should be accountable for following procedures and permit conditions, regardless of the outcome. A worker who achieves record throughput by taking short cuts must be treated in the same way as a worker who causes a spill by taking a short cut. Both cases represent an unacceptable deviation from approved standards, and both workers must be held accountable.

The most successful accountability programs involve the workers as well as management. Peer pressure can be much more powerful than management edict. The workers also know what rewards they value, and they should have a voice in deciding who has earned recognition for outstanding performance.

Strive to continuously improve. The metrics and management reviews associated with this element should be used not only to correct performance deficiencies, but also to improve ongoing operations and support a reliability-minded culture. To accomplish this, managers must be knowledgeable about personnel performance, facility activities, and facility conditions within their area of responsibility. Managers should observe work activities and assess overall performance. Good performance should be promptly acknowledged and rewarded with positive feedback. Best practices should be shared with other workers so everyone can deliver superior performance. To improve performance during abnormal situations, workers should be challenged with "What-if" scenarios randomly, at opportune times, and the workers should walk through their response. Lessons learned from investigations (see Chapter 19, Incident Investigation) should be shared widely so that all workers can perform better in similar situations.

Maintain fitness for duty. Workers are expected to be both physically and mentally fit to perform their required duties. Management must be clear in its expectation that workers take primary responsibility for their own fitness – that they get adequate sleep and rest when they are off duty and that they avoid using substances that would adversely affect their performance. To encourage compliance, management should offer an employee assistance program that will help workers adopt healthier lifestyles, deal with personal issues, and recover from substance abuse. To enforce compliance, management should develop specific policies that address fitness for duty issues, such as limiting secondary employment and randomly testing for substance abuse. In addition, management must assign workers to only those tasks for which they have been trained and have demonstrated competency to perform, and management should limit demands for overtime or unusual work schedules that could degrade workers' performance.

Conduct field inspections. Frequently conducting unannounced inspections on all shifts is one way to help ensure that operations are conducted in accordance with best practices. Peer inspections are particularly effective, but supervisors and upper-level managers must also be visibly involved on a routine basis. Supervisors should ensure that rounds are conducted as scheduled and that logs and other paperwork are properly completed. Persons conducting routine safety and housekeeping audits should also make a point of checking ongoing maintenance work activities to ensure that (1) the proper permits are in place, (2) the permits specify appropriate conditions, and (3) workers are following procedures and conforming to permit requirements. The frequency of inspections should be based on perceived risk. For example, during normal operations a few spot checks each week may be sufficient; during turnarounds, unusual conditions, or high-risk activities, several inspections each day are probably warranted.

Correct deviations immediately. Another means to help prevent complacency and ensure effectiveness is management adherence to the same rules and values they espouse. Workers notice managers' actions, including actions that managers choose to not take. For example, if someone is not following a prescribed procedure and a manager walks by without stopping to correct the situation, this demonstrates to all witnesses that management's commitment to following proper procedures, and perhaps to overall safety, is not as strong as the facility claims it to be. Even if the manager was too preoccupied with a pressing issue to notice the situation this time, repeated indifference will undermine the *operations* program. All levels of the organization must be mindful of this trap, and everyone, whether employee or contractor, must take the time to stop, investigate, and correct unsafe actions or unsafe conditions whenever they are observed.

17.3 POSSIBLE WORK ACTIVITIES

The RBPS approach suggests that the degree of rigor designed into each work activity should be tailored to risk, tempered by resource considerations, and tuned to the facility's culture. Thus, the degree of rigor that should be applied to a particular work activity will vary for each facility, and likely will vary between units or process areas at a facility. Therefore, to develop a risk-based process safety management system, readers should perform the following steps:

1. Assess the risks at the facility, investigate the balance between the resource load for RBPS activities and available resources, and examine the facility's culture. This process is described in more detail in Section 2.2.
2. Estimate the potential benefits that may be achieved by addressing each of the key principles for this RBPS element. These principles are listed in Section 17.2.
3. Based on the results from steps 1 and 2, decide which essential features described in Sections 17.2.1 through 17.2.5 are necessary to properly manage risk.
4. For each essential feature that will be implemented, determine how it will be implemented and select the corresponding work activities described in this section. Note that this list of work activities cannot be comprehensive for all industries; readers will likely need to add work activities or modify some of the work activities listed in this section.
5. For each work activity that will be implemented, determine the level of rigor that will be required. Each work activity in this section is followed by two to five implementation options that describe an increasing degree of rigor. Note that work activities listed in this section are labeled with a number; implementation options are labeled with a letter.

Note: Regulatory requirements may specify that process safety management systems include certain features or work activities, or that a minimum level of detail be designed into specific work activities. Thus, the design and implementation of process safety management systems should be based on regulatory requirements as well as the guidance provided in this book.

17.3.1 Maintain a Dependable Practice

Define Roles and Responsibilities
1. Develop a written policy for managing the *operations* element.
    a. General guidance applies to all RBPS elements.

     b. Detailed guidance specifically addresses *operations* requirements.

     c. Detailed guidance addresses specific *operations* requirements for each job function.

2. Include specific roles and responsibilities in the management system governing the *operations* element.

     a. Roles and responsibilities are not written, but are generally understood.

     b. Responsibilities are assigned on an ad hoc basis and generally controlled by the work group.

     c. A written description of the requirements and lines of authority exists, but assigned responsibilities and enforcement are not included.

     d. A formal written document addresses the scope, roles, responsibilities, accountabilities, and enforcement of the program.

3. Develop procedures, permits, checklists, and other written standards governing the *operations* element.

     a. *Operations* procedures and practices are generally not written; they are generally based on standing practice at the facility.

     b. The *operations* procedures are documented and included in worker training.

     c. The *operations* procedures are documented, included in worker training, and occasionally revised by management.

     d. The *operations* procedures are documented, included in worker training, and workers regularly offer suggestions for improvement.

4. Train all employees and contractors in the *operations* element.

     a. Employees are expected to learn *operations* procedures from coworkers during on-the-job training.

     b. A written description of *operations* policies and procedures is provided to new employees during initial training.

     c. New employee and contractor training describes the *operations* policies and procedures, and trainees must demonstrate their understanding of it before being allowed to work independently.

     d. Item (c), and regular refresher training is required.

5. Describe the interfaces between the *operations* element and other RBPS elements, especially the *training, procedures, safe work, readiness*, and *asset integrity* elements.

     a. Information is provided if requested by other element owners.

     b. Interfaces are learned during on-the-job training.

     c. A written description of the interfaces exists, but assigned responsibilities for performing required tasks are not provided.

     d. A formal written program document exists that defines the interfaces, and its implementation is routinely monitored.

6. Develop special procedures to control operations for units that process extremely toxic or otherwise hazardous chemicals.
   a. Corporate policy addresses this issue, and key personnel at the facility are mindful of special hazards presented by some operations (e.g., sampling, catalyst addition).
   b. Procedures are based on regulatory and corporate requirements.
   c. The facility's program-level governing procedure provides a comprehensive set of specific requirements.

Establish Standards for Performance

7. Develop goals for the *operations* element.
   a. General goals are established without worker input.
   b. Specific goals are established with worker input.
   c. Specific goals are established with worker input. Goals are challenging, but achievable, and are under the workers' direct control.
8. Communicate progress toward operational goals.
   a. Communication significantly lags performance.
   b. Communication is timely and shows progress toward the goal.
   c. Communication is timely and shows progress toward the goal. Intermediate milestones are noted and celebrated when achieved.

Validate Program Effectiveness

9. Identify metrics by which operations effectiveness will be judged.
   a. Effectiveness is judged by managers.
   b. Effectiveness is judged by production metrics or other lagging indicators.
   c. Effectiveness is judged by personnel job results, such as lower error rates, better process stability, higher productivity, first-time right, and so forth.
10. Periodically evaluate whether workers understand the *operations* program, and provide refresher training as necessary.
    a. Job performance is monitored during initial training.
    b. Item (a), and job performance is reviewed after an incident.
    c. Performance is periodically monitored by supervisors.
    d. Performance is continuously monitored by supervisors and co-workers, and behavior deficiencies are corrected.
11. Review incident investigation results, and correct any root causes related to *operations* program deficiencies.

    a. Incident investigation results are occasionally reviewed with workers.

    b. Item (a), and operational discipline is improved to reduce human errors.

    c. Incident investigation results are regularly reviewed with workers. Both operational discipline and engineered systems are improved to provide defense-in-depth against human errors.

    d. Item (c), and the *operations* program is revised to address deficiencies identified as root causes of incidents.

17.3.2 Control Operations Activities

Follow Written Procedures

(See the list of work activities in Chapter 10, Operating Procedures.)

12. Define the expectations for a worker.

    a. No expectations are defined beyond the employee handbook.

    b. Workers are expected to follow procedures, but compliance is not rigorously enforced.

    c. Workers are expected to follow procedures and are held accountable for deviations, regardless of the outcome.

    d. Item (c), with peer checking and continuous improvement.

13. Develop and use checklists for critical operations and maintenance activities.

    a. Workers develop personal checklists as needed.

    b. Standard checklists are developed, but their use is generally not enforced.

    c. Standard checklists are developed, and their use is occasionally reviewed.

    d. Standard checklists are developed, and their proper use is independently audited.

14. Perform independent verifications properly.

    a. Independent verifications are performed by supervisors on an ad hoc basis.

    b. Verifications are required, but their performance is generally not enforced.

    c. Verifications are required, and their performance is enforced.

    d. Verifications are required, and safety-critical verifications are performed by an independent work group.

Follow Safe Work Practices

(See the list of work activities in Chapter 11, Safe Work Practices.)

Use Qualified Workers

(See the list of work activities in Chapter 14, Training and Performance Assurance.)

15. Coordinate on-the-job training activities with current operating conditions.

    a. On-the-job training is conducted based on a calendar-driven schedule.

    b. On-the-job training is coordinated with current conditions.

    c. On-the-job training is coordinated with current conditions and suspended if conditions warrant.

Assign Adequate Resources

16. Provide adequate operating staff.

    a. Mandatory overtime is required to cover routine job positions.

    b. Mandatory overtime is required to cover job positions during unusual situations (e.g., unit turnaround, construction).

    c. Staffing is occasionally supplemented with overtime.

    d. Staffing exceeds minimum requirements. Those not needed for daily operations use their time to receive or conduct training, review and update procedures, conduct peer observations, assist with special projects, and so forth.

17. Provide adequate support staff.

    a. The operating staff performs all job functions.

    b. The operating staff performs all job functions, but some centralized support staff members are available (engineering, maintenance planning, human resources).

    c. Support staff time is assigned to specific units or facilities.

    d. Support staff can specialize and become expert in specific areas.

18. Provide adequate facilities.

    a. Facilities are crowded.

    b. Facilities are adequate, but are considered temporary.

    c. Facilities are permanent and adaptable to different uses.

    d. Facilities are built to support specific missions.

19. Provide adequate equipment, for example, radios, meters, computers, vehicles, and so forth.

    a. Equipment is available, but must often be shared.

    b. Equipment is available, well maintained, and reasonably reliable.

    c. Item (b), and equipment is among the best available for its purpose.

    d. Item (c), and ample spares and backup systems are available.

Formalize Communications between Workers

20. Develop and use verbal communication protocols.

   a.   Communication protocols are developed by workers on an ad hoc basis.
   b.   Communication protocols are developed, but their use is not enforced.
   c.   Communication protocols are developed, and their use is loosely enforced.
   d.   Communication protocols are developed, and their proper use is enforced.
21. Issue timely orders to operators.
   a.   Orders are issued in an ad hoc manner.
   b.   Orders are communicated in a shift briefing.
   c.   Orders are verbally communicated in a shift briefing; written orders are posted and maintained.
   d.   Item (c), and a prioritized required reading list is maintained for each operator.

Formalize Communications between Shifts

22. Develop and use protocols for shift changes.
   a.   Shift-to-shift turnovers are happenstance.
   b.   Written logs are transferred between shifts.
   c.   Written logs are signed off, transferred, and reviewed in a pre-shift briefing.
   d.   Written logs are kept for each job position in a standard format. Logs are transferred and reviewed in a pre-shift briefing.

Formalize Communications between Work Groups

23. Develop and use protocols for communications between work groups.
   a.   Communications are ad hoc.
   b.   Communications are written on standard forms.
   c.   Communications are written on standard forms that are accounted for when completed or closed out.
   d.   Item (c), with specific accountability for execution and closure.

Adhere to Safe Operating Limits and Limiting Conditions for Operation

24. Require adherence to safe operating limits and limiting conditions for operation.
   a.   Safe operating limits and limiting conditions are not defined; operators must determine when to shut down a process based on judgment and experience.
   b.   Procedures include safe operating limits, and operators are trained to shut down the process if they feel they cannot control the situation.

    c.   Procedures specify safe operating limits and actions to take if limits are exceeded.

    d.   Item (c), and limiting conditions for operation, such as critical support systems and resources, are also specified. Operators are empowered to shut down the process any time they feel the situation is unsafe. Exceeding a safety limit or violating a limiting condition is considered a near miss and is investigated.

25. Establish operating procedures for abnormal situations.

    a.   Operators use their best judgment in abnormal situations.

    b.   Procedures specify a generic set of actions for any abnormal situation.

    c.   Procedures specify steps for responding to defined abnormal situations.

    d.   Item (c), and additional resources are immediately available to help diagnose and control the situation.

26. Ensure that those authorizing deviations from standard procedures are well aware of the risks and have a sense of vulnerability.

    a.   Those authorizing deviations rely on their experience.

    b.   General hazards identification training is provided for those authorizing deviations.

    c.   Authorizers generally participate in risk analyses, and they have a sound appreciation of how to stay within safe operating limits during most routine and nonroutine operations.

    d.   Item (c), and lessons learned from other facilities are routinely reviewed. Engineering is involved in analyzing novel situations before authorization is granted.

27. Ensure that persons authorized to approve abnormal operations have the (1) training and experience to understand a wide range of hazards and (2) knowledge of well-established methods to manage the risk associated with the hazards.

    a.   Deviations can be authorized by any supervisor or manager, and this responsibility can be delegated within a department.

    b.   Deviations can be authorized by any supervisor or manager, with the concurrence of the area safety representative or the shift representative on the central safety committee.

    c.   Deviations can be authorized only by experienced personnel designated by management who have completed specified training (and sometimes refresher training) requirements.

Control Access and Occupancy

28. Establish a system to control access to central control rooms.

    a.   The central control room is a gathering area for everyone authorized to enter the facility.

    b. The central control room is a gathering area for the operations staff.

    c. The central control room is segregated from other administrative areas. Workers only enter the control room on official business with the operators.

    d. Item (c), and measures are in place to limit noise and distractions.

29. Establish a system to control access to process areas by nonoperations employees, including a means to ensure verbal communication between the operators and nonroutine employees who need to enter the controlled area.

    a. Signage on doors or walkways leading to process areas say, "Authorized Personnel Only", but employees generally consider themselves authorized if they contact an operator.

    b. Contractors must check in with the control room prior to entering a process area; visitors are not allowed in process areas unless an employee escorts them.

    c. Procedures require all nonoperations personnel (e.g., managers, maintenance, technical, laboratory, and other support personnel) to notify operators prior to entering a process area, but additional communication is not required.

    d. All nonoperations personnel must comply with a procedure that (1) requires approval from operations prior to entering a process area, (2) documents the scope of the activity and the personnel involved, and (3) requires notification of operations when the entry is concluded. The area operator should maintain periodic contact with, or oversight of, the workers and their activities.

30. Control access to areas of the facility where special hazards exist.

    a. At the discretion of a supervisor or project engineer, particularly hazardous areas may be barricaded with caution tape.

    b. Particularly hazardous areas are barricaded and the hazard is identified. Workers rarely cross these boundaries.

    c. Item (b), and anyone entering the barricaded area must notify operations before entry.

    d. Item (c), and hard barricades (e.g., temporary metal walls or fences) are used to seal off particularly hazardous or sensitive areas.

17.3.3 Control the Status of Systems and Equipment

Formalize Equipment/Asset Ownership and Access Protocols

31. Develop procedures and/or permits to control work activities.

    a. Control of operating equipment is not formalized. Supervisors and technical staff may adjust settings and alignments at will. Maintenance that can be performed without requiring shutdown, such as instrument calibration, lubrication, or tightening flanges or packing, is performed whenever personnel are available.

  b. Any work by the maintenance staff or contractors requires a permit. Supervisors and technical staff may adjust settings and alignments at will.

  c. All work activities are under the direct control of the responsible operator. Anyone else wanting to work on or near the equipment must get a permit in accordance with the *safe work* element or receive specific authorization from the operator.

Monitor Equipment Status

32. Develop specific inspection logs for operator rounds.

  a. Operators note any abnormal equipment conditions on an unstructured log sheet.

  b. General log sheets are developed that list the type of equipment to be inspected.

  c. Specific log sheets are developed for surveillance rounds. Different logs are used for different types of inspections (e.g., routine patrols, daily inspections, weekly inspections).

  d. Item (c), and log sheets specify acceptable ranges for parameters.

33. Check local equipment conditions frequently.

  a. Operators inspect equipment if they suspect a problem.

  b. Operators inspect equipment at the beginning of each shift.

  c. Operators inspect equipment periodically and record local readings on a standard log sheet. Supervisors check logs occasionally.

  d. Item (c), and recorded readings are crosschecked against panel instruments. Supervisors check logs regularly.

34. Develop a list of tests and testing procedures that operators should exercise when the opportunity arises.

  a. Operators note the "test" results when equipment fails, for example, whether the alarm sounded when the compressor tripped.

  b. Operators devise their own tests for troublesome equipment.

  c. Operators have a list of tests and testing procedures that they exercise when the opportunity arises. Work orders are submitted if tests fail.

  d. Item (c), and all tests are recorded and submitted to the reliability engineer.

Maintain Good Housekeeping

35. State housekeeping expectations and enforce them.

  a. Trash is routinely removed, and major spills are cleaned up.

    b.   Operators keep work areas generally clean of trash and spillage, but equipment surfaces are visibly stained or deteriorated.

    c.   Individual operators are responsible for housekeeping in their work areas.   Management considers housekeeping in performance reviews.

    d.   Management has implemented a formal program for housekeeping, provides resources, and monitors ongoing compliance.

Maintain Labeling

36. Develop standards for labeling and color-coding equipment and posted operator aids.

    a.   Operators label equipment using facility terminology.

    b.   Labels conform to applicable codes, but no company standards are developed.

    c.   Company standards for labeling are developed, considering applicable codes, local language(s), and expectations.

37. Apply consistent labeling and color-coding to all equipment.

    a.   Labeling is left to the discretion of the vendor or project team.

    b.   All major equipment is labeled, but the labeling is not consistent between areas and operators add handwritten labels as they see fit.

    c.   All equipment referenced in procedures is clearly and consistently labeled.

    d.   Item (c), and appropriate labels are part of the specifications for purchased equipment packages.

38. Apply special labeling and color-coding to all safety-critical equipment.

    a.   All items use similar labeling.

    b.   Safety-critical items are specially labeled, but the labeling is not consistent between areas.

    c.   All safety-critical equipment is clearly and consistently labeled.

    d.   Item (c), and these labels are audited as part of annual emergency procedure reviews.

Maintain Lighting

39. Operating areas should be well lit, and lighting failures should be promptly corrected.

    a.   Working areas are generally lit, but some areas have poor lighting.

    b.   Working areas are well lit, but burned out lights are not promptly replaced.

    c.   Working areas are well lit, and burned out lights are promptly replaced when operators report them.

    d.   Item (c), and operators can adjust lighting levels in key areas, such as control rooms.

Maintain Instruments and Tools

(See the list of work activities in Chapter 12, Asset Integrity and Reliability.)

  40. Develop a program for ensuring that instruments and tools are kept in good working order.

    a.   Equipment is repaired if operators report a problem.

    b.   Equipment is inspected on a preventive maintenance schedule and repaired as needed.

    c.   Item (b), and operators inspect equipment before each use and report deficiencies for repair.

    d.   Item (c), and a formal system is in place for tagging out of service equipment. Supervisors check equipment status regularly and expedite repair or replacement.

17.3.4 Develop Required Skills/Behaviors

Emphasize Observation and Attention to Detail

  41. Encourage workers to notice and report abnormalities in processes and equipment.

    a.   Workers report major problems and equipment failures.

    b.   Workers note any problems requiring maintenance and try to avoid unplanned shutdowns.

    c.   Workers note conditions adverse to safety in their logs and corrective actions are promptly initiated.

    d.   Workers note all discrepancies in their logs and corrective actions are promptly initiated.

Promote a Questioning/Learning Attitude

  42. Establish and promote an environment that welcomes questions regarding the safety of all aspects of the operation, including nonroutine activities, even if the activities are planned and executed by experts.

    a.   A questioning attitude is recognized as a positive attribute, but only when a valid question is posed.

    b.   The facility's culture promotes a "safety first" approach, and all employees are encouraged to question the basis for safety of all operations.

    c.   The facility's culture promotes a "safety first" approach, and workers are empowered to shut down any operation that they believe to be unsafe until their concerns are addressed.

43. Establish and promote an environment that encourages workers to develop a thorough understanding of their process.

    a.   Worker training includes basic information on process dynamics.

    b.   A questioning attitude is recognized as a positive attribute, but experts tend to limit the answers to just what they think the workers need to know.

    c.   The facility's culture promotes a questioning attitude and recognizes workers who develop good troubleshooting skills.

    d.   The facility's culture promotes a questioning attitude and recognizes workers who develop good troubleshooting skills. Workers routinely challenge each other to explain the conditions they observe and develop optimum response strategies.

Train Workers to Recognize Hazards

44. Train workers on how to recognize hazards, and how to recognize when unknown hazards may be present.

    a.   Workers receive general training on hazard identification.

    b.   Workers receive specific training in risk analysis.

    c.   Workers receive specific training in risk analysis applied to their areas and ongoing training in lessons learned from incidents.

45. Train workers when to involve others in risk analyses.

    a.   Workers rely on their own judgment of risks.

    b.   Workers involve managers if they perceive unusually high risks.

    c.   Workers routinely involve managers and technical experts in risk analyses as long as the job is not delayed.

    d.   Workers routinely involve managers and technical experts in risk analyses until they are fully satisfied that they understand the hazards and can manage the risks.

Train Workers to Self-Check and Peer-Check

46. Encourage workers to act deliberately and stop if conditions do not match their expectations.

    a.   Workers execute tasks in conformance with procedures and training.

    b.   Workers execute tasks in conformance with procedures and stop if a procedural condition cannot be met.

    c.   Item (b), and workers confirm the expected response before executing the next step.

    d.   Item (c), and workers confirm that they have correctly identified the specific component before taking action.

Establish Standards of Conduct

47. Define management expectations for worker behavior consistent with reliable performance.
    a. Only prohibited behaviors are defined, such as fighting, horseplay, or substance abuse, and violators are punished.
    b. Item (a), with training in interpersonal relations (e.g., diversity training, conflict resolution).
    c. Item (b), with peer observation and coaching to maintain standards.
    d. Item (c), and participation extends to contractors.

48. Encourage open communication.
    a. Worker input can be offered through a suggestion box.
    b. Workers can bring any suggestion or concern to their manager for resolution.
    c. Item (b), and alternate communication channels have been established if workers feel an issue needs the attention of upper management.

49. Encourage teamwork.
    a. Each work group/shift is independent; however, they cooperate with official requests to collaborate on specific projects.
    b. All work groups/shifts recognize that their goal is to support reliable, efficient operations, and they align themselves to support that goal.
    c. Item (b), and the workers are rewarded for cooperation.

17.3.5 Monitor Organizational Performance

Maintain Accountability

50. Hold workers accountable for their performance.
    a. Workers are punished if their behavior causes a loss.
    b. Workers are coached or disciplined if their behavior causes a near miss.
    c. Workers are coached or disciplined if their behavior deviates from approved procedures and practices, even if the outcome is good.
    d. Item (c), and the lessons learned are shared throughout the organization.

51. Reward workers for good performance.
    a. Workers' continued paychecks are their reward for good performance.
    b. Token rewards are offered to a few individuals.
    c. Significant recognition and financial rewards are offered to individuals and teams when goals are met.
    d. Item (c), and the workers help choose the award recipients.

Strive to Continuously Improve

52. Define performance metrics and progress toward improvement goals.

    a. Performance metrics are defined, but management is primarily interested in negative trends.
    b. Performance metrics are defined, and management rewards positive accomplishments.
    c. Item (b), and management solicits worker ideas for further improvement.

Maintain Fitness for Duty

53. Ensure that workers are both physically and mentally fit to perform their required duties.

    a. Supervisors observe the fitness for duty of their work group.
    b. Workers must pass regular physicals and are subject to random testing for substance abuse.
    c. Item (b), and the company has an employee assistance program.
    d. Item (c), and work hours and schedules have strict limits.

Conduct Field Inspections

54. Periodically audit work practices in the field to verify that they are consistent with training.

    a. Field inspections are performed ad hoc.
    b. Field inspections are performed after incidents.
    c. Field inspections are performed occasionally on day shift.
    d. Field inspections are performed on all shifts, routinely and after significant changes.

55. Establish a system to routinely inspect work areas to determine if (1) best practices are being followed, (2) abnormal activities are controlled by appropriate permits, and (3) good housekeeping is being maintained.

    a. Inspections are left to the discretion of the supervisor for the area where the work is taking place.
    b. Supervisors, are required to periodically conduct field inspections, and the inspection protocol includes checking on nonroutine work to ensure that required permits are in place.
    c. All managers, supervisors, engineers, and other technical staff members are required to conduct field inspections on a defined schedule, and the results are reviewed in shift meetings.
    d. Item (c) using a well-defined method, and the results are trended and used to help identify where modifications are needed.

56. Review completed logs and reports, and based on the results of the review, take steps to improve their accuracy and completeness.

a. Written logs are kept if work activities span multiple shifts.
b. Written logs are kept.
c. Written logs are kept and regularly reviewed for accuracy and completeness.
d. All written logs, permits, work orders, and so forth are regularly reviewed for accuracy and completeness. If discrepancies are noted, the worker is coached on how to properly complete required documents.

Correct Deviations Immediately

57. Correct deviations from practices and procedures whenever they are noticed.
a. Management only corrects non-performance.
b. Management corrects only those deviations that present a safety hazard.
c. Management corrects departures from practices and procedures whenever observed.
d. Management and co-workers correct departures from practices and procedures by employees or contractors whenever observed.

17.4   EXAMPLES OF WAYS TO IMPROVE EFFECTIVENESS

This section provides specific examples of industry tested methods for improving the effectiveness of work activities related to the *operations* element. The examples are sorted by the key principles that were first introduced in Section 17.2. The examples fall into two categories:

1. Methods for improving the performance of activities that support this element.
2. Methods for improving the efficiency of activities that support this element by maintaining the necessary level of performance using fewer resources.

These examples were obtained from the results of industry practice surveys, workshops, and CCPS member-company input. Readers desiring to improve their management systems and work activities related to this element should examine these ideas, evaluate current management system and work activity performance and efficiency, and then select and implement enhancements using the risk-based principles described in Section 2.1.

17.4.1 Maintain a Dependable Practice

Use gain sharing to encourage superior performance. Individuals more readily embrace organizational goals when they see direct benefit to themselves. Rewarding behavior that leads to improvements in safety, environmental, and operational performance is one of the surest ways to motivate adoption of and adherence to *operations* principles.

17.4.2 Control Operations Activities

Provide workers with portable electronic checklists and procedures. Electronic versions of procedures and checklists can be easily loaded onto personal digital assistants (PDAs). If field instruments and components are labeled with bar codes, the worker can use the PDA to confirm that the correct device or material is selected and can directly enter readings and notes. If a operator has any questions, she can call up procedural details and diagrams. Once the surveillance round or procedure is complete, the operator can download the data to the central computer for recordkeeping, trending, comparison with remote instrument readings, work orders, notifications, and so forth.

Use a computer-based required reading list. The need to inform workers of changing conditions, procedures, and policies and to provide information that will help workers do their jobs better is ongoing. However, the flow of information does not stop when an individual worker is away from the facility. Electronic systems are far more reliable than paper-based systems for managing document inventories, providing documents on demand, and maintaining records of when documents were looked at by specific individuals.

Have supervisors conduct pre-job briefs. Anyone performing nonroutine activities should be informed of their work assignment in the context of work conditions at the facility. A responsible person should describe the purpose of the work; describe any special considerations; provide a contact point in case of question, abnormality, or emergency; and inspect the work area to ensure that actual conditions match expectations.

Have supervisors conduct job-planning briefs. One way to ensure reliable operation is to eliminate surprises. Supervisors should frequently communicate the current status of planned and ongoing work to the operating staff. Supervisors should also conduct regular planning meetings to coordinate any upcoming work or special activities that will take place that day, during the next week, or over the coming month.

Carefully schedule high hazard work to minimize risk. Some days are more hectic than others; high hazard work should be scheduled when workers can devote extra attention to detail. To minimize risk, schedule high hazard work when workers are most alert, when other activities will be least distracting, and when other work groups can provide the best support.

Designate a shift technical advisor. Operators are not expected to have the educational background of an engineer or chemist, yet circumstances occur when a thorough technical understanding of the process is vital to managing unusual or abnormal situations. Having a designated technical advisor on call for each shift helps ensure that operational decisions are based on solid technical information.

Designate a building/area manager. Sometimes the activities in one part of a building or area may have safety implications in another location. For example, an operator may be unaware of planned work on the building fire sprinkler when approving a hot work permit in the unit. Having a designated facility manager who is responsible for every activity in a specific area will help ensure that unexpected interactions do not occur. This manager should control access to the facility, approve all facility maintenance or modification, and coordinate all work activities in the facility.

Define limiting conditions for operation. In addition to the safe process operating limits documented in the procedures, other critical resources and systems are usually necessary for a safe startup or continued normal operations. In aviation, for example, a plane may be safely flown with one failed engine, but the required minimum equipment list demands that all engines be operating before takeoff. Similarly, management should identify critical systems, such as the waste incinerator or emergency shortstop system, which must be operable before startup and if they fail during operation, require immediate compensatory action, such as reducing throughput or shutting down. Some resources, such as a fire brigade, must also be available. Collectively, these make up the limiting conditions for operation that must be satisfied to start the process or continue normal operation.

17.4.3 Control the Status of Systems and Equipment

Implement a formal program for housekeeping. Formal systems for organizing plant assets and maintaining good housekeeping have shown to improve both productivity and safety. Workers should be responsible for their work areas, and management must provide the resources to sustain the program.

Install and maintain clear labels, signs, and operator aids. Workers should be able to walk into any area of the facility and identify what hazardous materials are present, as well as identify specific pieces of equipment. Workers are often oblivious to deteriorating, missing, obsolete, or misleading labeling in familiar areas; therefore, on a regular basis, management should have workers inspect areas outside their normal duty station for labeling deficiencies.

17.4.4 Develop Required Skills/Behaviors

Randomly challenge workers' diagnostic skills. Having workers regularly think about unusual situations helps to ensure reliable performance when those situations arise. Management should challenge workers with "What-if?" scenarios on a regular basis and reward those who identify the best way to handle the situation. Scenarios should be realistic and can be derived from actual situations in the facility, lessons learned from incidents, and hypothetical scenarios that could challenge the plant's safe operating limits. Afterwards, best solutions can be published as part of workers' required reading, along with recognition or reward for those individuals or teams who performed well.

Expand work observations to include peers. Too often, enforcement of best practices fails because it adds one more burden on supervisors and technical personnel who are already working very long hours. One alternative is to increase worker involvement by providing time and incentive for workers to observe and coach peers (see Chapter 6, Workforce Involvement). Obviously, to work well, the program must be properly structured so that it focuses more on recognizing workers doing their job well than on catching a worker making a mistake. However, with time, workers will recognize that everyone must perform their jobs well to achieve the team goals and rewards for all.

Train supervisors to recognize workers with diminished capacity. Workers with diminished capacity, perhaps due to illness, emotional stress, or substance abuse, can be quite adept at covering their weakness. However, an astute supervisor can be trained to notice certain telltale signs. Management should provide supervisors with training that will help them detect when workers may not be capable of performing their required duties as a result of such problems.

Vary work assignments. When workers do the same job every day, they may become complacent and inattentive. By varying work assignments, management cultivates a more flexible workforce and helps maintain worker interest in the job. In particular, rotating workers between control room and field assignments helps ensure that the board operators maintain their knowledge of field conditions and the work activities they must coordinate remotely when they are on the board.

Allow work teams flexibility in work assignments. When a worker recognizes that he may not be able to work at peak performance, for example, because he is taking cold medication, the worker should have the opportunity to coordinate with others on the work team to take less demanding assignments that day. Management should encourage such self-reporting and coordination and not intervene unless an individual is abusing the system to avoid difficult or unpleasant work duties.

17.4.5 Monitor Organizational Performance

Reward workers who spot known defects. People are poor observers of things that do not happen. Thus, in high reliability systems, workers are often less attentive precisely because the system tends to run quite well. To encourage careful ongoing observations, managers can introduce observable (but harmless) abnormalities into the system, such as a small oil puddle near a pump, and reward those workers who observe and report or correct the condition.

Take a daily snapshot of permitted work. Most nonroutine work is planned one or more days ahead of time. Some nonroutine work places a strain on the operations group (e.g., they are often tasked to closely monitor contractors), but the increase in operational risk resulting from this strain is typically tolerable. However, overburdening the system with nonroutine work is altogether possible. If a number of nonroutine jobs are proceeding concurrently, the risk increase could be significant. This is particularly true if other, unrelated factors are simultaneously placing a strain on the organization; for example, the field operator is on vacation, the loader is covering for the field operator, and the senior operator is off sick. Under these conditions, the risk associated with otherwise typical nonroutine work activities may become unacceptable. By reviewing all of the planned and unplanned work for each day, along with the staffing, training, and experience of workers in both the production and maintenance departments, managers can at least be aware of the intensity of higher risk and presumably make better decisions regarding what activities are acceptable given the current situation.

Provide an anonymous employee-reporting program. Workers are sometimes reluctant to report operational issues for fear of reprisal from managers or co-workers. Management should provide an independent communication channel through which workers can anonymously report any concerns. Depending on the organization's culture, a third party may be required to assure the workers that their anonymity will be maintained.

Provide a process safety observer. During the hectic hours of a process startup, the operating crew can easily get caught up in resolving one issue and lose track of others. An independent observer whose primary responsibility is avoiding a process incident can provide an objective, safety-oriented perspective during complex startup and shutdown activities. This observer can also help ensure effective crew resource management so that some workers can maintain vital functions while others troubleshoot problem areas.

17.5 ELEMENT METRICS

Chapter 20 describes how metrics can be used to improve performance and when they may be appropriate. This section includes several examples of metrics that could be used to monitor the health of the *operations* element, sorted by the key principles that were first introduced in Section 17.2.

In addition to identifying high value metrics, readers will need to determine how to best measure each metric they choose to track. In some cases, an ordinal number provides the needed information, for example, total number of workers. Other cases, such as average years of experience, require that two or more attributes be indexed to provide meaningful information. Still other metrics may need to be tracked as a rate, as in the case of employee turnover. Sometimes, the rate of change may be of greatest interest. Since every situation is different, the reader will need to determine how to track and present data to most efficiently monitor the health of RBPS management systems at their facility.

17.5.1 Maintain a Dependable Practice

- *Number of incidents with operations issues as root causes.* An upward trend may indicate deficiencies in the operations element.
- *Number of qualified personnel in defined roles.* Decreases in these numbers below minimum staffing requirements may indicate imminent breakdowns due to lack of resources.
- *Progress toward performance goals.* A lack of progress may indicate that the *operations* program is not being implemented effectively, that aging equipment is not being maintained, or that required management support is lacking.
- *Staff turnover rates.* An increase may indicate that workers are disillusioned and cannot meet expectations with the resources provided. Even if the rate is low, a plan for filling key job roles in a timely manner should be established.

17.5.2 Control Operations Activities

- *The incidence of shortcuts identified by near misses and incidents.* Increases in these numbers may indicate failure to enforce best practices, overly aggressive performance goals, inadequate staffing, or rewards favoring results over behavior.
- *Number of incidents during which safe operating limits were exceeded.* An upward trend may indicate inadequate safety margins or poor control of abnormal conditions.
- *Number of human-machine deficiency and near miss reports.* A declining number may indicate that management has resolved many error-likely situations or that workers are frustrated that reported conditions are not corrected and have begun reporting fewer instances.
- *Number of incidents attributed to trainees.* An upward trend may indicate poor control of on-the-job training activities.

- *Number of visitors to the control room.* A high number may correspond to frequent distracting conditions.
- *Number of labor hours per unit of product.* A declining number may indicate improving organizational efficiency or inadequate staffing.

17.5.3 Control the Status of Systems and Equipment

- *Number of nuisance and always-on alarms.* An increase may indicate excessive operator workloads or distraction, poorly defined operating limits, or inadequate resources to repair faulty devices.
- *Number of nonroutine and emergency maintenance work orders.* An increase may indicate that the operators are not identifying incipient conditions or that routine maintenance work is being neglected.
- *Number of audit findings related to inoperable instruments and tools.* Increases may indicate that operators are not attentive to the human-machine interface or that management is not providing resources to promptly repair deficiencies.
- *Number of incomplete shift logs or reports.* An increase may indicate excessive operator workloads or distractions.
- *Number of missed surveillance rounds.* An increase may indicate excessive operator workloads or lack of management attention.
- *Number of work orders attributed to equipment abuse.* An increase may indicate that the equipment is poorly designed or maintained or that operators are not following best practices.
- *Average time to complete repairs on the human-machine interface.* An increase may indicate inadequate resources or an inability to find spare parts for obsolete equipment.
- *Number of housekeeping audits and their scores.* Decreases in the number of audits and declining audit scores may indicate declining interest of management and workers; a stable or increasing number of audits and stable or improving scores normally indicates that good housekeeping is ingrained as part of *operations* practices.
- *Average time to resolve off-normal findings.* An increase may indicate inadequate resources or an increasing tolerance for process deviations.
- *Number of inspection deficiencies related to labeling.* Increases may indicate that labeling is not being maintained on an ongoing basis.
- *Number of inspection deficiencies related to lighting.* Increases might indicate that lighting is not being maintained on an ongoing basis.
- *Number of access permits issued.* An increase may indicate excessive operator workloads or distraction, even if the increase is to the result of planned activities, such as an upcoming turnaround; a decrease may indicate that workers are not bothering to obtain a permit before working on process equipment.

17.5.4 Develop Required Skills/Behaviors

- *Number of incidents caused by a lack of self-checking or peer-checking.* An upward trend may indicate poor attention to detail.
- *Number of times workers are challenged to solve "what-if" scenarios.* Low numbers or a decreasing trend may indicate a lack of management interest in improving worker skills.
- *Percentage of overtime hours.* An upward trend or unexpected spike may indicate an imminent breakdown in fitness for duty.
- *Absenteeism.* An upward trend or unexpected spike may indicate serious deterioration in fitness for duty.
- *Average time required to complete required reading.* An upward trend may indicate that workers are not maintaining current awareness of plant conditions and may indicate inadequate staffing.
- *Number of incidents involving disruptive personal behavior.* An upward trend may indicate poor control of the work environment or failure to enforce the company's progressive discipline policy.

17.5.5 Monitor Organizational Performance

- *Number of manager inspections of work locations.* Low numbers or a decreasing trend may indicate a lack of visible management support.
- *Number of unplanned shutdowns.* An increasing number may indicate that the *operations* program is failing or that equipment is wearing out.
- *Number of unplanned safety system activations for valid reasons.* An increasing number may indicate that the *operations* program is failing, for example, operators are not noticing and correcting incipient problems, or that the program is being widely embraced – operators are more willing to trip the process in response to an upset. A decreasing number may also indicate that the *operations* program is failing, for example, operators are bypassing critical interlocks, or that it is succeeding – operators are noticing and correcting incipient problems.
- *Number of unplanned safety system activations for invalid reasons.* An increasing number may indicate that the *operations* program is failing; for example, workers are not careful when testing or working near safety systems or workers are using safety systems for non-safety reasons. An increase could also mean that the *asset integrity* program is failing; for example, equipment malfunctions are triggering spurious alarms.
- *Frequency of communication of progress toward goals.* A declining communication frequency may indicate declining management interest or a reluctance to confront bad news. Declines also correlate with lower worker morale because they see no progress as a result of their efforts.

- *Percentage of manager inspections delegated to subordinates.* An increasing trend may indicate that management gives operational discipline a low priority.
- *Number of disciplinary actions.* A decreasing number may indicate that *operations* initiatives are effective or that management is tolerating poor performers. High numbers may indicate inappropriate reliance on individual attention and administrative controls instead of eliminating error-likely situations.
- *Percentage of workers failing random substance abuse tests.* An upward trend or unexpected spike may indicate serious deterioration in fitness for duty.

17.6 MANAGEMENT REVIEW

The overall design and conduct of management reviews is described in Chapter 22. However, many specific questions/discussion topics exist that management may want to check periodically to ensure that the management system for the *operations* element is working properly. In particular, management must first seek to understand whether the system being reviewed is producing the desired results. If the organization's *operations* performance is less than satisfactory, or it is not improving as a result of management system changes, then management should identify possible corrective actions and pursue them. Possibly, the organization is not working on the correct activities, or the organization is not doing the necessary activities well. Even if the results are satisfactory, management reviews can help determine if resources are being used wisely – are there tasks that could be done more efficiently or tasks that should not be done at all? Management can combine metrics listed in the previous section with personal observations, direct questioning, audit results, and feedback on various topics to help answer these questions. Activities and topics for discussion include the following:

- Discuss roles and with workers to verify their understanding of their responsibilities and the lines of authority.
- Discuss possible upsets and incidents with workers to verify their understanding of notification responsibilities.
- Discuss performance goals and current plant performance with operators to verify their understanding.
- Verify that current practices match policies and expectations, for example:
  o Radio traffic is free of idle chatter and nonstandard communication language.
  o Shift turnover logs are being kept contemporaneously and transferred in an organized manner.

- o Car seals, drain plugs, and hatch covers are in place.
- o Pressure between rupture disks and relief valves is being checked routinely.
- o Work areas are free of distractions or unauthorized entertainment devices.
- Review the number of overtime hours worked by individuals and departments to determine if adequate resources are being provided to perform necessary tasks.
- Discuss working conditions to determine whether shortcuts are necessary to get the job done in time.
- Determine whether required crosschecks are actually being done or are simply signed off.
- Check maintenance work orders to determine the percentage of "emergency" work.
- Review unit logs to verify that work groups are coordinating activities with responsible operators.
- Review the process for authorizing nonroutine activities for evidence of complacency.
- Monitor the number of visitors and administrative duties that distract operators from their primary tasks.
- Identify the number of nuisance and always-on alarms.
- Investigate whether maintenance is being deferred to meet production goals.
- Tour the work area to assess housekeeping and the status of required safety devices, such as chocks at truck stations, locks on critical valves, charged and available fire extinguishers and so forth.
- Investigate the reasons for any significant differences in the performance of different shifts, teams, areas, or departments.
- Determine how often supervisors observe work in the field.
- Determine whether peers are observing and coaching peers on an ongoing basis outside of formal training settings.
- Verify that the organization chart is up to date and that clear responsibilities and lines of authority are being maintained.

Regardless of the questions that are asked, the management review should try to evaluate the organization's depth on several levels. Is the operational discipline sufficient? Does the formality of operations extend beyond routine tasks? When things go wrong, is it likely that key personnel understand the chain of command that must be followed to make prudent decisions? Also, is there sufficient depth in key personnel? If the entire organization depends on a single individual to make all of the really tough risk judgments, what will happen if that person suddenly resigns or falls ill? Management reviews provide an opportunity for the organization to honestly assess its depth, and take action to address any concerns before experiencing a loss event as a result of a breakdown in operational discipline, and before losing key drivers in the unrelenting quest for excellence in human performance. Any weaknesses revealed by the management review should be resolved as described in Chapter 22.

17.7   REFERENCES

17.1   *EPA Chemical Accident Investigation Report – Tosco Avon Refinery, Martinez, California*, EPA550-R-98-009, U.S. Environmental Protection Agency Chemical Emergency Preparedness and Prevention Office, Washington, 1998.

17.2   *Conduct of Operations at Nuclear Power Plants*, DS347, International Atomic Energy Agency, Vienna, 2005.

17.3   *Conduct of Operations Requirements for DOE Facilities*, DOE Order 5480.19, U.S. Department of Energy, Washington, 2001.

17.4   Galsworth, Gwendolyn D., *Visual Systems: Harnessing the Power of the Visual Workplace*, American Management Association, New York, 1997.

17.5   Hirano, Hiroyuki, *5 Pillars of the Visual Workplace – The Sourcebook for 5S Implementation*, Productivity Press, Portland, Oregon, 1995.

18

EMERGENCY MANAGEMENT

Around 8 a.m. on April 16, 1947, a fire was detected in a cargo hold aboard the freighter *Grandcamp* while it was moored near Texas City, Texas, loading ammonium nitrate fertilizer. At the time the fire started, the ship had been loaded with 1,400 tons of ammonium nitrate (in bags) in one hold and 800 tons of the same material in another hold. Over the next hour, the ship's captain decided to not use water to extinguish the fire, fearing that some of the cargo would be lost. Instead, the captain ordered sailors to close the hold and use high pressure steam to displace the oxygen. Unfortunately, once set afire, ammonium nitrate coated with paraffin does not need oxygen to burn. Worse yet, the heat from the steam increased the rate of burning. The ship exploded at 9:12 a.m., killing at least 468 people, which in terms of fatalities is the worst industrial accident that has ever occurred in the U.S. The explosion aboard the *Grandcamp* also caused several other large explosions and fires over the next 16 hours involving a nearby chemical plant, nearby oil terminals, and ultimately led to a large explosion early in the morning of April 17 involving 1,000 tons of ammonium nitrate aboard the freighter *Highflyer*, which was moored in a slip adjacent to the *Grandcamp*.

According to one author who studied the Texas City disaster, "Safety and emergency preparedness . . . were grossly deficient, considering the enormity of the dangers. . . . Without preparations, little chance existed that anyone could cope with the effects of an accident quickly enough to prevent it from escalating into a disaster." Clearly, many failures occurred in the emergency response effort, including (1) the decision to use steam to displace oxygen rather than water to remove heat and (2) the failure to evacuate bystanders from the dock adjacent to the *Grandcamp*. In fact, hundreds of people gathered in the dock area between 8:00 a.m. and 9:12 a.m. that morning to get a better view of the brightly colored smoke and flames (Ref. 18.1).

18.1   ELEMENT OVERVIEW

Developing appropriate emergency management and response capabilities is one of nine elements in the RBPS pillar of *managing risk*. This chapter describes the attributes of a risk-based management system for emergency management. Section 18.2 describes the key principles and essential features of a management system for this element. Section 18.3 lists work activities that support these essential features, and presents a range of approaches that might be appropriate for each work activity, depending on perceived risk, resources, and organizational culture. Sections 18.4 through 18.6 include (1) ideas for improving the effectiveness of management systems and specific programs that support this element, (2) metrics that could be used to monitor this element, and (3) issues that may be appropriate for management review.

18.1.1 What Is It?

Emergency management includes (1) planning for possible emergencies, (2) providing resources to execute the plan, (3) practicing and continuously improving the plan, (4) training or informing employees, contractors, neighbors, and local authorities on what to do, how they will be notified, and how to report an emergency, and (5) effectively communicating with stakeholders in the event an incident does occur.

The scope of the *emergency* element extends well beyond "putting out the fire." This chapter focuses on three aspects of emergency planning and response:

- Protecting people, including people who are onsite, offsite, and emergency responders.
- Responding to catastrophic accidents involving explosions, large releases of chemicals, or other large releases of energy.
- Communicating with stakeholders, including neighbors and the media.

This chapter does not specifically address accidents caused by natural disasters or malevolent actions (e.g., intentional attack, public demonstrations, sabotage), although managing the consequences of many of these events will be similar to what is done in the event of a process safety incident. This chapter also does not address related issues such as business continuity planning, recovery, or requirements to preserve forensic evidence that may be useful in an incident investigation. Reference 18.2 provides a thorough treatment of these issues.

18.1.2 Why Is It Important?

The consequences of any particular incident can be significantly reduced with effective emergency planning and response. Failure to establish and enforce a perimeter to keep bystanders and nonresponders at a safe distance from the *Grandcamp* on April 16, 1947, directly contributed to several hundred fatalities. Even if no one is killed or seriously injured by an incident, the facility's license to operate within the community may come into question, and the answer will be strongly influenced by the public's perception of the competence of emergency response activities. Effective emergency management saves lives, protects property and the environment, and helps reassure stakeholders that, in spite of the incident, the facility is well managed and should be allowed to continue to operate.

18.1.3 Where/When Is It Done?

Emergency management activities typically occur at the facility and in the community where the accident might occur. These activities include (1) planning and training, which occur frequently, (2) drills and exercises, which typically occur once or more each year, and (3) actual responses, which should occur rarely if other RBPS elements are effectively implemented. Activities also include coordination with local authorities, for example, by attending monthly meetings of the local emergency planning committee (LEPC).

18.1.4 Who Does It?

Emergency planning is typically performed by specialists, both within and external to the facility. Planners consult with the operations group and review work products from the *risk* element to identify and select planning scenarios. Emergency response plans should be developed in concert with potentially involved or affected work groups, and they should be frequently reviewed with all potentially involved or affected workers. The operations group is typically responsible for immediate emergency response activities, such as shutting down the process and isolating hazardous material inventories, and they are assisted as quickly as possible by specially trained teams whose activities are coordinated by an incident commander. These teams often include facility-sponsored response teams, outside agencies, including fire departments, medical responders, hazardous material (HAZMAT) teams, and, in some locations, mutual aid response teams from nearby facilities. Crisis management, which is beyond the scope of this book, is normally led by a senior manager and focuses on issues beyond mitigating the immediate effects of the incident.

18.1.5 What Is the Anticipated Work Product?

Effective *emergency management* should reduce the magnitude of effects of an incident, including any loss of good will with stakeholders. An intermediate, more tangible work product is effective and tested emergency response plans, trained and equipped response teams, and effective methods of protecting (1) personnel who could otherwise be harmed by the incident (including emergency response personnel), (2) the environment, and (3) property, both offsite and onsite.

Emergency management is closely linked to the *risk* element. In fact, risk was considered in emergency planning long before risk was proposed as a basis for developing any management system. When developing emergency plans, one intuitively asks the three fundamental risk questions:

- What can go wrong – What types of emergencies should we plan for?
- How bad could it be – Will operators be able to put out the fire with portable extinguishers or do we need professional firefighters with a pumper truck?
- How often might it happen – Is the likelihood of a particular accident scenario high enough to justify in-house response capability or should we depend on local authorities?

Effective emergency response also requires trained personnel and dependable equipment; it is very dependent on the *training* and *asset integrity* elements.

18.1.6 How Is It Done?

Emergency management activities are largely "done" well in advance of an incident. They include:

- Thorough planning.
- Effective training.
- Realistic drills.
- Effective two-way communication with stakeholders.
- Establishing the culture and operational discipline needed to ensure that personnel adhere to emergency plans and procedures.

If all of these pieces are in place when an incident occurs, emergency management activities will be based on carefully developed plans, proper training, and well defined roles and responsibilities. Failure to plan for emergencies, or failure to execute the plan when required, can quickly transform an accident into a disaster.

18.2 KEY PRINCIPLES AND ESSENTIAL FEATURES

Large fires, explosions, and toxic releases are rare events. However, facilities must carefully plan for emergencies, train personnel, and provide necessary equipment. Decay in the capability or readiness of emergency responders can go unnoticed for a very long time. Therefore, drills must truly test the emergency response plans and procedures. Every response must be as effective as possible – lives often depend on it.

Policy issues for the *emergency* element extend one step further than for most other RBPS elements. After identifying what can go wrong, emergency planners start by asking, "How much should we do?" In many cases, "very little" or "nothing" is an appropriate answer. Those facilities leave emergency response activities to local authorities, choosing instead to invest their resources in activities and systems to (1) reduce the frequency of accidents, (2) safely evacuate personnel from the area affected by an accident, and (3) mitigate the consequences of accidents. Once management decides on the balance between emergency response and other possible strategies, it then establishes a policy of defining roles and responsibilities, the scope of response activities, and other aspects addressed in this chapter.

The following key principles should be addressed when developing, evaluating, or improving any management system for the *emergency* element:

- Maintain a dependable practice.
- Prepare for emergencies.
- Periodically test the adequacy of plans and the level of preparedness.

The essential features for each principle are further described in Sections 18.2.1 through 18.2.3. Section 18.3 describes work activities that typically underpin the essential features related to these principles. A facility should evaluate the risks and potential benefits that may be achieved as a result of improvements in this element, decide to adopt some or all of the essential features, and execute some or all of the work activities, depending on perceived risk and/or process hazards that it identifies. However, none of this will be effective if the accompanying *culture* element does not embrace the use of reliable management systems. Even the best management system, without the right culture, will not ensure a highly capable and effective emergency management capability.

18.2.1 Maintain a Dependable Practice

An incident triggers emergency management activities. People will do (or intentionally decide to not do) something in an emergency, even if it is wrong. Planning for emergencies greatly increases the likelihood that people will do the right things. However, plans must be tested regularly and rigorously because emergency response activities seem to be more prone to erosion from lack of attention than any other RBPS element. For example, verifying that an effective evacuation plan is in place, or that the plan is understood, is very difficult unless the plan is periodically tested and updated. Because emergency response activities are infrequent, and facilities normally don't get a second chance to address systemic weaknesses, maintaining a dependable practice for this element is critical.

Most facilities in the process industries do not lack for emergency plans. More likely, the facility will have multiple 3-ring binders dealing with incidents ranging from a major spill that threatens the environment to an armed employee intent on killing co-workers to a large release that could form a toxic vapor cloud. The challenge is not developing more plans; rather it is in ensuring that plans that already exist (1) cover the range of credible scenarios and (2) are likely to work if needed.

Develop a written program. The written program is not the facility's emergency response plan. Rather, it is a document that places all of the activities that support this element into an understandable framework.

The program should be jointly developed by representatives of (1) process safety, (2) environmental, health, and safety, (3) emergency response, (4) operations, and (5) local emergency response agencies. Each of these functions (which may include a high degree of overlap, particularly at smaller facilities) normally holds part of the knowledge needed to develop a comprehensive emergency response program, and all of this knowledge should be considered when developing the emergency response program.

The written emergency response program, or a subordinate document, should clearly delineate between those incidents that operators should respond to and those that should be handled by the facility's emergency response team (ERT) or equivalently trained outside responders. Failure to address this issue will require that operators and other front line employees make instant decisions on how to respond to an emergency, likely leading to unacceptable risk because people generally do a poor job of considering a wide range of factors when making decisions in a crisis.

No single right answer exists as to what level of responsibility should be placed on operators when faced with a small spill, incipient fire, or other unplanned event. Operators normally have a much deeper understanding of the process and its hazards than do emergency responders and can normally act more quickly in terms of both (1) formulating a response action and (2) executing the action. However, operators are generally not as well equipped as emergency responders (in terms of response and personal protective equipment), and are almost certainly not as well trained in emergency response activities. Because they feel a certain ownership of the process, some operators may accept more personal risk while responding to an emergency than the company might endorse. This ownership issue, and the need to understand tolerable risk when responding to emergencies, can best be managed if the range of acceptable actions is documented, reviewed, and communicated to operators and other affected workers in writing, and reinforced in training situations.

Designate an owner and define roles and responsibilities. As in any other management system, the *emergency* program will consist of activities that ultimately support essential features and key principles of the system. Unless roles are defined and responsibility is clearly assigned at the work activity level, some activities are likely to be overlooked.

Facilities normally designate an owner for the *emergency* element. This is often the leader of the ERT, typically a person very skilled in directing emergency response activities. This assignment can be a good fit, and usually works well as long as this individual is aware of the full range of emergency management activities and actively champions issues that extend beyond the ERT.

In the event of a major emergency, at least two distinct types of activities must be managed. First, someone must direct all of the tactical activities performed by emergency responders; this responsibility is assigned to the incident commander. This individual has direct operational control of all aspects of the response. This is never a shared responsibility. Additional responders or assets arriving on the scene will be assigned to this individual, or incident command will be formally transferred to another individual. For example, the emergency medical technician, the fire chief, and the leader of the HAZMAT response team are not peers charged with making decisions via consensus. One of these individuals, or some other person designated as the incident commander, decides how best to coordinate the actions of the fire brigade, medical responders, and HAZMAT team members, often based on advice and input from specialists in each area.

The second activity involves all tasks except those being executed by the responders under the direction of the incident commander, ranging from accounting for personnel to responding to questions from the media. One person would not be able to manage these divergent activities while also being the incident commander. Therefore, the *emergency* element should address the roles and responsibilities of teams or persons assigned to (1) respond to the incident and (2) execute supporting actions.

A member of the facility's management team is sometimes designated as the owner for the *emergency* element, mainly to ensure that the full range of emergency management activities receives proper attention. This person is often assigned to manage all of the activities that are not the direct responsibility of the incident commander. Also, this person is often heavily involved in emergency planning; participates in outreach programs, such as belonging to the LEPC; and often chairs the community advisory panel described in Chapter 7.

Define the scope of the program. Many facilities divide their emergency response plans primarily by the cause of the emergency and associated hazards, rather than by the area of the facility. The response to a bomb threat generally does not depend on which department receives the call. Plans to respond to fires also generally apply throughout the facility, although plans to respond to fires will be more detailed in units that process flammable materials, and the specific firefighting tactics may differ greatly from one unit to the next. The program scope normally includes:

- The physical areas (although many programs address the entire facility).
- The general types of emergencies, and if multiple plans exist, a cross reference between type of emergency and response plan to help personnel quickly locate information.
- Specific actions for different types of emergencies. For example, will offensive actions be undertaken (e.g., will the fire brigade enter the process area to extinguish a large pool fire), and who will be assigned to undertake the actions?
- Integration with other responders, including mutual aid agreements between nearby facilities, command and control for response teams made up of personnel from different companies at a multi-company facility, and agreements with local authorities.

Involve competent personnel. Given that emergencies are rare events, very few people at a facility ever develop the depth of expertise needed to understand both (1) the range of credible facility-specific accident scenarios and their potential impact and (2) emergency response operations, including operating without utilities and infrastructures that are normally very reliable. Therefore, the emergency management program and specific emergency response and related contingency plans should be developed and reviewed by a team that collectively has (1) a broad understanding of the facility and its hazards and (2) training and experience in emergency response operations.

18.2.2 Prepare for Emergencies

By definition, emergencies are come as you are events. There is little time or opportunity to develop, update, or revise plans. Responders are faced with choosing a course of action based on a range of preplanned response options or derivatives of those options. The response options are typically limited by personnel, their training, equipment, communication protocols, and external support. In general, providing each of these resources has a cost. Planners often have to make a risk-based decision, balancing the potential need with the cost of obtaining and maintaining these resources.

Identify accident scenarios based on hazards. The first step in planning for emergencies is to identify the range of emergencies that may occur. To a degree, this is a natural thought process: facilities along the gulf coast plan for hurricanes; facilities in the midwest plan for tornados. In a similar manner, facilities should determine the range of credible process incidents as a starting point in their planning. Large-scale incidents involving chemicals generally present three classes of hazards: fires (thermal effects), explosions (pressure effects), and toxic vapor clouds (physiological effects). Often, the response plan can address a few scenarios involving each hazard class and cover the range of credible large-scale accident scenarios.

A second class of incidents involves acute danger to workers in the immediate vicinity of the hazard. Hazards in this category include nitrogen asphyxiation in an enclosed space, thermal or chemical burns, falls, electrocution, or other workplace safety hazards that do not have distant effects. In most cases, facilities address these hazards in their emergency response plans with the objectives of (1) protecting responders from injury and (2) providing medical treatment as quickly as possible. The remainder of this chapter will focus on large-scale process safety incidents that are likely to have distant effects – fires, explosions, and toxic material releases.

A list of hazardous materials stored or processed at the facility, along with the maximum intended inventory and hazard information for these chemicals, is a readily available starting point for identifying accident scenarios. In addition, credible accident scenarios are normally identified in risk analysis reports. A combination of expert opinion, local (unit or facility) incident history, industry wide incident history, hazard review documents, and hazardous material inventory records are also normally used to identify potential accident scenarios for further consideration.

If hazards at nearby facilities could potentially affect your workers, include those scenarios in emergency management planning. For example, decide in advance how nearby facilities will notify each other in the event of an emergency, such as a toxic gas release, and include actions that your workers should take based on various types of notifications (e.g., activate your facility's shelter-in-place plan).

Unfortunately, another potential cause of emergencies is intentional attacks (including acts of sabotage by facility workers). Although safeguards protecting against malicious incidents are well beyond the scope of this book, the *emergency* element should logically address these events as well. Planning for these scenarios can be particularly complex as well-informed attackers may simultaneously (1) cause an explosion or a release of a hazardous material and (2) disable part or all of the facility's emergency response capabilities. Thus, close coordination with local emergency response agencies is a vital part of preparing for intentional attacks.

Assess credible accident scenarios. Evaluate the list of credible accident scenarios to determine the types and range of potential effects. Response activities that focus on protecting people and the environment normally depend on (1) the type of effect, such as a flammable gas release, toxic gas release, or explosion and (2) the range of effects, in other words, the distance to a defined end point. Often, facilities use computer models to predict the range of effects for each type of credible release. If the list of potential accident scenarios is long, or if it contains scenarios that would result in similar effects, further analysis or/and modeling is often used to screen the possible scenarios.

In addition, planners need to know specific details about the potential source of a release to properly equip and train the ERT. For example, if one planning scenario is a release of chlorine, planners need to know what types of containers will be present at the facility (e.g., cylinders, ton containers, or rail cars) so that the ERT will have the proper types of seal kits stocked in appropriate locations.

Select planning scenarios. Select planning scenarios based on the types of releases, the footprint or potentially affected area, and the incident history within the industry. Intended actions, such as evacuation versus shelter-in-place, should also be considered when selecting scenarios for inclusion in the emergency response plan. Planners should try to strike a balance between high-consequence/low-frequency scenarios and lower-consequence/higher-frequency scenarios. For example, fires that are limited to a direct-fired heater might occur once in 100 years per heater, whereas a large release of a flammable gas with the potential for distant effects might be expected to occur less than once in 10,000 years per process unit. These scenarios may represent equivalent risk and both should be addressed when planning emergency response activities.

In addition, the *emergency management* element is often tasked with addressing scenarios that are not specific to a process. For example, many facilities have invested in the equipment and training needed to perform confined space rescue or vertical rescue.

Plan defensive response actions. The primary objective of emergency planners is to protect life and health, both for onsite personnel and offsite neighbors. Immediate protection is normally provided by defensive actions.

Specifically, defensive response actions include:

- Emergency recognition and reporting.
- Authority and methods for raising a standby or evacuation alarm.
- Personal protective equipment to assist in evacuation.
- Safe havens or shelter-in-place locations (and when this strategy should be used).

- Evacuation routes, assembly points, and actions that should be taken at the assembly points.
- Alternate evacuation routes, shelter-in-place locations, and assembly points.
- Headcount/personnel accountability procedures.
- Medical response, particularly if immediate first aid or special antidotes are needed.
- Notification of management and regulatory authorities (if required).
- Establishment of an emergency operations center (EOC), including command and control of defensive actions and policies regarding transition of incident command authority.

Emergency plans should address defensive actions, such as evacuation and shelter-in-place, describing (1) the specific actions to be taken by employees who do not respond to the event, (2) the means that will be used to notify these employees of which action(s) to take, and (3) the basis on which the plan should be implemented. These plans should be reinforced through actual drills. Physically moving to the designated safe location and conducting a headcount (versus simply evaluating the plans on paper) will test the system and help reinforce in workers' minds what the evacuation signal sounds like and where their assembly point and shelter-in-place areas are located.

Another important aspect of defensive operation is quickly putting operating process units in a safe condition prior to evacuation. Operator actions necessary to do this should be included in the operating procedures, and reviewed or practiced as often as required to provide a high degree of assurance that, in the event of an emergency, operators can quickly execute the required actions without having to look them up in the procedures manual.

Facility emergency response plans should be integrated with community emergency response plans. LEPCs often develop emergency response plans that correspond to facility plans. The LEPC plans describe specific actions to be taken (community evacuation, shelter-in-place, etc.), and the means that will be used to notify people of the emergency, which action(s) to take, and when it is safe to resume normal activities.

Plan offensive response actions. Written plans are a good way to transmit technical knowledge and coordinate response activities. Reflecting on the 1947 Texas City disaster described at the start of this chapter, a fire-response plan for ammonium nitrate that informed the ship's captain to use water instead of steam to extinguish an ammonium nitrate fire might have limited the event to a minor incident with little property damage and no loss of life.

For example, plans for taking offensive actions in response to a toxic chemical release should specifically address (Ref. 18.3):

- Firefighting preplans and consequence model results that help predict the range of effects.
- The boundaries for the "hot" and "warm" zones.
- Control of access into and out of these zones.
- Response team communications, with particular emphasis on communications between responders, operations, and supporting groups (e.g., if facility security will be used to control access, how will they communicate with the ERT?).
- Staffing and specific duties, including:
    o Incident commander
    o Safety officer
    o Responders
    o Decontamination team
    o Technical specialists supporting the response team
- Guidance for PPE selection.
- Use of the buddy system.
- Decontamination procedures.

Employers have a fundamental responsibility to protect their employees and the public from harm. However, many facilities rely on local fire departments and offsite HAZMAT teams (supported by emergency medical services and local police) to take offensive response actions and/or take actions to protect the public. Relying on local response agencies does not necessarily indicate lack of commitment to this element. In fact, if fires are very rare events, the risk associated with using facility personnel to fight fires or perform fire rescue may be much higher than using well-trained full-time firefighters who put out fires and perform rescue activities on a daily basis. Factors that should be considered when deciding whether or not to establish an ERT include:

- The costs and benefits of investing in fixed/unmanned systems rather than an ERT. Normally, it costs much less to maintain fixed systems than to train and support an ERT, but an ERT is much more agile. Trained personnel can respond to a wide range of emergencies and quickly change tactics when needed.
- Staffing. Smaller facilities are unlikely to have the staff required to mount a timely emergency response effort.
- Willingness to turn over control to local authorities. If the local fire department is called to the facility for a response, facility personnel will

likely have to turn over incident command to the senior firefighter on the scene.

- Concern that, once onsite, local authorities may prohibit facility employees from taking actions that might mitigate the effects of the emergency. A lack of understanding on the part of local emergency responders could prohibit facility staff members from taking beneficial actions.

- Understanding that the primary responsibility of fire departments and other responders is to prevent harm to people, including firefighters, not to prevent property damage. In fact, firefighters may decide to point their hoses at nearby facilities to prevent the spread of the fire rather than use their resources to try to save parts of the facility that is on fire.

- Estimates of the response capability of the local fire department and other local emergency responders. These can be based on staffing and training (a fully staffed municipal fire and medical response agency versus a volunteer fire department that has no paid staff), budget, equipment, class rating, or similar criteria.

If a facility decides to depend primarily on outside agencies for emergency response services, close links with such agencies are essential. Some facilities may be reluctant to invite the local fire department to visit their facility because they are concerned that they may be cited for a fire code violation. Although this is possible, failing to familiarize offsite emergency responders with your facility could significantly impede the effectiveness of response efforts. The risk of a potentially hindered emergency response (because of unfamiliarity with the premises) should outweigh facility fears of a citation.

If a facility decides to have internal emergency response capability, either using its own employees or via a mutual aid agreement with nearby facilities, it will need to establish objectives for emergency responders. These typically include (1) approaching the hazard to rescue a person who is unable to escape without aid, (2) stopping a toxic release, (3) containing a fire, or (4) taking other actions to mitigate the consequences of the incident. These actions are often very specific to the hazards present at a facility and should be planned by trained emergency responders who are very familiar with operations at the facility.

Another aspect of planning offensive operations involves negotiating mutual aid or other cooperative agreements. These agreements often involve multiple facilities that are located near each other, but they can also involve mutual aid agreements with community responders, such as the local fire department, or formal agreements among several companies that operate units within a facility and share a single ERT.

Develop written emergency response plans. Written emergency response plans spell out offensive response actions and form the basis for determining what facilities, equipment, staffing, training, communication, coordination, and other resources or activities are required. Many state and national regulations specify minimum content for emergency response plans. (Several of the references listed in this chapter address the required content for emergency response plans, as well.) Emergency response plans normally address all of the essential features identified in this key principle, for example, communications, physical facilities and equipment, training, and so forth. These plans also normally include annexes or plans for each scenario (or group of similar scenarios) that requires offensive actions by emergency responders. Emergency response plans and associated training should also emphasize the importance of preserving and securing the area after the incident until the incident investigation team has an opportunity to document the condition of the equipment and record other data that may be relevant to the investigation.

Provide physical facilities and equipment. The resources needed for emergency response should be described in the response plan(s). Facilities normally apply the same sort of decision analysis to the purchase of fixed or durable emergency response equipment that they would to other capital assets, basing decisions on the cost and expected benefit. Facilities should also give considerable thought to the location of equipment: siting it too far from likely response areas will increase response time, while siting it too close to locations where hazardous materials may be released can make it difficult or impossible for the ERT to reach its equipment in the event of an incident.

Maintain/test facilities and equipment. Emergency response activities employ a wide range of facilities and equipment, much of which is rarely (or never) used. Regardless, it must be maintained, periodically inventoried, and tested to ensure that it will function when it is needed.

Most facilities routinely test emergency evacuation alarms. Fewer facilities have formal preventive maintenance tasks assigned to (1) test and maintain emergency lighting, (2) replace batteries in radios and bullhorns stored at assembly points (or similar communication equipment), and (3) inventory 5-minute escape packs. Maintaining both emergency evacuation and emergency response equipment is important. In most cases, failure to do this is a management system issue – responsibility for maintaining equipment that is required for emergency evacuation is not well defined.

Determine when unit operator response is appropriate. Plans should address (1) small spills and releases that are typically contained and cleaned up by operators, maintenance employees, or other unit personnel without activating the emergency response program and (2) incipient fires that are typically put out using a fire extinguisher. Specifically, the plans should explain (1) when it is appropriate for unit personnel to address these types of situations, (2) the proper PPE to be worn and other precautions to take, (3) the intended means to contain and collect the spill (including the intended means to dispose of sorbents containing spilled process materials), (4) training requirements, and (5) alert/notification requirements.

Train ERT members. Poor decisions on the proper course of action or poorly executed emergency response activities increase the risk for personnel at the facility, the public, and the responders themselves. Emergency response is like most skills – it is learned through training and reinforced through experience. Fortunately, major incidents are rare, but this means that most emergency responders have limited real life experience with major emergencies. Therefore, competence must be developed through realistic training. Such training is available at highly specialized training facilities, such as the fire schools sponsored by several major universities.

Plan communications. Effective two-way communication is a critical element to emergency management. Potentially affected people must be informed of the plan and brought on board. This is normally a very simple task for facility employees, and becomes progressively more difficult with contractors, local authorities, neighbors, and other stakeholders. Just as plans are routinely updated to reflect changes to the facility or to response tactics, the communication effort must also be an ongoing process.

Although risk communication is not explicitly included in this element, if neighbors perceive that a facility presents a significant risk to their safety, addressing concerns about risk is necessary before describing emergency response actions that neighbors may need to take to protect themselves. After all, when neighbors are told how bad things might be, they will naturally want to know how the incident might happen and what the facility is doing to protect their safety, which is a key objective for the *outreach* element.

In the event of an emergency, a reliable means must be available to alert those in danger that they should take action, and specifically what action to take. Facilities normally implement some sort of audible system to alert workers, sometimes coupled with a visual signal in high noise areas. Community alert systems vary widely, including police cars driving up and down streets using their PA systems to notify residents, specialized audible alarms (similar to tornado sirens), Reverse 911® systems, and emergency warnings carried on local radio and TV stations. Regardless of the method used to alert people to an emergency, the alarm must be be recognizable, and clearly distinguishable from normal message traffic carried on whatever media is used.

Inform and train all personnel. Written plans are of little or no value if they are not widely understood. Individuals who might be affected by an emergency should be trained or notified on (1) how they will be alerted of an emergency, (2) what actions they may be asked to take, and (3) what to do to protect themselves. This practice is relatively straightforward for onsite workers to whom the plan can be explained periodically in a safety meeting or similar forum and reviewed when it changes. For the same reasons, training emergency responders on the emergency response plan and notifying them of minor changes to the plan is also a straightforward action. However, if a facility anticipates using offsite responders, such as the local fire department, consideration should be given to how to train these organizations on facility-specific hazards, as well as on changes to the facility that might affect response efforts.

In addition to being trained on what to do, all personnel should understand management's intent for emergency response. What actions are acceptable and what are not? Many workers, when faced with an emergency, will pursue unsafe actions to mitigate the consequences, particularly if a co-worker could be injured. For example, a high incidence of double fatalities occurs for confined space incidents: first the person working in the confined space collapses, and shortly afterward, a co-worker collapses while attempting to rescue the first victim. Policy issues such as, "Do not attempt to take offensive actions such as . . . " should be clearly communicated to all personnel and reinforced periodically through training or other communication modes.

If any accident scenarios can affect the public, response activities should be coordinated with local emergency planners. LEPCs are normally aware of nearby facilities that cannot be easily evacuated or present other special challenges, for example, schools, hospitals, nursing homes, jails, and can provide invaluable help when planning for emergencies at these locations.

If any of the accident scenarios can affect nearby residents, the residents should be told how they will be notified of an emergency and provided with literature explaining the range of possible actions they may have to take, such as shelter in place or evacuate. Local authorities, politicians, and the local media can also be an invaluable ally in efforts to reach out to the public.

Periodically review emergency response plans. Just as the accuracy of operating procedures decays rapidly once people stop paying attention to them, the accuracy and utility of emergency response plans will likewise decay. For example, most facilities provide a list of emergency contact phone numbers near the front of a plan. All too often, this list is inaccurate because of personnel changes. Emergency response plans are particularly prone to this sort of decay, because unlike operating procedures, workers are less likely to get them out and read them on a routine basis.

18.2.3 Periodically Test the Adequacy of Plans and Level of Preparedness

If emergency communications are inadequate, the only indications of a possible problem may be a drill or an emergency response activity that exposes this weakness.

Conduct emergency evacuation and emergency response drills. There is no substitute for practice. First, the plans may simply not work as designed.

Second, individuals may misunderstand their role, creating a critical gap in execution. Third, practice reinforces skills, transforming rules-based actions into skill-based reactions (see Chapter 14, Training and Performance *Assurance*, for further discussion of the effect of training on performance). Effective practice, resulting in skill-based execution, dramatically reduces the response time and the likelihood of human error.

Conduct drills under realistic conditions, such as on short notice, at night, or in inclement weather. Too often, drills are scheduled months in advance for mid-afternoon on a weekday in a warm (but not too hot) time of the year. Thus, facilities may fail to learn that the ERT does not have supplemental lighting that they would need for a nighttime response or that the ERT's protective suits are so hot that responders must be relieved after 15 minutes on a mid-August afternoon.

Finally, facilities should conduct evacuation drills periodically, including accounting for personnel. These types of drills are often not scheduled for production workers because facilities do not want to interrupt operations or increase risk by executing a startup after an emergency shutdown. Some facilities conduct these drills at shift change to mitigate any impacts, often as part of a safety meeting. Although this completely eliminates any element of surprise, it does cause all personnel to periodically walk to the assembly point, which helps reinforce the assembly point location in their mind, and it also may uncover weaknesses in the headcount procedures.

Conduct tabletop exercises. Even if the plan is up to date, in many cases managers and others who are not members of the ERT are not fully aware of their responsibilities. An efficient method for addressing this potential gap is to hold table top exercises – practicing a response without the expense of assembling the response team and staging a full-blown drill. In most cases, persons assigned to critique drills spend their time with the responders, thereby missing opportunities to critique the management team. Tabletop exercises focus attention on management, communications, coordination, and logistics. Gaps in any of these areas could impact the effectiveness of response efforts.

Practice crisis communication. Crisis communication is unlike most other forms of communication. The company spokesperson will be under extreme pressure to release information that is simply not available, and in some cases, reporters' questions can be very badgering or confrontational. Some companies integrate crisis communication and media participation into drills. Footage of fire trucks rolling to the mock emergency tends to air on the 6 o'clock news and can help with community outreach; however, such drills do not train the company spokesperson to answer hard questions under extreme pressure. Many facilities choose to train their designated spokespersons in specialized short courses taught by experienced news reporters using realistic accident scenarios.

Critique exercises, drills, and actual responses. One critical element for each of these events is feedback from experienced observers. Without constructive criticism, facilities will miss many of the lessons that can help improve the effectiveness of the emergency management program. The critique should result in specific actions to improve overall performance.

Conduct assessments and audits. Audits are addressed in detail in Chapter 21. Audits of the *emergency management* element are particularly important because less-than-adequate training or poor performance may go unnoticed for years, particularly at a facility that does not conduct regular and effective emergency response drills.

Address findings and recommendations. Exercises, drills, assessments, and audits all offer insight on the emergency management program. However, to provide value, the observations need to be documented as findings or recommendations (depending on the source), and these findings and recommendations need to be translated into actions that will improve the *emergency management* element. Some companies will also periodically share the lessons learned at each facility, thereby leveraging their value across the organization.

18.3 POSSIBLE WORK ACTIVITIES

The RBPS approach suggests that the degree of rigor designed into each work activity should be tailored to risk, tempered by resource considerations, and tuned to the facility's culture. Thus, the degree of rigor that should be applied to a particular work activity will vary for each facility, and likely will vary between units or process areas at a facility. Therefore, to develop a risk-based process safety management system, readers should perform the following steps:

1. Assess the risks at the facility, investigate the balance between the resource load for RBPS activities and available resources, and examine the facility's culture. This process is described in more detail in Section 2.2.
2. Estimate the potential benefits that may be achieved by addressing each of the key principles for this RBPS element. These principles are listed in Section 18.2.
3. Based on the results from steps 1 and 2, decide which essential features described in Sections 18.2.1 through 18.2.3 are necessary to properly manage risk.
4. For each essential feature that will be implemented, determine how it will be implemented and select the corresponding work activities described in this section. Note that this list of work activities cannot

be comprehensive for all industries; readers will likely need to add work activities or modify some of the work activities listed in this section.

5. For each work activity that will be implemented, determine the level of rigor that will be required. Each work activity in this section is followed by two to five implementation options that describe an increasing degree of rigor. Note that work activities listed in this section are labeled with a number; implementation options are labeled with a letter.

Note: Regulatory requirements may specify that process safety management systems include certain features or work activities, or that a minimum level of detail be designed into specific work activities. Thus, the design and implementation of process safety management systems should be based on regulatory requirements as well as the guidance provided in this book.

18.3.1 Maintain a Dependable Practice

Develop a Written Program

1. Develop a written description of the emergency management program that addresses each of the essential features listed in this subsection.
   a. Some programmatic activities, such as roles and responsibilities, can be found in the emergency response plan, but no separate description of the emergency response program exists.
   b. A written description of the emergency response program has been developed by the ERT leader; it mainly addresses emergency response actions.
   c. A written description of the emergency management program has been developed based on input from a number of departments. It addresses a wide range of emergency management issues, including tactical response activities, crisis communications, and business continuity.

Designate an Owner and Define Roles and Responsibilities

2. Designate a single person who has overall responsibility for the facility's emergency management program.
   a. Facility personnel have an understanding that the safety department is responsible for all aspects of the emergency response program.
   b. The overall responsibility for the emergency response program has been formally assigned to the ERT leader.
   c. Overall responsibility for emergency response, along with related responsibilities such as crisis communications, has been assigned to a senior manager at the facility. Other aspects, such as

business continuity planning, are formally assigned to corporate personnel. An integrated set of facility and corporate emergency management plans has been established that address a wide range of crisis management issues.

3. Include programmatic roles and responsibilities in the written program description.

   a. Although no formal responsibilities are assigned for activities such as emergency planning, training, or conducting drills, these and other emergency response planning and training activities do occur. The ERT leader oversees all ERT training; all other activities are the responsibility of the facility's safety department.

   b. The roles and responsibilities have been defined for emergency responders and departments that operate high hazard units.

   c. Overall responsibility for emergency management has been integrated into appropriate business processes. For example, the manufacturing director is responsible for recovery planning; the business manager is responsible for business continuity planning; the facility's safety manager is responsible for evacuation, personnel accountability, and facility security; the ERT leader is responsible for drills, ERT member training, coordinating with local emergency planners, and tactical response activities.

Define the Scope of the Program

4. Define the scope of the emergency response program, including:

   - The physical scope to which the emergency response program applies, for example, a single unit, a contiguous multi-unit facility, all areas within a contiguous facility and remote warehouses, transportation incidents, and so forth.

   - The types of emergencies considered or excluded, for example, fires, explosions, toxic exposure hazards resulting from accidental release of process materials, hazards related to natural disasters, hazards related to intentional attacks such as bomb threats, and so forth.

   - Types of actions to be taken by facility personnel, for example, use of facility personnel in fighting fires.

   - Integration with other response agencies.

   a. With the exception of evacuation and personnel accountability, the emergency response program is informal. Local authorities will be asked to protect workers and the public from harm in the event of an emergency.

   b. Linkage between the defined scope and the emergency response plan is clear.

  c. In addition to item (b), the emergency response plan is regularly reviewed with outside agencies that could be called on to respond. Efforts have been made to align response plans and resolve possible jurisdictional conflicts.

Involve Competent Personnel

5. Ensure that the overall emergency response program and key aspects of the emergency response plan are developed and/or reviewed by persons with the proper blend of facility knowledge and emergency management experience.

  a. With the exception of evacuation and personnel accountability, the emergency response program is informal. Neither the overall program nor emergency plans are reviewed by "experts."

  b. Emergency response plans are audited by corporate or third party personnel, who check to see that the plans include all of the content that is mandated by corporate policy and regulatory requirements. However, auditors do not normally have the experience needed to spot facility-specific deficiencies.

  c. Emergency response programs and plans are jointly prepared by unit personnel and trained emergency management personnel, and periodically reviewed by independent experts.

18.3.2 Prepare for Emergencies

Identify Credible Accident Scenarios Based on Hazards

6. List accident scenarios that represent the range of consequences identified in previous hazard identification and risk assessment work activities.

  a. The emergency response plan consists of a basic evacuation plan, personnel accountability procedure, and emergency contact phone numbers. It does not address any specific accident scenarios.

  b. Work products from the *risk* element are used to bound the range of accidents that might occur.

  c. The link between information generated by the *risk* element and the scenarios considered when developing emergency management and response plans is clear.

7. Expand the list of accident scenarios based on expert opinion.

  a. The emergency response plan consists of a basic evacuation plan, personnel accountability procedure, and emergency contact phone numbers. It does not address any specific accident scenarios.

  b. The emergency response plan addresses common accident scenarios such as structural fires, severe weather, security incidents, and so forth.

    c.   When considering contingencies, planners review work products from the *risk* element to help identify unique high-risk scenarios, such as the presence of water-reactive materials in an area, as well as generally applicable and credible accident scenarios such as fires and explosions.

    d.   In addition to item (c), emergency management planners consider a wide range of issues, including transportation incidents and business continuity planning.

8.   Critically review the accident scenario list and (1) remove scenarios that are not credible or are very unlikely to be severe enough to warrant emergency response, (2) consolidate scenarios that appear to be very similar in terms of effects and tactics that might be used for response, and (3) ensure that the list includes both worst credible accident scenarios and more likely, less severe, scenarios.

    a.   The emergency response plan addresses generic types of accident scenarios, such as fires, explosions, natural disasters, and security events.

    b.   The emergency response plan addresses fire and explosion in a general manner, but it provides specific guidance for a release of each highly toxic material at the facility and provides scenario-specific plans for certain other events (tornado warnings, bomb threats, etc.).

    c.   The entire range of credible accident scenarios is reviewed and scenarios are ranked based on risk. Based on this, emergency management plans are prepared in a comprehensive, yet efficient, manner.

Assess Credible Accident Scenarios

9.   Assess the range of accident scenarios in terms of types of consequences, such as fire, explosion, toxic release, and so forth, and the footprint where these consequences might be felt.

    a.   Emergency management planners only consider onsite emergencies, such as structural fires.

    b.   Accident scenarios are addressed by type (fire, explosion, etc.), but no effort is made to determine the degree or range of effects that might result from credible accidents.

    c.   Accident scenarios are evaluated, and consideration is given to the potential severity of the consequences. For example, evaluation of fire scenarios includes the potential for pool fires, jet fires, and vapor cloud explosions.

    d. Consequence modeling is used to predict accident effects. Models are used to predict the distance to specified endpoints for representative high-consequence scenarios.

Select Planning Scenarios

10. Based on the results of work activities to identify and assess planning scenarios, select a group of scenarios that span the range of effects and types of consequences (fires, explosions, toxic vapor releases, etc.). Include credible worst case scenarios as well as more likely (and less severe) accident scenarios, and span the gamut of different tactical response activities.

    a. Plans are based on hazards rather than accident scenarios.

    b. Plans are based on a few limiting case scenarios, but personnel recognize that facility emergency responders and local emergency response agencies cannot provide the required resources for some of the "worst case" scenarios. Relatively little effort is applied to more likely and less severe planning scenarios.

    c. Plans are based on a wider range of scenarios, and if planners determine that the resources available to respond are less than adequate, additional effort is applied to either increasing response capabilities or otherwise reducing risk associated with these limiting case scenarios.

11. Model the expected impacts from the planning scenarios to determine the geographical area that might be affected by each scenario.

    a. Plans are based on hazards present at the facility, such as fire/explosion resulting from a release of flammable materials, but the specific scenarios that could lead to such an event have not been analyzed.

    b. A few worst case scenarios have been modeled, but limited effort has been applied to identifying the range of credible planning scenarios, and personnel generally agree that these worst case scenarios are not credible. Therefore, these scenarios are typically not used to plan emergency response activities.

    c. A range of credible planning scenarios have been modeled, and the results of this effort are (1) used in planning response activities and (2) compiled for use by emergency responders to estimate distances to certain end effects and, on that basis, specify appropriate actions to protect the public.

12. Expand the range of accident scenarios to include events that are not process specific, such as confined space rescue or vertical rescue, if (1) the ERT will be called on to respond in these situations and (2) special training or equipment is required to execute the response.

a. Plans are based solely on process hazards.

b. Emergency response plans and capabilities are limited by the budget. Scenarios that require unbudgeted training or equipment are generally relegated to municipal fire departments and other local emergency response agencies.

c. Emergency response planners have evaluated a wide range of accident scenarios, along with the required training and equipment, and have made risk-informed decisions regarding which scenarios warrant additional equipment and training.

Plan Defensive Response Actions

13. Develop a written emergency management plan that specifies actions to protect and account for employees, contractors, and visitors.

   a. Plans have been developed, but little consideration is given to implementing those plans under adverse conditions.

   b. Plans are designed to work under adverse conditions, such as in severe weather, at night, or after a severe accident.

   c. Plans are developed based on minimum unit staffing and training level, and planners closely monitor changes to any constraints, such as the number of persons assigned to assist in evacuation and headcount on each shift, to help ensure that plans continue to be viable.

14. Practice the emergency management plans.

   a. Emergency evacuation drills are held annually. The drills are announced and occur during daylight hours.

   b. A variety of announced drills (evacuation, shelter-in-place, etc.) are conducted periodically.

   c. In addition to item (b), unannounced drills are periodically conducted at random hours, with particular emphasis on the accuracy and speed of reporting personnel accountability.

Plan Offensive Response Actions

15. Decide if emergency response offensive actions will be executed by a facility-sponsored ERT, via mutual aid agreements with nearby facilities, or solely by local authorities.

   a. Management intends for operators to respond to emergencies in the unit until the supervisor decides to terminate the response and evacuate to a safe location.

b. The decision to field an ERT capable of taking offensive action is based primarily on the perceived benefit of maintaining control of emergency response actions (versus turning over incident command to local authorities).

c. The decision to field an ERT capable of taking offensive action is based on perceived risk.

16. Establish unit or building preplans that address the range of accident scenarios that have been identified.

a. Planning is based on hazards rather than potential accident scenarios.

b. Planning scenarios are based on perceived risk, and capabilities are provided to address planning scenarios in the highest risk group.

c. Planning scenarios are intentionally selected to span the range of risk and required capabilities. For example, a confined space rescue capability is provided even though the risk associated with confined space entry is perceived to be low.

Develop Written Emergency Response Plans

17. Develop a written plan or suite of plans that address emergency management; if multiple plans are developed, provide links or references to relevant parts of other documents so that a user can quickly find information in the event of an emergency.

a. Emergency response plans are a combination of customized generic plans (e.g., for facility fires) and ad hoc plans that have been issued as policies, procedures, and memos over the life of the facility.

b. Emergency response plans have been consolidated into a manual (or multi-volume set of manuals) with a table of contents.

c. Emergency response plans have been consolidated into a manual, and appropriate cross references exist to help the user quickly locate particular sections, forms, contact lists, and so forth.

d. In addition to the written manual, the current version of the emergency response plan is stored in a secure location on the facility's computer network, and a current copy is also maintained on a laptop computer that is kept in the emergency response van. The plan includes hyperlinks to facilitate navigation between documents.

Provide Physical Facilities and Equipment

18. Based on the plans that are developed, provide the facilities and equipment necessary to execute the plans.

    a. Equipment and facility needs are generally understood, but some needs remain unfunded.

    b. Equipment and facility needs are periodically reviewed. Budget approval is based on risk and an assessment of how the proposed investment will affect risk.

19. Store the emergency response equipment where it will be accessible.

    a. Emergency response equipment is generally stored in a trailer near the process area.

    b. Considerable thought has been given to selecting storage locations, including how the proposed location(s) might become inaccessible in the event of an incident.

    c. Considerable thought has been given to selecting storage locations, and any changes to buildings or roads that might affect traffic patterns are reviewed via the *management of change* element to help ensure that the changes do not impact access to emergency response equipment or block access to process units by emergency responders.

Maintain/Test Facilities and Equipment

20. Identify emergency response equipment, including required inspections, tests, and other preventive maintenance or replacement activities, and establish a system to ensure that equipment is properly maintained and tested.

    a. Emergency response equipment is identified, and the ERT leader is generally recognized as the person responsible for maintaining the equipment.

    b. Emergency response equipment is identified, and the ERT leader has established a paper-based system to plan and execute maintenance activities. Records are kept in a file in the ERT leader's office.

    c. Emergency response equipment is identified, and maintenance of the equipment is tracked in a CMMS that also ensures that records are maintained.

21. Periodically inventory consumable emergency response equipment, paying particular attention to expiration dates.

    a. The ERT or team leader periodically performs a visual check to determine if additional supplies are needed.

    b. In addition to item (a), the check includes an examination of expiration dates for critical supplies.

    c. All critical ERT supplies and consumable materials are periodically inventoried using a written checklist. Expiration dates are checked, and the records of the inventory are kept on file for a designated period of time.

22. Identify emergency evacuation equipment, including required inspections, tests, and other preventive maintenance or replacement activities, and establish a system to ensure that equipment is properly maintained and tested.

   a. Emergency alarms are routinely tested. However, no procedures are in place to support testing of the evacuation alarm.

   b. Emergency alarms are routinely tested in accordance with written procedures. The procedures include steps to (1) confirm that the alarm can be heard through all potentially occupied areas of the facility and (2) use a variety of activation means, so that all of the various activation means are tested over time.

   c. In addition to item (b), the facility has identified all of the equipment that is required for orderly emergency evacuation and routinely tests or maintains this equipment.

Determine When Unit Operator Response is Appropriate

23. Include in the emergency response plans, either directly or by reference, procedures used by operators and other unit personnel to address small spills/releases and incipient fires.

   a. Operators rely on generic instructions and training, such as information contained in MSDSs or fire extinguisher training, to address a small spill/release or to put out an incipient fire.

   b. Operator response actions are included in written procedures and are emphasized in initial and refresher training, including guidance on when it is appropriate to discontinue the response action and evacuate to a safe location.

   c. In addition to item (b), initial and refresher training includes hands-on training for fighting incipient fires and for cleaning up mock spills.

Train Emergency Response Team Members

24. Train incident commanders and all ERT members on all of the skills needed to effectively and safely mount an emergency response or rescue effort.

   a. ERT members receive initial and refresher training, mostly conducted by third-party trainers in a classroom environment. The training primarily covers basic skills such as donning and checking protective gear.

    b.   In addition to classroom training, all ERT members participate in at least one drill or significant hands-on exercise or training event, such as "fire school," each year.

    c.   In addition to classroom and hands-on training, emergency response drills are as realistic as possible. Drills often occur in process areas under realistic conditions, for example, at night without benefit of facility lighting.

Plan Communications

25. Provide a means to alert all personnel to the emergency and what actions they should take to protect themselves.

    a.   A single plant-wide emergency alarm is used.

    b.   The plant-wide alarm includes a code (e.g., the code may be a combination of long and short horn blasts) that indicates that an emergency has been declared, provides the location of the emergency, and instructs workers in what actions they should take

    c.   In addition to the code, the facility has a loudspeaker system that is used to describe the nature of the emergency and provide verbal instructions to workers.

    d.   The facility has several layers of alarm systems with many automated, planned announcements so that the announcements can be made quickly and accurately.

26. Provide equipment for emergency communications, including (1) employee notification, (2) communication protocols for ERT members, (3) communication protocols with external authorities/ responders, and (4) communications that are not directly part of the response effort, for example, communications with the media and other stakeholders.

    a.   Emergency responders intend to use the facility's telephone and radio systems.

    b.   Some radios are equipped with a second high-priority channel, and all ERT members carry these specially equipped radios.

    c.   The communication plan extends beyond the ERT; the emergency operations center is equipped with radios that can communicate with external responders, such as local authorities, mutual aid groups.

    d.   The communication plan addresses the full gamut of communications issues, including communication needs of emergency responders, communication among facility management, and communication with stakeholders, corporate officers, and the media.

Inform and Train All Personnel

27. Ensure that all personnel (1) are aware of the emergency response program, (2) understand the facility's policy governing actions to take in an emergency, and (3) know how to recognize and report an emergency.

    a. Emergency recognition and response is covered in new employee orientation.

    b. Unit-specific emergency recognition and response training is provided to new operators and maintenance employees.

    c. All facility personnel are trained to the "first responder – awareness level." Anyone who has not completed this training cannot enter a process area without an escort. All operators are trained to the "first responder – operations level."

    d. In addition to item (c), first responder refresher training is provided annually.

28. Ensure that all personnel at the facility, including contractors, can recognize emergency alarms and know what actions to take for each type of alarm.

    a. Alarm response is addressed in initial training provided to all new employees and contractors. It is reinforced by visual aids posted in conspicuous locations that show evacuation routes and assembly areas.

    b. In addition to item (a), alarm response is further emphasized during evacuation drills and periodic tests of the alarm systems.

    c. In addition to item (b), worksite inspection checklists include an item to ask a random selection of employees and contractors to describe the various alarms and state what they should do in the event that the alarm is sounded. Immediate retraining is provided to anyone who does not answer the question properly, and training is reviewed if a consistent pattern of incorrect answers emerges.

29. Train managers and technical personnel who take an active role in emergency management, but who do not direct or actively participate in the tactical response.

    a. Managers and technical personnel assigned to perform specific activities are provided with written procedures and must annually sign a form stating that they have read and understood the procedures.

    b. Initial training is provided to managers and technical personnel who are assigned to take an active role in emergency management.

c.   Initial and refresher training is based on plant procedures. Some drills and exercises are designed to measure the performance of persons who are not part of the ERT but who are assigned to take an active role in emergency management activities.

30. Ensure that neighbors know what to do in the event they are notified of an emergency.

a.   Proper community response to alarms is described in a mailing or community meeting.

b.   Proper community response to alarms is covered frequently at public events, and emphasized as part of the *outreach* element (see Chapter 7).

c.   Efforts to continuously improve the capability to alert neighbors are pursued and supported by the facility.

Periodically Review Emergency Response Plans

31. Maintain emergency response plans current and accurate, and periodically review the plans.

a.   Plans are reviewed and recertified annually.

b.   In addition to item (a), plans are reviewed before and after drills or exercises and response actions determine if the plans need to be improved or expanded.

c.   In addition to item (b), a formal link or step is included in the *management of change* element to determine if a proposed change will affect any aspect of the emergency response plans, and this step often results in (1) an update to the emergency plans or (2) revision or denial of the requested change.

18.3.3 Periodically Test the Adequacy of Plans and Level of Preparedness

Conduct Emergency Evacuation and Emergency Response Drills

32. Periodically conduct drills to assess the (1) effectiveness of the plan and (2) state of readiness of the ERT.

a.   Drills are limited to actions by the ERT with little or no integration with operations.

b.   Drills periodically involve joint exercises with external emergency response agencies, and are sometimes covered by the media. Independent observers are assigned to assess performance.

c.   In addition to item (b), drills include operations employees. In advance of the drill, the drill planners provide specific instructions to operators to perform certain credible actions that may complicate the emergency response activities. Doing this helps test the capability of the ERT to recognize and respond to the unexpected.

Conduct Tabletop Exercises

33. Periodically conduct tabletop exercises or other actions to train managers and other personnel who would help manage the crisis but who are not directly involved in the tactical emergency response activities.

    a. Drills are limited to actions by the ERT and facility evacuations.

    b. Management actions are integrated into major facility emergency response drills. Independent observers are assigned to assess performance.

    c. The emergency management program includes both (1) drills that are designed primarily to test actions by emergency responders and (2) tabletop exercises that are designed to test the actions of senior managers who are not directly involved in tactical response activities. Independent subject matter experts are assigned to assess performance of this group.

Practice Crisis Communication

34. Provide training and refresher training, as needed, on crisis communication.

    a. Managers and employees are instructed to not communicate with the media under any circumstances; this responsibility is assigned to the corporate spokesperson.

    b. Senior managers at the facility participate in crisis communications training and are specifically authorized to communicate with the media if the corporate spokesperson is not available.

    c. Senior managers at the facility are trained in crisis communication and other types of media communication, and are expected to be the local spokesperson for the company.

Critique Exercises, Drills, and Actual Responses

35. Conduct a formal critique using independent, experienced observers.

    a. The ERT gathers at the conclusion of each drill (or actual response) and discusses what worked well, and what did not go as planned. The ERT leader takes notes and decides what, if any, changes are needed.

    b. Key members of the ERT meet with members of the management staff after each drill or major response activity and discuss what changes should be made. Recommendations are resolved and tracked to conclusion using a formal management system.

c. Experienced responders who do not play an active role in the facility's emergency response program observe each drill or exercise and facilitate a formal critique. The discussion is documented; recommendations are resolved and tracked to conclusion using a formal management system. The same sort of critique and follow up is applied to response actions (except independent observers are not involved).

Conduct Assessments and Audits

36. In addition to exercises and drills, periodically evaluate the emergency management program to ensure that all of the elements required to maintain a dependable practice are in place and remain effective.

a. The emergency management program is included in the scope of periodic regulatory-driven audits. Program elements are compared to regulatory requirements by auditors who are knowledgeable of the requirements but who are not subject matter experts in emergency management.

b. The emergency management program is reviewed by subject matter experts within the company during periodic visits to the facility.

c. The emergency management program is periodically audited by experts within the company or experienced consultants. The audit protocol includes regulatory requirements, corporate standards, and best practices.

Address Findings and Recommendations

37. Record any deficiencies or recommendations for improvement resulting from exercises, drills, assessments, and audits; resolve them into action plans; and implement the action plans.

a. No formal debriefing takes place after drills, but the ERT leader and others involved in the drill take action to address any deficiencies that they noticed.

b. A formal debriefing takes place, actions are developed to address deficiencies, individuals are assigned to implement the actions, dates are established, and actions are communicated to the responsible person.

c. Actions resulting from critiques of drills/exercises, assessments, and audits are entered into a well-established action tracking system, and management periodically reviews the progress on each action.

38. Publish findings or recommendations that may hold wide interest to other facilities and might prod other facilities to reflect on, "Could that happen here?"

   a. A debriefing is held after drills, but no formal list of actions is recorded.

   b. The list of actions is kept on a company-wide intranet page, and key personnel at other facilities have read only access to this information.

   c. When evaluating deficiencies and developing action plans, facility personnel involved in emergency management activities, along with counterparts at the corporate level, consider if any issues might represent systemic weaknesses throughout the company. A conscious decision is made regarding the need to publicize any lessons learned on a wider basis.

18.4 EXAMPLES OF WAYS TO IMPROVE EFFECTIVENESS

This section provides specific examples of industry tested methods for improving the effectiveness of work activities related to the *emergency* element. The examples are sorted by the key principles that were first introduced in Section 18.2. The examples fall into two categories:

1. Methods for improving the performance of activities that support this element.
2. Methods for improving the efficiency of activities that support this element by maintaining the necessary level of performance using fewer resources.

These examples were obtained from the results of industry practice surveys, workshops, and CCPS member-company input. Readers desiring to improve their management systems and work activities related to this element should examine these ideas, evaluate current management system and work activity performance and efficiency, and then select and implement enhancements using the risk-based principles described in Section 2.1.

18.4.1 Maintain a Dependable Practice

Build a strong relationship with external response agencies and other organizations that will be part of the response or recovery team in the event of an emergency. For example:

- Take on a leadership role with the LEPC to increase influence in decision making by local authorities, for example, influence local emergency planners to fund a Reverse 911® service for the community.
- Consider donating equipment (foam, special cameras to help locate people in the event of a fire inside a large building, etc.) to the local fire department or other agency that would help with response activities at your facility, and encourage other nearby facilities to do the same.
- Establish contracts and plans with an environmental remediation contractor to quickly respond and mitigate any spread of contaminants.

18.4.2 Prepare for Emergencies

Develop unit-specific preplans. In many cases, emergency responders are unfamiliar with the layout of each specific unit. Unit-specific preplans that depict the general equipment layout along with the location of emergency response equipment (standpipes, fire extinguishers, etc.), egress routes, and special hazards on each level of the process area can be used both to familiarize responders with the area and to quickly review critical information prior to entering the area under emergency conditions.

Include emergency actions in job-specific training and refresher training. Facilities often focus their refresher training on standard operating procedures, possibly at the expense of the procedures that operators will need to follow in an emergency. In reality, operators are normally well aware of how to perform their jobs during most routine modes of operation, and may be less likely to remember all of the tasks associated with emergency actions or emergency shutdowns, such as starting and monitoring the emergency generator or diesel-powered fire water pump.

Address facility security in emergency plans. During an emergency, a reporter may sneak into a facility to obtain a better photo or closer look. This can (1) place the reporter in harm's way, (2) potentially interfere with emergency response activities or burden responders, for example, add to the work of the decontamination or medical response teams, or (3) result in embarrassing or trade secret information being published. If spectators are a potential issue, establishing facility security becomes even more important. As discussed at the start of this chapter, a large number of people who were killed in the 1947 disaster at Texas City were spectators who had gathered to watch the fire aboard the *Grandcamp*.

Coordinate emergency response plans for accidents and intentional attacks. The effects and response actions for accidents are often identical to those for intentional attacks. Moreover, the return on investment (ROI) for applying resources to emergency response may far exceed the ROI for additional investments in facility security because investments in facility security traditionally address a limited number of attack modes, whereas investments in emergency response will likely reduce the expected losses for a wide range of accident and intentional attack scenarios.

Apply a "train-the-trainer" approach. Providing training and periodic refresher training to emergency responders and others, such as first responders at the operations level, can be very expensive. Many large facilities find it cost effective to designate a few personnel to become trainers who can then train other employees at convenient times, reducing the cost for external trainers and potentially reducing overtime costs for trainees.

18.4.3 Periodically Test the Adequacy of Plans and the Level of Preparedness

Involve multiple agencies in emergency response drills. Many facilities intend to execute emergency response activities in concert with local response agencies and mutual aid partners. Conducting joint training will help uncover issues with communication, command and control, and different or incompatible response tactics. Just as important, joint training normally fosters improved teamwork.

18.5 ELEMENT METRICS

Chapter 20 describes how metrics can be used to improve performance and when they may be appropriate. This section includes several examples of metrics that could be used to monitor the health of the *emergency* element, sorted by the key principles that were first introduced in Section 18.2.

In addition to identifying high value metrics, readers will need to determine how to best measure each metric they choose to track. In some cases, an ordinal number provides the needed information, for example, total number of workers. Other cases, such as average years of experience, require that two or more attributes be indexed to provide meaningful information. Still other metrics may need to be tracked as a rate, as in the case of employee turnover. Sometimes, the rate of change may be of greatest interest. Since every situation is different, the reader will need to determine how to track and present data to most efficiently monitor the health of RBPS management systems at their facility.

18.5.1 Maintain a Dependable Practice

- *Number of meetings or other contacts with local emergency responders or the LEPC regarding emergency response plans.* These types of work activities can simply stop if nobody is watching, and facilities may have no indication of the failure to plan and coordinate activities unless an actual emergency occurs.

- *Number of meetings or other contacts with the community regarding how they will be notified of an emergency and what they should if they are notified.* If these activities stop, facility neighbors may not know how to react to an emergency, and facility management may not be aware of this until an actual emergency occurs.

18.5.2 Prepare for Emergencies

- *Number of trained ERT members on each shift.* This metric should be closely tracked, particularly if the facility's intent is to have a fully functional team onsite at all times without having to call people in to the facility.
- *Number (or percent) of units that have up-to-date plans.* High values indicate that unit-level emergency plans are being maintained.
- *Number of errors/omissions in the emergency response plan or its annexes discovered during drills and training.* A high error rate indicates that the system to periodically review plans is ineffective.
- *Percent of preventive maintenance work orders for emergency response equipment that are past due.* Maintaining emergency response equipment is very important; however, because it is seldom used, failure to maintain the equipment may go unnoticed or unreported.
- *Number (or percent) of emergency response plans or annexes to the plan that are past due for periodic review.* High values indicate that weaknesses in these plans may go unnoticed until an emergency occurs.
- *Percent of failed tests or inspections of emergency response equipment.* A high rate of test failures indicates that the maintenance program needs to be enhanced.
- *Supply status for ERT consumable supplies.* Consumable supplies should be present and not out of date. If the stock of ERT consumable supplies is chronically low, more rigor needs to be applied to this work activity.

18.5.3 Periodically Test the Adequacy of Plans and the Level of Preparedness

- *Percent of ERT members who are up to date on emergency responder training requirements.* Training is an essential part of preparing for emergencies. Poor compliance with the training schedule indicates that execution of emergency response plans may be unreliable and puts the personnel involved at risk.

- *Fraction of drills that are conducted as scheduled.* Drills are a vital part of the *emergency* element. Poor compliance with the drill schedule indicates that execution of emergency response plans may be unreliable.
- *Number of changes to the emergency response tactics or logistics based on critiques of drills or other exercises.* No emergency response plan is perfect. If the critiques find no areas for improvement, it is likely that the critiques need to be more thorough.
- *Results of opinion surveys among operators regarding their perception of the unit's or facility's state of preparedness for emergencies.* Opinion surveys will often show if emergency response capabilities are improving or diminishing.

18.6   MANAGEMENT REVIEW

The overall design and conduct of management reviews is described in Chapter 22. However, many specific questions/discussion topics exist that management may want to check periodically to ensure that the management system for the *emergency* element is working properly. In particular, management must first seek to understand whether the system being reviewed is producing the desired results. If the organization's level of *emergency* is less than satisfactory, or it is not improving as a result of management system changes, then management should identify possible corrective actions and pursue them. Possibly, the organization is not working on the correct activities, or the organization is not doing the necessary activities well. Even if the results are satisfactory, management reviews can help determine if resources are being used wisely – are there tasks that could be done more efficiently or tasks that should not be done at all? Management can combine metrics listed in the previous section with personal observations, direct questioning, audit results, and feedback on various topics to help answer these questions. Activities and topics for discussion include the following:

- Review the range of unit-, building-, or scenario-specific preplans that have been developed to determine if they adequately address recent changes that have occurred at the facility, new regulatory requirements, and any other issues that would affect the scope of the *emergency* element.
- Compare worst credible accident scenarios that have been identified in unit process hazard reviews or other documents to the preplans in the emergency response plan to determine if (1) they are addressed and (2) the preplans adequately reflect the potential magnitude of the consequences associated with the accident scenario.

- If the facility depends primarily on local agencies for emergency response, review any changes that have recently occurred in any of their capabilities. Do any of the changes pose unacceptable risk to the facility?
- If the facility is part of a mutual aid response organization, review changes in the capabilities of the mutual aid group. Have any members recently dropped out, and if so, does this pose unacceptable risk to the facility?
- Examine ERT staffing and any other resource needs for the *emergency* element. In particular, confirm that the facility has enough trained ERT members to respond in the intended manner (without having to call anyone in to work). If not, develop contingency plans until the ERT can be fully staffed. If this is a chronic problem, determine why employees are reluctant to join the ERT.
- Review the results of drills and tabletop exercises that were conducted since the previous management review. Also examine lessons learned from each drill and tabletop exercise and the status of improvement plans that resulted from these activities.
- If any situations occurred that required a significant emergency response since the previous review, review the performance of the ERT. If the event uncovered any systemic weaknesses in the management system for the *emergency* element, review the status of actions for addressing these issues.
- Review major training events that have taken place since the last review and training events that are planned for the upcoming year. Are any alternatives available that might provide better value, for example, better training at the same cost or equivalent training at a lower cost?
- Review lessons learned from response activities at other facilities within the company or industry. Are the lessons being effectively applied?

At most facilities that manufacture, process, or store hazardous materials, seriously considering the worst credible accident that might occur is uncomfortable. However, a workable plan must be in place for responding to incidents. This capability is developed and maintained through careful planning, effective training, well planned drills/exercises, and an ongoing commitment to the *emergency management* element. There is no time for planning, training, or other preparation once an explosion, toxic release, or fire occurs. Regardless of what questions are posed, managers should attempt to determine if the entire facility, including ERT members and all other personnel, is prepared for the range of credible incidents that might occur. Does the answer change if the incident occurs at 3 a.m. on Sunday? What about if no power or city water is available at the facility? The answers to these and other difficult questions can provide useful insight regarding the facility's level of emergency preparedness.

18.7 REFERENCES

18.1   Stephens, Hugh W., *The Texas City Disaster, 1947*, University of Texas Press, Austin, Texas, 1997.

18.2   Stringfield, William H., Emergency Planning and Management: Ensuring Your Company's Survival in the Event of a Disaster, Government Institutes, Rockville, Maryland, 1996.

18.3   *Guidelines for Technical Planning for On-Site Emergencies*, American Institute of Chemical Engineers, New York, New York, 1995.

Additional reading

Lees, Frank P., *Loss Prevention in the Process Industries, Hazard Identification, Assessment, and Control*, 2nd edition, Butterworth-Heinemann, Oxford, England, 1996.

Theodore, Louis, Reynolds, Joseph P., and Taylor, Francis B., *Accident and Emergency Management*, John Wiley and Sons, New York, New York, 1989.

IV. LEARN FROM EXPERIENCE

Risk-based process safety (RBPS) is based on four pillars: (1) committing to process safety, (2) understanding hazards and risk, (3) managing risk, and (4) learning from experience. To learn from experience, facilities should focus on:

- Investigating incidents that occur at the facility to identify and address the root causes.
- Applying lessons from incidents that occur at other facilities within the company and within the industry.
- Measuring performance and striving to continuously improve in areas that have been determined to be risk significant.
- Auditing RBPS management systems as well as the performance of work activities that make up the management system.
- Holding periodic management reviews to determine if the management systems are working as intended and if the work activities are helping the facility effectively manage risk.

This pillar is supported by four RBPS elements. The element names, along with the short names used throughout the *RBPS Guidelines*, are:

- Incident Investigation (*incidents*), Chapter 19
- Measurement and Metrics (*metrics*), Chapter 20
- Auditing (*audits*), Chapter 21
- Management Review and Continuous Improvement (*management review*), Chapter 22

The management systems for each of these elements should be based on the company's current understanding of the risk associated with the processes with which the workers will interact. In addition, the rate at which personnel, processes, facilities, and products change, or incidents occur (placing demands on resources), along with the process safety culture at the facility and within the company, can also influence the scope and flexibility of the management system required to appropriately implement each RBPS element.

Chapter 2 discussed the general application of risk understanding, tempered with knowledge of resource demands and process safety culture at a facility, to the creation, correction, and improvement of process safety management systems. Chapters 19 through 22 describe a range of considerations specific to the development of management systems that support efforts to learn from a wide variety of sources. Each chapter also includes ideas to (1) improve performance and efficiency, (2) track key metrics, and (3) periodically review results and identify any necessary improvements to the management systems that support commitment to process safety.

19

INCIDENT INVESTIGATION

The in-flight failure of the Columbia Space Shuttle on January 16, 2003, which resulted in the deaths of the seven-member crew, is a classic example of a failure to adequately investigate and address abnormal system performance. To protect the shuttle's fragile thermal protection system, the shuttle design specifications required no shedding of foam insulation. Despite this requirement, at least 65 missions had experienced foam loss, including six instances of foam shedding from the bipod ramp, the same location that ultimately caused the loss of the Columbia. For example, on October 7, 2002, during launch of STS-112, foam loss caused 707 dings in the thermal protection system, 298 of which were greater than an inch in one dimension. One piece of debris knocked off a thermal protection tile, exposing the orbiter's skin to the heat of re-entry. Fortunately, the missing tile on STS-112 happened to be at the location of a thick aluminum plate, so a burn-through did not occur. Despite these near miss incidents and extensive experience that was contrary to the design requirements of the shuttle, NASA did not thoroughly investigate the root causes of foam separation. Experience with successful missions had lulled NASA into believing that the shuttle was immune from damage from shedding debris. Proper investigation of these near miss incidents could have prompted NASA to change the shuttle design to prevent the shedding of the foam, to decrease the shuttle's susceptibility to a foam strike, or both (Refs. 19.1 and 19.2).

19.1　ELEMENT OVERVIEW

Developing, sustaining, and enhancing the organization's incident investigation competency is one of four elements in the RBPS pillar of *learning from experience*. This chapter describes the characteristics of an effective incident investigation program and how organizations might begin to enhance their own incident investigation process. Section 19.2 describes the key principles and essential features of a management system for this element. Section 19.3 lists work activities that support these essential features, and presents a range of approaches that might be appropriate for each work activity, depending on perceived risk, resources, and organizational culture. Sections 19.4 through 19.6 include (1) ideas for improving the effectiveness of management systems and specific programs that support this element, (2) metrics that could be used to monitor this element, and (3) issues that may be appropriate for management review.

19.1.1　What Is It?

Incident investigation is a process for reporting, tracking, and investigating incidents that includes (1) a formal process for investigating incidents, including staffing, performing, documenting, and tracking investigations of process safety incidents and (2) the trending of incident and incident investigation data to identify recurring incidents. This process also manages the resolution and documentation of recommendations generated by the investigations. Figure 19.1 shows an overview of the *incidents* element activities.

At some facilities, the *incidents* element is used to assign blame to personnel involved in an incident. This approach results in ineffective recommendations being implemented. A more effective approach is to develop recommendations that address the systemic causes of the incidents. The *incidents* element is not a process to assign blame, but a process to develop effective recommendations to address the underlying, system-related causes of incidents.

19.1.2　Why Is It Important?

Incident investigation is a way of learning from incidents that occur over the life of a facility and communicating the lessons learned to both internal personnel and other stakeholders. Depending upon the depth of the analysis, this feedback can apply to the specific incident under investigation or a group of incidents sharing similar root causes at one or more facilities.

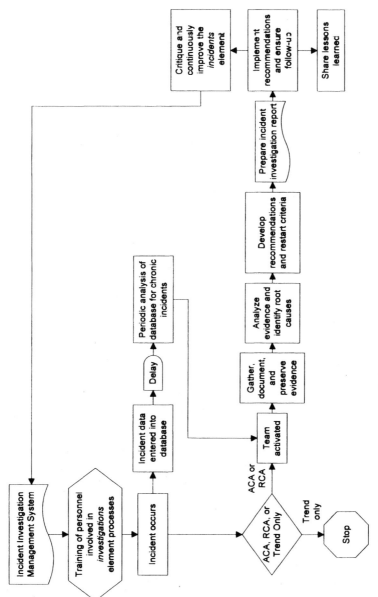

FIGURE 19.1. Incident Investigation Flowchart

Limited investigations may only provide feedback concerning the specific incident scenario by only focusing on the causal factors (sometimes referred to as direct causes) of an incident. Addressing the causal factors can reduce the frequency of the direct causes of the incident and/or reduce the consequences of future similar incidents.

Thorough investigations can provide feedback on the performance of the RBPS elements. By identifying and addressing the root causes of equipment failures and personnel errors, leveraged solutions can be developed that will reduce the frequency of entire categories of incidents and/or reduce the consequences of entire categories of incidents.

19.1.3 Where and When Is It Done?

Incident investigations are conducted whenever and wherever incidents occur; because performing investigations remotely is rarely effective. The investigation team is usually based near the incident scene to allow them to more efficiently collect data by conducting interviews, gathering physical evidence, and so forth. Analysis of the root causes of incidents may occur anywhere, but being near the personnel involved in the management of the facility is preferable, as it helps foster discussions with facility personnel.

Companies decide what combinations of consequences and frequencies are appropriate to trigger each type of incident feedback process: (1) formal investigation (root cause analyses [RCAs]), (2) less formal investigation (apparent cause analyses [ACAs]), and (3) trending of incident data with no immediate investigation performed. Figure 19.2 shows the relationship of these different levels of analysis.

The term apparent cause analysis was originally defined by the Department of Energy. Other terms that are typically applied to these different level of analyses include Level 1 and Level 2 analyses or basic and detailed analyses (Ref. 19.3).

19.1.4 Who Does It?

Personnel who have formal training in incident investigation or RCA techniques typically perform investigations and lead investigation teams. A multidisciplinary team is appropriate for incident investigations with greater consequences and risks. An individual or a two-person investigation team may investigate incidents with lower consequences and risks. Personnel throughout the company provide assistance to the investigation team. For incidents involving significant human injuries or incidents with potential regulatory impact, the legal department will often assume overall management of the investigation. Investigation teams should engage the company's public affairs group as the focal point for communications to the media and other external organizations (see the *outreach* element).

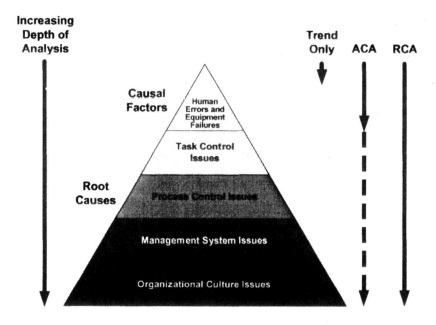

FIGURE 19.2. Incident Investigation Levels of Analysis

19.1.5 What Are the Anticipated Work Products?

The main products of an incident investigation system are: (1) basic data for all recorded incidents, including those that do not qualify for immediate investigation (trend-only incidents), (2) analysis reports for ACAs and RCAs, (3) identification of the causes of the incidents that are investigated, and (4) implemented recommendations (lessons learned and action items) that will reduce the risk of similar incidents. The results of the investigation process should be documented in a standard incident report form. Other types of work products may include up-to-date incident summary lists, action item tracking summaries, and trending data such as incident characteristics, causes, and recommendations. Data gathered during the review or generated by the investigation are usually kept at least until the action items and lessons-learned are implemented and communicated, and usually longer in accordance with company and regulatory requirements.

Outputs of the *incidents* element should be used to facilitate the performance improvement of other elements. In most cases, recommendations aimed at the root causes of an incident will involve modifications to other RBPS elements.

19.1.6 How Is It Done?

A company selects its incident investigation approach and documents the approach in its investigation procedures. When an incident occurs, it is categorized, investigated, and trended according to the procedures. Data, such as interviews, paper/electronic records, physical analyses, position observations, photos, videos, and so forth, are gathered; causal factors and root causes are identified; and recommendations are made. The incident database is periodically analyzed for trends that may indicate risk-significant recurring incidents that merit formal investigation. Once identified, analyses may be performed to identify the root causes of these trends.

19.2   KEY PRINCIPLES AND ESSENTIAL FEATURES

The *incidents* element is a vital process for learning from experience for facilities that manufacture, store, or otherwise use hazardous chemicals. Without an investigation program, a company typically suffers from repeated failures and incidents because less formal feedback mechanisms are not sufficient to identify effective recommendations. Incident investigation is a key process in the continuous improvement of the other RBPS elements addressed in this book.

Risk is a significant driver for the *incidents* element. For serious accidents, the actual consequences usually drive the level of effort expended in the investigation of the incident(s), but for less serious accidents, the actual and potential severity of the loss, combined with the likelihood of recurrence, influence the level of effort. For near misses, the potential consequences combined with the anticipated frequency of recurrence are used to determine the level of investigation effort. For chronic incidents, the actual consequences combined with the frequency of recurrence are used to determine the level of investigation effort.

Incident investigations are performed when (1) a single incident with large actual or potential consequences occurs or (2) an analysis of data indicates a large risk from a group of more frequent but less severe incidents. An investigation team typically consists of a leader and one to five additional members with varying backgrounds. Teams may include front-line personnel (operators and maintenance personnel), contractors, forensic experts, and system or process experts.

The company culture has a strong influence on the effectiveness of the investigation process. Performing investigations is usually only one of many duties assigned to facility personnel. Therefore, they often have a vested interest in completing the analysis quickly to allow them to get back to their regular job. If not countered by management pressure to dig deeply into the underlying causes of the incident, cursory analyses will often result. However, managers themselves are often reluctant to dig deeply because the team will probably discover weaknesses in the management systems. As a result, many companies experience the same failures over and over because the root causes are never identified and corrected.

The following key principles should be addressed when developing, evaluating, or improving any management system for the *incidents* element:

- Maintain a dependable incident reporting and investigation practice.
- Identify potential incidents for investigation.
- Use appropriate techniques to investigate incidents.
- Document incident investigation results.
- Follow through on results of investigations.
- Trend data to identify repeat incidents that warrant investigation.

The essential features for each principle are further described in Sections 19.2.1 through 19.2.6. Section 19.3 describes work activities that typically underpin the essential features related to these principles. Facility management should evaluate the risks and potential benefits that may be achieved as a result of improvements in this element. Based on this evaluation, management should develop a management system, or upgrade an existing management system, to address some or all of the essential features, and execute some or all of the work activities, depending on perceived risk and/or process hazards that it identifies. However, these steps will be ineffective if the accompanying *culture* element does not embrace the use of reliable management systems. Even the best management system, without the right culture, will not produce effective incident investigations.

19.2.1 Maintain a Dependable Incident Reporting and Investigation Practice

Implement the program consistently across the company. Investigations are a responsibility that is typically shared across many personnel in the company. In order to achieve consistency, investigators need a defined process and clear expectations. In addition, because most investigation personnel only perform investigations on an occasional basis, they typically need the guidance provided by a policy or procedure and/or element expert. The more detailed the guidance provided to the teams through the program documentation and through an element expert, the greater the level of consistency that will be achieved.

Providing a champion and coordinator for the program will enhance consistency. The champion, usually an upper level manager within the company, helps to set management expectations for the investigation. The coordinator supports the team; helps to overcome routine obstacles encountered by the team, such as obtaining supplies, coordinating logistics, and processing reports; and provides continuity from one investigation to the next.

Define an appropriate scope for the incident investigation element. What does the program cover? What types of incidents does the company want to be reported and investigated? Which should simply be trended? Specific examples of the types of incidents that trigger a detailed investigation (an RCA), a lower level investigation (an ACA), or trending of the incident will help personnel understand the expectations of the company. Abstract criteria, such as challenging the last line of defense in a system, are less desirable because they result in wide variations in interpretation, and therefore, wide variations in implementation. If the program includes other types of losses besides safety (reliability, quality, productivity, security), providing specific examples of these types of incidents is also appropriate.

Involve competent personnel. RCAs require a greater level of expertise in the analysis techniques than ACAs. The higher skill level is usually acquired through additional training and experience in applying the techniques. Because RCAs make up only a fraction of the incident investigations typically performed in a company, they are often assigned to a core group of investigators. ACAs are performed more frequently and do not require as much expertise in the investigation techniques. Therefore, ACAs may be performed by a larger group of personnel, with oversight and assistance from the RCA investigators.

Everyone in the company performs incident reporting. Often, only one person, the individual involved, knows about a near miss. If that person doesn't report it, no one will. Therefore, all personnel must be aware of the types of incidents that the company wants to be reported and how to report them.

Monitor incident investigation practices for effectiveness. Implementation of the *incidents* element provides a feedback mechanism for most of the other RBPS elements. However, a poorly performing *incidents* element will probably not trigger an assessment of the *incidents* element itself. To determine if the program is achieving its goals, metrics are needed to monitor the effectiveness of the *incidents* element activities in reducing repeat incidents and improving company performance. When the metrics point to ineffective implementation of the *incidents* element, intervention is required to identify recommendations.

19.2.2 Identify Potential Incidents for Investigation

Monitor all sources of potential incidents. In order to investigate an incident, it must first be identified. While some incidents are obvious, others, particularly near misses, may go unnoticed. Several methods can be used to identify incidents, including verbal reports from personnel and reviews of documents, such as logs, work orders, emergency response activations, and data trends. Careful review of these data sources allows near misses and incident precursors to be identified.

Ensure that all incidents are reported. The most important source of data for the identification of incidents is front-line personnel. Periodic assessment of the effectiveness of incident reporting by front-line personnel and others in the company assists in identifying barriers to reporting incidents.

Initiate investigations promptly. The faster the investigation gets started, the easier it will be to collect accurate data. Clear responsibilities for initiating an investigation will allow for the rapid initiation of an investigation. Having a list of currently qualified investigators will aid in getting the investigation team formed quickly. Packing an incident investigation kit with basic tools and equipment will help the team start investigating quickly, rather than wasting time searching for routine items.

19.2.3 Use Appropriate Techniques to Investigate Incidents

Collect appropriate data during the investigation. Performing an investigation without collecting all of the required data often results in the identification of incorrect causes and the development of ineffective recommendations. Developing a list of the types of data typically available and/or required for an analysis will help the team to efficiently collect all of the data it needs. Listing the data collected in the investigation report also helps the reviewers understand the depth of the analysis.

Interface with the emergency management element. Interfaces with the *emergency* element are required to ensure preservation of evidence and to determine if emergency response is within scope of an RCA.

Use effective data collection methods. Investigations usually require extensive interviews of personnel. Because these interviews often focus on errors or performance gaps involving the individual being interviewed, or a co-worker, they can be emotionally charged. Proper interviewing techniques can greatly increase the effectiveness of the interviews. Preserving physical data also presents a number of challenges; preparing a test plan can greatly improve the usefulness of the data collected from testing.

Use appropriate techniques for data analysis. Consistent use of analysis techniques is a common struggle for most companies. Providing one or two acceptable techniques allows the company to achieve proficiency in the application of the technique(s). A step-by-step procedure for using each analysis technique may not be possible because of the variety of types of incidents investigated. However, detailed guidance that addresses the typical analysis process, along with the common problems faced by the team, can be provided. Unusual issues often require the advice of the program coordinator or senior investigators to resolve.

Investigate causes to an appropriate depth. ACAs and RCAs need to consistently dig to the appropriate depth of causes. Guidance provided by the investigation techniques and by the *incidents* element procedures can assist in achieving this goal. For example, the company needs to ensure that root causes are identified when RCAs are performed. Providing sufficient guidance to the investigators to determine if the causes identified by the analysis are deep enough to be considered root causes will assist in achieving this goal. Some incident investigation methods use predefined lists of root causes to assist in this task. CCPS's *Guidelines for Investigating Chemical Process Safety Incidents* (Ref. 19.4) can provide additional guidance and examples on this topic.

Demand technical rigor in the investigation process. Technical rigor refers to the thoroughness of the arguments used to support the causes and recommendations identified by the investigation team. In most cases, the team will have some preconceived notions regarding the causes of an incident. In some cases, team members may have recommendations in mind before the analysis even begins. As a result, the team often examines data with a tainted view. Techniques that encourage the team to not only look at the data that support its preconceived ideas but also consider data that refute or disprove its theories tend to result in more rigorous investigations. This approach improves acceptance of the recommendations generated by the team because it allows management to assess the validity of the conclusions in light of all the available data, not just the data that support the team's preconceived ideas.

Provide investigation personnel with appropriate expertise and tools. The team that performs the investigation needs to have the technical knowledge to perform the investigation or at least the ability to understand the technical issues being investigated. This means the technical capabilities can either be provided as part of the team or as a resource to the team.

When contractor organizations, suppliers, or union labor are involved in an incident, having representatives from these organizations on the team is often desirable. This provides additional technical resources to the team and also makes the investigation more understandable and visible. This can lead to a more collaborative working relationship among the company, the facility, the union, the contractors, and the suppliers as promoted by the *involvement* and *contractor* elements.

Develop effective recommendations. Recommendations are developed for each cause (causal factor and root cause) identified by the analysis. Recommendations are tied to the causes and supporting data identified by the analysis. Without specific, actionable wording, recommendations can be implemented in less effective ways than anticipated by the investigation team.

19.2.4 Document Incident Investigation Results

Prepare incident investigation reports. A basic outline or report template can be provided to the investigation teams to provide report consistency across different investigations. However, teams will have to adapt the outline to the specifics of the current investigation. One approach that reduces the amount of time required to generate the report is to use the results of the analysis techniques as the core of the report. For example, a logic tree, such as a cause and effect tree, or a time-based cause and effect chart, such as a causal factor chart, that was used to analyze the data provides a wealth of information regarding the incident. Its inclusion in the report can significantly reduce the incident description and cause discussions in the report.

Provide clear linking between causes and recommendations. A clear linkage between causes and recommendations in the analysis report will make it easier for the reader to understand the connections between the incident, the causes identified, and the recommendations. The supporting and refuting data associated with each identified cause will also be readily apparent to the reader. A clear linkage between causes and recommendations also allows management to assess the thoroughness and validity of the investigation. In certain investigations, the legal department may want to control any report documenting a clear connection between causes and effects in an effort to reduce the company's liability exposure. More information regarding the legal issues surrounding investigations can be found in CCPS's *Guidelines for Investigating Chemical Process Safety Incidents* (Ref. 19.4).

19.2.5 Follow Through on Results of Investigations

Resolve recommendations. The investigation process produces positive results for the company only when the recommendations generated by the team are resolved and any action items are implemented. The most significant recommendations usually focus on long-term solutions. As such, the implementation of these recommendations is normally not needed to get the system restarted, but is intended to prevent recurrence of the incident over the next weeks, months, and years. Failure to implement the action items resulting from these recommendations may not be obvious until it is too late (when the next incident occurs) unless recommendations are tracked to completion. A database, spreadsheet, or other computer-based application is the most practical method for tracking recommendation resolution.

Periodically, the status of investigation recommendations is reviewed with responsible managers. This review facilitates the identification and resolution of obstacles to the implementation of the investigation recommendations.

Communicate findings internally. For the program to remain effective, participation is needed from a wide variety of personnel, primarily in the data-gathering step. To encourage their continued support and participation in the program, these workers need some feedback on the performance of the program. Their primary need is to know that the time they spent in providing data to the investigation team led to improvements in the company. In other words, they need to know that their investment in the investigation was worthwhile.

Communicating investigation results internally can be accomplished in a variety of ways. Typical methods include (1) incorporating results into formal presentations that are specific to an incident, (2) incorporating results into periodic presentations that may incorporate lessons learned from multiple analyses, (3) presenting the lessons learned during tailgate safety meetings, (4) issuing the results as required reading, (5) distributing them through e-mail, or (6) posting the investigation results on a facility bulletin board. The method(s) used should be appropriate to the significance of the recommendations. Too many presentations on issues with lower risk significance issues can lead to overload.

Communicate findings externally. Incidents that occur at one facility often provide opportunities to strengthen systems at other facilities. For example, the team investigating a pump seal failure at one facility may develop recommendations that would effectively control risk at other facilities. However, the investigation team may not be able to identify the generic implications of the incident for other facilities. As a result, external communications of the incident investigation findings are vital for other facilities and companies to perform this assessment.

Maintain incident investigation records. Once the investigation recommendations are approved, the company's attention normally shifts to implementing the investigation team's recommendations. However, the facility should maintain permanent, retrievable records of the investigation to allow future investigations to take advantage of the information compiled and generated by the team. Investigation reports are also helpful when performing risk analyses. The information in these reports can help identify new hazards and risks. There may also be legal issues related to retention of investigation documents. The company's legal department should be consulted when structuring the document retention policy to limit the company's potential liability.

During an investigation, the team concentrates on assessing the effectiveness of the RBPS elements. To achieve this goal, the team must know if prior investigations have been effective in improving the management systems of the company. Records from prior investigations are fundamental to this assessment.

19.2.6 Trend Data to Identify Repeat Incidents that Warrant Investigation

Would tread separation on one vehicle's tire make the manufacturer think about potential generic problems at the manufacturing facility? How about multiple tread separations, on one particular vehicle model, most associated with a single brand and type of tire manufactured at a Decatur, Illinois, plant (Ref. 19.5)? The only way to see this link is through trending of data. Trending facilitates looking across investigations performed by a variety of personnel to identify common underlying threads between the incidents. Trending is particularly useful in identifying lower consequence incidents that have a medium to high frequency of occurrence. These incidents usually do not justify an analysis based on the consequences of a single occurrence. However, they collectively represent a significant risk to the company that warrants an investigation.

Log all reported incidents. A database of incident characteristics is key to identifying repeat incidents that warrant an investigation. The database should be structured to be consistent with the incident reporting requirements and the database analysis requirements. If management has no plans to analyze the data for trends, entering it into the database is probably not beneficial. Only collect data that will be analyzed.

Analyze incident trends. The incident database should be periodically analyzed for trends to identify recurring causes of incidents and repeat incidents. The incidents can also be sorted from highest to lowest consequence or potential consequence. Then, commonalities between the higher consequence incidents may be identified. Sorting by incident characteristics and causes can also reveal significant common causes of incidents. This periodic analysis of the database should be assigned to a specific individual, typically the program coordinator, and the results should be reviewed with management. Trends typically result in triggering an analysis of these chronic issues. However, data trends must be carefully interpreted to understand their underlying causes. For example, a sharp increase in reported incidents may be the result of a change in company culture, where personnel become more conservative in their reporting, rather than a change in the performance of the facility. A new individual may simply have a different interpretation of the reporting requirements. An investigation may be required to understand this type of relationship.

19.3 POSSIBLE WORK ACTIVITIES

The RBPS approach suggests that the degree of rigor designed into each work activity should be tailored to risk, tempered by resource considerations, and tuned to the facility's culture. Thus, the degree of rigor that should be applied

to a particular work activity will vary for each facility, and likely will vary between units or process areas at a facility. Therefore, to develop a risk-based process safety management system, readers should perform the following steps:

1. Assess the risks at the facility, investigate the balance between the resource load for RBPS activities and available resources, and examine the facility's culture. This process is described in more detail in Section 2.2.
2. Estimate the potential benefits that may be achieved by addressing each of the key principles for this RBPS element. These principles are listed in Section 19.2.
3. Based on the results from steps 1 and 2, decide which essential features described in Sections 19.2.1 through 19.2.6 are necessary to properly manage risk.
4. For each essential feature that will be implemented, determine how it will be implemented and select the corresponding the work activities described in this section. Note that this list of work activities cannot be comprehensive for all industries; readers will likely need to add work activities or modify some of the work activities listed in this section.
5. For each work activity that will be implemented, determine the level of rigor that will be required. Each work activity in this section is followed by two to five implementation options that describe an increasing degree of rigor. Note that work activities listed in this section are labeled with a number; implementation options are labeled with a letter.

Note: Regulatory requirements may specify that process safety management systems include certain features or work activities, or that a minimum level of detail be designed into specific work activities. Thus, the design and implementation of process safety management systems should be based on regulatory requirements as well as the guidance provided in this book.

19.3.1 Maintain a Dependable Incident Reporting and Investigation Practice

Implement the Program Consistently Across the Company
1. Establish and implement written procedures to report on, collect data related to, investigate, and learn from incidents
   a. A general policy statement exists, but it contains no specific guidance.

b. In addition to item (a), an investigation form is used during investigations.

c. In addition to item (b), an investigation procedure is used.

d. In addition to item (c), detailed guidance and specific examples are provided to investigation personnel.

2. Assign a job function as the champion of the *incidents* element to:
   - Provide guidance to the RCA analysts.
   - Monitor the effectiveness of the program.
   - Assist in removing roadblocks encountered by the investigation teams.

   a. The safety department champions the *incidents* element.

   b. A champion of the *incidents* element is formally designated.

   c. In addition to item (b), a specific time commitment is dedicated to the function, with clearly defined roles and responsibilities.

3. Assign an incident investigation coordinator to:
   - Assist incident investigators with the details of performing an investigation.
   - Perform a review of cause trends (categories or codes) to increase consistency across the company.

   a. The safety manager is the *incidents* element coordinator.

   b. An *incidents* element coordinator is formally designated.

   c. In addition to item (b), a specific time commitment is dedicated to the function, with clearly defined roles and responsibilities.

Define an Appropriate Scope for the Incident Investigation Element

4. Define the technical scope of the *incidents* element by specifying the risk and consequence thresholds that trigger different levels of investigations.

   a. Only catastrophic incidents trigger an analysis.

   b. Catastrophic incidents trigger RCAs, and less risk significant incidents trigger ACAs.

   c. Serious incidents trigger RCAs, and minor incidents trigger ACAs.

   d. Minor incidents trigger RCAs, and incidents with operational impacts trigger ACAs.

5. Communicate the technical scope of the *incidents* element by providing facility-specific examples of the types of incidents that are investigated, and the appropriate level of effort for each type or level of analysis.

    a. General guidance identifies the types of incidents that are investigated, for example, those with large releases or significant injuries.

    b. Specific examples clarify the incidents that are investigated, such as death; injury; severe property damage; actuation of designated safeguards, including safety critical systems, evacuations, rescues, and shelter-in-place; releases; and environmental damage.

    c. In addition to item (b), guidance on the investigation effort and outcomes for each type of investigation is provided to investigators.

6. Identify facility-specific examples of near miss incidents and train workers to appropriately report near misses.

    a. Worker training includes general near miss reporting and investigation guidance (e.g., challenging the last line of defense, challenging two or more safety barriers).

    b. Worker training includes facility-specific examples of near misses that are investigated.

    c. In addition to item (b), guidance is provided on the investigation effort and outcomes for each type of incident.

Involve Competent Personnel

7. Provide awareness training and refresher training on *incidents* element processes to all employees and contractors, focusing on the appropriate reporting of incidents, including near misses, and the basic approach of the incident investigation program.

    a. Only selected personnel are trained.

    b. All employees are trained.

    c. All personnel, including contract personnel, receive initial and refresher training.

8. Provide RCA and forensics training to incident investigation leaders, focusing on the skills needed to lead an investigation team and the use of RCA techniques.

    a. Basic knowledge training (e.g., ask Why? repeatedly) is provided, but no practical training.

    b. In addition to item (a), training is provided on the administrative processes used in the company.

    c. In addition to item (b), investigators receive hands-on training with the analysis techniques.

9. Provide RCA review training to managers responsible for championing, guiding, and reviewing incident investigations, focusing

on the essential elements and products of an RCA (less skill training is required for this group).

   a. Basic knowledge training (e.g., ask Why? repeatedly) is provided, but no details on what to look for during an RCA review.

   b. In addition to item (a), training is provided on the specific issues and items to address during an RCA review.

   c. In addition to item (b), a company-specific checklist is provided, and training includes practice in reviewing actual analyses.

Monitor Incident Investigation Practices for Effectiveness

10. Establish a formal investigation review process for teams to use at the conclusion of each investigation.

   a. Teams are given guidance on performing post-investigation reviews.

   b. In addition to item (a), formal submittal of the post-investigation review is required.

   c. In addition to item (b), periodic assessment of the post-investigation review is performed.

11. Establish and collect incident investigation performance metrics.

   a. Formal performance metrics are established.

   b. In addition to item (a), metrics are monitored regularly.

   c. In addition to item (b), a status report is periodically provided to management.

12. Perform self-assessments of the management systems and practices for the *incidents* element.

   a. Self assessments of *incidents* element performance are performed by company personnel.

   b. In addition to item (a), self-assessments of the *incidents* element performance are performed by external personnel.

13. Perform management reviews of the incident investigation process.

   a. High visibility incidents are reviewed by management.

   b. In addition to item (a), a review is performed of *incidents* element trends.

   c. In addition to item (b), most investigation reports are reviewed.

19.3.2 Identify Potential Incidents for Investigation

Monitor All Sources of Potential Incidents

14. Perform active assessments of data, such as logs, data trends, and emergency response actuations, that could identify near misses and incident precursors.

a. Data sources are monitored.

b. In addition to item (a), the results of the data monitoring are periodically reported.

Ensure that All Incidents Are Reported

15. Identify and eliminate barriers for reporting incidents.

   a. Periodic surveys of personnel are performed to assess near miss reporting effectiveness, but no specific actions are implemented to eliminate barriers.

   b. In addition to item (a), trend graphs and statistics are developed showing the number and type of incidents reported.

   c. In addition to item (b), personnel are assigned formal action items to eliminate significant reporting barriers.

Initiate Investigations Promptly

16. Initiate investigations as promptly as possible, but within a specified time following the incident.

   a. Formal time requirements are established for each type of investigation (i.e., ACAs and RCAs).

   b. The time difference between the incident and the beginning of the investigation is formally tracked to identify potential investigation initiation issues.

   c. In addition to item (b), kits containing the basic tools and equipment that the investigation team will need are assembled, and a formal surveillance task is assigned to maintain the kit inventory.

17. Establish time limits for notifying the appropriate personnel of incidents.

   a. Formal time requirements are established for reporting incidents to regulatory authorities.

   b. In addition to item (a), formal time requirements are established for reporting incidents to internal personnel, such as the legal department or upper management.

   c. In addition to item (b), a current call list and call procedures for incident notifications are maintained.

19.3.3 Use Appropriate Techniques to Investigate Incidents

Collect Appropriate Data During the Investigation

18. Develop a list of information, data, interviews, and records that incident investigators typically consider collecting during investigations.

    a. A generic industry list of potential data sources is available.

    b. A facility-specific list of potential data sources is developed and maintained.

    c. Area- or incident-type-specific lists of potential data sources are developed and maintained.

Interface with the Emergency Management Element

19. Develop a formal interface between the *emergency* element and the *incidents* element.

    a. Informal agreements govern interactions between *emergency* element and *incidents* element personnel.

    b. A general written guideline governs the interface between the *emergency* element and *incidents* element personnel.

    c. A specific written procedure governs the interface between the *emergency* element and *incidents* element personnel.

Other work activities associated with planning for and responding to emergencies, including protecting evidence, are addressed in the *emergency* element in Section 18.2.2.

Use Effective Data Collection Methods

20. Use consistent and effective methods, such as interviewing techniques and test plans, to collect data.

    a. Guidance is provided on interviewing techniques and questions, along with test plan guides and examples.

    b. The use of structured interviewing techniques and the development of formal test plans for physical data analysis are required.

21. Use consistent and effective methods to preserve data.

    a. The incident site is temporarily secured, but not under investigation team control.

    b. The incident site is secured until released by the investigation team.

    c. In addition to item (b), physical evidence is securely stored as long as needed.

Use Appropriate Techniques for Data Analysis

22. Provide data collection guidance and methods for performing incident investigations to facilitate rigorous analysis of the data collected. Guidance and methods address the following key issues:
    - Collection of all relevant data.
    - Assessment of both supporting and refuting data.
    - Documentation of the source of each statement and conclusion to allow an assessment of the validity of each piece of data and each conclusion.
    - All causes are identified.
    - Development of recommendations for each cause identified.
    - Consistent depth of analysis for each type of investigation.
    - Consistent identification of causes across multiple investigations.
    a. Teams use whichever structured techniques that they feel are appropriate.
    b. One or two analytical techniques are specified because they meet most of the above criteria.
    c. The use of analytical techniques that meet all of the above criteria is specified.

Investigate Causes to an Appropriate Depth

23. Analyze each incident in accordance with the analysis level defined in the investigation program.
    a. Specific criteria that must be met for each ACA and RCA, including guidance on depth of analysis, are provided.
    b. A checklist is used for each analysis to assess the investigation against the established depth-of-analysis criteria.

Demand Technical Rigor in the Investigation Process

24. Ensure that the investigation team approaches the investigation with an open mind and considers all evidence.
    a. This issue is addressed during incident investigation training.
    b. All investigation reports are required to specifically document (1) credible causes that were not investigated and (2) data that do not support or refute the team's findings and conclusions.

Provide Investigation Personnel with Appropriate Expertise and Tools

25. Assign personnel who have expertise in investigation methodologies to perform investigations.

a. Investigators are assigned based on who knows the most about the equipment, or the area supervisor is assigned, regardless of his investigation training and skills.

b. An informal process is used to assign investigators, who are generally chosen based on their investigation experience and knowledge of the process.

c. Investigators are assigned from a controlled, pre developed list of personnel who are qualified to perform ACAs and RCAs.

d. In addition to item (c), external personnel, such as personnel from other company facilities or contractors, are used to supplement facility personnel when expertise is needed in a specific area, or to provide independence from the company.

Develop Effective Recommendations

26. Develop appropriate recommendations for each cause.

a. Recommendations are tied directly to a cause of the incident.

b. All causes have associated recommendations.

c. All causes have associated recommendations that address the immediate cause, as well as the generic implications of the cause across the facility.

d. In addition to item (c), recommendations are developed that address the generic implications of the cause across the company.

19.3.4 Document Incident Investigation Results

Prepare Incident Investigation Reports

27. Provide specifications for the content of reports.

a. Each team is required, as a minimum, to complete a standard investigation form.

b. In addition to item (a), a model investigation report is used as the basis for developing reports.

c. In addition to item (b), guidance is provided on how to complete each field.

Provide Clear Linking Between Causes and Recommendations

28. Require the investigation team to specifically show the relationship between the incident, the causes, and the recommendations identified.

a. Investigation reports list causes and recommendations.

b. Investigation reports include a table or matrix that shows the relationship between direct causes, underlying causes, and recommendations. This linkage can be incorporated into the report template specified by work activity 27.

19.3.5 Follow Through on Results of Investigations

Resolve Recommendations

29. Establish a system to promptly address and resolve the incident report recommendations. Resolutions and recommendations are documented and tracked. A database is usually required to track the status of each recommendation.

   a. A database is used to track the status of recommendations from incident investigations. This database can be an integrated action tracking system that is used to track actions generated from all RBPS elements.

   b. In addition to item (a), the database is integrated with an e-mail system so that automatic status reports and reminders can be sent to responsible individuals.

30. Review the status of each action item on a periodic basis, including review and approval of changes to the implementation plans and schedules. Ensure that the reasons for slips in the implementation schedule or significant changes to the scope of actions are valid and well documented. Periodically review the status of action items and conformance to the plan and schedule with upper management.

   a. The database created in work activity 29 or 36 is used as the basis for a formal, periodic status review of recommendation status with management and the individuals assigned to resolve each recommendation.

   b. In addition to item (a), documentation of schedule changes is maintained.

Communicate Findings Internally

31. Review incident investigation reports with all affected personnel whose job tasks are relevant to the incident findings, including contract employees where applicable.

   a. Informal processes are used to communicate investigation results.

   b. Investigation results are communicated via e-mails, newsletters and postings within the facility.

   c. In addition to item (b), investigation results are communicated during safety meetings and tailgate sessions within the facility.

   d. In addition to item (c), interactive communication within the facility is used to communicate report findings.

    e. In addition to item (d), feedback is solicited on the information presented within the facility.

Communicate Findings Externally

32. Review incident investigation reports with all relevant facilities.
    a. Data are posted in a centrally located file with limited availability.
    b. Data are posted on a web site or centrally located file available to all company personnel.
    c. Information is distributed via e-mails or other one-way communication methods to other company facilities.
    d. Interactive sessions are held with other facilities to share incident investigation experience.
33. Share findings and lessons learned with industry peer groups, respecting business confidentiality.
    a. Selected information is distributed via news releases.
    b. Information is distributed via informal discussions at conferences.
    c. In addition to item (b), data are input into industry databases.
    d. In addition to item (c), formal conference presentations are made and technical papers are written.
34. Assess incidents and incident recommendations from other facilities for their impact on the reader's facility.
    a. Informal assessments are performed of external incidents.
    b. Outside organizations, such as CCPS, CSB, and OSHA perform impact assessments for external incidents.
    c. A formalized program assesses the potential applicability of external incidents.

Maintain Incident Investigation Records

35. Develop a document retention policy with regard to data logs and summaries, investigation reports, and investigation files.
    a. Individual analysts maintain their own investigation records.
    b. In addition to item (a), a document retention policy for investigation records is used.
    c. In addition to item (b), a central archive of investigation reports is maintained.
    d. In addition to item (c), a central archive of backup information for each investigation is maintained.

19.3.6 Trend Data to Identify Repeat Incidents that Warrant Investigation

Log All Reported Incidents

36. Use an incident database to trend incident characteristics and track recommendations.

    a. Individuals' memories are relied upon to identify dominant incident and investigation trends (no formal database).

    b. A database is used to identify dominant incident characteristics and is integrated with the requirements from work activities 29 and 30.

    c. In addition to item (b), a database is maintained that contains complete records for all incidents.

    d. In addition to item (c), criteria that define the acceptable content for each field are developed.

37. Populate the database established in work activity 36 from all sources (see work activity 14).

    a. "Interesting" incidents are added to the database.

    b. All incidents reported through the *incidents* element are added to the database.

    c. In addition to item (b), incidents reported from other sources (see work activity 34) are included in the database.

Analyze Incident Trends

38. Perform a periodic analysis of the incident database to identify adverse trends.

    a. Adverse trends are identified by informal analyses.

    b. Adverse trends are identified by periodic assessment of the incident data created in work activity 36.

    c. Incident trending is assigned to a specific individual.

    d. In addition to item (c), specific guidance is provided on what constitutes an adequate trend analysis.

39. Periodically report adverse trends to management.

    a. Incident data created in work activity 36 are periodically assessed and the results are reported.

    b. In addition to item (a), the results are formally presented to facility management.

    c. In addition to item (b), the results are formally presented to corporate management.

40. Perform an analysis of historical incident data during each incident investigation to determine any prior similar instances.

    a. Each analyst performs an ad hoc historical analysis.

b. Each analyst is required to perform historical analyses.

c. In addition to item (b), the results of the historical analyses are incorporated into the investigation reports.

19.4 EXAMPLES OF WAYS TO IMPROVE EFFICIENCY AND EFFECTIVENESS

This section provides specific examples of industry tested ways to improve the effectiveness of work activities related to the *incidents* element. The examples are sorted by the key principles that were first introduced in Section 19.2. The examples fall into two categories:

1. Methods for improving the performance of activities that support this element.
2. Methods for improving the efficiency of activities that support this element by maintaining the necessary level of performance using fewer resources.

These examples were obtained from the results of industry practice surveys, workshops, and CCPS member-company input. Readers desiring to improve their management systems and work activities related to this element should examine these ideas, evaluate current management system and work activity performance and efficiency, and then select and implement enhancements using the risk-based principles described in Section 2.1.

19.4.1 Maintain a Dependable Incident Reporting and Investigation Practice

Integrate health, safety, environmental, reliability, quality, security, and customer service investigations. Most incidents have multiple types of consequences, yet share the same root causes. For example, a typical equipment failure can have quality, reliability, and safety impacts. Using one program to investigate all types of incidents, regardless of their impact, eliminates the expense and inefficiency of separate investigation programs and also allows recommendations to be developed in an integrated manner. Improvements in safety performance often occur when reliability and quality improvements are made. As a result, the *incidents* element and the *asset integrity* element often have strong ties.

Use software to manage the information generated by the incidents element. Documenting and maintaining incident reports, monitoring recommendation implementation, and trending are all tasks that can be performed with software tools. However, to experience efficiency gains, the software must be easy to use and fully integrated with other company information systems.

Some companies have implemented an incident reporting and tracking system based on the capabilities of their e-mail system (e.g., Microsoft® Outlook™ or Lotus™ Notes™). This integration builds in the rapid and efficient dissemination of incident reports and action item tracking. Data can be exported to standard spreadsheet programs (e.g., Microsoft® Excel™) for analysis. Other implementations have been based on the computerized maintenance management systems (CMMS) used by many companies to track maintenance and modification work. CMMSs typically have a built-in equipment hierarchy that makes trending of incidents by equipment type and location easier. The incident report can be handled as a modified work order, so most personnel who enter maintenance reports, such as maintenance, operations, and engineering workers, require no additional training on how to use the system.

Adopt a commercially available investigation method and/or software. Commercially available investigation methods can be adopted by the company. This can speed the adoption process and reduce some of the startup problems. Customization of the methodology is usually required to facilitate acceptance of the program by facility personnel. This customization should be completed during pilot testing before rolling out the program facility wide. Dedicated software packages are also available that tend to have greater functionality than the homegrown systems. However, they may not be as efficient overall if they are not well integrated with the company's other data systems and require the users to learn a new program. As the number of people involved in the implementation process increases, this issue becomes more significant.

Perform pilot testing. Normally, incident investigation programs must be customized to effectively meet the specific needs of the facility or company and its culture. In most cases, the best approach is to perform pilot testing in phases. This allows the process to be tested a few times and adjusted as needed before rollout to the entire facility.

19.4.2 Identify Potential Incidents for Investigation

Use the risk matrices used for proactive risk assessments to classify the actual and potential losses from incidents. Because accidents are relatively rare in most companies, trending of accidents and losses tends to be an unreliable performance indicator for the *incidents* element. Trending of actual and potential losses is a better leading indicator of company performance. Determining the potential losses is inherently speculative, so the use of loss categories is more appropriate. Using the same categories that are used in proactive analyses is most efficient as it allows the team to use the same matrix used in the risk element. This also facilitates comparison between the proactive estimates of risk and actual experience.

Focus on near miss reporting. For companies that truly embrace a goal of zero accidents, aggressive reporting and investigation of near misses is vital. Near misses provide low cost opportunities to discover weaknesses in the management systems before disasters occur. For continual improvement, management should broaden the definition of a near miss over time to ensure a steady stream of reports. Near miss reporting is also an important metric because it is an indication of the willingness of personnel to participate in the program. Incidents cannot be investigated if they are not reported.

Use a standard method for estimating the potential consequences of an incident. Incident trending is only effective if consistent methods are used to determine the parameters that are trended. Consistent estimation of property damage, business interruption, environmental clean-up, and other costs can be facilitated through the use of a cost calculator tool. This tool can be executed in a spreadsheet format and made available on the corporate intranet. This allows the assumptions used in incident investigation calculations to be consistent with those used as inputs to other business decisions.

Use the investigation program to investigate abnormally positive incidents. Incidents are typically defined as negative deviations from desired performance. Normally, *incidents* activities are performed to identify system improvements to eliminate undesirable behaviors. However, the same investigation system can be used to identify causes of very desirable incidents; for example, a facility turnaround that went exceptionally well or a project that was completed ahead of schedule. The causes for this better-than-anticipated performance can be identified, and the recommendations can help ensure these causes are strengthened and reinforced. In other words, the incident investigation process is also an effective way to learn from, and replicate, positive experiences. These investigations often have a very uplifting effect on morale as attention is focused on the positive versus the negative.

19.4.3 Use Appropriate Techniques to Investigate Incidents

Develop a library of generic analyses. Incident investigation teams often encounter similar types of equipment failures or human errors. Maintaining templates of generic failure modes encountered in these situations can speed up the analysis of the causes of the incident. For example, fault tree or why tree templates for pump failures, piping failures, relief valve failures, rupture discs, generic human errors, and so forth can be made available on the company's intranet. These generic trees provide a starting point for the analysis, and they can be modified based on the facts of the specific failure under investigation. Development of generic equipment failure trees can be based initially on proactive analysis activities, such as failure modes and effects analyses, fault tree analyses, layer of protection analyses, or reliability-centered maintenance analyses. Generic why trees can not only be developed at the equipment failure and human error level, but also at the root cause level. Root cause listings, tables, charts, and trees represent a generic why tree at the root cause level of the analysis. Like any other checklist methodology, the efficiency gained must be balanced against the potential that the team will overlook issues not included in the list or tree.

Use an incident scoring system to determine the level of effort expended during the investigation and in communicating the results. A scoring system that incorporates multiple factors can be used to determine the level of effort to be expended during the investigation and in communicating the investigation results. Factors that could be considered in the system might include the type of material released, the size of a release, degree of control over the consequences, effectiveness of safety devices, onsite and offsite impacts, cost impacts, reputation impacts, and regulatory impacts. Using the scoring system during the initial stages of the investigation will be challenging because much of the needed information may not available. The incident score can be used to help companies sort external incident data flowing into their facility (see work activity 34 in Section 19.3.4). For example, a red alert would prompt greater and more immediate attention than a yellow, green, or white alert.

Ensure recommendations can be resolved. Ensure that the recommendations are specific enough so that those responsible can determine when the recommendation has been resolved. For example, consider the recommendation: "The company should stress the use of management of change processes with personnel." What must be done to resolve this? Vague recommendations will often lead to minimal effort in their resolution and no change in behavior. Instead, specific action items such as, "Modify the XYZ procedure to include ...", "Revise the purchasing specifications for ABCs to specify that the drums must be blue", or "Include a discussion of this incident and the proposed changes to the XYZ process in a tailgate session for all mechanics" state exactly what needs to be done and how. Because these recommendations are specific, they have a greater likelihood of being implemented.

Develop a formal restart policy that involves the investigation team. Develop a formal process for the restart of systems affected by incidents that involves the investigation team in the decision making process. This activity often involves the development and implementation of interim recommendations that reduce the immediate risk of operation. Long-term recommendations are usually needed to address long-term and broad risk reduction.

19.4.4 Document Incident Investigation Results

Use a two-phase report review process. A two-phase report review process can be used to ensure that the facts presented in the report are correct before the team spends time working on the development of recommendations. Initially the reviewers focus on the sequence of events and other factual data included in the report. Based on reviewers' comments, the investigation team then clarifies the report facts and develops recommendations based on the reviewed data. This approach reduces the burden on the investigation team, but increases the level of effort required by the report reviewers.

Provide ready access to incident investigation data and recommendation status. To use the information and lessons learned throughout the incident investigation program and as inputs to other RBPS elements (*procedures, risks*, etc.), ready access to the information is essential. To provide ready access to the information, data from incident investigations should be integrated into other operational data and management systems. This will increase the usage of the data and reduce the training time required because personnel will be able to use a system they are already familiar with.

19.4.5 Follow Through on Results of Investigations

Use a software program to track the status of recommendations. As noted above, the use of a database will reduce the effort needed to track the status of recommendations to completion. Because many other RBPS elements also require that recommendations be tracked to completion, using the same, integrated database to track all recommendations would be most effective. Integration of the database with the company's e-mail program would also reduce the effort required to notify personnel of the status of their recommendations.

Prioritize recommendations. Recommendations can be prioritized using a cost/benefit analysis. Quantitative cost/benefit ratios can be calculated for each recommendation. However, this is usually a resource-intensive approach. A qualitative method of assessing costs and benefits is to just list descriptions of the costs and benefits, such as "reduces potential for double batching" or "requires periodic painting of markings on the floor," in a two-column table. This is often sufficient to prioritize the recommendations and requires much less effort than a quantitative analysis.

Track recommendations to completion. A common problem in companies is the failure to resolve recommendations (i.e., implement the recommendations or justify alternative actions) that are generated by work activities associated with the *incidents* element. This is a relatively common issue because the majority of the recommendations generated by effective *incidents* element implementation are long term in nature. In other words, they are not required in order to run the process in the short term. Instead, they are methods to reduce risk in the near future and long term. As a result, implementation of the action items is often delayed. Tracking recommendations to completion, combined with accountability for those responsible, is essential to getting them resolved. Using a central database facilitates this tracking.

Perform proactive assessments of the effectiveness of recommendations. Trending typically is focused on the overall effectiveness of the investigation process. To determine if individual recommendations are effective usually requires the monitoring of a leading indicator associated with the recommendation. By monitoring leading indicators, problems with recommendation effectiveness can be identified long before the next incident occurs. This monitoring can also help identify unanticipated detrimental effects of implementing the recommendation.

19.4.6 Trend Data to Identify Repeat Incidents that Warrant Investigation

Use a software program to perform trend investigation data. As noted above, the use of a database will facilitate trending. The data can be analyzed in standard spreadsheet software, such as Microsoft® Excel™, if the investigation data can be easily exported from the investigation database.

19.5   ELEMENT METRICS

Chapter 20 describes how metrics can be used to improve performance and when they may be appropriate. This section includes several examples of metrics that could be used to monitor the health of the *incidents* element, sorted by the key principles that were first introduced in Section 19.2.

In addition to identifying high value metrics, readers will need to determine how to best measure each metric they choose to track. In some cases, an ordinal number provides the needed information, for example, total number of workers. Other cases, such as average years of experience, require that two or more attributes be indexed to provide meaningful information. Still other metrics may need to be tracked as a rate, as in the case of employee turnover. Sometimes, the rate of change may be of greatest interest. Since every situation is different, the reader will need to determine how to track and present data to most efficiently monitor the health of RBPS management systems at their facility.

19.5.1 Maintain a Dependable Incident Reporting and Investigation Practice

- *Repeat causes.* Repeat causes indicate a weakness in the investigation process. It generally means that either the recommendations that are being developed are ineffective, or the recommendations are not being implemented.

- *Actual and potential losses from incidents.* Losses typically decrease over time. However, the number of reported incidents may actually increase as personnel become more aware of what the company wants reported.
- *Facility performance.* Facility performance data, such as product production volumes, unplanned shutdowns, actuations of safety systems, actuations of protection devices, and use of emergency modification processes can also serve as leading indicators. See the *operations* element for suggestions regarding specific system performance metrics.
- *Number of qualified investigation leaders.* A decrease in the number of investigation leaders can introduce a resource constraint that affects the overall program effectiveness.

19.5.2 Identify Potential Incidents for Investigation

- *Ratio of accident to near miss reports by shift, department, or unit.* An increase in this ratio may indicate an issue with the incident reporting process.
- *Number of incidents reported per unit time.* The number of incidents reported per unit time is driven by two underlying factors: the number of accidents and near misses that actually occur and the number that are reported.
- *Average time to initiate investigation.* This metric will assist in the identification of barriers to the start of an investigation.

19.5.3 Use Appropriate Techniques to Investigate Incidents

- *Ratio of low/moderate/high level of effort investigations.* This ratio can indicate that lower levels of analysis are being used when higher levels of effort are appropriate.
- *Average effort expended per investigation.* This metric will help management calibrate investigation personnel to the expected level of effort for each type of analysis (i.e., ACAs and RCAs).

19.5.4 Document Incident Investigation Results

- *Average time to complete investigation reports.* Delays in completing the reports can indicate that the investigation and reporting process are not given sufficient priority in the company.

19.5.5 Follow Through on Results of Investigations

- *Average time to resolve recommendations.* Leveraged recommendations that address fundamental company processes generally take longer to implement, and therefore, can adversely impact this metric.
- *Number of times recommendation completion dates are revised.* This metric indicates how committed the company is to implementing recommendations as originally planned.
- *Number of lessons learned communications.* This metric indicates how often the investigation results are being formally communicated throughout the facility.
- *Number of times presentations and review dates are revised.* This metric also indicates how committed the company is to completing reports on schedule.

19.5.6 Trend Data to Identify Repeat Incidents that Warrant Investigation

- *Number of investigations triggered by trend analysis.* This metric can indicate how often trending analyses are performed.

Other factors that are generally not suitable for quantification as formalized metrics are proposed in the next section, Management Review. These topics may serve as the basis for qualitative judgments of the health of the *incidents* element.

19.6   MANAGEMENT REVIEW

The overall design and conduct of management reviews is described in Chapter 22. However, many specific questions/discussion topics exist that management may want to check periodically to ensure that the management system for the *incidents* element is working properly. In particular, management must first seek to understand whether the system being reviewed is producing the desired results. If the organization's level of *incidents* is less than satisfactory, or it is not improving as a result of management system changes, then management should identify possible corrective actions and pursue them. Possibly, the organization is not working on the correct activities, or the organization is not doing the necessary activities well. Even if the results are satisfactory, management reviews can determine if resources are being used wisely – are there tasks that could be done more efficiently or tasks that should not be done at all? Management can combine metrics listed in the previous section with personal observations, direct questioning, audit results, and feedback on various topics to help answer these questions. Activities and topics for discussion include the following:

- Review a list of major incidents that have occurred at the facility since the last management review for this element.
- Select a group of incident investigation reports representing a range of severity, different units, and so forth, and for each investigation report:
  - o Review team compositions to be sure that the teams have been staffed appropriately for the incidents under investigation.
  - o Verify that the incident team leaders were trained in the methodology used during the investigation.
  - o Evaluate the overall quality of the work product, paying particular attention to determining if (1) the methodology chosen by the team seems appropriate, (2) the team properly applied the method that was selected, (3) the causes identified by the team make sense, and (4) the recommendations appear to address the root causes that the team identified.
  - o Review the status of open recommendations for these incidents.
- Review the list of all open recommendations and how the number of open recommendations is changing with time.
- Review the timeliness of completion of recommendations. If recommendations are not completed in a timely manner, determine the causes of the delays in implementation.
- Review the list of ongoing incident investigations, with particular focus on the level of effort applied to completing the investigation; determine if the level of effort appears to coincide with the level of risk associated with the incident.
- Review the amount of time that elapsed between the actual occurrence of incidents and the time that documented investigation reports were issued.
- Examine the current status of actions to address management system issues for the *incidents* element that were identified by the *audits* element, or other RBPS element.
- Identify lessons that should be communicated to other units at the facility or other facilities within the company.
- For significant lessons that were communicated to other units and facilities, review what actions (if any) were implemented.

In addition to identifying specific areas for improvement, an effective management review process educates the entire leadership team on the importance of reporting near misses, honestly investigating incidents with the intent to learn, challenging conventional thinking, and focusing on management system improvements.

19.7   REFERENCES

19.1   Columbia Accident Investigation Board, Report Volume 1, U.S. Government Printing Office, August 2003.

19.2   Vaughan, D.; *The Challenger Launch Decision,* Chicago: University of Chicago Press, 1996.

19.3   U.S. Department of Energy, *Accident/Incident Investigation Manual*, DOE/SSDC 76-45/27, 2nd edition. Washington, DC: U.S. Department of Energy, 1985.

19.4   Center for Chemical Process Safety (CCPS), *Guidelines for Investigating Chemical Process Incidents, Second Edition*, New York: American Institute of Chemical Engineers, 2003.

19.5   Inside the Ford/Firestone Fight, Time Magazine, May 29, 2001.

20

MEASUREMENT AND METRICS

On September 23, 1999, the Mars Climate Orbiter was lost as it attempted to enter orbit (Ref. 20.1). During its 9-month journey, propulsion maneuvers were required ten times more often than were expected by the navigation team. After the spacecraft loss, investigators discovered that the trajectory errors were introduced by a software module that had been coded in the wrong measurement units. Had the unexpected deviations in trajectory been investigated during the flight, the loss of the orbiter might have been avoided. Like flight trajectories, management system metrics provide data about actual system performance versus intended system performance. If the underlying causes of deviations are not investigated, understood, and corrected, the "flight" of the management system may continue to appear normal for months or years until a critical maneuver, such as a unit startup or retirement of a key worker, results in a catastrophic system failure.

20.1   ELEMENT OVERVIEW

Identifying and using relevant process safety metrics over the life of a process is one of four elements in the RBPS pillar of *learning from experience*. This chapter describes a process for establishing and maintaining process safety leading indicators (metrics) to aid the near-real-time monitoring of RBPS management system effectiveness and provide input to continuous improvement. Section 20.2 describes the key principles and essential features of a management system for this element. Section 20.3 lists work activities that support these essential features and presents a range of approaches that might be appropriate for each work activity, depending on perceived risk, resources, and organizational culture. Sections 20.4 through 20.6 include (1) ideas to improve the effectiveness of management systems and specific programs that support this element, (2) metrics that could be used to monitor this element, and (3) management review issues that may be appropriate for the *metrics* element.

20.1.1 What Is It?

The *metrics* element establishes performance and efficiency indicators to monitor the near-real-time effectiveness of the RBPS management system and its constituent elements and work activities (Ref. 20-2). This element addresses which indicators to consider, how often to collect data, and what to do with the information to help ensure responsive, effective RBPS management system operation.

A combination of leading and lagging indicators is often the best way to provide a complete picture of process safety effectiveness (Ref. 20-3). Outcome oriented lagging indicators, such as incident rates, are generally not sensitive enough to be useful for continuous improvement of process safety management systems because incidents occur too infrequently. Measuring process safety management performance requires the use of leading indicators, such as rate of improperly performed line breaking activities.

20.1.2 Why Is It Important?

Fortunately, serious process safety accidents occur relatively infrequently. However, when they do occur, they usually involve a confluence of root causes, some of which involve degraded effectiveness of management systems or, worse, complete failure of management system activities. Facilities should monitor the real-time performance of management system activities rather than wait for accidents to happen or for infrequent audits to identify latent management system failures. Such performance monitoring allows problems to be identified and corrective actions to be taken before a serious incident occurs.

20.1.3 Where/When Is It Done?

One or more metrics can be established for each RBPS element, or a few can be created for the entire system. *Metrics* can address performance issues, efficiency issues, or both (effectiveness) in all operating phases. Once data gathering/refreshing systems are in place, metrics can be viewed anywhere, although proposed corrective actions generally occur at the subject facility. The frequency for refreshing the individual metrics may range from daily, to weekly, to monthly or longer, depending upon the dynamic nature of the metrics, the anticipated costs of data collection, and the local needs.

Higher risk situations usually dictate a greater need for formality and thoroughness in scope and level of detail, for example, a larger number of sensitive metrics. Situations requiring a smaller number of more global metrics covering a smaller scope of the RBPS activities may be adequate for lower risk processes.

Facilities whose operations are very dynamic may result in safety management system element metrics that change frequently, resulting in a need to gather and report metric data more frequently. Facilities with sound process safety cultures can generally rely on taking snapshots of metrics at less frequent intervals and using fewer metrics to gauge the effectiveness of the system activities. Facilities with an evolving or undetermined process safety culture may require more numerous metrics and greater "command and control" management system features to ensure good performance.

20.1.4 Who Does It?

Metrics data are usually collected by personnel involved in the operation of the RBPS management system element work activities. Users of the metrics can range from those personnel, to element owners, to facility or corporate management.

20.1.5 What Is the Anticipated Work Product?

The output of this activity is a set of metrics that are sensitive enough to help facility management monitor the performance and efficiency of the RBPS management system on a near-real-time basis. Metrics may be placed into a scorecard format and provided to various members of facility and corporate management for routine monitoring, or issued in a bulletin format when there are significant or abrupt changes in performance. Outputs of the *metrics* element can also be used to facilitate the performance of other elements. For example, metrics that identify dysfunctional RBPS element activities can help target *auditing* activities. The ultimate product is using metrics to (1) identify evolving management system weaknesses and (2) make adjustments to RBPS element work activities before the activities degrade into a failed state (performance or efficiency).

20.1.6 How Is It Done?

Metrics can be established as a facility designs, corrects, or improves its process safety management system (Ref. 20-4). Establishing metrics (and in particular, the data gathering and refreshing mechanisms) is simpler to do during the initial design and implementation of the system. Each RBPS chapter in this book has a section that contains a list of possible metrics proposed for that element's key principles (Section X.5, where X is the chapter number). Readers can select from these examples or develop their own ideas. Typically, a small set of metrics is proposed, data are gathered, and the set is pilot tested to see if tracking the metric data helps identify management system degradation. This metrics experiment should last a minimum of several "metric refresh cycles" and, at most, until the next formal RBPS audit is conducted. At that time, the audit can show whether the metrics have been correctly projecting the performance of the process safety management system.

20.2   KEY PRINCIPLES AND ESSENTIAL FEATURES

Safe operation and maintenance of facilities that manufacture, store, or otherwise use hazardous chemicals requires robust process safety management systems. The primary objective of the *metrics* element is to provide a means for near-real-time monitoring of the performance and efficiency of a process safety management system. Metrics generate leading indicators of performance that are critical to a facility's ability to determine if process safety incidents are likely to occur. Such leading indicators are also important inputs for achieving continuous improvement of process safety.

The following key principles should be addressed when developing, evaluating, or improving any system for the *metrics* element:

- Maintain a dependable practice.
- Conduct *metrics* acquisition.
- Use *metrics* to make corrective action decisions.

The essential features for each principle are further described in Sections 20.2.1 through 20.2.3. Section 20.3 describes work activities that typically underpin the essential features related to these principles. Facility management should evaluate the risks and potential benefits that may be achieved as a result of improvements in this element. Based on this evaluation, management should develop a management system, or upgrade an existing management system, to address some or all of the essential features.

Some or all of the identified work activities, depending on perceived risk and/or process hazards should be undertaken. However, none of these work activities will be effective if the accompanying *culture* element does not embrace the use of reliable management systems. Even the best *metrics* activities, without the right culture, will not produce sustainable process safety performance and efficiency over the life of a process.

20.2.1 Maintain a Dependable Practice

A written program that documents the intentions of the *metrics* element is key to long-term success and continuous improvement. Defining roles and responsibilities, which *metrics* data should be collected and how often, and the necessary technical expertise of personnel is critical to having an effective *metrics* system. Records should be maintained concerning *metrics* activities so that performance and efficiency can be periodically evaluated.

Establish consistent implementation. Like all RBPS elements, a written description of the *metrics* practices will help maintain institutional memory for the practice. Establishing and implementing formal procedures to develop and maintain process safety management performance and efficiency metrics should include designating the owner of the *metrics* system. This person has the responsibility to monitor the effectiveness of the *metrics* element on a routine basis.

Determine triggers for metrics collection and reporting. A variety of metrics may be defined by the metrics practice. Each metric may have its own data collection frequency or refresh rate. Metrics should be compiled and communicated to management at regular intervals, with special updates as needed. An overall schedule for collecting appropriate metrics for each element and each facility area should be defined.

Ensure that the scope of the metrics is appropriate. Facilities implementing a metrics system should identify the process areas in which the metrics data will be collected. In addition, not all RBPS elements may be selected for development of metrics. Facilities should determine those elements for which near-real-time effectiveness information will be most helpful. Finally, facilities should determine whether performance metrics or efficiency metrics, or both, are needed.

Involve competent personnel. Roles and responsibilities of various personnel involved in the metrics system should be defined. Awareness and refresher training on the metrics system should be considered for all employees. Some training on industrial statistics may be necessary for key personnel involved in actual metric collection and use.

Keep metrics practices effective. To have an effective, long lasting system, companies should maintain an archive of all metrics. Lessons learned from using the *metrics* element to monitor the effectiveness of RBPS elements can provide valuable input to the *auditing* element.

20.2.2 Conduct Metrics Acquisition

Too many metrics can overwhelm an organization and too few will not provide sufficient real-time monitoring of RBPS system effectiveness. Facilities should define the appropriate number, scope, and refresh rate of metrics. Using a practical format and selecting the best media for users is as important as the technical content of the metrics.

Implement appropriate element metrics. Metrics should be defined for any RBPS elements for which real-time effectiveness monitoring is judged to be important (Refs. 20-3 and 20-4). Ensure that appropriate input data and means of collection are established for selected RBPS elements. Be careful not to select too many metrics to avoid overwhelming both the collection and analysis processes.

Collect and refresh metrics. Provide appropriate resources to collect and refresh metrics data at appropriate intervals. If the interval is longer than a year, the metric may not be necessary.

Summarize and communicate metrics in a useful format. Summarize metrics data in a format that facilitates identification of performance deficiencies. Graphs are generally more informative than tabulated data, but graphs that are "smoothed" may mask early indications of unacceptable deviations. A scorecard or dashboard format can be designed to integrate and summarize information for decision makers.

20.2.3 Use Metrics to Make Element Corrective Action Decisions

Metrics should drive correction or improvement, otherwise they are a waste of resources. Facilities may need to gain experience with monitoring certain *metrics* to learn what movements in the metrics mean and when action is indicated.

Use the metrics element to improve RBPS elements. Establish calibration limits for action for the metrics program. Some metrics may require a "high" limit, some a "low" limit, and some may require both. Calibration limits should be thought of as the process safety equivalent to statistical process quality control limits. For example, a high limit may be set for the percentage of work orders that are improperly classified as a replacement-in-kind. The number of training sessions (e.g., 2) per quarter is an example of a low limit. As a metric trends toward a calibration limit, action is suggested prior to reaching the limit. Calibration limits may need to be set and reset periodically as the facility needs change. Using these limits, the metrics element owner should create a process and a tool for communicating the RBPS health status and potential corrective actions/adjustments to the RBPS elements.

20.3 POSSIBLE WORK ACTIVITIES

The RBPS approach suggests that the degree of rigor designed into each work activity should be tailored to risk, tempered by resource considerations, and tuned to the facility's culture. Thus, the degree of rigor that should be applied to a particular work activity will vary for each facility, and likely will vary between units or process areas within a single facility. Therefore, to develop a risk-based process safety management system, readers should perform the following steps:

1. Assess the risks at the facility, investigate the balance between the resource load for RBPS activities and available resources, and examine

the facility's culture. This process is described in more detail in Section 2.2.

2. Estimate the potential benefits that may be achieved by addressing each of the key principles for this RBPS element. These principles are listed in Section 20.2.

3. Based on the results from steps 1 and 2, decide which essential features described in Sections 20.2.1 through 20.2.3 are necessary to properly manage risk.

4. For each essential feature that will be implemented, determine how it will be implemented and select the corresponding work activities described in this section. Note that this list of work activities cannot be comprehensive for all industries; readers will likely need to add work activities or modify some of the work activities listed in this section.

5. For each work activity that will be implemented, determine the level of rigor that will be required. Each work activity in this section is followed by two to five implementation options that describe an increasing degree of rigor. Note that work activities listed in this section are labeled with a number; implementation options are labeled with a letter.

Note: Regulatory requirements may specify that process safety management systems include certain features or work activities, or that a minimum level of detail be designed into specific work activities. Thus, the design and implementation of process safety management systems should be based on regulatory requirements as well as the guidance provided in this book.

20.3.1 Maintain a Dependable Practice

Establish Consistent Implementation

1. Establish and implement procedures to develop and maintain process safety management performance and efficiency metrics.
   a. The written practice is informal with limited metrics.
   b. A simple written procedure defines a few metrics.
   c. A detailed written program defines a formal scorecard of performance metrics.
   d. A detailed written program defines a formal scorecard of performance and efficiency metrics.

2. Assign an owner of the metrics system to monitor the program's effectiveness on a routine basis.
   a. An informal metrics owner occasionally monitors effectiveness.
   b. A part-time metrics owner, local or offsite, occasionally monitors effectiveness.
   c. Multiple metrics owners across the facility monitor metrics in their respective areas.

     d.   A single, full-time metrics owner regularly monitors metrics.

Determine Triggers for Metrics Collection and Reporting

3. Determine when metrics data are gathered.
   a.   *Metrics* data are gathered "for cause."
   b.   *Metrics* data are gathered on an ad hoc basis.
   c.   *Metrics* data are gathered on a regular schedule.
   d.   *Metrics* data are gathered and refreshed on a regular schedule.

Ensure that the Scope of Metrics Is Appropriate

4. Determine the facility areas in which the metrics element should be applied.
   a.   The scope is based upon where/when people elect to generate metrics.
   b.   A few metrics are applied in selected facility areas.
   c.   Comprehensive metrics are applied in selected areas.
   d.   Comprehensive metrics are applied in all facility areas.
5. Determine the RBPS elements for which metrics would be useful.
   a.   Element metrics are generated on an ad hoc basis.
   b.   Metrics are established for selected elements.
   c.   Metrics are established for regulatory elements.
   d.   Metrics are established for all RBPS elements.
6. Determine whether performance indicators, efficiency indicators, or both are desired.
   a.   Metric types vary based upon perceived need.
   b.   Efficiency indicators are established.
   c.   Performance indicators are established.
   d.   Both performance and efficiency indicators are established.

Involve Competent Personnel

7. Define the metrics roles and responsibilities for various groups of personnel.
   a.   Roles/responsibilities are accepted on an ad hoc basis.
   b.   *Metrics* is an informal duty for several people.
   c.   *Metrics* roles/responsibilities are assigned to job functions/ departments.
   d.   A single person has overall responsibility for all *metrics* duties.
8. Provide training on the metrics system.
   a.   Informal training is provided.
   b.   *Metrics* practice is broadcast (e.g., through e-mail) to everyone one time.
   c.   Initial *metrics* awareness training is provided once to affected personnel.

    d. Initial and refresher training on *metrics* awareness are provided to affected personnel.

9. Provide detailed training to those who are assigned specific roles within the metrics system.

    a. Informal detailed *metrics* training is provided.

    b. *Metrics* practice is broadcast (e.g., through e-mail) to everyone one time.

    c. Detailed *metrics* training is provided once to key *metrics* personnel.

    d. Detailed and refresher training for the *metrics* element is provided to key *metrics* personnel.

Keep Metrics Practices Effective

10. Maintain records of metrics system data.

    a. *Metrics* data are kept informally for a brief period.

    b. *Metrics* data are kept for several years.

    c. A *metrics* log is kept by the *metrics* coordinator and is accessible to all affected personnel.

    d. Item (c), and data are accessible via the company network.

11. Establish and collect metrics data on the metrics element.

    a. Data are collected informally.

    b. Basic *metrics* activity data are collected.

    c. *Metrics* performance indicators are collected regularly.

    d. *Metrics* performance and efficiency indicators are collected regularly.

12. Provide input to internal audits of metrics practices.

    a. Informal internal reviews "for cause" are performed.

    b. *Metrics* performance information is provided to internal reviewers annually.

    c. *Metrics* performance and efficiency information is provided to reviewers for regular monthly/quarterly management reviews.

20.3.2 Conduct Metrics Acquisition

Implement Appropriate Element Metrics

13. Develop appropriate metrics for each selected RBPS element.

    a. Informal metrics are identified.

    b. A few performance indicators are selected for several elements.

    c. A formal process is used to determine performance indicators.

    d. A formal process is used to determine performance and efficiency indicators.

14. Ensure that an appropriate means exists for collecting data on selected RBPS elements.

    a. Facility personnel collect data on an informal basis.

    b. A manual data collection procedure is established.

c.   An electronic data collection and metrics refresh process is established.

Collect and Refresh Metrics

15. Collect and refresh metrics data at appropriate intervals.
    a.   Metrics are collected on an informal schedule.
    b.   Metrics are refreshed on an informal schedule.
    c.   Data collection and metric refreshing occurs at a standard interval.
    d.   Metrics are collected and refreshed at an appropriate interval for each indicator.

Summarize and Communicate Metrics in a Useful Format

16. Summarize metrics data in a useful format.
    a.   Informal *metrics* summarizing is done.
    b.   The *metrics* format is informally established by each data collector.
    c.   *Metrics* data are collected in a spreadsheet format.
    d.   *Metrics* data are summarized in a scorecard or dashboard format.
17. Communicate metrics data.
    a.   *Metrics* reporting is established by each data collector.
    b.   *Metrics* are reported to a few management personnel.
    c.   *Metrics* are communicated widely within the facility.
    d.   *Metrics* are available via the company network.

20.3.3 Use Metrics to Make Corrective Action Decisions

Use Metrics to Improve RBPS Elements

18. Establish action limits for the metrics.
    a.   Action is left to the discretion of the user.
    b.   Limits are provided for some *metrics*.
    c.   Limits are provided for efficiency *metrics*.
    d.   Limits are provided for both efficiency and performance *metrics*.
19. Create a communication process for the RBPS health status and for potential corrective actions/adjustments to the RBPS elements.
    a.   *Metrics* and element corrective action logs/summaries are available upon request.
    b.   *Metrics* and element actions are communicated to RBPS element personnel.
    c.   *Metrics* and element actions are communicated regularly to all facility personnel.

20.4   EXAMPLES OF WAYS TO IMPROVE EFFECTIVENESS

This section provides specific examples of industry tested methods for improving the effectiveness of work activities related to the *metrics* element.

These examples were obtained from the results of industry practice surveys, workshops, and CCPS member-company input. Readers desiring to improve their management systems and work activities related to this element should examine these ideas, evaluate current management system and work activity performance and efficiency, and then select and implement enhancements using the risk-based principles described in Section 2.1.

Use existing data collection systems to gather RBPS metrics. Distributed control systems, computerized maintenance management systems, and other data historian systems are capable of handling data collection, archiving, and display functions to support the *metrics* element. Since a significant investment has already been made in these systems, they should be used as much as practical in the gathering of data for RBPS metrics.

Coordinate data collection with the facility reliability effort as much as possible. Because both reliability and RBPS management systems are intended to prevent failures, the data needed for both efforts will be similar. The economic need for reliability data is often better understood; this viewpoint can facilitate the gathering of RBPS data.

Use existing RBPS element software and databases to gather RBPS metrics. Databases used for the *incidents*, *management of change*, *asset integrity*, and other RBPS elements can often be easily mined for data using available programs or by using special programs to filter and evaluate data.

20.5 ELEMENT METRICS

This chapter describes how metrics can be used to improve performance and when they may be appropriate. This section includes several examples of metrics that could be used to monitor the health of the *metrics* element, sorted by the key principles that were first introduced in Section 20.2.

In addition to identifying high value metrics, readers will need to determine how to best measure each metric they choose to track. In some cases, an ordinal number provides the needed information, for example, total number of workers. Other cases require that two or more attributes be indexed to provide meaningful information, such as averaging employees' length of service to obtain a unit's average years of experience per employee. Still other metrics may need to be tracked as a rate, as in the case of employee turnover. Sometimes, the rate of change may be of greatest interest. Since every situation is different, the reader will need to determine how to track and present data to most efficiently monitor the health of RBPS management systems at their facility.

> *Note: the use of leading indicator performance metrics and efficiency indicators is relatively new. The idea of developing metrics to monitor the effectiveness of a metrics element may, on the surface, seem a little convoluted. However, any process safety activity that is worth doing is worth monitoring, and companies should consider appropriate ways to monitor and improve the effectiveness of its metrics activities.*

20.5.1 Maintain a Dependable Practice

- *Number of RBPS elements for which metrics are maintained.* A large percentage of elements indicates that the facility's use of metrics is thorough.
- *Evidence that metrics use has caused improvement.* Changes in practices brought about because metrics scorecards highlighted a potential problem indicate that *metrics* use is meeting its purpose.
- *Number of people trained on metrics element.* A high number indicates an active program.
- *The number of audit findings dealing with the metrics element.* A high number indicates poor performance of the *metrics* program.
- *Results of audits or management reviews indicating that metrics are in consistent use.* Interviews that indicate prevalent use of metrics indicates good system performance.

20.5.2 Conduct Metrics Acquisition

- *Number of metrics for which data are collected.* A high percentage indicates the program is active.
- *The refresh rate for metrics.* A high number indicates that the program is active.
- *Staff-hours required to develop metrics.* A high or increasing number might indicate that efficiency of *metrics* work activities needs improvement.
- The number of metrics communications tools developed. A high number indicates an active program.
- The frequency of communicating metrics. A high rate indicates an active program.

20.5.3 Use Metrics to Make Element Corrective Action Decisions

- Percentage of management personnel that use metrics for decision making. A high percentage indicates that the program is useful.
- Percentage of employees who have seen metrics. A high percentage indicates that employees see the program in use and this reinforces the process safety culture element.

- Frequency of metrics use in management review meetings. A high rate indicates that the program is useful to management.
- The number of problems avoided/discovered through the use of metrics. A high number indicates that the metrics element is meeting its objective.

20.6 MANAGEMENT REVIEW

The overall design and conduct of management reviews is described in Chapter 22. However, many specific questions/discussion topics exist that management may want to check periodically to ensure that the management system for the *metrics* element is working properly. In particular, management must first seek to understand whether the system being reviewed is producing the desired results. If the organization's level of activity in the use of *metrics* is less than satisfactory, or it is not improving as a result of management system changes, then management should identify possible corrective actions and pursue them. Possibly, the organization is not working on the correct activities, or the organization is not doing the necessary activities well. Even if the results are satisfactory, management reviews can help determine if resources are being used wisely – are there tasks that could be done more efficiently or tasks that should not be done at all? Management can combine the metrics listed in the previous section with personal observations, direct questioning, audit results, and feedback on various topics to help answer these questions. Activities and topics for discussion include the following:

- The use of metrics for each operating area.
- Employee opinions about the use of key process safety performance metrics. Are workers aware of the metrics, or know where to find information about them?
- Data collection methods.
- Frequency of metrics updates.
- Any recent audit findings that addressed the metrics element.
- Status of corrective actions and whether they are overdue.
- Any incidents or trends that found metrics failure to be a root cause or contributing factor.
- Employee opinions on the effectiveness of new employee and contractor training on metrics.
- Resource utilization on the collection and refreshing of metrics.

In addition, the element owner normally makes a special effort to understand the reasons for any trends or anomalies in the *metrics* element.

Finally, if any major projects to address known *metrics* gaps are ongoing, a briefing should be prepared for the management review committee.

The results of a management review of *metrics* activities should demonstrate that leadership at the facility (1) is aware of and values *metrics* and (2) is intent on ensuring that useful metrics are collected. Management reviews should focus their efforts on selecting, collecting, and maintaining *metrics* information, including ways for improving the efficiency of work activities that support this element. In addition, an effective management review process educates the entire leadership team on the importance of *metrics* and the role this element can play in helping to identify RBPS weaknesses, manage risk, and sustain the business.

20.7 REFERENCES

20.1   Mars Climate Orbiter Mishap Investigation Board – Phase I Report, National Aeronautics and Space Administration, Washington, November 10, 1999.

20.2   ProSmart – The Tool You Need to Improve Process Safety, Center for Chemical Process Safety, American Institute of Chemical Engineers, New York, www.aiche.org/ccps/prosmart/

20.3   *Guidance on Safety Performance Indicators – Guidance for Industry, Public Authorities and Communities for Developing SPI Programmes related to Chemical Accident Prevention, Preparedness and Response*, Organization for Cooperation and Economic Development Environment, Health and Safety Publications, Series on Chemical Accidents No. 11, Danvers, Massachusetts, 2005.

20.4   *Step-by-Step Guide to Developing Process Safety Performance Indicators*, U.K. Health and Safety Executive, ISBN 0 7176 6180 6, Norwich, England, 2006.

21

AUDITING

The explosion at the Longford gas plant is described in Chapter 9. An audit conducted by a corporate team six months prior to the explosion had determined that the gas plant was successfully implementing its process safety management system. However, a Royal Commission subsequently investigated the explosion and found significant deficiencies in the areas of (1) risk identification, analysis, and management, (2) training, (3) operating procedures, (4) documentation, and (5) communications. These long-standing problems had not been detected by the prior audit.

21.1 ELEMENT OVERVIEW

Critical evaluation of the RBPS management system is one of four elements in the RBPS pillar of *learning from experience*. This chapter covers formal methods for performing periodic RBPS management system audits, which should reduce risk by proactively identifying and correcting weaknesses in management system design and implementation. Section 21.2 describes the key principles and essential features of a management system for this element. Section 21.3 lists work activities that support these essential features, and presents a range of approaches that might be appropriate for each work activity, depending on perceived risk, resources, and organizational culture. Sections 21.4 through 21.6 include (1) ideas for improving the effectiveness of management systems and specific programs that support this element, (2) metrics that could be used to monitor this element, and (3) issues that may be appropriate for management review.

21.1.1 What Is It?

The *audits* element is intended to evaluate whether management systems are performing as intended. It complements other RBPS control and monitoring activities in elements such as *management review* (Chapter 22), *metrics* (Chapter 20), and inspection work activities that are part of the *asset integrity* and *conduct of operations* elements (Chapters 12 and 17). The *audits* element comprises a system for scheduling, staffing, effectively performing, and documenting periodic evaluations of all RBPS elements, as well as providing systems for managing the resolution of findings and corrective actions generated by the audits.

The following terms, as used in these *Guidelines*, have these meanings:

An audit is a systematic, independent review to verify conformance with prescribed standards of care. It employs a well-defined review process to ensure consistency and to allow the auditor to reach defensible conclusions.

An RBPS management system audit is the systematic review of RBPS management systems and is used to verify the suitability of these systems and their effective, consistent implementation.

Standards of care are established guidelines, standards, or regulatory requirements against which an audit is conducted. Standards of care also typically include the self-imposed requirements of the organization being audited.

A finding is a conclusion reached by an auditor based upon data collected and analyzed during the audit. While findings could be either positive or negative, the usage here will be consistent with that adopted by most organizations; in other words, findings will indicate a deficiency in the construction or implementation of an RBPS element based on the requirements established by the standard(s) of care.

An observation is a conclusion reached by the auditor that is not directly related to compliance with the standard of care. Some organizations use the term observation to cite good programs, procedures, or practices identified during the audit. Other organizations use the term to indicate conclusions that, while the requirements established by the standard(s) of care have been met, opportunities remain for improving the construction or implementation of an RBPS element. The latter context is used in these *Guidelines*.

A recommendation is a proposed remedial activity intended to correct a deficiency that resulted in a finding.

21.1.2 Why Is It Important?

The *audits* element evaluates RBPS management systems to ensure that they are in place and functioning in a manner that protects employees, customers, communities, the environment, and physical assets against process safety risks.

Audits are important control mechanisms within the overall management of process safety. In addition, audits can provide other benefits such as the identification of opportunities for improved operability, increased safety awareness, and greater confidence regarding compliance with regulatory requirements.

21.1.3 Where/When Is It Done?

Audits are conducted throughout the development and implementation of the RBPS management system. The nature and frequency of the audits will be governed by factors such as the current life cycle stage of the facility, the maturity (degree of implementation) of the RBPS management system, past experience (e.g., prior safety performance and audit results) and applicable facility, corporate, legal, regulatory, or code requirements.

The *audits* element applies during all life cycle stages, but primarily focuses on operating facilities. The audits should be performed where access to process safety management records and subject matter experts is most convenient. Depending on the life cycle stage, that might be the pilot plant, the engineering offices, or the facility.

While they can be scheduled on an as-needed basis, audits of a particular RBPS management system are typically conducted at some predetermined interval, for example, frequencies ranging from once per year to once every three years are common.

21.1.4 Who Does It?

Audits can be conducted by qualified personnel selected from a variety of sources, depending upon the scope, needs, and other aspects of the specific situation. Audits are typically conducted by teams. Team members might be selected from staff at the facility being audited, from other company locations (e.g., from another operating facility or from corporate staff functions), or from outside the company, for example, a consulting firm.

21.1.5 What Is the Anticipated Work Product?

The short-term product of the *audits* element is a review of the implementation of one or more RBPS elements against identified standards of care, addressing the manner, degree, quality, and effectiveness of the implementation. A report of observations, findings, and recommendations for any needed improvements is typically prepared to document the results of an audit.

Longer-term results include the implementation of the remedial activities necessary to address the findings of the audit, the consequent enhancement in the effectiveness of the management system for other RBPS elements, and the resultant improvements in process safety performance.

21.1.6 How Is It Done?

An audit involves a methodical, typically team-based, assessment of the implementation status of one or more RBPS elements against established requirements, normally directed by the use of a written protocol. The *audits* effort will be primarily focused on operating facilities; however, companies may choose to augment the audit function with information generated by other processes. This information may have been obtained during different life cycle stages, such as research and development or design; during other, related functions, such as vessel fabrication or inspection; or at such non-operational locations as a company corporate office or supplier. This additional information can greatly increase the effectiveness of the audit by taking advantage of the full range of systematic, independent evaluation processes within an organization, minimizing the duplication of effort.

Data are gathered through the review of program documentation and implementation records, direct observations of conditions and activities, and interviews with individuals having responsibilities for implementation or oversight of the element(s) or who might be affected by the RBPS management system. The data are analyzed to assess compliance with requirements, and the conclusions are documented in a written report.

Recommendations for addressing any performance gaps identified by the audit are proposed, responsibilities and schedules for addressing the recommendations are assigned, and recommendations are tracked to their resolution. The report and documentation of the resolution of the recommendations are maintained as required to meet programmatic needs and any regulatory requirements.

Audit results should be trended over time to determine whether or not RBPS performance is improving, with program adjustments made as necessary.

21.2   KEY PRINCIPLES AND ESSENTIAL FEATURES

Safe operation and maintenance of facilities that manufacture, store, or otherwise use hazardous chemicals requires reliable management systems, human performance, and equipment operation. Implementing the *audits* element provides a periodic, independent check on all three aspects, with particular emphasis on management systems. The *audits* element is an important quality control function for the implementation of RBPS management systems, and significant process incidents have graphically demonstrated the consequences of ineffective audit systems. Common problems include audits that are not conducted as required, audits that fail to detect serious management system flaws, and failure to properly follow-up on audit findings.

The following key principles should be addressed when developing, evaluating, or improving any management system for the *audits* element

- Maintain a dependable practice.
- Conduct element work activities.
- Use *audits* to enhance RBPS effectiveness.

The essential features for each principle are further described in Sections 21.2.1 through 21.2.3. Section 21.3 describes work activities that typically underpin the essential features related to these principles. Facility management should evaluate the risks and potential benefits that may be achieved as a result of improvements in this element. Based on this evaluation, management should develop a management system, or upgrade an existing management system, to address some or all of the essential features, and execute some or all of the work activities, depending on perceived risk and/or process hazards that it identifies. However, these steps will be ineffective if the accompanying *culture* element does not embrace the use of reliable management systems. Even the best management system, without the right culture, will not ensure a thorough, discerning audit.

21.2.1 Maintain a Dependable Practice

When a company identifies or defines an activity to be undertaken, that company likely wants the activity to be performed correctly and consistently over the life of the facility. The *audits* element should be documented to an appropriate level of detail in a procedure or a written program addressing the general management system aspects discussed in Section 1.4.

Ensure consistent implementation. The written program governing the *audits* element should establish (1) the scope and objectives of the *audits* element implementation, (2) criteria for determining the frequency and depth of scrutiny for audits as a function of the perceived risk associated with the audited facilities or activities, (3) organizational responsibilities for the conduct of the audits, (4) requirements for the resolution of audit findings and recommendations, (5) documentation requirements, and (6) any other considerations necessary to establish the foundation for an effective *audits* program.

The scope of an *audits* program refers to the facilities and units to be covered (physical scope), the subject areas and criteria against which the audit is to be conducted (analytical scope), and the period of RBPS implementation that is to be reviewed (temporal scope). Failure to clearly define the scope of the *audits* program can lead to (1) misunderstandings among the groups being audited, the auditors, and the management recipients of the audit report,

(2) inconsistent or inaccurate audit results, (3) findings being missed, or (4) the inclusion of inappropriate conclusions in audit reports.

When defining the physical scope of an *audits* program, management must determine whether the scope includes all facilities, only manufacturing facilities, or only facilities handling certain hazardous materials. Similarly, the scope may cover only wholly owned facilities or it may cover joint ventures and partnerships, and some companies may choose to also extend the scope to contract manufacturing operations. Among the parameters considered in defining the physical scope of the *audits* program are:

- Type of facility (manufacturing, terminals, etc.).
- Ownership (wholly owned, joint ventures, etc.).
- Geographical location.
- Facility coverage (all units versus selected units).
- Nature of operations and risks.

When defining the analytical scope of an RBPS management system audit program, the factors to be considered include:

- Facility or company policies and standards.
- Regulatory requirements.
- Other management control mechanisms.

The temporal scope will commonly extend back to the previous audit, but may be narrowed or expanded to meet special needs. For example, an audit might be more narrowly focused on the performance of a management system element subsequent to the implementation of particular remedial measures intended to improve element performance.

Other considerations in defining the scope of an *audits* program include practical limitations on the availability of resources; however, such limitations must not be permitted to impact the integrity of the *audits* program. No single correct approach exists to defining the scope of an *audits* program, and decisions on audit scope should be made within the context of the overall process safety management program and the perceived risk associated with the audited facilities or activities.

The frequency with which RBPS management systems audits are conducted (i.e., the maximum interval between audits) is dependent on the objectives of the *audits* program, the nature of the operations involved, and any applicable regulatory requirements. Corporate programs will typically establish a maximum permissible interval between audits, allowing for more frequent audits as circumstances dictate (e.g., for higher risk operations, or when particularly negative results in past audits indicate a need for more frequent scrutiny).

Many organizations find it appropriate to assign one individual to oversee the *audits* program. This role is typically assigned at the corporate level to ensure a suitable degree of objectivity and independence from the organizational structure of the facility being audited. However, many organizations have an *audits* element owner at the facility level who performs functions such as coordinating audit schedules, tracking the resolution of recommendations, and monitoring and responding to trends in audit results.

The various roles involved in the implementation of the *audits* element must be defined and responsibilities clearly assigned. In addition to the roles described above, another particular consideration relates to the staffing of audit teams. As noted below, some organizations develop lead auditors who, in addition to their day-to-day responsibilities, assist in audits at other facilities across the company. The availability of such personnel must be formalized and coordinated with their home departments.

Many organizations find that the development and use of standardized audit protocol(s) helps to ensure the dependable focus of audit activities. Specialized protocols can address the unique aspects of regulations and other requirements against which the RBPS management system must be audited.

The program description should also document the intended format and content for audit reports. The audit report should document the results of the audit, indicating where and when the audit was done, who performed the audit, the audit scope, the manner in which it was conducted, and the audit findings.

The facility's written description of its *audits* element should also address the following issues related to audit reports:

- Should the audit report include both findings and recommendations, or just findings? Some facilities prefer the audit team to document recommendations separately, or not at all. This allows the facility to select an action for addressing the finding that makes the most sense and frees the facility from justifying why management chose an action that differs from an auditor's recommendation.
- To what degree should the audit report summarize the auditor's basis for the conclusions that were drawn? For example, how much detail needs to be provided regarding the amount and nature of the documentation reviewed?
- Should the audit team report on opportunities to improve the effectiveness of the process safety management activities, either through improved efficiency or performance? If so, should these opportunities for improvement be reported separately?

- Should the audit team document particularly effective practices so that these can be shared with other facilities within the company?

Involve competent personnel. Auditing, particularly serving as a lead auditor, requires a special mix of familiarity with applicable requirements; strong, focused organizational abilities; dogged curiosity; tact; and often the ability to confidently stand one's ground in the face of determined rebuttal.

The organization should identify the skills necessary to successfully function as an auditor, and then seek suitable candidates for that role. Some organizations, recognizing the specialized expertise required of lead auditors, develop a cadre of trained, experienced auditors across the corporation. These auditors typically lead audits at facilities other than their own to help ensure a suitable degree of objectivity for the audit team. This role is typically an adjunct to the individual's normal responsibilities.

Training for auditors should address topics such as (1) developing an intimate familiarity with the corporate and regulatory requirements that are to be audited, (2) the specifics of the facility's *audits* program procedure, (3) legal considerations associated with conducting and documenting audits, (4) the mechanics of conducting an audit, and (5) specialized topics such as interviewing skills.

Curiosity and classroom training do not automatically make an effective auditor. Fortunately, audits are typically a team-based activity. This provides an opportunity for new auditors to work under the guidance of more experienced auditors as they develop their skills. Team staffing efforts should take this into consideration through the inclusion, where feasible, of a less experienced auditor-in-training to ensure that a sufficiently large pool of experienced auditors exists to meet organizational needs.

The audit team must have sufficient objectivity and independence from the function being audited to ensure a credible evaluation of the RBPS management system element(s) being audited. The following terms are often used to describe increasing levels of potential audit team objectivity: 1st party, 2nd party, and 3rd party.

First party auditors are from the facility being audited and, potentially, provide the lowest degree of objectivity, but likely provide the greatest flexibility with timing. Also, 1st party auditors are likely to have the most in-depth knowledge regarding the actual workings of the unit. When 1st party auditors are used, management should attempt to identify auditors who are knowledgeable of the RBPS element(s) and requirements, but who are also as objective as possible regarding the implementation of the specific RBPS element(s) being audited. Often, the requirement for objectivity results in choosing 1st party auditors who have relatively little direct involvement with the element(s) being audited. Otherwise, 1st party auditors have the potential for being accustomed to the status quo, and may be less likely to recognize problems or gaps as such.

Second party auditors come from another location within the company, perhaps from another facility or from a corporate group such as a centralized safety or process safety function. Auditors from other facilities are often selected, understandably, from facilities making the same, or a similar, product. One advantage of this approach is that it fosters cross-learning among facilities. However, highly knowledgeable personnel at other facilities (who likely make the best audit team members) may not be readily available for this activity. Also, if this other facility is part of the same business unit, a perception of lessened independence can exist. For this reason, auditors from a central staff function are typically perceived to have the greater degree of objectivity. They also generally have (or quickly gain) more auditing experience than an auditor from a facility who might be called on to audit outside their facility only infrequently.

Third party auditors (typically, consulting companies who can provide experienced auditors) potentially provide the highest degree of objectivity. However, 3rd party auditors do not have the first-hand familiarity with company programs and policies, and may not have the familiarity with the process, that a facility employee has. Finally, this potentially higher objectivity may come with an increased out-of-pocket cost, and some of the knowledge gained by the audit team walks out the door at the end of the audit.

Clearly, the selection of the audit team is a complex issue that requires careful consideration. Some companies have successfully developed hybrid approaches that capture many of the benefits from more than one of these models. For example, a facility may use a combination of 2nd and 3rd party auditors to ensure independence, promote cross-learning, and train personnel at sister facilities to become effective auditors. Some facilities assign knowledgeable facility employees, such as senior operators or maintenance personnel, engineers, or quality control inspectors, who do not play a management role in the process safety program, as members of a predominantly 2nd or 3rd party audit team. Doing this provides facility-specific 1st party perspective while maintaining audit team independence. Some companies use predominately 1st party teams, but require one 2nd or 3rd party member. If properly designed and supported by management, these hybrid approaches can improve both the transparency of the audit activity and the effectiveness of the audit, while simultaneously supporting other goals, such as *workforce involvement*.

Identify when audits are needed. The *audits* element description will specify the maximum interval between audits of a particular RBPS management system. The *audits* coordinator should establish an ongoing audit schedule that ensures that each program is audited at least as frequently as required by these criteria. However, circumstances may warrant that a particular RBPS program be audited more frequently or on an episodic basis. Among the factors to consider in determining audit frequency are:

- Degree of perceived risk. More frequent audits may be considered for operations that pose higher levels of risk. Higher risks may result from the particularly hazardous nature of the materials present; the type of process involved, for example, one that operates at extreme pressures or in a highly corrosive environment; or the proximity of potentially exposed populations or resources.
- Frequency of process or equipment changes. Many details of RBPS management system implementation are closely tied to the process or equipment details. For areas in which frequent changes in the process or equipment occur, such as multi-purpose batch processing environments, more frequent audits may be warranted to ensure that the more likely impacted RBPS elements (e.g., *risk assessments, procedures, training, asset integrity*) properly reflect such changes.
- Process safety culture maturity. Any substantive evidence of a weak process safety culture should be considered in a decision to increase the frequency of audits.
- RBPS management system maturity. More frequent audits may be warranted for operations that have new or evolving RBPS programs.
- Results of prior evaluations. Significant gaps in RBPS management system implementation revealed in a prior audit may indicate the need to perform the next audit sooner than the program schedule would normally dictate.
- Incident or regulatory history. When a location has experienced a significant trend of incidents or near misses, or has experienced or expects regulatory scrutiny, increasing the frequency of audits may be appropriate. In addition to identifying possible management system deficiencies, more frequent audits may increase awareness of process safety within the facility staff.

Particularly problematic results with the last two items on the above list may warrant special out-of-cycle audits (i.e., even more frequently than normally scheduled) for a particular RBPS program, or one or more elements within the program.

21.2.2 Conduct Element Work Activities

Note: This narrative assumes that a 2nd or 3rd party audit team is conducting an audit of the full RBPS management system. Many of these same activities would be required for a 1st party audit; however, some of the logistical issues and preparatory tasks would be simpler, or nonexistent, for a team auditing its own facility.

Prepare for the audit. Pre-audit tasks include (1) scoping and scheduling the audit, (2) assembling the audit team, (3) assigning activities to each audit team member, (4) gathering information so that the audit team will be better prepared when it arrives on site, and (5) planning onsite audit activities. Each activity is discussed in further detail below. Based upon the scope of the audit, modifying the standard audit protocols may be necessary. As part of the initial planning activity, the audit team leader, perhaps with a representative of the facility, should review the protocols, make any necessary revisions, and factor the changes into the audit planning.

Determine the audit scope and schedule. The first step for the audit team leader in preparing for an audit is to contact the designated representative at the facility to be audited to establish the precise schedule and scope for the audit. The physical, analytical, and temporal scope of a particular audit is defined consistent with the considerations outlined in Section 21.2.1. Some flexibility in schedule may be required to address issues, such as resource conflicts, caused by maintenance turnarounds and other unique circumstances. Audits of particularly broad scope, or of large facilities, may need to be broken into multiple facility visits. For these reasons, the audit team leader should initiate discussions sufficiently early to ensure that any imposed audit deadline can be met.

Assemble the team. A comprehensive RBPS management systems audit normally requires a team effort. Using a multi-person team for the audit (1) provides more than one perspective, (2) provides an opportunity for intra-team discussion of issues, (3) allows involvement of personnel with a variety of disciplines, skills, and experiences, and (4) reduces the onsite time and associated disruption of facility routine required to conduct the audit.

The ideal team for an RBPS management systems audit will include individuals who have:

- Familiarity with the process.
- Experience in process safety management.
- Experience in audit techniques.

More than one of these characteristics may be found in a single team member and, collectively, the team should provide all of these characteristics. Predominately 2nd or 3rd party audit teams may include a representative of the facility being audited to provide the required familiarity with the process, if this is lacking in the designated team.

The size of the team should be determined based upon the scope of the audit, while maintaining a perspective on the additional burden that the audit will impose on facility personnel. Narrowly focused audits or audits of small, lower risk facilities with streamlined RBPS management systems can be accomplished in several days, while a comprehensive audit of a large facility with a detailed RBPS program might take several weeks. Audit durations of more than two weeks should be given careful consideration with respect to whether the team size is adequate to the task and, if so, whether the audit may need to be broken up into more than one trip.

On the other hand, excessively large teams can provide an undue administrative burden for the team leader and may be difficult for the audited facility to support. From a practical standpoint, teams of six or fewer auditors are typical, even for large facilities.

After the audit schedule has been confirmed, the team leader should select the team members who will satisfy the needs for the particular audit, considering availability constraints. Some iteration may be required to assemble the proper team that satisfies the requirements of both schedule and team expertise.

Assign responsibilities. Responsibility for auditing the various RBPS elements should be assigned to individual team members based upon their expertise, past auditing experience, and interest in particular elements.

Gather advanced information. Certain background information will be required by the audit team to allow them to properly prepare for the audit. To increase the efficiency of the audit team, such information should be obtained and distributed to the audit team members for their review before the start of the audit, when logistical issues and other concerns (such as controls over proprietary information) allow. Such information might include:

- Descriptions of the processes operated at the facility to be audited.
- Information on the organizational structure at the facility, including identification of the various RBPS element owners.
- Program descriptions and procedures for the implementation of the RBPS element.
- Past audit reports and documentation of the resolution of recommendations.
- Other common records related to RBPS element implementation that are practicable to be provided ahead of the audit.
- Summary of any local regulations.

Such information should be available as a normal facet of designing and implementing a management system. It should be uncommon that a facility should have to develop such information solely for the benefit of the audit team.

Upon receipt of their audit assignments and background information, team members should review the supplied information, familiarize themselves with the audit protocol(s), and strategize their approach to their audit responsibilities. Lack of time to do pre-audit preparation is sometimes a problem for 2nd party auditors, who often also have extra work to do at their own facility prior to leaving for the period of the audit. Audit team leaders need to be realistic about the amount of preparation time those individuals will have.

Plan onsite activities. Once the audit schedule, scope, and team membership have been finalized, the team leader, with input from the facility and audit team members, can establish an agenda for the onsite audit effort. Where feasible, this agenda should include preliminary interview schedules for those representatives of the facility staff whom the various team members will want to interview (identified by role, if not by name).

Conduct the audit. The audit team should conduct a kick-off meeting prior to the start of the audit. This meeting should include facility staff responsible for RBPS element implementation, managers of the departments that will be audited, and other interested or affected parties. This meeting should be used to confirm the overall schedule of the audit, explain the mechanics of the audit, and generally attempt to address any curiosities and anxiety on the part of the facility staff.

The bulk of the onsite time will be devoted to the auditors gathering data about the implementation of the RBPS program. Generally, three approaches are used to gather this information: (1) records reviews, (2) direct observations, and (3) interviews.

While different auditors have different preferences, one approach is to interview those parties responsible for RBPS element implementation to hear their perspective on how this is being accomplished. Next, based upon this input and the auditors' familiarity with the RBPS procedures (from their advance preparation), the auditors gather information from the review of records and the observation of conditions and practices in the field. Additional interviews with the RBPS element owners, and others who are involved in implementing the elements or who are affected by the elements, are used to broaden the auditor's understanding of the implementation of the element and to clarify inconsistent information.

The audit team should immediately apprise facility management of any activities or conditions encountered during the audit that appear to pose an imminent hazard to personnel or facilities.

Reviewing every record associated with a particular RBPS element requirement is not necessary, and, for that matter, is often impractical. Valid sampling approaches should be used to identify a representative number of records for review, sufficient to serve as a valid basis for conclusions without overworking the analysis. The perceived risk associated with the facility or activity being audited, and the role in mitigating that risk expected of the RBPS element requirement under review, should be factored into the depth of the scrutiny given to that requirement.

As each auditor gathers data, the focus should be on assessing the design and implementation of the RBPS element(s) against the established requirements addressed in the audit protocol. Strengths, weaknesses, and gaps in the implementation should be noted. The auditor should seek additional data when unambiguous conclusions cannot be drawn. Detailed notes should be maintained to support preparation of the subsequent audit report. Depending upon the context, the audit team (especially a 3rd party team) may elect to retain detailed documentary evidence to substantiate the basis for the audit findings. This is typically not necessary if the audit team presents a detailed summary of specific findings at the conclusion of the onsite activities, and if the facility has been provided ample opportunities to review and rebut audit findings during the course of the audit. Such documentation is needed only until the final audit report is issued by the team and accepted by the facility.

Preliminary findings and observations should be periodically shared with facility staff through brief meetings held during the course of the audit. For shorter audits, these meetings might be scheduled daily. Such meetings allow facility staff to provide commentary and additional information on the perspectives shared by the auditors. Without this activity, auditors can pursue an avenue of inquiry that may be based on misleading or incomplete information, or a false premise. Clearing up misconceptions as quickly as possible saves time both during the audit and in audit follow-up activities (i.e., the facility will not have to justify why an audit finding is erroneous and, therefore, no action is required).

A final closeout meeting should be held with facility staff to summarize preliminary audit findings before the audit team leaves the facility. All significant issues should be addressed in this meeting. While the intent should be to ensure that the written report will not contain any issues that were not previously discussed, the audit team should clearly communicate that these preliminary findings neither constitute a final nor comprehensive summary of the findings as they will appear in the report. This meeting provides another (but not final) opportunity for the facility staff to provide input to the auditors' deliberations, and allows the team leader to confirm the anticipated schedule for the delivery of the draft report, if any, and the final report.

Document the audit. Based on information gathered during the audit, the team members should complete and document their evaluation of the implementation of those RBPS elements that they were assigned to audit. The audit team leader has the responsibility for assembling each team member's contributions into the comprehensive audit report in a timely fashion. The format and content of the report should conform to what is stated in the facility's written description of the *audits* element. At a minimum, the report should document the results of the audit, indicating where and when the audit was done, who performed the audit, the audit scope, the manner in which it was conducted, and the audit findings. The team should be aware that an audit report can potentially become a public document (e.g., as a consequence of a legal or regulatory proceeding). Consequently, auditors should use care to ensure that the report is factual, while avoiding potentially inflammatory content such as sweeping generalizations, editorial commentary, and unsupported speculation.

The team leader often prepares a draft report and forwards it to appropriate facility staff for their review and comment. Any feedback is shared with the responsible audit team member(s) to determine if the audit report content will be affected. After all such issues have been resolved, the team leader prepares and issues a final audit report.

Address audit findings and recommendations. Upon completion, the audit report must be distributed to appropriate parties for follow-up action. The required distribution of the audit report may be determined by the *audits* program document. Typically, the appropriate recipients of the audit report include the manager of the facility that was audited and at least one level of supervision above that manager. Corporate legal counsel may provide input with regard to the breadth and manner of distribution of the report. Failure to distribute the audit report to appropriate individuals may compromise the value of the audit.

Facility staff members responsible for audit follow-up must develop an action plan to address the audit report by first determining what they view to be the appropriate response to each audit finding. The audit report may or may not include specific recommendations for addressing the audit findings, depending upon the agreed upon content. If it does, facility staff will need to review these recommendations, judging their likely effectiveness and practicality. If the report does not include specific recommendations, facility staff will have to propose their own solutions for addressing the gaps.

Ultimately, facility management will receive a list of proposed recommendations for their consideration. Typical management decisions for a particular recommendation might be:

- Adopt as proposed.
- Adopt in principle (i.e., achieve the stated intent through another, equivalent means).
- Reject based upon the assertion that the recommendation was based upon an error in fact or judgment.
- Reject because of changed circumstances that make the recommendation no longer required.

The process of reviewing the recommendations and determining whether to adopt or reject them is termed the resolution of the recommendations, and should be accomplished by facility management in a timely fashion. The resolution of the recommendations will yield a list of action items that must be implemented to achieve the intent of the adopted recommendations. (Note that resolution of a single recommendation may require multiple action items; conversely, a single action item may address multiple recommendations.)

Each resulting action item must be assigned to a responsible party and should be given a due date. The management system should (1) track the status of all actions until they are implemented and (2) monitor the system periodically to ensure compliance. The completion date of each action item should also be documented. Successful implementation of all associated action items permits the closure of the recommendation.

Some organizations require the facility to report proposed resolutions and final closure of recommendations to a central corporate authority that endorses the proposed resolution and tracks each recommendation to closure. Such controls are more common in organizations that coordinate the audit function at the corporate level.

The *audits* program document should provide guidance on retention practices for audit records. At a minimum, the documentation maintained should include the final audit report and records of the means and date of closure for each recommendation (draft documents and working papers are typically destroyed once the final audit report is issued). Records should be maintained at least until the next audit, or longer as determined by regulatory requirements or corporate policies and needs. Master copies of records should be protected against unauthorized modification or loss. For example, records maintained in workers' personal files are prone to loss if the individual retires, resigns, or moves to another position in the company.

21.2.3 Use Audits to Enhance RBPS Effectiveness

Monitor RBPS maturation over time for each facility. While audits, in their narrowest sense, focus on verifying conformance with established standards for the implementation of RBPS, many organizations will aspire to levels of performance beyond mere compliance. By implementing the *audits* element, the organization can create a system for monitoring performance over time, which allows the organization to track the maturation of the RBPS management system. In this sense, the *audits* element complements the *metrics* and *management review* elements described in Chapters 20 and 22, respectively.

Relevant performance metrics should be identified for each RBPS element. Such metrics must bear a direct relationship to the effectiveness of the RBPS element, and should be clearly quantifiable. These metrics should be updated for each audit. Where performance is particularly problematic, more frequent or specially targeted audits may be appropriate. Results should be trended to monitor whether or not performance is improving.

Continued poor performance is an indication that weaknesses exist in the program for implementing that particular RBPS element, or that the element is not being consistently implemented in accordance with the established program. Implementation problems may be due to a variety of causes, such as: (1) a failure to clearly define roles and responsibilities, (2) inadequate training provided to those responsible for element implementation, (3) failure to commit resources, or (4) failure to hold staff accountable for their performance in support of RBPS objectives. Each of these causes may indicate a deficiency in the organization's process safety culture.

A root cause analysis may be required to determine the underlying causes of continuing performance problems including, if pertinent, process safety culture weakness (see Chapter 3). Once the causes are identified, remedial actions can be proposed, selected for implementation, and tracked for their effectiveness.

Share best practices. Implementation of the *audits* element can also allow the identification of those RBPS elements whose implementation practices are particularly effective. While such best practices may not be universally applicable, noting them during audits and properly documenting them allows these successes to be shared across the company.

The individual coordinating the *audits* element at the corporate level should be alert to such potential best practices. A system should be provided for researching the details, codifying the practice, and sharing this information with RBPS element owners across the company for their consideration.

21.3   POSSIBLE WORK ACTIVITIES

The RBPS approach suggests that the degree of rigor designed into each work activity should be tailored to risk, tempered by resource considerations, and tuned to the facility's culture. Thus, the degree of rigor that should be applied to a particular work activity will vary for each facility, and likely will vary between units or process areas at a facility. Therefore, to develop a risk-based process safety management system, readers should perform the following steps:

1. Assess the risks at the facility, investigate the balance between the resource load for RBPS activities and available resources, and examine the facility's culture. This process is described in more detail in Section 2.2.
2. Estimate the potential benefits that may be achieved by addressing each of the key principles for this RBPS element. These principles are listed in Section 21.2.

3. Based on the results from steps 1 and 2, decide which essential features described in Sections 21.2.1 through 21.2.3 are necessary to properly manage risk.
4. For each essential feature that will be implemented, determine how it will be implemented and select the corresponding work activities described in this section. Note that this list of work activities cannot be comprehensive for all industries; readers will likely need to add work activities or modify some of the work activities listed in this section.
5. For each work activity that will be implemented, determine the level of rigor that will be required. Each work activity in this section is followed by two to five implementation options that describe an increasing degree of rigor. Note that work activities listed in this section are labeled with a number; implementation options are labeled with a letter.

> *Note: Regulatory requirements may specify that process safety management systems include certain features or work activities, or that a minimum level of detail be designed into specific work activities. Thus, the design and implementation of process safety management systems should be based on regulatory requirements as well as the guidance provided in this book.*

21.3.1 Maintain a Dependable Practice

Ensure Consistent Implementation

1. Develop an *audits* program that addresses issues such as: the scope of application, scheduling, team staffing, recommendation resolution, and documentation of audits.
   a. An informal *audits* program exists, but has not been documented.
   b. A written program document has been developed that addresses the basic regulatory mandates for the *audits* program.
   c. Written program documentation has been expanded to more comprehensively include auditing against facility or company standards and other non-regulatory requirements.
2. Establish an *audits* element owner.
   a. Oversight of the *audits* program is provided on an ad hoc basis.
   b. Oversight of the *audits* program is conducted as a part-time, formal activity.
   c. A permanent staff position is dedicated to oversight of the *audits* program.
3. Define roles and responsibilities for the *audits* element.
   a. Roles and responsibilities are informally defined.

      b. Roles and responsibilities are formally defined and documented.

      c. Item (b) and roles and responsibilities are reinforced through an active accountability system.

4. Develop audit protocols.

      a. Audit protocols have been developed to address general considerations related to the conduct of audits, but they do not address specific requirements.

      b. Audit protocols explicitly address regulatory requirements.

      c. Audit protocols have been expanded to address facility or company-specific requirements.

Involve Competent Personnel

5. Define the skills needed for the audit team members.

      a. Audits are staffed in an ad hoc fashion.

      b. Basic skills, such as familiarity with regulatory requirements and auditing practices, have been defined.

      c. More advanced skills, such as interviewing techniques, sampling schemes, and written and oral communications have been added to the auditor profile.

      d. A structured system has been defined identifying skill sets for various roles, such as lead auditor, corporate audit expert, facility coordinator, team member, and so forth.

6. Provide the required training to the audit team members.

      a. Informal training is provided.

      b. Formal, documented training is provided on basic skills.

      c. Formal, documented training is provided on advanced skills.

7. Ensure appropriate objectivity for audit teams.

      a. Audits are typically staffed with 1st party membership; however, controls are in place to ensure that auditors do not review activities with which they have a direct association.

      b. Some 2nd party membership is included on many audits.

      c. Significant involvement of 2nd party membership is included on most audits; 3rd party members are not uncommon.

      d. Audits are coordinated at corporate level, with strong emphasis on team objectivity. 3rd party members constitute a significant fraction of the audit team membership.

Identify When Audits Are Needed

8. Establish baseline schedules for audits.

      a. Audits are scheduled in a reactive or episodic fashion.

    b.   Audits are scheduled strictly according to minimum frequencies determined from regulatory requirements.

    c.   Audit frequencies exceed basic regulatory requirements, based upon considerations such as the perceived level of risk associated with facility activities.

9.   Identify triggers for additional audits.

    a.   Additional audits are conducted only in the most extreme circumstances, at the discretion of management.

    b.   Additional audits are effectively used as tools to identify system performance problems, but only at the discretion of management.

    c.   Formalized protocols have been established to determine if audits are needed more frequently than currently scheduled.

21.3.2 Conduct Element Work Activities

Prepare for the Audit

10. Review audit protocols.

    a.   Standard protocols are reviewed to ensure that they are up to date.

    b.   Standard protocols are customized to match any unique scope considerations for the audit.

Determine the Audit Scope and Schedule

11. Confirm the audit schedule and define the scope.

    a.   The schedule and scope are mandated upon the facility by the corporate environment, health, and safety function.

    b.   The scope and schedule are negotiated in a manner that enhances the productivity of audit.

Assemble the Team

12. Select team members, assign RBPS management system elements to the respective team members, and confirm their availability.

    a.   The audit schedule, personnel availability, and other constraints may be significant determinants of team membership.

    b.   Careful consideration is given to required skill sets, background, and knowledge of team members; the audit schedule is adjusted, if required, to ensure the availability of proper team members.

    c.   Item (b) and consideration is given to using audits as an opportunity to train new auditors (in a fashion that does not detract from the quality of the audit).

Assign Responsibilities

Assign Responsibilities

13. Assign audit responsibilities to audit team members based on expertise, experience, and interest.

    a. The audit team leader assigns activities to audit team members based on knowledge of each auditor's expertise.

    h The audit team leader consults with each audit team member. Assignments are based on (1) not allowing audit team members to review areas in which they have previously provided services and (2) audit team member interest, experience, and expertise.

Gather Advanced Information

14. Assemble and distribute to the team members information that will assist in their advance preparation for the audit.

    a. Basic information, such as facility organization charts and RBPS responsibility assignments, is provided to the audit team.

    b. Additional information addressing the needs of the audit team members is provided. Such information could include process descriptions, RBPS program documentation and procedures, past audit reports, and so forth.

Plan Onsite Activities

15. Plan audit activities and prepare the onsite agenda.

    a. Basic logistical issues are addressed.

    b. Detailed plans are prepared, including preliminary schedules for interviews.

Conduct the Audit

16. Conduct the audit kick-off meeting.

    a. A brief introductory meeting is held.

    b. A detailed kick-off meeting is carefully planned and held, with proper attendance, to increase potential success of the audit.

17. Gather audit data though records sampling and reviews, observations, and interviews.

    a. Data are gathered in sufficient detail, using an appropriate balance between sources (records, observations, and interviews) to support valid conclusions regarding the degree of compliance and effectiveness of RBPS management system.

    b. Item (a) and more extensive sampling yields additional data, which provide greater validation of conclusions (e.g., as required for a settlement agreement audit).

18. Assess RBPS implementation strengths, weaknesses, and gaps relative to established requirements.

   a. General implementation problems are identified.

   b. A thorough evaluation of implementation strengths, weaknesses, and gaps results from a detailed analysis.

   c. Item (b) and further analysis is conducted to determine not just the problems, but the underlying causes of the problems.

19. Report preliminary audit observations and findings in periodic meetings during the audit.

   a. Meetings are held infrequently, or are brief in content.

   b. Meetings are sufficiently frequent and detailed to keep the facility closely advised of audit progress.

   c. Regular meetings are effectively used to enhance the value of the audit. Sufficient information is exchanged between the audit team and facility personnel to facilitate identification and resolution of problems, audit team misconceptions, necessary changes in audit schedule, and so forth.

20. Conduct an audit close-out meeting.

   a. The audit team shares high level, preliminary conclusions with facility representatives.

   b. The audit team shares detailed findings with facility representatives. While the findings are still preliminary, they are based upon thorough analysis, and substantive changes are not anticipated.

Document the Audit

21. Prepare a draft report and forward it to the appropriate facility personnel for review.

   a. A draft report is prepared, but it is given limited distribution to facility staff.

   b. Broader distribution of the draft report enhances the potential that issues are identified and resolved prior to issuing the final report.

22. Issue the final report and forward it to the appropriate facility personnel.

   a. A summary report is prepared, covering only the audit team's findings.

   b. A more detailed report is issued that documents the basis for audit findings, as well as recommendations for remediation.

   c. In addition to the content described in item (b), the report describes audit methodology, team qualifications, and other content appropriate to validate the quality of the audit. The report also addresses areas of exceptional performance.

Address Audit Findings and Recommendations

23. Develop an action plan to address report findings; assign responsibilities and establish deadlines.
    a. Audit follow-up is handled in ad hoc fashion.
    b. A detailed action plan is formulated for each recommendation, with clear assignment of responsibilities for implementation and required due dates.
24. Follow up to resolve, and document resolution of, audit recommendations.
    a. Recommendations are resolved in an ad hoc fashion.
    b. A formalized system is implemented for recommendation follow-up and tracking of recommendations to resolution.
25. Maintain required documentation of the audit, including resolution of recommendations
    a. An informal system is implemented for maintaining documentation.
    b. A formal system is implemented for storing and protecting audit documentation.
    c. An advanced document management system is implemented to maintain and protect information, which also enhances its availability to personnel having a need to access the information.

21.3.3 Use Audits to Enhance RBPS Effectiveness

Monitor RBPS Maturation over Time for Each Facility

26. Monitor RBPS performance over time.
    a. Data are collected and reviewed, but analysis is limited.
    b. Goals are established, and performance against these goals is analyzed over time.
27. Identify continuing RBPS management system or performance weaknesses.
    a. Analysis is sufficient to identify more substantive problems.
    b. Detail analysis yields a comprehensive awareness of systemic problems.
    c. Item (b) and analysis is sufficient to identify underlying problems, common causes, element interface issues, and so forth.
28. Implement RBPS management system enhancements to address weaknesses.

a. Corrective actions focus on the immediate problems or gaps.

b. Corrective actions include more substantive responses that address underlying problems with management systems.

Share Best Practices

29. Identify RBPS management system strengths.

   a. Strengths are noted if obvious.

   b. Audit teams make a concerted effort to identify significant best practices.

30. Disseminate information on RBPS management system best practices to other facilities.

   a. Best practices are shared informally between facilities.

   b. A concerted effort is made to disseminate best practices to all company facilities.

   c. Item (b) and, where appropriate, best practices are shared with the industry.

21.4   EXAMPLES OF WAYS TO IMPROVE EFFECTIVENESS

This section provides specific examples of industry tested methods for improving the effectiveness of work activities related to the *audits* element. The examples are sorted by the key principles that were first introduced in Section 21.2. The examples fall into two categories:

1. Methods for improving the performance of activities that support this element.

2. Methods for improving the efficiency of activities that support this element by maintaining the necessary level of performance using fewer resources.

These examples were obtained from the results of industry practice surveys, workshops, and CCPS member-company input. Readers desiring to improve their management systems and work activities related to this element should examine these ideas, evaluate current management system and work activity performance and efficiency, and then select and implement enhancements using the risk-based principles described in Section 2.1.

21.4.1 Maintain a Dependable Practice

Properly train audit team members. In addition to training on applicable regulations and basic auditing techniques, train auditors on softer topics, such as interviewing skills, negotiation skills, written communications skills, and so forth.

Implement an apprentice system during which new auditors have opportunities to work with seasoned auditors in order to develop their skills.

One CCPS sponsor has developed a virtual auditor refresher training module for experienced auditors. The module consists of conference calls and net meetings in three sessions over a one-month timeframe as well as off-line group breakouts and exercises. An electronic challenge test is administered once the training is completed. 2nd party auditors are required to complete this refresher training every four years.

Standardize audit procedures. Develop one *audits* system for the corporation, to be used at all facilities. This will prevent having to "reinvent the wheel" multiple times, and should help ensure consistency across the corporation. One CCPS sponsor utilizes a detailed 2nd party RBPS audit protocol checklist for all global audits that includes more than 300 individual audit items. This same checklist is also used at all facilities for 1st party self audits.

Make appropriate use of written audit protocols to help ensure comprehensive audits. Develop protocols unique to the regulations and other requirements being audited, but do not allow the literal content of the protocols to limit auditors in their inquiries. Protocols cannot anticipate every question or area of inquiry, and experienced auditors use protocols as guides, and not scripts, for audits.

Ensure that resources to support the audits program are properly budgeted. The *audits* program can require appreciable resources in terms of staff time and expenses at the facility and, perhaps, corporate level. Such resources should be budgeted ahead of time to ensure their availability in support of the *audits* program.

21.4.2 Conduct Element Work Activities

Be discriminating in the scheduling of audits. Audits are very time consuming, both for the organization being audited and the auditor. Consequently, initiatives to more effectively steward this investment in time should be sought, but without sacrificing the quality of the audit.

- Schedule audits sufficiently far ahead of time, to ensure the availability of needed audit team staff and to ensure that unanticipated circumstances do not prevent compliance with regulatory and company schedule deadlines.
- More frequently audit any RBPS management systems (or particular elements) that are undergoing transitions or, through analysis of

incidents element trends, are showing indications of performance problems.

- Effectively use management reviews (see Chapter 22) between audits to identify and address problems so that they do not become surprises during audits.
- Do parts of the audit in conjunction with other RBPS activities to minimize duplication of effort and optimize resource use. For example, audit process safety information as a part of the hazard assessment planning effort, when such information will be scrutinized anyway.
- Implement a program in which a certain number of RBPS elements are audited each quarter such that each element is audited at the required frequency. Avoid auditing all of the elements in the same quarter by staggering the audits throughout the year. This method also helps to reduce spikes in resource needs that accompany a cluster of audit recommendations coming due in a short time interval.
- Be aware that various corporate functions may be auditing the same facility at different times, and with potentially overlapping analytical scopes; for example, corporate fire protection and RBPS audits may each look at a number of common issues. Be respectful of the demands that audits place on facility staff and try to coordinate such audits whenever possible. Avoid redundant recommendations.

Give proper consideration to audit team composition. Use qualified, skilled auditors. An inexperienced team will take longer to find less, seriously affecting both the efficiency and performance of the audit activity. While more experienced auditors may be hard to break free from other responsibilities, the net trade-off for doing so will be favorable. Some CCPS member companies have found it helpful to:

- Use 2nd party auditors (from corporate groups, or other facilities) to enhance communications and sharing of best practices between facilities.
- Provide an appropriate mix of 1st party, 2nd party and, perhaps, 3rd party auditors on the team to ensure adequate familiarity with facility processes and systems, sufficient team objectivity, and auditing expertise. To enhance objectivity when selecting 2nd party auditors, choose those who are farthest removed from the organizational reporting structure for the facility being audited. Use qualified 3rd party auditors when internal audit resources do not exist or are not sufficient, or where a different perspective or specialized expertise is required.
- Develop a cadre of experienced lead auditors across the corporation who can serve as 2nd party auditors and mentors. Rotate different auditors through the team leader role to provide cross-training and the sharing of fresh viewpoints.

- Where possible, configure audit teams so that more than one team member is capable of auditing each element. Share, discuss, and test preliminary conclusions within the team before sharing them with the facility.

Pay careful attention to audit logistics. Request that the facility provide the audit team a conference room or other suitable work space for the duration of the audit, large enough for the entire audit team to meet together. The entire team should meet periodically during the day (e.g., at lunch and prior to the daily meeting with the facility representatives, if such is held) in order to share information and leads.

Unless a facility cafeteria is convenient to the team meeting room, have lunches delivered to the audit team to save the time that would otherwise be required to commute to and from offsite locations.

Use the evenings to organize notes, identify gaps in data gathering, and plan the next day's activities.

Make maximum use of computer databases when performing records reviews. Facilities with computerized systems for implementing certain elements (such as *management of change*) or maintaining relevant records (such as piping inspection records) will often be trending the very data that auditors are seeking. If possible, request the facility to provide team members access to such databases, or request that specific data searches be made for the team.

Ensure that an adequate amount of representative data is reviewed. When scheduling interviews, seek to include representatives from each operating area and each craft group. Conduct some interviews and observations on off-shifts or on the weekend, to ensure that a broader cross-section of personnel are surveyed, and to provide an opportunity to interface with workers with fewer day-shift supervisors present.

Use appropriate sampling schemes to ensure that sufficient numbers of documents are reviewed and that a suitable cross-section of documentation is sampled.

Where possible, retain copies of key documentary evidence until the final report has been issued.

Establish an integrated corrective action tracking system. Some organizations have developed and implemented an integrated corrective action tracking system to monitor all relevant RBPS action items, including recommendations from incident reports, process hazard analyses, emergency drills, audits, and so forth. These systems are transparent so that everyone in the organization can see open and overdue action items, and can electronically run reports to query the data. Such systems are commonly connected to e-mail systems to inform responsible personnel (and, often, their supervisors) of overdue actions, and to allow responsible parties to update the system when actions are completed. These systems also facilitate the re-assignment of responsibility for open items in the event of organizational changes.

21.4.3 Use Audits to Enhance RBPS Effectiveness

Prioritize follow-up activities. Some organizations use a relative risk ranking approach to prioritize the resolution of audit recommendations when establishing follow-up schedules.

Provide for a high level review of audit results. Facility management may be required to review the results of audits with corporate management in order to provide them perspective on the health of the local systems. Corporate management involvement reinforces the importance of RBPS program performance.

Some corporations track audit recommendation resolution at the corporate level, especially when the audits are 2nd party audits coordinated by a central staff department. The higher visibility given to recommendations tracked at this level helps reinforce their importance and the timeliness of their resolution.

Avoid making audits a competition. Corporate management should be very careful if they elect to score audits and use the results to give facilities relative ranks. Doing so creates the potential for a competitive environment in which facility managers become more focused on their scores than on the messages underlying the findings.

While there may be room to debate the interpretation of requirements, or of data, unreasonable pushback on audit findings from facility managers should not be tolerated by the organization.

21.5 ELEMENT METRICS

Chapter 20 describes how metrics can be used to improve performance and when they may be appropriate. This section includes several examples of metrics that could be used to monitor the health of the *audits* element, sorted by the key principles that were first introduced in Section 21.2.

In addition to identifying high value metrics, readers will need to determine how to best measure each metric they choose to track. In some cases, an ordinal number provides the needed information, such as total number of workers. Other cases, average years of experience, for example, require that two or more attributes be indexed to provide meaningful information. Still other metrics may need to be tracked as a rate, as in the case of employee turnover. Sometimes, the rate of change may be of greatest interest. Since every situation is different, the reader will need to determine how to track and present data to most efficiently monitor the health of RBPS management systems at their facility.

21.5.1 Maintain a Dependable Practice

- *Percentage of near miss and incident investigations identifying RBPS management system weaknesses that were not detected by prior audits.* A high value suggests that prior audits were not of sufficient rigor or depth to effectively identify the true problems in RBPS management system development or implementation.
- *Percentage of audits completed according to schedule.* A low value suggests that the organization may not assign sufficient value to the *audits* element.
- *Percentage of audits having few significant findings.* While a high value may indicate an exemplary RBPS management system, an alternate explanation could be that the audit teams may have become complacent or too accommodating.
- *Number of previous audits conducted by each audit team member.* A high value is one indicator (but not a guarantee) of audit team qualifications.

21.5.2 Conduct Element Work Activities

- *Average and maximum number of days overdue for open recommendations.* A high value calls into question the value that the organization places on the *audits* program, and the resources assigned to resolve issues identified by the audit.
- *Number or percentage of unresolved audit recommendations.* A high value calls into question the value that the organization places on the *audits* program, and the gravity that it assigned to problems identified by the audit.
- *Number of person-days required to complete an audit.* A low value calls into question the rigor of the audit. A high value, absent extenuating circumstances necessitating a particularly detailed audit, may point to inefficient audit practices.
- *Interval between completion of onsite work and completion of the audit report.* A high value calls into question the value that the organization places on the *audits* program.

21.5.3 Use Audits to Enhance RBPS Effectiveness

- *Percentage of audit findings that are repeat findings.* A high value suggests that audit results are not being effectively used to improve RBPS management system performance.

- *Trends in the number or significance of findings over a series of audits of the same facility.* An increasing trend may indicate that RBPS management system performance is degrading. Alternatively, auditing practices may be improving.
- *Percentage of recommendations that are rejected by the facility management.* A high value may indicate either too much pushback from management, or insufficient analysis by the audit team.

21.6   MANAGEMENT REVIEW

The overall design and conduct of management reviews is described in Chapter 22. However, many specific questions/discussion topics exist that management may want to check periodically to ensure that the management system for the *audits* element is working properly. In particular, management must first seek to understand whether the system being reviewed is producing the desired results. If the organization's implementation of the *audits* element is less than satisfactory, or it is not improving as a result of management system changes, then management should identify possible corrective actions and pursue them. Possibly, the organization is not working on the proper activities, or the organization is not doing the necessary activities well. Even if the results are satisfactory, management reviews can help determine if resources are being used properly – are there tasks that could be done more efficiently or tasks that should not be done at all? Management can combine metrics listed in the previous section with personal observations, direct questioning, audit results, and feedback on various topics to help answer these questions. Activities and topics for discussion include the following:

- Evaluate whether the scope of the audit program covers every relevant facility and activity, consistent with the organization's scope of application for the RBPS management system.
- Evaluate whether the established frequencies for the audit(s) reflect (1) the perceived risk associated with the activities and facilities being audited and (2) past audit results and incident histories. Do the results of the audits(s) confirm the risk assumptions underlying the audit frequency?
- Determine whether audits are conducted according to schedule. Even if audits are coordinated at corporate level, facility management is responsible for ensuring that certain regulatory driven audits are completed within the prescribed time period.
- Evaluate whether the audit team staffing is appropriate. Was the aggregate expertise on the team sufficient to provide the requisite familiarity with facility operations, auditing techniques, and the

standards of care against which the audit was performed? Was the audit team sufficiently independent of the organization being audited?

- Review the audit report(s) and determine if the documentation is sufficient to establish a basis for evaluating the quality of the audit.
- Evaluate whether the action items proposed to address the audit findings are appropriate to correct the deficiencies detailed by the report findings or recommendations.
- Determine whether responsibilities for the resolution of audit recommendations are clearly assigned and deadlines established. Are the proposed deadlines credible considering the gravity of the findings that the recommendations address?
- Determine whether recommendation action item deadlines are consistently met. If not, what is done in lieu of the required activity? Are delays the result of uncontrollable factors, or just a lack of proper follow-up?
- Confirm that recommendation action items are closed out only upon the implementation of the action explicitly recommended. Do not accept closure for responses such as "work order written" or "engineering request submitted", which do not explicitly implement the final intent specified in the action items.

RBPS management system implementation requires the dedication of appreciable time and resources by the organization. Effective auditing serves to protect this investment by helping ensure that the process safety results intended from the management system are actually being achieved.

21.7   REFERENCES

Additional reading

CCPS, *Guidelines for Auditing Process Safety Management Systems*, 1993.

Frank, W.; Hobbs, D.; and Jones, D., "What Are Process Safety Management Audits Telling Operators?", Hydrocarbon Processing, October 2000.

22

MANAGEMENT REVIEW AND

CONTINUOUS IMPROVEMENT

On July 17, 2001, an explosion occurred in a spent sulfuric acid storage tank at a refinery in Delaware (Ref. 22.1). Welding work to repair tank leaks ignited hydrocarbon vapors in the vapor space of the tank. One contract maintenance worker was killed, and eight others were injured. The accident resulted from a confluence of several factors: (1) the corrosivity and flammability hazards associated with changing the tank from fresh acid service to spent acid service were not identified and controlled, (2) repeated requests for tank inspections and repairs were deferred or ignored, and (3) the hot work permit failed to specify atmospheric monitoring despite previous permit denials because of toxic and flammable gas concentrations. Each of these contributing factors was the result of a management system breakdown (*management of change*, *asset integrity*, and *safe work*) that could have been identified and corrected by timely management review.

22.1 ELEMENT OVERVIEW

Routinely reviewing the organization's process safety systems to spur continuous improvement is one of four elements in the RBPS pillar of *learning from experience*. This chapter describes the meaning of management review, the attributes of a good management review system, and the steps an organization might take to implement management reviews. Section 22.2 describes the key principles and essential features of a management system for this element. Section 22.3 lists work activities that support these essential features and presents a range of approaches that might be appropriate for each work activity, depending on perceived risk, resources, and organizational culture. Sections 22.4 through 22.6 include (1) ideas to improve the effectiveness of management systems and specific programs that support this element, (2) metrics that could be used to monitor this element, and (3) issues that may be appropriate for management review.

22.1.1 What Is It?

Management review is the routine evaluation of whether management systems are performing as intended and producing the desired results as efficiently as possible. It is the ongoing "due diligence" review by management that fills the gap between day-to-day work activities and periodic formal audits. Management review is similar to a doctor giving a routine physical examination – even when no overt signs of illness are present, life-threatening conditions may be developing that are best addressed proactively. Management reviews have many of the characteristics of a 1st party audit as described in Chapter 21. They require a similar system for scheduling, staffing, and effectively evaluating all RBPS elements, and a system should be in place for implementing any resulting plans for improvement or corrective action and verifying their effectiveness.

22.1.2 Why Is It Important?

Effective performance is a critical aspect of any process safety program; however, a breakdown or inefficiency in a safety management system may not be immediately obvious. For example, if a facility's training coordinator unexpectedly departed, required training activities might be disrupted. The existing trained workers would undoubtedly continue to operate the process, so there would be no outward appearance of a deficiency. An audit or incident might eventually reveal any incomplete or overdue training, but by then it could be too late. The *management review* process provides regular checkups on the health of process safety management systems in order to identify and correct any current or incipient deficiencies before they might be revealed by an audit or incident.

22.1.3 Where/When Is It Done?

Management reviews should be conducted wherever RBPS elements are implemented. The depth and frequency of each management review should be governed by factors such as the current life cycle stage of the facility, the maturity or degree of implementation of the RBPS management system, the level of management performing the review, past experience (e.g., incident history, previous reviews, and audit results), and management's view of the risk posed by the activities to be reviewed. Most of the *management review* effort will be focused on operating facilities.

While they can be scheduled on an as-needed basis, management reviews of a particular RBPS element are typically conducted at a predetermined interval (e.g., frequencies ranging from monthly to annually are common), and they may be scheduled in conjunction with other regularly scheduled meetings, such as facility safety committee meetings.

22.1.4 Who Does It?

Strictly speaking, every level of management – from the process supervisor to the facility manager to the board of directors – should conduct periodic management reviews. Further discussion in this book focuses on program level reviews, which should be conducted by a manager who is one or two levels above the person responsible for the day-to-day execution of a specific RBPS element. This is usually the manager ultimately responsible for the proper functioning of the facility's overall process safety management system; however, larger facilities may charter a Process Safety Committee to conduct the reviews under the leadership of a senior manager.

22.1.5 What Is the Anticipated Work Product?

The output of a management review is generally an internal memorandum summarizing the review, any deficiencies or inefficiencies noted, and recommendations for improvement or corrective action. The recommendations should be given deadlines and then assigned to specific individuals. All outputs of the *management review* element are intended to facilitate the performance of other elements. In addition, the management reviews provide input that the *audits* element can use to focus its efforts.

22.1.6 How Is It Done?

Management reviews are conducted with the same underlying intent as an audit – to evaluate the effectiveness of the implementation of an entire RBPS element or a particular element task. However, because the objective of a management review is to spot current or incipient deficiencies, the reviews are more broadly focused and more frequent than audits, and they are typically conducted in a less formal manner.

Nevertheless, like an audit, a management review at least checks the implementation status of one or more RBPS elements against established requirements. The management review team meets with the individuals responsible for managing and executing the subject element to (1) present program documentation and implementation records, (2) offer direct observations of conditions and activities, and (3) answer questions about program activities. The team attempts to answer such questions as:

- What is the quality of our program?
- Are these the results we want?
- Are we working on the right things?

Organizational changes, staff changes, new projects or standards, efficiency improvements, and any other anticipated challenges to the subject element are also discussed so that management can proactively address those issues.

Recommendations for addressing any existing or anticipated performance gaps or inefficiencies are proposed, and responsibilities and schedules for addressing the recommendations are assigned. Typically, the same system used to track corrective actions from audit findings is used to track management review recommendations to their resolution. The meeting minutes and documentation of each recommendation's resolution are maintained as required to meet programmatic needs.

Management review results should be monitored over time, and more frequent reviews should be scheduled if persistent problems are evident.

22.2   KEY PRINCIPLES AND ESSENTIAL FEATURES

Safe operation and maintenance of facilities that manufacture, store, or otherwise use hazardous chemicals requires robust process safety management systems. The primary objective of the *management review* element is to ensure that the required RBPS activities produce the desired results over the life of the facility. The following key principles should be addressed when developing, evaluating, or improving any system for the *management review* element:

- Maintain a dependable practice.
- Conduct review activities.
- Monitor organizational performance.

The essential features for each principle are further described in sections 22.2.1 through 22.2.3. Section 22.3 describes work activities that typically underpin the essential features related to these principles. Facility staff should evaluate the risks and potential benefits that may be achieved as a result of improvements in this element and, on that basis, develop a management system (or upgrade an existing management system) to address some or all of the essential features. Some or all of the identified work activities, depending on perceived risk and/or process hazards should be undertaken. However, none of these work activities will be effective if the accompanying *culture* element does not embrace the use of reliable management systems. Even the best management reviews, without the right culture, will not produce reliable management systems.

22.2.1 Maintain a Dependable Practice

Documenting the *management review* program is the first step in maintaining a dependable practice. The procedures governing management reviews must be established, reviewers should be trained, and their effectiveness should be periodically verified. The management system should be designed to detect current or incipient weaknesses in RBPS elements so they can be corrected before a serious breakdown occurs.

Define roles and responsibilities. The written description of the *management review* system should identify the organization's objectives and the procedure that must be followed to achieve those objectives. The procedure should (1) identify who is responsible for, and accountable for, its execution and (2) provide clear lines of authority for making decisions. The person ultimately accountable for the performance of the process safety management system should lead the management review and be assisted by other senior managers. The owner of a specific RBPS element should be responsible for gathering any requested data on the performance of the element and present it to the reviewers. Others involved in implementing the element should be responsible for reporting their observations about how the element is being executed, what improvements should be made, and what challenges they anticipate.

Establish standards for performance. *Management review* is primarily focused on ensuring that other program elements are adhering to their standards of performance, yet the management reviews themselves should conform to their own established standards for performance. These standards typically define: (1) the scope and objectives of the *management review* element, (2) the criteria for determining the frequency and depth of scrutiny for management reviews as a function of the perceived risk associated with the facilities or activities, (3) the requirements for the resolution of findings and recommendations, (4) documentation requirements, and (5) any other considerations, such as field audit time, necessary to establish the foundation for an effective *management review* program.

Validate program effectiveness. Subsequent reviews, periodic auditing, and root cause incident investigation all provide insight into the effectiveness of the *management review* program. For example, if a deficiency is noted during a management review, the reviewers should attempt to determine if the deficiency existed, but was missed, during a previous management review. If audits discover program deficiencies, management should investigate why the management review program had not already discovered and corrected that deficiency. Each time an incident investigation identifies an issue related to the failure of an RBPS element, management should investigate whether it is

(1) an isolated deficiency or (2) indicative of a systemic problem that previous management reviews had overlooked. Any of these situations requires appropriate corrective action, because a fully effective *management review* program should have detected and corrected an RBPS element breakdown before it caused an incident or audit finding.

22.2.2 Conduct Review Activities

Once a *management review* system is in place, the reviews must be reliably performed. Reviews should be scheduled based on (1) the perceived risk of a breakdown in each element and (2) the consequences of that failure. Information must be gathered and summarized so the review can proceed efficiently, but the review process must be flexible enough to probe areas of perceived weakness with field verifications, as necessary. Any corrective actions deemed necessary by the review team should be implemented as swiftly as possible.

Prepare for the review. Facilities should maintain a regular schedule for conducting management reviews, as illustrated in Table 22.1. All elements should be reviewed at least annually, and some elements should be reviewed more frequently, depending on the process risks and factors that could change those risks, such as new equipment, new products, organizational changes, or staffing changes. In preparation for the next scheduled review, the person(s) responsible for the selected element(s) should inquire about any special topics that the reviewers may want to investigate. Examples of these can be found throughout this book – the description of each element includes a list of metrics and topics that might be included in a management review. Once the review scope has been established, the pre-review tasks include: (1) scheduling a specific time for the review, (2) gathering information and subject matter experts so that the management review team's questions can be efficiently answered, and (3) preparing a summary presentation of current activities.

TABLE 22.1. Example Schedule for Management Reviews

Month	Topic (Chapter)
January	Operations (17)
February	Knowledge (8), Management of Change (15)
March	Asset Integrity (12)
April	Procedures (10)
May	Outreach (7), Emergency (18)
June	Contractors (13)
July	Culture (3), Involvement (6)
August	Incidents (19)
September	Safe Work (11), Readiness (16)
October	Competency (5), Training (14)
November	Standards (4), Risk (9)
December	Metrics (20), Audits (21), Management Review (22)

Determine the review scope. The review leader must first decide which aspects of the subject element will be reviewed. At a minimum, the review will typically look for evidence that the basic program requirements are being met and that any previous review, audit, or incident investigation findings relevant to that element have been (or are being) satisfactorily resolved. Beyond that, the review leader should look for evidence that the management system is robust and can handle foreseeable challenges. For example, if it is known that a key worker is planning to retire, the review scope might include a discussion of the succession plans to preserve workforce competency (Section 5.2.2) and the organizational management of change (Section 15.2.3). Or if a new unit is scheduled to start up, the review scope might include a discussion of how the surge in training demand will be met (Section 14.2.2). The management review team leader should define the scope sufficiently early to facilitate the collection of necessary data and/or the scheduling of subject matter experts.

Schedule the review. Ideally, the management reviews will be scheduled at regular times each month and the element(s) due for review each meeting will be published well in advance. However, the exact review dates can be adjusted to accommodate current plant conditions, such as an unexpected shutdown. The objective is to schedule the review in a timely manner that will maximize participation of knowledgeable personnel without creating unnecessary resource conflicts.

Gather information. A management review normally involves gathering information about the subject element from several subject matter experts. To increase the efficiency of the management review team, current information on the element's metrics should be gathered and distributed to the reviewers before the start of the meeting, along with a summary of the corrective actions taken to resolve any previous review, investigation, or audit findings. Data should also be gathered to help answer any special topics the reviewers may have included in the scope.

Prepare a presentation. The element owner should prepare a presentation summarizing the state of the RBPS element based on the information gathered. The presentation should be designed to both inform the management team about current conditions and trends and also alert them to known weaknesses and upcoming challenges.

Conduct the review. The senior manager should convene the review meeting and have the owner briefly present his views on the current state of the RBPS element. The bulk of the meeting time should be devoted to reviewing data about the implementation of the element, focusing on those factors that most directly affect process safety. For example, when reviewing the *asset integrity* element, questions might include:

- Is our written program adequate?
  - When was it last revised?

- o Does it match current practice?
- o Is the scope appropriate?
- o How does it compare to industry practice?
- Is our program effective?
  - o Were there incidents that resulted from failures?
  - o What percentage of downtime was unplanned?
  - o Are there recommendations whose resolution is overdue or delayed?
  - o Are there tests and inspections that are overdue?
  - o Have critical safety systems or interlocks been removed or bypassed?
  - o Are chain locks, car seals, and drain/vent plugs being actively managed?

Generally, as the subject matter experts answer questions in the meeting, the review team probes for explanations of any anomalies. Records reviews and direct observations are occasionally necessary to resolve points of uncertainty or contention that arise during the discussions.

While different managers have different styles, the overriding objective of the review is to ascertain the real state of affairs: Is the management system broken? Is it superficially okay, but brittle and subject to failure at any moment? Or is it okay and robust enough to withstand foreseeable challenges? To be effective, the review must consider both the qualitative aspects of the activity as well as quantitative measures, and must produce quality results. A review style that seeks the truth and engages the team in identifying practical corrective actions when necessary will be far more successful than a review style that either cheerfully accepts appearances that everything is okay or "shoots the messenger" instead of constructively dealing with bad news. The organization's review style is both a potent influence on, and an accurate reflection of, its process safety culture. A healthy process safety culture should be more focused on finding and correcting weaknesses than on hiding weaknesses to placate management or avoid their ire.

The perceived risk associated with the facility or activity being reviewed, and the subject RBPS element's role in mitigating that risk, should be factored into the depth of the scrutiny given to that element. For example, a facility that frequently uses contractors to perform high hazard work would want to review the *contractors* element more thoroughly and frequently than a facility that only uses contractors occasionally for relatively routine tasks such as painting or installing insulation. Recommendations resulting from the management review should be resolved in a timely manner, but the management review team should insist on immediate action to correct any activities or conditions identified during the review, such as a gap in *safe work* practices, that appear to pose an imminent hazard to personnel or facilities.

The meeting should conclude with a summary of any noted gaps or inefficiencies in the existing practice and a list of individual assignments to either develop or implement corrective action plans. A timetable for these activities should also be included.

Document the review.   Because of its informal nature, management reviews are typically documented in the form of meeting minutes or internal memoranda. At a minimum, the report should document the subject and date of the review, the participants, and any review findings, recommendations, and assignments.

Upon completion, the review results should be distributed to appropriate parties for information and followup action.   Typically, the appropriate recipients of the report include the manager who led the review, the process safety program manager, the owner of the element that was reviewed, and others who could benefit from the information gleaned by the review.

Address review findings/recommendations. Facility staff responsible for followup should develop action plans to address each review finding. The review usually includes specific recommendations for addressing the findings: however, if not included, the facility staff will have to propose their own solutions for improving the system or filling the gaps. Facility staff must track each recommendation until it has been resolved. The date and manner in which each recommendation was resolved should be documented.   For example, if a recommendation was rejected, the reason for that rejection should be noted. If an action plan was accepted, tracking information should include the date the change was implemented. This documentation should be retained until the next review of that element, or longer if required by corporate policy.

22.2.3 Monitor Organizational Performance

A safety management system can be seriously deficient, yet appear satisfactory by superficial measures – the paperwork appears to be in place and no serious incidents have been recorded. Complacency replaces a sense of vulnerability, and the execution of program tasks becomes perfunctory. The purpose of management reviews is to monitor the organizational performance of other RBPS elements, but the *management review* element can itself fall victim to the same complacency when upper management attention is directed elsewhere. Thus, the objective of monitoring review performance is primarily to provide a gauge of its effectiveness in identifying program weaknesses.

Strive to continuously improve.   The information gained during a management review should be focused on continuously improving the system. The management review offers an opportunity to incorporate best practices that may have been discovered by other units, plants, or divisions; or that workers may be aware of from previous employers. If an RBPS element fails, upper management should question whether the deficiency existed at the time of the last management review, and if so, why it wasn't discussed and corrected then. Perhaps the reviews are not thorough enough, or perhaps the reviews are conducted in a manner that encourages workers to hide the truth from upper management. In any case, upper management should use such occasions as opportunities to improve the review process.

Conduct field inspections. Conducting occasional unannounced spot checks is one way to help ensure that conditions in the field are consistent with those reported in the management review. Management can use opportunities incidental to other work to conduct such field checks. For example, the manager might stop and read the work permits posted at a job site and ask the workers what they would do if a specific alarm sounded. Or the manager might observe how contract personnel are controlled when entering the plant gate. Such random checks not only validate the management reviews, they demonstrate upper management's active interest in RBPS elements, which inspires the workers to be similarly attentive.

22.3   POSSIBLE WORK ACTIVITIES

The RBPS approach suggests that the degree of rigor designed into each work activity should be tailored to risk, tempered by resource considerations, and tuned to the facility's culture. Thus, the degree of rigor that should be applied to a particular work activity will vary for each facility, and likely will vary between units or process areas within a single facility. Therefore, to develop a risk-based process safety management system, readers should perform the following steps:

1. Assess the risks at the facility, investigate the balance between the resource load for RBPS activities and available resources, and examine the facility's culture. This process is described in more detail in Section 2.2.
2. Estimate the potential benefits that may be achieved by addressing each of the key principles for this RBPS element. These principles are listed in Section 22.2.
3. Based on the results from steps 1 and 2, decide which essential features described in Sections 22.2.1 through 22.2.3 are necessary to properly manage risk.
4. For each essential feature that will be implemented, determine how it will be implemented and select the corresponding work activities described in this section. Note that this list of work activities cannot be comprehensive for all industries; readers will likely need to add work activities or modify some of the work activities listed in this section.
5. For each work activity that will be implemented, determine the level of rigor that will be required. Each work activity in this section is followed by two to five implementation options that describe an

increasing degree of rigor. Note that work activities listed in this section are labeled with a number; implementation options are labeled with a letter.

> *Note: Regulatory requirements may specify that process safety management systems include certain features or work activities, or that a minimum level of detail be designed into specific work activities. Thus, the design and implementation of process safety management systems should be based on regulatory requirements as well as the guidance provided in this book.*

22.3.1 Maintain a Dependable Practice

Define Roles and Responsibilities

1. Develop a written policy for managing the *management review* element.
   a. General guidance applies to all elements.
   b. Detailed guidance addresses specific management review requirements.
   c. Detailed guidance addresses specific management review requirements for each RBPS element.
2. Include specific roles and responsibilities in the management system governing the *management review* element.
   a. Roles and responsibilities are informally defined.
   b. Roles and responsibilities are formally defined and documented.
   c. Roles and responsibilities are reinforced through an active accountability system.

Establish Standards for Performance

3. Develop *management review* protocols.
   a. Management review protocols address general considerations, but they do not address specific requirements.
   b. Management review protocols explicitly address regulatory requirements.
   c. Management review protocols address best practices.
4. Establish baseline schedules for reviews.
   a. Reviews are scheduled annually.
   b. Reviews are scheduled more frequently, based upon the perceived level of risk controlled by the RBPS element.

Validate Program Effectiveness

5. Identify measures by which the effectiveness of management reviews will be judged.

    a. Effectiveness is judged by managers.
    b. Effectiveness is judged by audit findings or other lagging indicators.
    c. Effectiveness is judged by the lack of repeat findings in subsequent reviews.

22.3.2 Conduct Review Activities

Prepare for the Review

Work activities 6 through 9 address this essential feature.

Determine the Review Scope

6. Establish the scope of the review.

    a. Review scope focuses on regulatory requirements.
    b. Review scope focuses on regulatory requirements and known system failures.
    c. Review scope includes topics such as those listed in the sixth section of Chapters 3 through 21 for each element in this book.

Schedule the Review

7. Confirm the review schedule.

    a. Reviews are rescheduled for management's convenience.
    b. Reviews are not rescheduled, regardless of who can attend.
    c. Reviews are rescheduled to avoid conflict with unforeseen events, such as an unplanned shutdown.

Gather Information

8. Gather information necessary for the review.

    a. Current metrics are gathered for the review.
    b. Current metrics and trends are gathered for the review.
    c. Item (b), with additional data to support the scope of the review.

Prepare a Presentation

9. Prepare a presentation.

    a. Raw metrics are presented.
    b. Metrics are presented with an analysis.
    c. Item (b), with a summary of current issues and suggestions for improvement.

Conduct the Review

10. Conduct the review meeting.
    a. Meeting is combined with other management activities.
    b. Dedicated, but perfunctory, meeting is held.
    c. The meeting is carefully planned and held, with proper attendance and participation.
    d. Item (c), with creative suggestions for improvement.
11. Assess RBPS implementation strengths, weaknesses, and gaps relative to established requirements.
    a. A cursory analysis identifies only the most obvious deficiencies.
    b. Implementation deficiencies are thoroughly evaluated.
    c. The gaps and inefficiencies are thoroughly evaluated, with effective suggestions addressing their underlying causes.

Document the Review

12. Document the review and forward it to the appropriate facility personnel.
    a. Only verbal summaries are communicated.
    b. A summary memorandum is prepared, covering only the team's findings.
    c. A more detailed memorandum is issued, addressing observations and the basis for conclusions, as well as recommendations and assignments for corrective action.

Address Review Findings/Recommendations

13. Develop an action plan to address review findings; assign responsibilities and establish deadlines.
    a. Followup is handled in ad hoc manner.
    b. Responsibilities for followup are assigned, but no due dates are established.
    c. A detailed corrective action plan is formulated for each recommendation, with clear assignment of responsibilities for resolution, and required due dates
14. Followup to resolve review recommendations, and document resolution.
    a. Most recommendations are resolved, eventually.
    b. Most recommendations are resolved in a timely fashion, via an informal system.
    c. A formalized system is implemented for recommendation followup, and tracking of recommendations to resolution.

22.3.3 Monitor Organizational Performance

Strive to Continuously Improve

15. Identify relevant metric(s) for each RBPS element.
    a. Basic metrics are developed, but they primarily address lagging indicators.
    b. A balanced mix of metrics is developed. Qualitative and quantitative metrics address both leading and lagging indicators.
16. Trend RBPS performance over time.
    a. Data are collected but not trended.
    b. Data are trended, but analysis is limited.
    c. Goals are established and performance/improvement against these goals is analyzed over time.
17. Identify continuing RBPS management system or performance weaknesses.
    a. Analysis is sufficient to identify more substantive problems.
    b. Detail analysis yields a comprehensive awareness of systemic problems.
    c. Analysis is sufficient to identify underlying problems, common causes, element interface issues, and so forth.
18. Implement management system enhancements to address weaknesses.
    a. Corrections address only the immediate problems and rarely provide the intended remediation on a systemic basis.
    b. More substantive responses address underlying problems to system deficiencies.

Conduct Field Inspections

19. Periodically spot check work practices in the field to verify that they are consistent with RBPS element requirements.
    a. Field inspections are performed after incidents.
    b. Field inspections are performed occasionally on day shift.
    c. Field inspections are performed on all shifts, incidental to every senior manager's work activities.

22.4 EXAMPLES OF WAYS TO IMPROVE EFFECTIVENESS

This section provides specific examples of industry-tested methods for improving the effectiveness of work activities related to the *management review* element. The examples are sorted by the key principles that were first introduced in Section 22.2. The examples fall into two categories:

1. Methods for improving the performance of activities that support this element.

2. Methods for improving the efficiency of activities that support this element by maintaining the necessary level of performance using fewer resources.

These examples were obtained from the results of industry practice surveys, workshops, and CCPS member-company input. Readers desiring to improve their management systems and work activities related to this element should examine these ideas, evaluate current management system and work activity performance and efficiency, and then select and implement enhancements using the risk-based principles described in Section 2.1.

22.4.1 Maintain a Dependable Practice

Make appropriate use of written protocols to help ensure comprehensive reviews. Develop protocols unique to the RBPS element being reviewed, but do not allow the literal content of the protocols to limit reviewers in their inquiries. Protocols cannot anticipate every question or area of inquiry; experienced reviewers use protocols as guides, and not scripts, for reviews.

Look for trends across different elements. Common strengths and/or weaknesses may be observed as various elements are reviewed throughout the year.

22.4.2 Conduct Review Activities

Provide workers with advance notice of special areas of inquiry. Suggested topics for management review are included within each element chapter of this book. By notifying workers of special areas of inquiry, the data can be gathered and presented during the review and any issues can be addressed immediately.

Establish an integrated corrective action tracking system. Some organizations have developed and implemented an integrated corrective action tracking system to monitor all relevant RBPS action items, including recommendations from incident reports, risk analyses, emergency drills, audits, and so forth. These systems are transparent so that anyone in the organization can view open and overdue action items, and electronically run reports to query the data. Such systems are commonly connected to e-mail systems to inform responsible personnel (and, often, their supervisors) of overdue actions, and to allow responsible parties to update the system when actions are completed. These systems also facilitate the reassignment of responsibility for open items in the event of organizational changes.

22.4.3 Monitor Organizational Performance

Avoid making reviews a competition. Corporate management should avoid scoring reviews and using the results in a manner to rank or punish individuals. To do so risks (1) creating a competitive situation in which managers become more focused on their scores than on the messages underlying the findings or (2) creating an environment in which subject matter experts are reluctant to offer honest appraisals of system weaknesses.

Provide an anonymous employee-reporting program. Workers are sometimes reluctant to report management system issues for fear of reprisal from managers or co-workers. Management should provide an independent communication channel through which workers can anonymously report any concerns. Depending on the organization's culture, a third party may be required to assure the workers that their anonymity will be maintained.

22.5   ELEMENT METRICS AND INDICATIONS

Chapter 20 describes how metrics can be used to improve performance and when they may be appropriate. This section includes several examples of metrics that could be used to monitor the health of the *management review* element, sorted by the key principles that were first introduced in Section 22.2.

In addition to identifying high-value metrics, readers will need to determine how to best measure each metric they choose to track. In some cases, an ordinal number provides the needed information, for example, total number of workers. Other cases require that two or more attributes be indexed to provide meaningful information, such as averaging employees' length of service to obtain a unit's average years of experience per employee. Still other metrics may need to be tracked as a rate, employee turnover, for example. Sometimes, the rate of change may be of greatest interest. Since every situation is different, the reader will need to determine how to track and present data in a manner that most effectively monitors the health of RBPS management systems at their facility.

22.5.1 Maintain a Dependable Practice

- *Changes in performance goals.* This provides an indication of resource requirements and upper management interest.
- *The number of repeat findings in reviews.* Increases in these numbers may indicate failure to treat the findings seriously.

22.5.2 Conduct Review Activities

- *Number of management reviews per time period.* Decreases in these numbers may indicate a lack of management interest and support.

- *The number of deficiencies identified by management reviews.* Increases in these numbers may indicate failure to enforce best practices; sudden declines in these numbers may indicate a lack of depth or honesty in the reviews.
- *The time required to resolve deficiencies identified by management reviews.* Increases in these numbers may indicate failure to provide resources or failure to hold managers accountable for meeting their commitments.

22.5.3 Monitor Organizational Performance

- *The type and number of findings in audits.* Increases in the number or severity of findings, or repetitive types of findings, may indicate failure to implement effective corrective actions.
- *The number of incidents attributed to RBPS element failures.* Increases in these numbers may indicate failure to implement effective corrective actions.
- *Percentage of reviews delegated to subordinates.* An increasing trend may indicate that upper management gives the reviews a low priority.

22.6 MANAGEMENT REVIEW

The overall design and conduct of management reviews has been described in this chapter. Like all other elements, the *management review* element should be subject to review periodically to ensure that the program is working properly. However, rather than have the management reviewers attempt to review themselves, this review can best be conducted as part of the periodic external *audit* program. Any weaknesses revealed should be resolved as an audit finding.

22.7 REFERENCES

22.1 *Investigation Report – Refinery Incident at Motiva Enterprises LLC*, Report No. 201-05-I-DE, U.S. Chemical Safety and Hazard Investigation Board, Washington, October, 2002.

23

IMPLEMENTATION

Applying risk to process safety is by no means revolutionary. In the early 19th century, process safety primarily involved siting high-risk processes in places where a fire or explosion would do the least amount of damage.

With time, companies began to understand the causes of accidents and began using standards to manage process safety. In 1911, the American Society of Mechanical Engineers issued its landmark Boiler and Pressure Vessel Code. Technical standards continued to evolve throughout the 20th century, and they remain one of the foundations of process safety management. Major accidents in the 1970s and 1980s demonstrated that technical standards alone were insufficient, so industry began to develop management systems to (1) eliminate inconsistent/unreliable behavior (at all levels of the organization) and (2) provide multiple layers of safeguards against accidents. In some countries, safety management systems were mandated by law, starting with the Seveso Directive in 1982 and continuing to the present day. Yet 25 years later, serious accidents still occur.

Clearly, improvements to safety are still necessary, and risk-based process safety is the next step forward. The rest of this chapter addresses a series of questions that are likely to confront anyone who promotes using a risk-based approach:

1. What is a risk-based process safety management system?
2. Why implement a risk-based process safety management system?
3. What are the first steps toward implementation?
4. Should I start with a specific RBPS element or try to change my entire process safety management program in a single effort? If using an incremental approach is acceptable, where should I start?

5. How would I implement a risk-based process safety management system? For example, can this approach be used to:
   - Upgrade one or more elements of an existing process safety management system to further reduce risk and improve plant performance?
   - Implement just one of the "new" RBPS elements, effectively retrofitting the element into an existing process safety management system?
   - Address performance issues for one or more elements that are part of a compliance-based process safety management system?
   - Implement an effective risk-based process safety management system for a relatively low hazard facility?
6. If my organization is not ready to embrace a new approach to process safety, can the information presented in this book be applied in other ways?

Question 1 was addressed in Chapter 2. Questions 2 through 6 are addressed in Sections 23.1 through 23.5, respectively.

23.1   REASONS TO IMPLEMENT A RISK-BASED PROCESS SAFETY MANAGEMENT SYSTEM

First, consider the reasons to implement a process safety management program. The Center for Chemical Process Safety (CCPS) book titled, *Guidelines for Implementing Process Safety Management Systems* lists the following benefits (Ref. 23.1):

- Improved efficiency arising from consolidating a range of discrete safety-related activities.
- Cost savings from the systemic review of new projects and identifying safety enhancements early in the design process.
- Reduced downtime.
- Reduced maintenance costs.
- Improved operations.
- Improved customer satisfaction resulting from enhanced quality.
- Increased prestige within the industry and among shareholders.
- Improved employee recruitment and retention.
- Improved labor relations through involving all employees and, if applicable, their union representatives, in process safety management.

In addition, an effective process safety management program will help reduce the risk associated with incidents involving uncontrolled releases of hazardous materials or energy. Is the process safety management program at your facility delivering these benefits? If not, facility management should consider improving one or more elements.

Risk-based process safety builds on the foundation established by the CCPS over the past 20 years. The risk-based approach advocated in this book has the following attributes:

It incorporates state-of-the-art and experience-tested features in a holistic process safety management system. The RBPS approach specifically includes novel elements such as culture, outreach, conduct of operations, metrics, and management review, along with the more traditional elements that were initially introduced by the CCPS in its 1989 book titled, *Guidelines for Technical Management of Chemical Process Safety* (Ref. 23.2). These new elements address gaps in process safety management systems that have been widely observed over the past 15 years.

Its holistic application will tend to minimize aggregate risk. Since risk is used to design and implement the management system for each element, the work products should be more useful, and the intensity of work activities that support each element should be better tuned to the residual risk they are designed to manage.

It optimizes the use of resources. At many companies, resources are steadily shrinking in response to the increasingly competitive global economy. Each of the 20 chapters devoted to the individual RBPS elements lists several ideas for improving the effectiveness of the associated work activities, either through efficiency improvements (make do with less) or performance improvements (maximize the benefit from existing or slightly enhanced efforts).

It provides a management system that is better tuned to hazards and risks. As companies move to a risk-based approach to process safety management, they need to decide whether the level of effort applied to work activities for each element is inadequate, excessive, or optimal. The RBPS approach also requires that users overtly state the scope of the work activities and the required level of reliability, which helps focus effort and promote effective risk management practices.

23.2   FIRST STEPS TOWARD IMPLEMENTATION

Facilities contemplating upgrading to an RBPS management system can thoroughly examine their current situation and build a business case supporting implementation of RBPS. A sound business case requires that the benefits be stated in clear terms, be tangible, and be realistic. A good feeling or level of excitement will not, over the long term, provide the organizational energy needed to implement and sustain an effort to move to a new approach. Management support normally depends on facts, not feelings.

If an organization is not ready to embark on a large initiative right now, another option for using this book is to focus on incremental improvement. Use information in the fourth, fifth, and sixth sections of Chapters 3 through 22 to find the nuggets that will help improve the effectiveness of each element of the existing process safety management program. Find a new metric and convince management to try it. After all, what gets measured tends to get done. Persuade management to hold periodic management review meetings for weak elements of the existing process safety management system to drive specific improvements.

However, if the decision is made to completely revamp the management system for one or more elements, consider the following:

- *Obtain management commitment and support.* Identify a senior manager who will serve as the project champion.
- *Solicit assistance.* Recruit a project team staffed with diligent, respected, and knowledgeable employees representing a variety of functions and all levels of the organization. As described in Chapter 6, involving operators and maintenance craftspeople is critical in any effort to design and implement process safety programs.
- *Clearly define the scope and objectives for the task.* Develop a charter that clearly states the scope, objectives, budget, schedule, milestones, and deliverables.
- *Define roles and responsibilities.* Ultimate success will depend on the actions of many people. Management must define the roles and responsibilities for all participants, both during the implementation phase and throughout ongoing execution of work activities.
- *Manage expectations.* Compliance-based process safety management systems were not implemented overnight; expecting that a risk-based approach can be implemented without a few growing pains is unrealistic.
- *Communicate.* All of the issues in this list have a communication element. Effective communication is critical to efforts to make any change that has wide-ranging impact. The communication effort should keep potentially affected people informed of progress and prepare them for upcoming changes. Another equally important objective should be to promote the effort; make it clear why change is needed and advertise the expected benefits.
- *Move forward in discrete, manageable, and well-defined steps.* Adopt an incremental approach, guided by perceived baseline risk, potential risk reduction, and other potential improvements that could come from

a more effective management system. Consider also what resources are available and the organization's process safety culture.

- *Keep it simple.* A modest initial implementation project that works is better than an overly ambitious one that fails.
- *Obtain widespread commitment.* Elicit employee perspectives and address concerns.
- *Provide adequate training.* All affected personnel must be appropriately educated on the new or revised management system. Training should emphasize their roles and responsibilities, and this should be periodically reinforced via refresher training.
- *Field-test the system prior to its official implementation.* Debugging the proposed RBPS management system via early pilot testing will prevent wide-scale failure and rejection.
- *Develop tools to streamline future implementation activities.* Based on the results of pilot testing and initial full-scale implementation, develop generic implementation plans with templates, guidelines, and procedures that can be used by other facilities.
- *Apply proven project management tools and methods.* Keep the implementation project focused on the objectives, and manage the scope of the effort.
- *Be flexible.* The organization will change throughout the implementation and beyond; the implementation team, and the new system, must adapt to these changes.
- *Be aware.* All projects come with hidden agendas, which may be more important (or create more obstacles) than the stated objectives.
- *Continually reinforce the commitment.* Emphasize to the management team that their actions can do more to reinforce (or destroy) the work of the implementation team than any other factor.

23.3   START WITH RBPS ELEMENTS THAT PROVIDE THE GREATEST RISK BENEFIT TO YOUR FACILITY

One potential pitfall with any new management system is that the implementation team can get lost in the details, particularly if they are working within a very tight schedule. Even though OSHA mandated that almost all of the requirements in its process safety management (PSM) standard be addressed within 90 days of when the final standard was issued in February 1992, most U.S. facilities took several years to fully implement well designed PSM programs. Change is difficult; rapid change can be confusing, even mind-numbing.

Rest assured that a risk-based process safety management system does not need to be developed and implemented in 90 days. In fact, in almost all cases, attempting to move that quickly would be a mistake. An RBPS approach to process safety management can be implemented one piece at a time, based on what a facility decides will deliver the maximum benefit. Furthermore, there is no mandate to implement all 20 elements.

Figure 23.1 uses an event tree to graphically demonstrate a thought process that can be used to help decide which elements should receive the highest priority. The three risk questions that were first presented in Chapter 9 should be asked sequentially; these are shown across the top of the figure. If no (or relatively low) hazards are present at a facility (end point A in Figure 23.1), a minimal program is probably sufficient. In fact, the most important element for low hazard facilities is likely *management of change* (*MOC*). A relatively high level MOC program can help block the inadvertent introduction of new hazards to a facility. In addition to managing change, a low hazard facility often implements management systems to (1) ensure compliance with standards, codes, and regulations (the *standards* element), (2) control nonroutine work to protect against personal injury hazards (the *safe work* element), (3) prepare for emergencies (the *emergency* element), and (4) learn from near miss incidents and loss events (the *incidents* element).

Higher hazard facilities demand more extensive process safety management programs. For example, a comprehensive risk assessment may show that a high hazard facility is relatively low risk because of the low consequence and low frequency of the range of accident scenarios that have been identified (end point B in Figure 23.1). In this case, identifying the critical safeguards that help ensure that potential accident scenarios are low consequence and/or low frequency, and carefully maintaining or enhancing those safeguards, is very important. This often includes work activities associated with the *knowledge*, *risk*, *procedures*, *training*, and *asset integrity* elements, as well as the RBPS elements identified above for a low hazard facility.

As different answers to the three fundamental risk questions cause the endpoint to move from B toward E, more extensive and reliable systems are typically needed to manage risk. In fact, a facility that finds itself at endpoint E will probably fail to remain in business unless significant and rapid changes are made to better manage risk. Long-term operation at endpoint D requires either (1) high levels of performance coupled with very vigilant oversight of the process safety management program or (2) process changes to reduce the expected consequences of possible accident scenarios.

Despite efforts to reduce risk with inherently safer approaches, managing risk will remain critical to the success of any facility that manufactures, stores, or processes hazardous chemicals. Furthermore, each company must determine what risk level is tolerable for its operations. Figure 23.1 is not intended to recommend a particular risk management strategy; it merely demonstrates a thought process that readers may use when considering where to focus their initial RBPS implementation efforts.

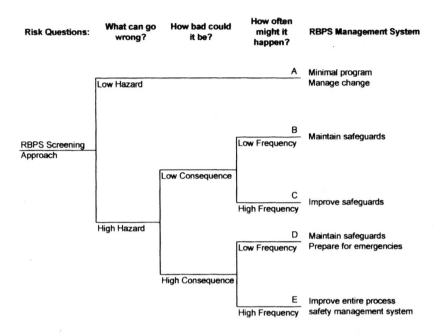

FIGURE 23.1. A Risk-based Approach to Identifying Which RBPS Elements to Implement

Facilities can improve their risk-reduction efforts by implementing additional elements, improving performance for existing elements, or both. For example, a facility that is concerned about human error may increase the effort applied to the *procedures* and *training* elements. This can be done by selecting those work activities that generally fall toward the bottom of the lists of implementation options presented in Sections 10.3 and 14.3, respectively, while simultaneously implementing the *operations* element. Another facility may determine that equipment failure dominates their overall risk and decide to apply more effort to the *asset integrity* and *incidents* elements.

Clearly, risk management has no "one size fits all" solution; no simple formula exists to help facilities determine exactly which work activities to adopt and what the appropriate level of detail for each work activity should be. However, one overriding theme has been presented in each chapter: *Efforts to implement any process safety management system will be less effective if the accompanying culture element does not embrace the pursuit of excellence.* Without a sound process safety culture, even the best RBPS management systems will not promote and help sustain safe and reliable operations.

23.4   IMPLEMENTATION EXAMPLES

This section describes four case studies involving major initiatives to use the risk-based approach. The four case studies address:

- Using RBPS to upgrade the *procedures* element (Section 23.4.1).
- Implementing the *conduct of operations* element (Section 23.4.2).
- Fixing a deficient *management of change* element (Section 23.4.3).
- Using RBPS to develop and implement a new process safety management system for an existing nonregulated, low hazard facility (Section 23.4.4).

For each case, the background section describes a hypothetical situation to provide context, then describes an example implementation plan. The first three case studies take the key principles and essential features for each element and examine how the facility chose to address these in developing their new management system. The tabular format allows a direct comparison between the key principles and essential features for the element and the implementation options chosen by the implementation team. In many cases, a brief commentary is provided for each key principle to explain the implementation team's specific thought process in addition to the summary description of the overall thought process.

The first three case studies conclude with a list of metrics and management review items. As described in Chapter 20, *metrics* should help management track the implementation effort and ongoing execution of the new or revised work activities. Chapter 22 describes the *management review* element. Since all of these case studies involve implementation activities, and the rate of change will be quite high during the implementation period, management will likely choose to hold frequent review meetings (e.g., monthly to quarterly). As the rate of change slows, management can increase the interval between review meetings.

In practice, readers charged with implementing a risk-based process safety management system should "drill down" one level further and closely examine the work activities associated with each essential feature. However, readers should not feel compelled to (1) implement or address every work activity presented for a particular element, (2) implement work activities at a consistent level of detail (e.g., choose all of the "c" options), or (3) limit their thinking to only the work activities and implementation options presented in any of the element chapters.

Section 23.4.4 is a case study involving development of a risk-based process safety management system where no formal process safety management system currently exists. The implementation strategy is described at the element level, but in practice, readers who are faced with an effort of this magnitude should first apply the thought process introduced in Section 23.3 to select which elements should be addressed in the initial implementation. Once the elements are selected, review Sections 23.4.1 through 23.4.3 to obtain ideas for implementing management systems for elements where (1) performance could be improved by upgrading existing practices, (2) a completely new element needs to be developed, or (3) the existing practice is ineffective. Finally, readers should closely review the key principles, essential features, and work activities in the second and third sections of the chapter for each element (Chapters 3 through 22) that they intend to include in their RBPS management system.

23.4.1 Using RBPS to Upgrade the Operating Procedures Element

Background

A medium size facility employs 150 people and produces specialty polymers using batch reactors. Existing operating procedures consist of a combination of (1) formulation sheets that specify the recipe and time/temperature conditions in the reactor for each product and (2) standard operating procedures that describe the proper operation of the equipment. Although the facility is not aware of any regulatory deficiencies related to its operating procedures, the standard operating procedures are poorly written and not well organized. Operators seldom use them. One consequence of the practice is that each of the four shifts operates the equipment in a slightly different manner. In fact, the production superintendent has observed that many operators and supervisors start sentences with "On our shift, we . . . ." The management team at the facility believes that this variability frequently leads to equipment failure, and sometimes impacts product quality, yield, and throughput. In addition, the management team is concerned that failure to closely follow operating procedures could someday lead to a severe accident.

Even though the operating procedures are not always followed to the letter, operators have a generally high regard for safety, product quality, and yield at the facility. However, most operators, and many others at the facility, believe that completely documenting how to run the equipment is not possible (or practical). Most believe that undocumented "operator technique," learned through many years of experience, is an essential part of producing first-quality product.

Resources at the facility have historically been adequate, but over the past five years, there has been steady pressure to control costs, and several jobs have been eliminated through attrition. Adequate time will be allotted to this project to upgrade procedures, but no additional money is available for this project to hire contract procedure writers or pay overtime, or for any capital expenditures.

Although four production areas are located at the facility, the facility operates as a single unit. That is, one shift supervisor is responsible for all four production areas, and all four shift supervisors report to one production superintendent. The facility has traditionally applied a single management system to all aspects of its operations, including its standards for operating procedures.

Implementation Plan

A well respected senior process engineer has agreed to lead this project. The project leader will be assisted by a core team representing the production, quality, and maintenance departments. The core team will establish standards for format and technical content of the new operating procedures.

The procedure content will be developed and validated by four subteams representing the four areas of the facility. Each subteam will include one operator from each shift along with one maintenance technician. These subteams will develop the technical content for the new/revised operating procedures. In addition to developing easy to use procedures that help reduce risk, each subteam is being asked to determine the best way to operate each process area, with a goal of significantly reducing dependence on operator technique.

The core team anticipates that this initiative will be met with some degree of skepticism among operators. However, the procedure content will be largely based on existing practices. In addition, involving an operator on each shift from each of the four production areas should help their colleagues accept the results and generate a number of good ideas to improve operations in each process area. Although support for this initiative is strong, management has expressed a desire to limit changes to the existing management system for the *procedures* element to items that (1) directly contribute to the stated objectives, (2) address a regulatory deficiency, or (3) otherwise provide significant risk benefit.

The left column in Table 23.1 lists the key principles and essential features for the *procedures* element along with a reference to Chapter 10 where each key principle and essential feature is fully described. The right column contains a description of the RBPS implementation options chosen. In general, the team based its approach on its perception that procedures needed to be easier to use. Comments in the right column provide a slightly broader view of the implementation team's approach to each key principle.

TABLE 23.1. RBPS Implementation Options for Upgrading Operating Procedures

Key Principles and Essential Features	Implementation Options for This Case
Maintain a Dependable Practice (Section 10.2.1) Establish management controls. Control procedure format and content. Control documents.	Existing written management systems regarding procedures only address control of changes to operating procedures and document control. The team plans to develop a complementary policy governing the *procedures* element that (1) defines the roles and responsibilities, (2) references new subordinate procedures that address the required content for each operating procedure and acceptable formats, and (3) references the two existing change control and document control procedures. The existing procedure on document control will be upgraded to address the control/use of checklists. *Comment:* The core team did not believe that the lack of a formal management system led directly to the issues described in this case study. However, they believed that establishing a more formal written management system will help (1) all affected personnel review and comment on proposed changes for procedure content (and corresponding changes to the management system for this element) and (2) sustain the proposed changes.
Identify What Operating Procedures are Needed (Section 10.2.2) Conduct a task analysis. Determine what procedures are needed and the appropriate level of detail. Address all operating modes.	The core team and each subteam reviewed a task analysis for each area that was performed about 5 years earlier. The subteams updated the task analyses and reviewed it with many of their colleagues. The subteams confirmed that existing written procedures address each task, but suggested dividing some of the existing procedures that addressed multiple, unrelated tasks while combining other procedures to arrive at a more logical alignment of tasks and procedures. The subteams noted that the procedures generally addressed normal startup, normal operation, and normal shutdown, but did not always address (1) emergency shutdown, (2) recovery from emergency shutdown, and (3) other nonroutine conditions frequently encountered by operators, such as recovery from a power outage and failure of other utilities, such as cooling water or instrument air. The subteams also noted that relatively few written procedures exist for nonroutine (but repetitive) activities such as preparing vessels for inspection. These activities are instead governed by safe work practices and job safety analyses (procedures that are developed and reviewed on an ad hoc basis and not retained). However, the core team decided that the lack of these procedures did not affect the stated objective for this project and judged that existing policies, procedures, and practices are adequate to manage the risk associated with these nonroutine activities. *Comment:* The project leader and the core team effectively managed the scope of this effort. Although some believed that specific written procedures should be prepared for each nonroutine, repetitive task, that effort was outside of the scope of the project. However, the core team did obtain management commitment to revisit this issue at the conclusion of the project, and anticipates that management will support development of these procedures on an incremental basis, for example, immediately prior to when each task is next performed.

TABLE 23.1. **RBPS Implementation Options for Upgrading Operating Procedures** *(cont'd)*

Key Principles and Essential Features	Implementation Options for This Case
Develop Procedures (Section 10.2.3) Use an appropriate format. Ensure that the procedures describe the expected system response, how to determine if a step or task has been done properly, and possible consequences associated with errors or omissions. Address safe operating limits and consequences of deviation from safe operating limits. Address limiting conditions for operation. Provide clear, concise instructions. Supplement procedures with checklists. Make effective use of pictures and diagrams. Develop written procedures to control temporary or nonroutine operations. Group the tasks in a logical manner. Interlink related procedures. Validate procedures and verify that actual practice conforms to intended practice.	The core team chose to use a combination of the outline, multi-column, and checklist formats. The outline format is used for standard operating procedures, which generally cover equipment operation. Multi-column procedures govern production of specific products. This format was chosen because it allowed the writer to document each step, the expected system response, and what the operator should do if the expected response is not observed. Checklists were used to augment procedures governing well defined, but potentially hazardous tasks, such as tanker truck unloading, reactor boilout, and several operations that are not directly part of producing specialty polymers. The core team reviewed a broad sampling of procedures to make sure that they consistently described the expected system response and addressed all of the other essential features listed under this key principle. A detailed summary of these essential features was reviewed with the four subteams. The subteams set standards for how the procedures would systematically address: • operating limits • consequences of deviation • steps to avoid or correct deviations • limiting conditions for operation The subteams made extensive use of digital photographs. The subteams made judicious use of hyperlinks to help operators quickly navigate between related procedures. All procedures, checklists, and related documents were initially reviewed by the subteam for the respective production area. After the subteam agreed on the procedures, each procedure was reviewed by at least four other operators who routinely work in the area, including one from each shift. ***Comment:*** Although each shift (and sometimes each operator) felt that some of the steps were less than perfect (or described in too much or too little detail), they did agree that there should be one way to run the plant. In fact, many operators appreciated that they would not have to "reset" their equipment at shift change. For example, they could depend on equipment being in a certain mode (e.g., the operating procedures specified which control loops should be in automatic mode and which should remain in manual mode). Over time, operators began to actively contribute to the effort by reviewing draft procedures and suggesting improvements. Shift supervisors also demonstrated leadership and initiative, actively participating in efforts to develop and review procedures for processes areas in which each had previously worked as an operator.

TABLE 23.1. **RBPS Implementation Options for Upgrading Operating Procedures** *(cont'd)*

Key Principles and Essential Features	Implementation Options for This Case
Use Procedures to Improve Human Performance (Section 10.2.4) Use the procedures when training. Hold the organization accountable for following procedures. Ensure that procedures are available.	As a result of this project, plant operation stabilized. In addition, the entire organization realized that: • Operator technique can largely be captured in words and pictures, and doing so can help improve the competency of the entire organization. • There *is* one best way to operate the process. Enforcing the discipline of consistent operation helped uncover opportunities to further improve the operation as results (product quality, yield, etc.) became more predictable. Suboptimal operating methods became quite apparent as they consistently produced suboptimal results. These situations were subsequently addressed in a systematic manner by process engineering and product development. • Operators are ultimately not responsible for producing first-quality product; rather, they are responsible for (1) following procedures, (2) recognizing when the procedures are failing to produce the desired result and either taking proper action or notifying the appropriate person of the potential problem, and (3) alerting the appropriate person when they believe the procedures should be changed (versus simply doing what they know to be right, better, or best). *Comment:* Prior to this project, "good operators" were defined by who had the "best technique." This was based on who normally produced the best quality product (considering yield and cycle time). This project built on the organization's sound leadership and culture and began paving the way for implementing the *conduct of operations* element.
Ensure that Procedures are Maintained (Section 10.2.5) Manage changes. Correct errors and omissions in a timely manner. Periodically review all operating procedures.	As a result of this project, many operators and supervisors were stakeholders in the operating procedures and took pride in ensuring that they remained current and accurate. In addition, management considered the operating procedures to be a valuable asset and developed a streamlined management of change process to help remove any roadblocks in the way of maintaining current and accurate procedures. The facility developed a rotating review schedule that included each procedure and checklist. The time interval between reviews was based on perceived risk. *Comment:* One benefit of this do-it-yourself approach to writing procedures is that several operators on each shift have a substantial personal investment in the operating procedures. The procedures are more likely to be maintained current and accurate because the entire production department, and in particular, the operators who were part of the four subteams, do not want to see them fall into disarray.

This example is of a very focused implementation effort. Note that both the core team and the subteams had wide latitude in how they addressed management's objective to improve the operating procedures and increase compliance with the operating procedures. Note also that the team successfully stuck to the task at hand, even though they noted that other aspects of the operation may be improved through application of some of the other work activities for this element. (These opportunities were communicated to management for future consideration.) Finally, it should be noted that very few of the RBPS elements can stand alone. They represent an integrated suite of management systems. In this case, success will hinge on familiarizing operators with the new procedures (*training*) and instilling the discipline in all levels of the organization to follow written procedures (*operations*).

Metrics

The implementation team suggested that four metrics be established. The first metric is derived from one of the metrics related to procedure review in Chapter 10 and will be used only during the implementation phase. This metric will be reported monthly until the project is complete.

1. Number of operating procedures updated (compared to the plan for the implementation project).

The second and third metrics will be used to track ongoing performance, but both are lagging indicators. The implementation team recommended that these two metrics be reported quarterly.

2. The rate that incident investigation reports (including investigations related to product quality, yield, and equipment failures) indicate that operating procedures were incomplete or confusing.
3. The rate that incident investigation reports (including investigations related to product quality, yield, and equipment failures) indicate that operators failed to follow procedures.

The fourth metric will be used to track ongoing performance and represents a leading indicator. This metric is derived from one of the ideas to improve effectiveness in Section 10.4.4. The implementation team plans to recommend that, in addition to simply reading procedures to complete the annual procedure review, one operator from each shift will be asked to take the procedure to the field sometime during the month that is scheduled to be reviewed and carefully note any errors or omissions in the procedure.

4. In addition to resolving comments and updating the procedure, the production superintendent will record (1) the number of substantial differences between the four reviews and (2) the number of instances in which the operator assigned to review the procedure believed that the procedure was incorrect, but the operator was wrong.

The data from these reviews will only be reported as totals to protect the privacy of each operator who participates in the review, and the number of discrepancies will be reported monthly.

Management Review

Frequent and effective management review is a critical driver for any change initiative. To help focus implementation efforts, management intends to meet quarterly to review metrics, examine the progress being made by the implementation team, and discuss current performance. Some of the questions that management intends to consider are:

*   Do workers routinely reference operating procedures? If not, is it because they are not user friendly, or because the tasks described in the procedures are very simple?
*   Are there significant differences in the performance of different shifts, teams, areas, or departments? If so, are some shifts deviating from approved procedures?
*   Does the entire organization feel accountable for following procedures, and how does line management display this accountability to operators? Have there been any instances where operators have been rewarded for increased productivity achieved through bypassing procedures?

23.4.2 Implementing the Conduct of Operations Element

Background

After reviewing all of the RBPS elements, a facility that produces specialty metals and related materials that are used to fabricate fuel rod assemblies for the commercial nuclear industry has decided to improve its existing process safety management system by implementing parts of the *conduct of operations* element. Facility management believes that this element will help it further reduce the likelihood of a severe accident and improve product quality, which is critical to its business and to its customers.

The facility employs approximately 300 people. The chemical operations at the facility produce three basic materials, which are used in a number of areas throughout the manufacturing process. Many of the raw materials and intermediates present toxic exposure, chemical burn, and fire/explosion hazards; no fissile materials are present at the facility, in other words, no materials at the facility are capable of a causing a nuclear reaction. Operations at the facility are split approximately 50/50 between a chemical conversion area and downstream metal forming/working operations.

Over the years, the facility has implemented and upgraded its programs to help reduce human error, placing significant emphasis on procedures and training. In general, workers at the facility place a high value on following established procedures, but the workforce is somewhat resistant to change. The facility's culture is tilted toward product quality, although safety is a close second in everyone's mind. The facility's injury rate is lower than most sectors of the manufacturing industry, but somewhat higher than the chemical industry.

Implementation Plan

Management has formed a team with representatives from the production, maintenance, engineering, safety, and human resources departments. The team's objective is to develop and implement the *conduct of operations* element. The intent is to reduce human error and therefore improve safety and product quality. Management is particularly interested in efforts to improve communication between production and support groups (e.g., maintenance, quality control, and shipping) and between the four shift teams.

The left column in Table 23.2 lists the key principles and essential features for the *conduct of operations* element along with a reference to Chapter 17 where each key principle and essential feature is fully described. The right column contains a description of the RBPS implementation options chosen. In general, the team based this risk-based approach on its perception that communication is a problem at the facility, particularly for special instructions that accompany custom orders or when the facility chooses to deviate from its normal methods as a result of equipment failure or some other unusual condition. Comments in the right column provide a slightly broader view of the implementation team's approach to each key principle.

This scenario is another example of a limited implementation of an RBPS element. Many essential features were not addressed at all. For example, broad consensus was that labeling and lighting were adequate at the facility and that the facility's existing culture did not tolerate missing labeling or poor lighting. The implementation team did not spend any time generating standards for labeling or specifying a specific minimum number of lumens in production areas; rather it focused on essential features that it believed would provide real benefits.

TABLE 23.2. RBPS Implementation Options for Implementing the Conduct of Operations Element

Key Principles and Essential Features	Implementation Options for This Case
Maintain a Dependable Practice (Section 17.2.1) Define roles and responsibilities. Establish standards for performance. Validate program effectiveness.	The facility chose to develop a written policy that formally defines roles and responsibilities with specific focus on the *competency, knowledge, risk, procedures, asset integrity, training, incidents, metrics,* and *management review* elements; in many cases, the policy simply formalized and documented existing *operations* element practices.

Responsibility for implementing and executing *operations* element work activities is directly integrated with line management responsibilities; the operations manager at the facility (whose direct reports include the two production superintendents, the maintenance superintendent, the shipping/receiving supervisor, and the quality control supervisor) is personally leading the effort to design and implement the *operations* element.

Because this element will involve implementation of several new practices and work activities, as well as changes to many existing procedures and work activities, it will be communicated to the entire workforce multiple times using a variety of communication modes.

Goals will be developed that include:

- process metrics (e.g., compliance with the implementation plan).

- execution metrics (e.g., percent of employees who review the required reading file each shift).

- output metrics (e.g., in-process and final-product test data).

Program effectiveness will be measured using a combination of lagging and leading indicators, such as quality rejection rates and process capability measurements, respectively.

When designing the program, the implementation team plans to review all incident reports for the past two years to estimate the current reliability of defense in depth systems, and to identify any of these systems that appear to fail on a regular basis. These systems will likely receive special attention when implementing this element.

Comment: Because this element is completely new, considerable front-end planning is being applied to the implementation. In addition, line management has full responsibility for design, implementation, and execution of this element. The objective is to design a program that is effective and can be sustained over time, based on the current staffing level.

TABLE 23.2.   RBPS Implementation Options for Implementing the Conduct of Operations Element *(cont'd)*

Key Principles and Essential Features	*Implementation Options for This Case*
Control Operations Activities (Section 17.2.2) Follow written procedures. Follow safe work practices. Use qualified workers. Assign adequate resources. Formalize communications between workers. Formalize communications between shifts. Formalize communications between work groups. Adhere to safe operating limits and limiting conditions for operation. Control access and occupancy.	Although the facility has traditionally embraced written procedures and had a rigorous training and qualification program, the need to follow (or suggest revisions to) procedures and insist on full compliance with the training schedule will be reemphasized as part of the implementation of the *operations* element. Broad consensus is that facilities are adequate. Staffing levels within the operations group are adequate; however, senior management has been reluctant to replace workers in technical and staff positions who have left over the past few years; this situation is being closely monitored. That is one reason that the *competency* element is included in this initiative. The implementation team intends to closely examine the work activities related to communication and to choose implementation options that will greatly increase the reliability of shift-to-shift and inter-department communications within the operations group, which includes production, maintenance, shipping, and quality control. No special emphasis will be placed on limiting conditions for operations or control of access/occupancy; work activities that address these essential features are well established at the facility. *Comment:* Note that the implementation team considered each of the essential features, but greatly emphasized the two related to communication. Since procedures (and to a lesser degree, training) are one form of communication, procedures and training are also considered important to this initiative. The *competency* element includes maintaining the facility's memory, which requires communication between "generations." This is a particular concern as the headcount for the technical departments is slowly creeping downward.
Control the Status of Systems and Equipment (Section 17.2.3) Formalize equipment/asset ownership and access protocols. Monitor equipment status. Maintain good housekeeping. Maintain labeling. Maintain lighting. Maintain instruments and tools.	Existing management systems and practices have been determined to be adequate for all of these essential features.

TABLE 23.2.   RBPS Implementation Options for Implementing the Conduct of Operations Element *(cont'd)*

Key Principles and Essential Features	Implementation Options for This Case
Develop Required Skills/ Behaviors (Section 17.2.4) Emphasize observation and attention to detail. Promote a questioning/learning attitude. Train workers to recognize hazards. Train workers to self-check and peer-check. Establish standards of conduct.	As part of the initiative to improve compliance with procedures and training, the facility intends to closely examine procedures to determine whether they include the expected response at appropriate points. If not, the procedures will be updated. In conjunction with this effort, the facility plans to emphasize self-checking and monitor department wide performance in this area. Methods to improve self-checking will be of particular interest for maintenance and shipping/receiving departments. The facility believes that its practices are adequate for all of the other essential features under this key principle. ***Comment:*** The implementation team believes that self-checking will deliver positive benefits, particularly in the maintenance and shipping/receiving departments, where very little peer-checking currently takes place. In addition, this will help include all hourly employees in the effort to introduce a higher level of discipline into all work activities.
Monitor Organizational Performance (Section 17.2.5) Maintain accountability. Strive to continuously improve. Maintain fitness for duty. Conduct field inspections. Correct deviations immediately.	As part of the implementation, all levels of supervision and management have committed to conduct frequent field inspections. Supervisors will be directed to pay close attention to written and verbal communication and to model the expected behavior. The facility manager, along with subordinate managers, constantly advocates the need to make spot corrections whenever unsafe or otherwise incorrect behaviors are observed. Through this behavior and continued emphasis, taking the time to stop and correct deviations will become the norm for all supervisors and technical staff at the facility. Concurrent with the implementation of the first phase of the *operations* element, facility management is reviewing programs that help empower all workers at the facility to stop and correct deviations from practices or procedures any time they notice them. ***Comment:*** The implementation team chose to focus on two essential features: conducting frequent inspections and correcting deviations immediately. In addition, it insisted that persons in leadership positions become personally involved in implementing these essential features by consciously modeling the behavior. It also encouraged all employees to do likewise.

Metrics

The implementation team suggested that four metrics be established. The first metric is derived from the implementation plan described above and will be used only during the implementation phase. It will be reported weekly.

1. Conformance with plans to communicate the goals of this effort, as well as specific changes to work practices, to the entire organization. Note that the plan that will be developed by the implementation team calls for communicating this information multiple times using a variety of communication modes (e.g., meetings, newsletter articles, posters, special emphasis by supervisors).

The last three metrics will be used to track ongoing performance. The implementation team recommends that these metrics be reported monthly.

2. Conformance with plans for managers and supervisors to hold frequent, scheduled field inspections and record the results on a standard checklist.
3. Number of incomplete shift logs or reports (an indicator related to shift-to-shift communication).
4. Ratio of planned work orders (including work orders to check equipment condition) to emergency repair work orders (an indicator of the level of communication and cooperation between the production and maintenance groups).

The implementation team noted that the facility already tracks several output metrics, such as overall equipment effectiveness, which is a metric that is derived from throughput, availability, and quality metrics. The team believes that implementing a *conduct of operations* program will help improve overall equipment effectiveness, but also noted that many other factors also affect this metric. The team was concerned that random failures unrelated to the *operations* element, as well as the low "signal to noise ratio" inherent in the overall equipment effectiveness metric (and subordinate metrics), make them poor choices for directly measuring the effectiveness of the *operations* element.

Management Review

Frequent and effective management review is a critical driver for any change initiative. To help focus implementation efforts, management intends to meet quarterly to review metrics, examine the progress being made by the implementation team, and discuss current performance. Some of the questions that management intends to consider are:

*   Do workers understand their roles and responsibilities? Do they know the roles and responsibilities of their co-workers and the lines of authority? Do workers understand their notification responsibilities in the event of an upset or incident?

- Are communications reliable? Are logs being kept contemporaneously and transferred in an organized manner? Have there been incidents as a result of misunderstood communications between workers, shifts, departments, or management?
- Are adequate resources available to perform necessary tasks? Is overtime mandatory or excessive? Are shortcuts necessary to get the job done in time? Do workers have time to perform specified crosschecks?
- Are most work activities planned? Do all work groups coordinate activities with the responsible operators? Does evidence of complacency toward hazards exist when authorizing unusual activities?
- Is the work environment conducive to reliable performance? Is the operator often distracted from process monitoring duties by administrative activities or visitors? Are there frequent nuisance alarms? Are noises/distractions common in the work area, or are there unauthorized entertainment devices?
- Are significant differences noted in the performance of different shifts/teams/areas/departments?
- Are risk takers rewarded for successful outcomes?
- Do supervisors model desired behaviors? Are supervisors frequently observing work in the field? Are peers observing and coaching peers on an ongoing basis outside of formal training classes?

23.4.3 Using RBPS to Fix a Deficient Management of Change System

Background

Based on results of a corporate safety, health, and environmental audit, a large multi-unit facility has discovered that its *management of change (MOC)* element is ineffective; many changes are being made without proper authorization. When examining the possible courses of action to address this finding, the facility's staff also determined that (1) the facility has an ad hoc method to review hazards associated with changes and (2) no management system has been established to help ensure that actions associated with changes are completed, such as updating drawings and procedures.

The facility employs approximately 400 people, using 20 different batch and continuous processes to manufacture a range of specialty chemicals that are formulated into a variety of water treatment chemicals. Some of the raw materials and intermediates are hazardous, presenting toxic exposure, fire, and/or chemical burn hazards; however, the hazards are generally of moderate severity. The manufacturing processes are subject to many changes; historically the facility processes 10 requests for change per month. The audit team believes that the facility actually makes closer to 20 to 30 changes each month that should be reviewed and approved via the *MOC* element.

The management team has recently experienced significant turnover as a result of an acquisition. The production and maintenance staff is quite stable, but will begin to retire at an increasing rate over the next three to five years. The facility's injury rate is high compared to other facilities in the chemical industry, and facility employees exhibit a somewhat complacent attitude toward process safety. Resources at the facility have historically been tight. However, management believes that issues exist in the *MOC* element that need to be addressed, so a sufficient amount of time (but limited capital) will be made available to the team charged with fixing the *MOC* element.

Implementation Plan

Management has formed a team comprising members of the production, maintenance, process engineering, product development, and quality control departments and has charged the team with developing and implementing a new MOC program. Management intends for the new MOC program to be risk-based, or at least more attuned to process hazards. However, management and the implementation team recognize that a complex, multi-tiered *MOC* system is likely to fail. The team elected to specifically avoid electronic systems at this point because implementation team members felt that the entire organization would benefit from discussing hazards associated with proposed changes versus having the capability to rapidly approve (or deny) change requests.

The implementation team quickly recognized one reason the previous *MOC* system failed: it was only applied to certain "covered chemicals." Thus, changes were reviewed and approved via the *MOC* system only if the product being produced at the time involved a covered chemical. Since many process areas are used to produce multiple products, permanent changes made to equipment when it fell outside of the scope of the *MOC* element eventually affected equipment that was later used to process covered chemicals. The implementation team has therefore obtained management commitment to apply the *MOC* element to all process areas regardless of the material that is being produced at the time that the change is made.

The team also noted that the *MOC* system was generally disliked at the facility and anticipates the decision to apply the system on a facility-wide basis will be very unpopular. Because of the failure of the previous system and the general negative feelings toward the current *MOC* system, the team elected to include a high degree of management control in the new system.

The left column in Table 23.3 lists the key principles and essential features for the *management of change* element, along with a reference to Chapter 15 where each key principle and essential feature is fully described. The right column contains a description of the RBPS implementation options chosen. In general, the implementation team based its approach on its perception that (1) some of the processes involved a high degree of hazard (and high risk), (2) people-based resources were adequate, but no capital was available to purchase software, and (3) the facility's culture was unlikely to embrace the proposed system. Comments in the right column provide a slightly broader view of the implementation team's approach to each key principle.

TABLE 23.3. RBPS Implementation Options for Fixing a Deficient MOC System

Key Principles and Essential Features	*Implementation Options for This Case*
Maintain a Dependable Practice (Section 15.2.1) *Establish consistent implementation.* *Involve competent personnel.* *Keep MOC practices effective.*	The team intends to rewrite the existing MOC procedure. The MOC procedure will clearly define roles and responsibilities, and it will include a RACI chart that graphically shows roles and responsibilities (RACI stands for Responsible, Accountable, Consult, and Inform). The RACI chart lists all work activities in the first column and job positions along the top row. Symbols in the intersecting cells show who (1) is responsible for, (2) is accountable for, (3) must be consulted on, or (4) must be informed of the results of each work activity.
	The facility's engineering and maintenance manager has agreed to champion the implementation effort and assume overall responsibility for day-to-day oversight of the new *MOC* system.
	The team plans to provide (1) training to inform all employees of the changes and (2) more detailed training to personnel who typically generate MOC requests, such as engineers, operations and maintenance supervisors, maintenance planners, and so forth. The facility manager has agreed to attend as many of these training sessions as possible, introduce the training, and clearly demonstrate support for the new program.
	The administrative assistant for the process engineering department will become the MOC coordinator; other duties assigned to this new position include tracking completion of (1) hazard review team recommendations (see the *risk* element), (2) incident investigation team recommendations (see the *incidents* element), and (3) actions to address audit findings (see the *audits* element).
	The MOC coordinator will also report data for inclusion in the facility's metrics system. These data, along with the questions listed below, will be reviewed during quarterly management review meetings. The corporate safety, health, and environmental group will audit the *MOC* system semiannually until the facility receives three consecutive "green" ratings, based on the corporate auditing standard.
	Comment: This relatively high degree of structure should help address the issues that allowed the existing program to lapse. Combining the responsibility for tracking actions related to changes, risk reviews, incident investigations, and audits should help prevent implementation of changes related to these elements without formal review.

TABLE 23.3.   RBPS Implementation Options for Fixing a Deficient MOC System
(cont'd)

Key Principles and Essential Features	Implementation Options for This Case
Identify Potential Change Situations (Section 15.2.2) *Define the scope of the* MOC *system.*	The scope of the revised *MOC* element will include all equipment, facility, process, and procedure changes that could affect any of the production processes at the facility.
Manage all sources of change.	The scope of the revised *MOC* element will also include changes that do not directly affect a process, but are not clearly replacement-in-kind. For example, MOC approval will be required for facility changes, such as locating a temporary office trailer anywhere on the site, relocating maintenance shop operations, or closing a road to traffic for more than one day.
	The MOC program will not control changes to equipment if it can be demonstrated that new equipment complies with the original design specification, for example, replacement with a newer model of a field instrument that is identical to the existing instrument in all ways except that it includes a local indication of the process value.
	Key personnel, such as operators, supervisors, maintenance planners, maintenance mechanics, instrument technicians, and purchasing agents, will be instructed to challenge anyone who appears to be requesting that a change be made to either (1) provide an approved change authorization or (2) demonstrate why requested work is outside of the scope of the MOC program. Management has agreed to (1) encourage personnel to have a questioning attitude toward changes, (2) affirm the process of challenging approval for work that appears to be a change, and (3) make it clear to these key personnel that they are accountable for challenging such work.
	To guide this effort, the new MOC procedure will include examples of changes that must be approved via the *MOC* system, along with corresponding examples of replacements-in-kind; this list will be (1) sorted by types of change, (2) maintained by the MOC clerk, and (3) updated whenever unclear or controversial situations arise.
	Initial and refresher training for operations, maintenance, and engineering employees will address the scope of the *MOC* element, including a practical exercise during which participants will be tested on their ability to determine if a series of hypothetical changes require MOC authorization.
	The scope of the *MOC* element will include permanent, temporary, and emergency (off hours) changes.
	Comment: This detailed statement of scope should help all employees understand management's intent to apply the *MOC* element to all changes at the facility.

TABLE 23.3.   RBPS Implementation Options for Fixing a Deficient MOC System
(cont'd)

Key Principles and Essential Features	Implementation Options for This Case
Evaluate Possible Impacts (Section 15.2.3) *Provide appropriate input information to manage changes.* *Apply appropriate technical rigor for the MOC review process.* *Ensure that MOC reviewers have appropriate expertise and tools.*	The revised MOC procedure will include a detailed checklist of what information is normally required, or must be updated, for each type of change (equipment change, procedure change, etc.) and for each type of review (requestor-completed checklist, formal team-based hazard review, etc.). A completed checklist will be included with each change request, and the MOC coordinator will not process the request without either (1) a complete set of documentation or (2) authorization from the engineering and maintenance manager to develop this information in parallel with the hazard review meetings. Prior to submission for formal review, the manager of the unit in which the change will be implemented must endorse each change request. The unit manager will recommend a change review method. The engineering and maintenance manager will also review the change and endorse all change requests, including the proposed method for reviewing hazards (or consult with the unit manager to resolve concerns). All changes must be reviewed and endorsed by at least two independent persons who have the appropriate knowledge and experience; more significant changes require review via methods described under the *risk* element. *Comment:* This relatively high level of detail applied to the change review process will help ensure that information is available for reviewing the impact of the proposed change, that detailed designs/proposed changes are independently and thoroughly reviewed, and that the hazard review is performed using an appropriate method.
Decide Whether to Allow the Change (Section 15.2.4) *Authorize changes.* *Ensure that change authorizers address important issues.*	Based on the perceived risk associated with the change, any of the endorsers or approvers may request a formal risk review using the methodology developed for the *risk* element. The MOC form in the revised MOC procedure specifically requires certain levels of approval for each type of change. In general, endorsement is required from the facility's safety, health, and environmental manager, the engineering and maintenance manager, and the manager of each potentially affected department; the operations manager must approve all changes. Endorsers and authorizers will specify who needs to be notified of the change or trained on the change. *Comment:* This relatively high level of approval should help the organization adjust to the much more rigorous *MOC* system. The team designing the new *MOC* system considered allowing unit managers to approve changes (and reducing the number of endorsements), but ultimately decided to adopt this more rigorous approach, at least for the first 2 to 3 years.

TABLE 23.3. RBPS Implementation Options for Fixing a Deficient MOC System *(cont'd)*

Key Principles and Essential Features	Implementation Options for This Case
Complete Follow-up Activities (Section 15.2.5) *Update records.* *Communicate changes to personnel.* *Enact risk control measures.* *Maintain MOC records.*	The MOC procedure will include a set of forms that are required for various types of changes and change review methods. These forms (and attached documents) will record (1) all technical information associated with the proposed change, (2) the results of hazard reviews, or other reviews, that are conducted, (3) actions associated with the change, including post-implementation actions, such as issuing as-built drawings, (4) documentation that each action was completed, and (5) documentation that potentially affected workers, including contractors, were trained or notified of the change. The section of the form that addresses actions will list (1) each action, (2) whether it must be completed before the change is implemented, and when it must be completed if it can be completed after the change is placed into service, and (3) who is responsible for completing each action. If the change is temporary, one of the actions must be to remove the change prior to the expiration date. Sign-off sheets will be required to ensure that all potentially affected personnel are informed of each change prior to (1) when it is implemented, or when they are assigned to work on/operate the changed process, and (2) when it is removed from service (for temporary changes). If training is required, the documentation of training will include the means used to verify understanding of each trainee, and this documentation will be copied and forwarded to human resources to be maintained in the official training files. The MOC coordinator maintains all documentation related to change requests and will notify action owners at least one week prior to the completion date for all open actions. The MOC coordinator will provide a list of open, past-due actions to the engineering and maintenance manager at the end of each month. The MOC coordinator will also remind change requestors prior to expiration of a temporary change and notify the engineering and maintenance manager if a temporary change is not removed from service by the expiration date. Past-due open action items and expired temporary changes will be reviewed during the quarterly management review meeting. ***Comment:*** This relatively high level of control has been selected because the same audit team that found that the *MOC* system was not functioning properly also found that recommendations related to process hazard analyses and incident investigations were not consistently being addressed in a timely manner. Although the audit team did not include a finding regarding timely completion of action items from change authorizations, the MOC implementation team determined that addressing action items (or, more accurately, failure to address action items) is a systemic issue at the facility and believes that the *management review* element can help improve performance in this area.

Note that a high degree of centralized command and control is described for this *MOC* system. Facility management intends to revisit this system after it has received three consecutive green ratings by the corporate audit team, as this degree of control will require additional resources and may, in some cases, adversely impact the business (e.g., slow down the rate that changes can be made because of the rigorous approval process). However, given the current state of the *MOC* element and the facility's process safety culture, management strongly supports this approach.

Metrics

The implementation team suggested that two metrics be established. The first metric is aimed at detecting changes that were not reviewed and approved via the *MOC* system. The second metric should help alert management if the action items related to MOC authorizations are not being completed. The implementation team decided to focus on the post-startup action items as the *operational readiness* element has historically identified and addressed pre-startup MOC actions that were not complete at the time of the readiness review. Each metric will be reported monthly and reviewed in a quarterly management review meeting. The MOC coordinator will be responsible for collecting the data for each metric.

1. The ratio of completed maintenance work orders (excluding preventive maintenance work orders) to approved change requests.
2. The percentage of post-startup MOC actions that are completed as scheduled. To gather data for this metric, the MOC coordinator will examine the master register of past-due MOC actions and spot-check at least 25 post-startup actions from the previous month. The spot check will confirm (1) that the action was complete (e.g., a new revision of the affected drawing was issued) and, to the extent possible, (2) the quality of the action (e.g., the shaded areas and the content in those areas appears to match the redlined drawing in the approved MOC package). Any quality discrepancies will be listed, and the details will be provided to the engineering and maintenance manager.

Management Review

Frequent and effective management review is a critical driver for any change initiative. To help focus implementation efforts, management intends to meet quarterly to review metrics, examine the progress being made by the implementation team, and discuss current performance. Some of the questions that management intends to consider are:

- Is the MOC process being used? Is there any evidence of circumvention?
- Have any incidents occurred or trends been noted that found MOC system failure to be a root cause or contributing factor?
- How effective is the new employee and contractor training on MOC? Is refresher training effective?

23.4.4 Using RBPS to Develop and Implement a New Process Safety Management System

Background

A small facility that manufactures paraffin-based products uses drum/tote quantities of flammable solvents for some of its mixing, packaging, and printing operations. Solvents are also frequently used for cleaning equipment. Solvent drums and totes are stored in a room that is designed for indoor flammable liquid storage (per applicable standards for these facilities). Other hazards present at the facility include a hot oil system, high-pressure steam, and high-speed cutting and packaging equipment.

The facility employs 65 people and produces hundreds of different products. Over the years, engineering support has generally been limited to design and installation of new equipment, construction of new buildings, or other major investments. The company has provided almost no onsite technical support for process safety or process improvements. The facility's injury rate is higher than average for the manufacturing sector, but the injury log is dominated by soft tissue injuries resulting from packaging and material handling operations.

The facility has never experienced an injury as a result of a fire involving process materials, but small flash fires have occurred in the vent header that have caused equipment damage. However, because of a recent flash fire at a sister plant that resulted in a serious injury (along with recent discovery of previously unreported flash fires involving similar operations), senior management has become very concerned about process safety management.

The serious injury involved a flash fire that occurred when mineral oil was added to a 2,500-gallon mixing vessel via the manway. The incident investigation team determined that, although the mixture contained no flammable liquids (as defined by NFPA standards), the mixture was well above its flash point. The incident investigation team speculated that the vapor was ignited by a static discharge that was likely caused by a temporary change made earlier in the shift. The night shift maintenance mechanic had used a rubber hose to bypass a faulty control valve, allowing the operator to add the mineral oil directly to the mixing vessel via the manway. The operator was just starting to add solvent to the mixing vessel when the fire occurred.

The investigation team also discovered that:

- Flash fires involving similar mixing vessels had occurred at least six times over the past five years within the company but only one was investigated (none of the previous fires resulted in an injury). In addition,
  - the results of that incident investigation were not shared with the corporate safety group or with sister facilities that operated identical processes.
  - only 2 of the 6 recommendations from the one investigation that was performed were addressed.
- Until the most recent fire, company personnel believed that only flammable liquids and gases could be involved in a flash fire.
- Throughout the company, teams charged with performing hazard reviews for new facilities and new products focused almost entirely on operability, maintainability, and product quality issues. In fact, a quick review of the documentation from a representative number of hazard reviews for similar mixing processes revealed that teams never once documented hazards associated with fires.

Implementation Plan

As a result of this incident and the information uncovered by the investigation team, senior management decided to implement a risk-based process safety management system at each facility. It chartered a team consisting of corporate and facility personnel and charged the team with the task of developing a set of standards and a model risk-based process safety management system for customization and adoption at each manufacturing facility.

The team anticipates that this initiative will be met with some degree of skepticism by the workforce, but the initiative is strongly supported by senior management and the management teams at each facility. Management understands that designing and implementing the new program will require considerable time and some unplanned expenses. However, facility managers have been directed to fund this effort from their existing budget by reprioritizing existing activities.

The implementation team used the RBPS structure to develop the corporate standard and to design its model process safety management system. As shown in Table 23.4, the team elected to not address some RBPS elements and to defer other RBPS elements to a "second wave" as it believes that these elements provide substantially less risk benefit and was concerned that a broader scope might dilute higher priority efforts.

TABLE 23.4. Using RBPS to Develop and Implement a New Process Safety Management System

RBPS Pillar and Elements	Implementation Options for This Case
Commit to Process Safety *Culture* *Standards* *Competency* *Involvement* *Outreach*	The facility plans to focus on developing the *culture* element first and foremost. Facility management is particularly concerned about the lack of a sense of vulnerability among workers. The workforce is very involved in most aspects of plant operation. The high level of commitment demonstrated by the workforce is viewed as a strength throughout the company, and each facility intends to leverage this asset as it builds its RBPS program. **Comment:** Note that although this group includes five elements containing a total of more than 100 essential features, the team chose to focus on just one essential feature – maintain (or in this case, develop) a sense of vulnerability. This is an example in which one essential feature might underpin many aspects of the rest of the RBPS program.
Understand Hazards and Risk *Knowledge* *Risk*	The team has decided to identify and address knowledge gaps, particularly with respect to fire hazards. The team has obtained agreement from the corporate research and development (R&D) group to test the flash point of all products that are either (1) produced more than 10 days per year at any single facility or (2) likely to be heated above the flash point of the mixture, regardless of production volume. For each product that is heated above its flash point, corporate R&D has been directed to try to reformulate the product using components with a higher flash point. If that is not possible, corporate engineering has been directed to investigate options for inerting the mixing vessels. A checklist-based hazard review will be required for all existing and new products. R&D will not be allowed to turn the product over for full-scale production without certain hazard data, including flash point and a diagram showing the flammability range for the material at various oxygen-to–nitrogen ratios at the maximum operating temperature (for materials that will be heated in vessels inerted with nitrogen). Training will be provided to all employees on hazard recognition with special emphasis on fire science. **Comment:** Note that this effort is narrowly focused on fire hazards associated with heating materials above their flash points. Although this is certainly a hazard, it may not represent the dominant risk or even the dominant process safety risk at the facility. For example, hot oil, high-pressure steam, and high-speed cutting equipment may present similar or greater risk of fatality or disabling injury. This sort of myopic focus on a single hazard often occurs following a major incident. However, increased emphasis on the *incidents* element will lead to reporting and investigation of near miss incidents involving some of these hazards. Although learning from incidents is not the optimal approach to realizing that a wide range of hazards must be understood, it is sometimes the only effective means to expand management's thinking about hazards and risk.

TABLE 23.4. Using RBPS to Develop and Implement a New Process Safety Management System *(cont'd)*

RBPS Pillar and Elements	Implementation Options for This Case
Manage Risk *Procedures* *Safe Work* *Asset Integrity* *Contractors* *Training* *Management of Change* *Readiness* *Operations* *Emergency*	The team intends to focus on the *procedures, training,* and *management of change* elements. These elements are likely to deliver the quickest benefit in terms of providing a safe work environment. All facilities have already implemented equipment-specific lockout/tagout procedures and confined space entry procedures; the team believes that other safe work procedures will deliver relatively little risk benefit. **Comment:** Note that the team elected to forego the *asset integrity* element. This is because (1) loss of containment does not present a significant hazard and (2) these facilities have few, if any, active safety systems that would help prevent fires inside process equipment. This approach will likely need to change over time. As hazard review teams identify the need for additional engineering controls, maintenance tasks will be needed to ensure that the new controls/systems are dependable. In addition, incident investigation teams and, at some point in the future process hazard analysis teams, may also identify the need for additional engineering controls, which will likewise need to be maintained.
Learn from Experience *Incidents* *Metrics* *Audits* *Management Review*	As a result of the most recent incident, the corporate safety, health, and environmental group is directing implementation of an incident investigation program at each facility. In general, this program will address all of the work activities listed in Chapter 19, and the management systems will be relatively rigorous (i.e., the "b" or "c" level implementation option is being chosen for most work activities). Management has agreed to implement a periodic audit program. At first, facility management was concerned that conducting audits and addressing audit findings would add too much cost to the operation. However, the corporate safety, health, and environmental group successfully argued that if an RBPS activity was worthy of including in the program, it should be audited periodically at a reasonable level of scrutiny. Facility management has elected to not track any metrics or conduct management reviews at this time. **Comment:** The team decided to delay implementation of the *metrics* and *management review* elements until procedures and practices are well established. However, these two elements will likely be included in the next phase of the company's effort to improve process safety.

In total, the implementation team has chosen to focus on parts of the *culture, knowledge, risk, procedures, training, incidents,* and *audits* elements. It is clear that the company failed to learn from previous incidents. Thus, considerable effort is being applied to the *incidents* element in hopes that future near miss events will be reported and investigated.

23.5   OTHER APPLICATIONS

Over the past 20 or more years, almost all facilities that manufacture, store, or otherwise use hazardous chemicals have established some form of process safety management program. Many of these facilities have well developed, widely understood, and time tested management systems. Simply discarding such programs and replacing them with risk-based process safety management systems seems imprudent. However, facilities can use the material presented in this book in many ways, even if they do not choose to implement an RBPS management system. Six such options include:

1. *Review the effectiveness ideas presented in the fourth section of each element chapter* (Chapters 3 through 22). Most of the ideas were provided by CCPS member companies and describe strategies that they have used to improve efficiency. Most of these ideas are likely to provide benefit to any well established and proven process safety management system.

2. *Compare your existing metrics to those listed in the fifth section of each element chapter* (Chapters 3 through 22). Measuring any aspect of RBPS element performance has an intrinsic valve by emphasizing management interest and encouraging worker attention to the measured parameter. If most of a facility's current metrics are "lagging" indicators, adding some "leading" indicators may help drive improved performance.

3. *Consider implementing a management review process.* Facilities that do not have a management review process should consider the management review issues/questions in the sixth section of each element chapter (Chapters 3 through 22) and select key issues to assess periodically.

4. *Conduct a self assessment of your process safety management program.* Facilities can use the list of work activities in the third section of each element chapter (Chapter 3 through 22) to conduct self assessments. If current facility programs fall mostly in the implementation option "a" or "b" range, if the hazards present at the facility are limited and well understood, and if the perceived risk at the facility is relatively low, the likelihood is high that the process safety management program is properly scoped. However, if the self

assessment reveals a mismatch between actual and perceived risk, the facility should consider upgrading its process safety management system using the ideas in this book.

5. *Provide key personnel with a detailed overview of one or more RBPS elements.* The second section of each element chapter (Chapters 3 through 22) contains a detailed overview of how a management system might be structured for that element. This information can provide a broad overview to newly assigned process safety personnel or others who play an important role in executing the work activities associated with a particular element.

6. *Upgrade process safety overview training and other communication tools.* The first section of each element chapter (Chapters 3 through 22) is a summary description of that element. This information can help a facility develop content for process safety overview training, process safety management intranet web pages, posters, newsletters, or communication tools that support community outreach efforts.

23.6 CONCLUSIONS

Many more permutations of implementation exist than could be described in a single book. The four case studies presented in Section 23.4 span a wide range of situations that provide readers with ideas of how to use the concepts provided in Chapters 3 through 22.

The RBPS elements contained in this book are not presented in random order. The first five elements, collectively called the commit to process safety pillar require strong leadership and firm commitment. Virtually all initiatives can only survive with (and die without) proper leadership and commitment. Effective implementation and sustained high performance depends first and foremost on management commitment.

Management must understand hazards and risk. The level of effort applied to work activities should correspond to the level of on risk (or perceived risk). As the magnitude of risk increases, management is responsible for identifying hazards, analyzing risk, and, based on that information, managing risk effectively. Stakeholders, including employees, contractors, the community, investors, customers, and the broader industry, should hold management particularly accountable for this pillar. Thus, the *knowledge* and *risk* elements play a pivotal role in implementing RBPS. In fact, efforts to reduce costs by reducing the level of effort applied to process safety, without first understanding the impact on risk, are simply irresponsible.

The manage risk pillar, comprising the nine elements described in Chapters 10 through 18, is oriented toward managing risk on a day-to-day basis. These elements are largely interdependent. For example, if much of the training provided to operators were based on operating procedures, ensuring that the procedures are current and accurate before embarking on a major retraining initiative would be prudent. However, these links are normally easy to recognize, and most of these elements are widely understood and applied.

The final four elements, known collectively as the *learn from experience* pillar, are integral to any effort to improve. Failure to investigate incidents and audit the design and implementation of process safety management systems can lead to degradation of the process safety management system at a facility. Entropy will inevitably degrade systems and practices. *Metrics* and *management review* are well established methods for improving other aspects of operations. What gets measured normally gets done, and generally gets done well. This is particularly true if management pays attention to the metrics. Effective use of metrics and application of management review requires some moderation – too many metrics or management review topics will dilute their focus. Narrowly focused management review processes may cause an organization to become myopic, as described in the case study described in Section 23.4.4.

Finally, although implementation of a RBPS element or program has been described in this chapter as a "project," the resulting management system is a process – it does not have a defined end point. Thus, implementing a risk-based approach to process safety management is part of a continuing journey to better managing risk. Just as the processes within the industry continue to evolve over time, the optimal strategy for managing process safety will evolve as well. However, two facts are certain: (1) stakeholders' tolerance of catastrophic accidents will continue to decline and (2) meeting or exceeding stakeholders' expectations for safe operations will continue to be critical to long-term business success.

23.7   REFERENCES

23.1   Center for Chemical Process Safety, *Guidelines for Implementing Process Safety Management Systems*, American Institute of Chemical Engineers, New York, New York, 1994.

23.2   Center for Chemical Process Safety, *Guidelines for Technical Management of Chemical Process Safety*, American Institute of Chemical Engineers, New York, New York, 1989.

Additional reading

Center for Chemical Process Safety, *Plant Guidelines for Technical Management of Chemical Process Safety*, American Institute of Chemical Engineers, New York, New York, 1992 (and revised edition, 1995).

The American Chemistry Council's Responsible Care Management System, American Chemistry Council, Alexandria, Virginia.

API RP 750, *Management of Process Hazards*, American Petroleum Institute, Washington, DC, 1990.

24

THE FUTURE

Expectations for process safety have always been defined by stakeholders, both internal and external. Modern process safety professionals shudder when they recall that the basis for safety in many 19th century gun powder production facilities was simply locating particularly sensitive parts of the process in places where an explosion would do the least amount of damage. However, 200 years ago, more sophisticated accident prevention tools were not available, space was plentiful, and the process safety practice of proper facility siting was born. Stakeholder expectations, including the often divergent expectations of owners, stockholders, employees, neighbors, and the general public, will likely continue to shape process safety management for the foreseeable future, demanding still better tools and more effective practices.

Societal pressure to improve performance will continue to increase. Activist members of the public, and the governments representing them, will continue to push for safer processes that place less of a burden on their communities, in terms of both continuous effects, such as noise and nuisance odors, and risks associated with catastrophic releases of hazardous materials or energy. In particular, society will likely continue to push for more stringent standards for performance in developing countries. New acquisitions and businesses in other regions of the world will present challenges at many levels. Proven and emerging process safety practices that are effective in one country may fail elsewhere. In particular, efforts to establish and continuously improve process safety culture and operating discipline must recognize cultural differences.

Internal pressure to improve effectiveness will continue to increase. Globalization has led to a smaller world, one in which capital and goods move more quickly and freely than ever before. This trend is not likely to slow or reverse; in fact, it is highly likely that it will accelerate. Thus, companies will face continued pressure to do more with less to hold down costs and boost performance so they remain competitive and attract new investment.

The rate of change will continue to increase. Technology drives change, and the rate of change in many different technologies continues to accelerate unabated. Thus, management systems, including ones used to manage process safety, will need to be agile and flexible if they are going to keep up with the demands they will face.

Process safety management will continue to spread to more industries. Many of the principles that support the modern practice of process safety management can be traced to the commercial nuclear and aviation industries. Stakeholders have long held these industries to very high safety standards. Over the past 40 years, many proven risk analysis and risk management tools have been adapted and widely implemented within the chemical process industries. In some cases, this change was thrust upon the industry by regulation. However, in many cases, proven methods and management systems were voluntarily adopted because they simply made good business sense. Other industries will realize that management systems are a key element in efforts to ensure high levels of human performance and equipment reliability. RBPS methods will be adopted by a wide range of industries, ranging from underground mines to hospitals.

OUR HOPE AND VISION

The CCPS hopes that these *RBPS Guidelines* will catalyze thinking about effective ways to improve the process safety management activities at a facility. Although management system failures will continue to occur, we hope that implementing a risk-based approach to process safety will help practitioners design management systems that are more fault-tolerant, building in redundancy at critical points.

We propose the following seven statements as reasonably achievable goals for the next generation of process safety leaders. Years from now, when the CCPS sets out to write the next *Guidelines* for process safety management, perhaps the following "wish list" will be part of the foundation for the path forward.

Sustainable process safety management systems, particularly some of the thought-leading RBPS elements such as culture, competency, outreach, operational readiness, conduct of operations, metrics, and management review, will mature.

Developing and nurturing a first-rate culture will emerge from its current status as one of the hot topics in the field to one of many tools that are used widely throughout the industry. Engineers and managers have traditionally led efforts to develop and implement process safety management systems. For many years, their efforts focused on management systems and the technical tools that support the practice of process safety. Discussion of hard-to-quantify issues such as "organizational culture" fell outside of their comfort zone; these issues were simply left off the agenda. In the future, efforts to understand and nurture process safety culture will be viewed as fundamental to any process safety management system and critical to helping the business survive in the increasingly competitive global economy.

Management will maintain process safety competency and resolve the "loss of corporate memory" problem that is so prevalent in industry today. Experience is a powerful teacher, yet the painful lessons from watershed events are all too quickly forgotten if the organization does not deliberately decide to remember the past, prepare for transitions, and effectively learn. Companies will expand their outreach activities to embrace all stakeholder groups, and facilities will actively monitor and nurture community trust.

Management and workers alike will embrace operating discipline as an essential feature of improving and assuring human performance. Operational readiness, as a discipline, will be embraced, fully integrated with the concepts of change management, pre-startup review, and inspections.

Layered, effective management system control functions using metrics, management review, and audits will be used to wring value out of every unit of resource invested in process safety management. Metrics, management review, and audits will become the tools for driving and measuring change, rather than for pinpointing errors and shortcomings.

Fully integrated management systems and widespread use of the RBPS approach will grow in both regulated and nonregulated industries around the world.

The business case for process safety will be better understood and widely adopted. Safety, health, and environmental issues will be managed in much the same way that sales, raw materials, inventories, and capital are – industry will see the clear link between inputs such as work activities and outputs such as safe and reliable operation. The value will be much more measurable. RBPS practices will permeate the complete life cycle of equipment, processes, and facilities. In addition, RBPS practices will become commonplace as they are adopted by the entire range of industries that manufacture or use hazardous chemicals or energy.

Root cause analysis will be more creatively applied, not only to a wider range of incidents and near misses, but also to the "implementation mishaps" of the RBPS elements themselves.

Recognizing that good things happen through planning, while bad things happen all by themselves, the process safety community will apply one of its strongest diagnostic tools, root cause analysis (RCA), to its own processes, procedures, and practices. Just as an incident RCA seeks to identify specific management system root causes, an RBPS-RCA will identify management issues that cut across multiple elements, shining a bright light on systemic failures and improvement opportunities. Root cause analysis will be used for evaluating a wide range of undesired outcomes, including plant efficiency problems, chronic audit findings, safety culture issues, and equipment or human unreliability. In addition, root cause analysis will be used to better understand why things sometimes "go terribly right," providing the organization with a better understanding of what it needs to do more often, or do better, to be more successful.

Use of electronic, virtual environment, real-time tools for process safety management will become widespread.

Risk analysis tools will improve, be more available, and be easier to use. Expert systems will emerge to assist with real-time risk decisions. The CCPS's process safety management web community will grow, affording seamless virtual connectivity between workforces, facilities, companies, industries, and countries so that all can share the benefits of lessons learned and benchmark practices in real-time. Highly efficient electronic tools for controlling work flow, collecting metric data, and handling the more mundane aspects of process safety management will become the norm. Computers will assist workers with some of the technical aspects of process safety management: computer systems will communicate directly with the basic process control system, the process data historian, and the process operators and will identify subtle "near miss" events that currently go unnoticed by all but the most alert operator.

Hierarchical process safety performance measurement systems will be adopted throughout industries and countries.

Accident precursors and near misses will be routinely and consistently evaluated across the entire industry, and lessons learned and performance metrics will be shared. The CCPS's Lessons Learned Database will be expanded, counting hundreds of companies among its global membership.

Benchmarking will be widespread with transparent RBPS metrics, including both performance and efficiency indicators. Industries and countries will collaborate on common vocabulary, definitions, interpretations, and methods for measuring process safety performance, and publish their results so that the public can be better informed about risk and have greater confidence in the safety of the facilities that operate in their communities.

Lessons learned will occur "lower on the pyramid."

Widespread use of RBPS metrics will enable companies to correlate off-normal events to management system failures and dysfunctional behaviors, ultimately focusing on accident prevention by improving the culture at the base of the accident/incident pyramid. Companies will understand that "learning lower on the pyramid" leads directly to safer, more productive, and more profitable operations.

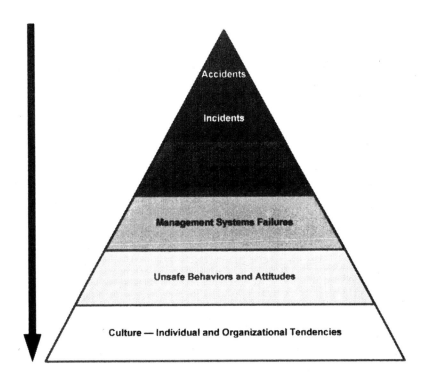

Fewer and less severe incidents will occur.

The final measure of all of our collective efforts, these *Guidelines* included, is, "Did we reduce the number and severity of process safety incidents, injuries, and fatalities?" Whatever we do and wherever the application of RBPS concepts takes us, we should always keep this ultimate goal in mind. Inherent safety tools will become the normal course of business – even for existing processes. Metrics will emerge that track the number of hazards that have been eliminated, not just the rate at which incidents are reduced.

As we strive for zero incidents, we will continue to extract and apply the lessons that we derive from those that do occur. Commercial aviation embraced this concept more than 40 years ago, resulting in a vast reduction in the number of commercial aviation incidents. In the same timeframe, the aviation industry has grown dramatically and become much more cost competitive, demonstrating that improved safety performance can coexist with economic competitiveness. In fact, in both the commercial aviation and chemical process industries, safe operation has proven to be vital to maintaining a viable business. The safety record in the commercial aviation industry, and the outstanding safety record for many chemical companies today, demonstrate that the goal of zero injuries and accidents is reasonable, and that safety improvements can directly support the goal of remaining economically competitive.

WE SHOULD CONTINUALLY STRIVE TO DO BETTER!

In thinking about the future, the one certainty is that change will be continual. New businesses will emerge, some existing business will fail, mergers, acquisitions, and divestitures will occur. All of these events will require that risk management systems be improved or revamped. New technologies, perhaps in biotechnology, nanotechnology, or some other new innovation, will benefit from the application of process safety management, but at the same time may require that different approaches be developed. Process safety management will need to become more agile to keep pace.

Standing still, congratulating ourselves on the successes of the past 20 years, and celebrating accidents that did not occur because of all of our hard work, will not prevent the next accident. Improvement will always be necessary. We must choose between moving forward, standing still, or slipping backward. We need not debate which direction to choose, only embrace the opportunity for each company to make a risk-informed decision regarding which forward path leads more directly to the ultimate goal of safe, effective, and economically competitive operation.

INDEX